Bayesian Filtering and Smoothing

Second Edition

Now in its second edition, this accessible text presents a unified Bayesian treatment of the state-of-the-art filtering, smoothing, and parameter estimation algorithms for non-linear state space models. The book focuses on discrete-time state space models and carefully introduces fundamental aspects related to optimal filtering and smoothing. In particular, it covers a range of efficient non-linear Gaussian filtering and smoothing algorithms, as well as Monte Carlo-based algorithms.

This updated edition features new chapters on constructing state space models of practical systems, the discretization of continuous-time state space models, Gaussian filtering by enabling approximations, posterior linearization filtering, and the corresponding smoothers. Coverage of key topics is expanded, including extended Kalman filtering and smoothing, and parameter estimation.

The book's practical, algorithmic approach assumes only modest mathematical prerequisites, suitable for graduate and advanced undergraduate students. Many examples are included, with the MATLAB and Python code available online, enabling readers to implement the algorithms in their own projects.

SIMO SÄRKKÄ is Associate Professor in the Department of Electrical Engineering and Automation at Aalto University, Finland. His research interests center on state estimation and stochastic modeling, and he has authored two books (2013 & 2019) on these topics. He is Fellow of ELLIS, Senior Member of IEEE, a recipient of multiple paper awards, and has been chair of MLSP and FUSION conferences.

LENNART SVENSSON is Professor in the Department of Electrical Engineering at Chalmers University of Technology, Gothenberg. His research focuses on non-linear filtering, deep learning, and tracking in particular. He has organized a massive open online course on multiple object tracking and received paper awards at the International Conference on Information Fusion in 2009, 2010, 2017, 2019, and 2021.

INSTITUTE OF MATHEMATICAL STATISTICS TEXTBOOKS

IMS Textbooks give introductory accounts of topics of current concern suitable for advanced courses at master's level, for doctoral students and for individual study. They are typically shorter than a fully developed textbook, often arising from material created for a topical course. Lengths of 100–290 pages are envisaged. The books typically contain exercises.

In collaboration with the International Society for Bayesian Analysis (ISBA), selected volumes in the IMS Textbooks series carry the "with ISBA" designation at the recommendation of the ISBA editorial representative.

Other Books in the Series (*with ISBA)

1. *Probability on Graphs*, by Geoffrey Grimmett
2. *Stochastic Networks*, by Frank Kelly and Elena Yudovina
3. *Bayesian Filtering and Smoothing*, by Simo Särkkä
4. *The Surprising Mathematics of Longest Increasing Subsequences*, by Dan Romik
5. *Noise Sensitivity of Boolean Functions and Percolation*, by Christophe Garban and Jeffrey E. Steif
6. *Core Statistics*, by Simon N. Wood
7. *Lectures on the Poisson Process*, by Günter Last and Mathew Penrose
8. *Probability on Graphs (Second Edition)*, by Geoffrey Grimmett
9. *Introduction to Malliavin Calculus*, by David Nualart and Eulália Nualart
10. *Applied Stochastic Differential Equations*, by Simo Särkkä and Arno Solin
11. **Computational Bayesian Statistics*, by M. Antónia Amaral Turkman, Carlos Daniel Paulino, and Peter Müller
12. *Statistical Modelling by Exponential Families*, by Rolf Sundberg
13. *Two-Dimensional Random Walk: From Path Counting to Random Interlacements*, by Serguei Popov
14. *Scheduling and Control of Queueing Networks*, by Gideon Weiss
15. *Principles of Statistical Analysis: Learning from Randomized Experiments*, by Ery Arias-Castro
16. *Exponential Families in Theory and Practice*, by Bradley Efron

Bayesian Filtering and Smoothing

SECOND EDITION

SIMO SÄRKKÄ
Aalto University, Finland

LENNART SVENSSON
Chalmers University of Technology, Gothenberg

CAMBRIDGE
UNIVERSITY PRESS

CAMBRIDGE
UNIVERSITY PRESS

Shaftesbury Road, Cambridge CB2 8BS, United Kingdom

One Liberty Plaza, 20th Floor, New York, NY 10006, USA

477 Williamstown Road, Port Melbourne, VIC 3207, Australia

314–321, 3rd Floor, Plot 3, Splendor Forum, Jasola District Centre, New Delhi - 110025, India

103 Penang Road, #05–06/07, Visioncrest Commercial, Singapore 238467

Cambridge University Press is part of Cambridge University Press & Assessment, a department of the University of Cambridge.

We share the University's mission to contribute to society through the pursuit of education, learning and research at the highest international levels of excellence

www.cambridge.org
Information on this title: www.cambridge.org/9781108926645

DOI: 10.1017/9781108917407

First published 2013
Second edition 2023

A catalogue record for this publication is available from the British Library

Library of Congress Cataloging-in-Publication Data
Names: Särkkä, Simo, author. | Svensson, Lennart, 1976- author.
Title: Bayesian filtering and smoothing / Simo Särkkä and Lennart Svensson.
Description: Second edition. | New York : Cambridge University Press, 2023. |
Series: Institute of Mathematical Statistics textbooks | Revised
edition of: Bayesian filtering and smoothing / Simo Sarkka. 2013. |
Includes bibliographical references and index.
Identifiers: LCCN 2022058412 | ISBN 9781108926645 (paperback)
Subjects: LCSH: Bayesian statistical decision theory. | Filters
(Mathematics) | Smoothing (Statistics)
Classification: LCC QA279.5 .S27 2023 | DDC 519.5/42–dc23/eng20230323
LC record available at https://lccn.loc.gov/2022058412

ISBN 978-1-108-92664-5 Paperback

Contents

Preface

The aim of this book is to give a concise introduction to non-linear Kalman filtering and smoothing, particle filtering and smoothing, and to the related parameter estimation methods. Although the book is intended to be an introduction, the mathematical ideas behind all the methods are carefully explained, and a mathematically inclined reader can get quite a deep understanding of the methods by reading the book. The book is purposely kept relatively short for quick reading.

The book is mainly intended for advanced undergraduate and graduate students in applied mathematics, computer science, and electrical engineering. However, the book is also suitable for researchers and practitioners (engineers) who need a concise introduction to the topic on a level that enables them to implement or use the methods. Readers are assumed to have a background in linear algebra, vector calculus, and Bayesian inference, and MATLAB or Python programming skills.

As implied by the title, the mathematical treatment of the models and algorithms in this book is Bayesian, which means that all the results are treated as being approximations to certain probability distributions or their parameters. Probability distributions are used both to represent uncertainties in the models and to model the physical randomness. The theories of non-linear filtering, smoothing, and parameter estimation are formulated in terms of Bayesian inference, and both the classical and recent algorithms are derived using the same Bayesian notation and formalism. This Bayesian approach to the topic is far from new and was pioneered by Stratonovich in the 1950s and 1960s – even before Kalman's seminal article in 1960. Thus the theory of non-linear filtering has been Bayesian from the beginning (see Jazwinski, 1970).

The main additions to the second edition of the book are the chapters on how to construct state space models of practical systems along with coverage of the iterated extended Kalman filters and smoothers, generalized statistical linear regression based filters and smoothers, and posterior linearization filters and smoothers. These additions have also resulted in a slight reordering of the material related to the coverage of Gaussian filters and smoothers. Methods for Bayesian estimation in discrete state systems, including, for example, the Viterbi algorithm, are now also covered.

Chapter 1 is a general introduction to the idea and applications of Bayesian filtering and smoothing. The purpose of Chapter 2 is to briefly review the basic concepts of Bayesian inference as well as the basic numerical methods used in Bayesian computations. Chapter 3 starts with a step-by-step introduction to recursive Bayesian estimation by demonstrating how to recursively solve a linear regression problem. The transition to Bayesian filtering and smoothing theory is explained by extending and generalizing the problem. The first Kalman filter of the book is also encountered in this chapter.

Chapters 4 and 5 are concerned with practical modeling with state space models. In particular, Chapter 4 is concerned with transforming continuous-time models of tracking models into discrete-time state space models that are compatible with the discrete-time estimation methods considered in this book, as well as examples of dynamic models. Chapter 5 proceeds to augment the models with linear, non-linear, Gaussian, and non-Gaussian measurement models and explains how certain classes of machine learning and signal processing models can be recast as state space models.

The Bayesian filtering theory starts in Chapter 6 where we derive the general Bayesian filtering equations and, as their special case, the celebrated Kalman filter, along with discrete state Bayesian filters. Taylor series-based non-linear extensions of the Kalman filter, the extended Kalman filter (EKF), and iterated extended Kalman filter (IEKF) are presented in Chapter 7. After that, Chapter 8 starts by introducing the moment matching-based general Gaussian filter algorithm, and the Gauss–Hermite Kalman filter (GHKF), cubature Kalman filter (CKF), unscented Kalman filter (UKF), and higher order cubature/unscented Kalman filters are then derived as special cases of it.

Chapter 9 introduces a different perspective and reformulates all the Gaussian filters in terms of enabling linearizations. The presentation starts with statistical linearization and the statistically linearized filter (SLF), and proceeds to statistical linear regression (SLR) and the related filters, which turn out to recover and extend all the Gaussian filters covered in the previous chapters. By further extending the concept of enabling linearizations, Chapter 10 introduces the posterior linearization filter (PLF), which generalizes the concept of iterated Gaussian filtering. Sequential Monte Carlo

(SMC)-based particle filters (PF) are explained in Chapter 11 by starting from the basic SIR filter and ending with Rao–Blackwellized particle filters (RBPF).

Chapter 12 starts with a derivation of the general (fixed-interval) Bayesian smoothing equations and then continues to a derivation of the Rauch–Tung–Striebel (RTS) smoother as their special case. In that chapter, we also present methods for smoothing in discrete state systems, including the Viterbi algorithm. The extended RTS smoother (ERTSS) and the iterated extended RTS smoother (IERTSS) are presented in Chapter 13. The general Gaussian smoothing framework is presented in Chapter 14, and the Gauss–Hermite RTS smoother (GHRTSS), cubature RTS smoother (CRTSS), unscented RTS smoother (URTSS), and higher order cubature/unscented RTS smoothers are derived as its special cases. The chapter then proceeds to the iterated posterior linearization smoother (IPLS), which generalizes the concept of iterated Gaussian smoothing.

In Chapter 15 we start by showing how the basic SIR particle filter can be used to approximate the smoothing solutions with a minor modification. We then introduce the numerically superior backward-simulation particle smoother and the reweighting (or marginal) particle smoother. Finally, we discuss the implementation of Rao–Blackwellized particle smoothers.

Chapter 16 is an introduction to parameter estimation in state space models concentrating on optimization and expectation-maximization (EM)-based computation of maximum likelihood (ML) and maximum a posteriori (MAP) estimates, as well as on Markov chain Monte Carlo (MCMC) methods. We start by presenting the general methods and then show how Kalman filters and RTS smoothers, non-linear Gaussian filters and RTS smoothers, and finally particle filters and smoothers, can be used to compute or approximate the quantities needed in the implementation of parameter estimation methods. This leads to, for example, classical EM algorithms for state space models, as well as to particle EM and particle MCMC methods. We also discuss how Rao–Blackwellization can sometimes be used to help parameter estimation.

Chapter 17 is an epilogue where we give general advice on selecting different methods for different purposes. We also discuss and give references to various technical points and related topics that are important but did not fit into this book.

Each of the chapters ends with a range of exercises that give the reader hands-on experience in implementing the methods and selecting the appropriate method for a given purpose. The MATLAB and Python source code needed in the exercises as well as much other material can be found on the book's web page.

We are grateful to many people who carefully checked the book and gave many valuable suggestions for improving the text. It is not possible to include all of them, but we would like to specifically mention Arno Solin, Robert Piché, Juha Sarmavuori, Thomas Schön, Pete Bunch, Isambi S. Mbalawata, Adrien Corenflos, Fatemeh Yaghoobi, Hany Abdulsamad, Jakob Lindqvist, and Lars Hammarstrand. We are also grateful to Jouko Lampinen, Aki Vehtari, Jouni Hartikainen, Ville Väänänen, Heikki Haario, Simon Godsill, Ángel García-Fernández, Filip Tronarp, Toni Karvonen, and various others for research co-operation that led to improvement of our understanding of the topic as well as to the development of some of the methods that now are explained in this book. We would also like to thank the editors of Cambridge University Press for their original suggestion for the publication of the book. We are also grateful to our families for their support and patience during the writing of this book.

Simo and Lennart

Symbols and Abbreviations

General Notation

$a, b, c, x, t, \alpha, \beta$	Scalars
$\mathbf{a}, \mathbf{f}, \mathbf{s}, \mathbf{x}, \mathbf{y}, \boldsymbol{\alpha}, \boldsymbol{\beta}$	Vectors
$\mathbf{A}, \mathbf{F}, \mathbf{S}, \mathbf{X}, \mathbf{Y}$	Matrices
$\mathbb{A}, \mathbb{F}, \mathbb{S}, \mathbb{X}, \mathbb{Y}$	Spaces or sets

Notational Conventions

$\mathbf{A}^\mathsf{T}$	Transpose of matrix
$\mathbf{A}^{-1}$	Inverse of matrix
$\mathbf{A}^{-\mathsf{T}}$	Inverse of transpose of matrix
$[\mathbf{A}]_i$	ith column of matrix $\mathbf{A}$
$[\mathbf{A}]_{ij}$	Element at ith row and jth column of matrix $\mathbf{A}$
$\|a\|$	Absolute value of scalar a
$\|\mathbf{A}\|$	Determinant of matrix $\mathbf{A}$
$\mathrm{d}\mathbf{x}/\mathrm{d}t$	Time derivative of $\mathbf{x}(t)$
$\frac{\partial g_i(\mathbf{x})}{\partial x_j}$	Partial derivative of g_i with respect to x_j
$(a_1, \ldots, a_n)$	Column vector with elements $a_1, \ldots, a_n$
$(a_1 \; \cdots \; a_n)$	Row vector with elements $a_1, \ldots, a_n$
$(a_1 \; \cdots \; a_n)^\mathsf{T}$	Column vector with elements $a_1, \ldots, a_n$
$\frac{\partial g(\mathbf{x})}{\partial \mathbf{x}}$	Gradient (column vector) of scalar function g
$\frac{\partial \mathbf{g}(\mathbf{x})}{\partial \mathbf{x}}$	Jacobian matrix of vector-valued function $\mathbf{x} \mapsto \mathbf{g}(\mathbf{x})$
$\mathrm{Cov}[\mathbf{x}]$	Covariance $\mathrm{Cov}[\mathbf{x}] = \mathrm{E}[(\mathbf{x} - \mathrm{E}[\mathbf{x}])\,(\mathbf{x} - \mathrm{E}[\mathbf{x}])^\mathsf{T}]$ of the random variable $\mathbf{x}$
$\mathrm{Cov}[\mathbf{x}, \mathbf{y}]$	Cross-covariance $\mathrm{Cov}[\mathbf{x}, \mathbf{y}] = \mathrm{E}[(\mathbf{x} - \mathrm{E}[\mathbf{x}])\,(\mathbf{y} - \mathrm{E}[\mathbf{y}])^\mathsf{T}]$ of the random variables $\mathbf{x}$ and $\mathbf{y}$
$\mathrm{diag}(a_1, \ldots, a_n)$	Diagonal matrix with diagonal values $a_1, \ldots, a_n$

$\sqrt{\mathbf{P}}$	Matrix such that $\mathbf{P} = \sqrt{\mathbf{P}}\,\sqrt{\mathbf{P}}^\mathsf{T}$
$\mathbf{P}^{-1/2}$	Alternative notation for $[\sqrt{\mathbf{P}}]^{-1}$
$\mathrm{E}[\mathbf{x}]$	Expectation of $\mathbf{x}$
$\mathrm{E}[\mathbf{x} \mid \mathbf{y}]$	Conditional expectation of $\mathbf{x}$ given $\mathbf{y}$
$\int f(\mathbf{x})\,\mathrm{d}\mathbf{x}$	Integral of $f(\mathbf{x})$ over the space $\mathbb{R}^n$
$\int_a^b g(t)\,\mathrm{d}t$	Integral of $g(t)$ over the interval $t \in [a, b]$
$p(\mathbf{x})$	Probability density of continuous random variable $\mathbf{x}$ or probability of discrete random variable $\mathbf{x}$
$p(\mathbf{x} \mid \mathbf{y})$	Conditional probability density or conditional probability of $\mathbf{x}$ given $\mathbf{y}$
$p(\mathbf{x}) \propto q(\mathbf{x})$	$p(\mathbf{x})$ is proportional to $q(\mathbf{x})$, that is, there exists a constant c such that $p(\mathbf{x}) = c\,q(\mathbf{x})$ for all values of $\mathbf{x}$
$\operatorname{tr}\mathbf{A}$	Trace of matrix $\mathbf{A}$
$\mathrm{Var}[x]$	Variance $\mathrm{Var}[x] = \mathrm{E}[(x - \mathrm{E}[x])^2]$ of the scalar random variable x
$x \gg y$	x is much greater than y
$x_{i,k}$	ith component of vector $\mathbf{x}_k$
$\mathbf{x} \sim p(\mathbf{x})$	Random variable $\mathbf{x}$ has the probability density or probability distribution $p(\mathbf{x})$
$\mathbf{x} \triangleq \mathbf{y}$	$\mathbf{x}$ is defined to be equal to $\mathbf{y}$
$\mathbf{x} \approx \mathbf{y}$	$\mathbf{x}$ is approximately equal to $\mathbf{y}$
$\mathbf{x} \simeq \mathbf{y}$	$\mathbf{x}$ is assumed to be approximately equal to $\mathbf{y}$
$\mathbf{x}_{0:k}$	Set or sequence containing the vectors $\{\mathbf{x}_0, \ldots, \mathbf{x}_k\}$
$\dot{\mathbf{x}}$	Time derivative of $\mathbf{x}(t)$

Symbols

α	Parameter of the unscented transform or a pendulum angle
α_i	Acceptance probability in an MCMC method
$\bar{\alpha}_*$	Target acceptance rate in an adaptive MCMC method
β	Parameter of the unscented transform, or a parameter of a dynamic model
γ	Step size in line search
$\Gamma(\cdot)$	Gamma function
$\delta(\cdot)$	Dirac delta function, or steering angle
$\delta\mathbf{x}$	Difference of $\mathbf{x}$ from the mean $\delta\mathbf{x} = \mathbf{x} - \mathbf{m}$
Δt	Sampling period
Δt_k	Length of time interval $\Delta t_k = t_{k+1} - t_k$

ε_k	Measurement error at time step k
$\boldsymbol{\varepsilon}_k$	Vector of measurement errors at time step k
θ	A parameter or heading angle
$\boldsymbol{\theta}$	Vector of parameters
$\boldsymbol{\theta}_k$	Vector of parameters at time step k
$\boldsymbol{\theta}^{(n)}$	Vector of parameters at iteration n of the EM-algorithm
$\boldsymbol{\theta}^{(i)}$	Vector of parameters at iteration i of an MCMC-algorithm
$\boldsymbol{\theta}^*$	Candidate point in an MCMC-algorithm
$\hat{\boldsymbol{\theta}}^{\text{MAP}}$	Maximum a posteriori (MAP) estimate of parameter $\boldsymbol{\theta}$
κ	Parameter of the unscented transform or auxiliary variable
λ	Parameter of the unscented transform or the Poisson distribution, or regularization parameter
λ_0	Parameter of the Poisson distribution
$\boldsymbol{\Lambda}$	Noise covariance in (generalized) statistical linear regression
$\boldsymbol{\Lambda}^{(i)}$	Noise covariance on the ith iteration of posterior linearization
$\boldsymbol{\Lambda}_k$	Noise covariance in a Gaussian enabling approximation of a dynamic model
$\boldsymbol{\mu}$	Mean of Student's t-distribution
$\boldsymbol{\mu}^{+,(i-1)}$	Predicted mean from iteration $i-1$
$\boldsymbol{\mu}^{(i)}$	Predicted measurement model mean at iteration i
$\boldsymbol{\mu}_k^-$	Predicted dynamic model mean in a statistical linear regression filter
$\boldsymbol{\mu}_k^+$	Predicted measurement model mean in a statistical linear regression filter
$\boldsymbol{\mu}_k$	Predicted mean of measurement $\mathbf{y}_k$ at time step k
$\boldsymbol{\mu}_k^-(\mathbf{x}_{k-1})$	Conditional mean moment of a dynamic model
$\boldsymbol{\mu}_k(\mathbf{x}_k)$	Conditional mean moment of a measurement model
$\boldsymbol{\mu}_k^{-(i)}$	Predicted dynamic model mean for sigma point i at time step k
$\boldsymbol{\mu}_k^{+,(i-1)}$	Predicted mean from iteration $i-1$
$\boldsymbol{\mu}_k^{(i)}$	Predicted measurement model mean for sigma point i at time step k or predicted measurement model mean at ith iteration
$\boldsymbol{\mu}_{\text{G}}$	Mean in generalized statistical linear regression approximation
$\boldsymbol{\mu}_{\text{L}}$	Mean in the linear (Taylor series-based) approximation
$\boldsymbol{\mu}_{\text{M}}$	Mean in the Gaussian moment matching approximation
$\boldsymbol{\mu}_{\text{Q}}$	Mean in the quadratic approximation
$\boldsymbol{\mu}_{\text{R}}$	Mean in the statistical linear regression approximation
$\boldsymbol{\mu}_{\text{R}}(\mathbf{x})$	Conditional mean moment in statistical linear regression

$\boldsymbol{\mu}_{\mathrm{S}}$	Mean in the statistical linearization approximation
$\boldsymbol{\mu}_{\mathrm{U}}$	Mean in the unscented approximation
ν	Degrees of freedom in Student's t-distribution
ξ	Unit Gaussian random variable
$\xi^{(i)}$	ith scalar unit sigma point
$\boldsymbol{\xi}$	Vector of unit Gaussian random variables
$\boldsymbol{\xi}^{(i)}$	ith unit sigma point vector
$\boldsymbol{\xi}^{(i_1,\ldots,i_n)}$	Unit sigma point in the multivariate Gauss–Hermite cubature
π	Constant $\pi = 3.14159265358979323846\ldots$
$\pi(\cdot)$	Importance distribution or linearization distribution
$\pi_k^{\mathbf{f}}(\mathbf{x}_k)$	Linearization distribution of a dynamic model
$\pi_k^{\mathbf{h}}(\mathbf{x}_k)$	Linearization distribution of a measurement model
$\boldsymbol{\Pi}$	Transition matrix of a hidden Markov model
$\Pi_{i,j}$	Element (i, j) of transition matrix $\boldsymbol{\Pi}$
ρ	Probability
σ^2	Variance
σ_i^2	Variance of noise component i
$\boldsymbol{\Sigma}$	Auxiliary matrix needed in the EM-algorithm
$\boldsymbol{\Sigma}_i$	Proposal distribution covariance in the Metropolis algorithm
τ	Time
φ	Direction angle
$\varphi_k(\boldsymbol{\theta})$	Energy function at time step k
$\Phi(\cdot)$	A function returning the lower triangular part of its argument or cumulative density of the standard Gaussian distribution
$\boldsymbol{\Phi}$	An auxiliary matrix needed in the EM-algorithm
ω	Angular velocity
$\boldsymbol{\Omega}_k$	Noise covariance in a Gaussian enabling approximation of a measurement model
$\boldsymbol{\Omega}_k$	Noise covariance at the ith iteration of a posterior linearization filter
$\mathbf{a}$	Action in decision theory or a part of a mean vector
$\mathbf{a}_{\mathrm{o}}$	Optimal action
$\mathbf{a}_k$	Offset in an affine dynamic model or constant term in a statistical linear regression approximation
$\mathbf{a}(\cdot)$	Non-linear drift function in a stochastic differential equation or acceleration
A	Resampling index
A_i	Resampling index

$\mathbf{A}$	Dynamic model matrix in a linear time-invariant model, the lower triangular Cholesky factor of a covariance matrix, the upper left block of a covariance matrix, a coefficient matrix in statistical linearization, or an arbitrary matrix
$\mathbf{A}^{(i)}$	Coefficient matrix at the ith iteration of posterior linearization
$\mathbf{A}_k$	Dynamic model matrix (i.e., transition matrix) of the jump from step k to step $k+1$, or approximate transition matrix in a statistical linear regression approximation
$\mathbf{A}_{\mathbf{x}}$	Jacobian matrix of $\mathbf{a}(\mathbf{x})$
b_k	Binary value in the Gilbert–Elliot channel model
$\mathbf{b}$	The lower part of a mean vector, the offset term in statistical linearization, or an arbitrary vector
$\mathbf{b}^{(i)}$	Offset term at the ith iteration of posterior linearization
$\mathbf{b}_k^{(i)}$	Offset term at ith iteration
$\mathbf{b}_k$	Dynamic bias vector or offset in an affine measurement model or constant term in a statistical linear regression approximation
$\mathrm{Be}(\cdot)$	Bernoulli distribution
$\mathbf{B}$	Lower right block of a covariance matrix, an auxiliary matrix needed in the EM-algorithm, or an arbitrary matrix
c	Scalar constant
c_k	Clutter (i.e., outlier) indicator
C	Arbitrary scalar constant
$C(\cdot)$	Cost or loss function
$\mathbf{C}$	The upper right block of a covariance matrix, an auxiliary matrix needed in the EM-algorithm, or an arbitrary matrix
$\mathbf{C}_k$	Cross-covariance matrix in a non-linear Kalman filter
$\mathbf{C}_{\mathrm{G}}$	Cross-covariance in the generalized statistical linear regression approximation
$\mathbf{C}_{\mathrm{L}}$	Cross-covariance in the linear (Taylor series-based) approximation
$\mathbf{C}_{\mathrm{M}}$	Cross-covariance in the Gaussian moment matching approximation
$\mathbf{C}_{\mathrm{Q}}$	Cross-covariance in the quadratic approximation
$\mathbf{C}_{\mathrm{R}}$	Cross-covariance in the statistical linear regression approximation
$\mathbf{C}_{\mathrm{S}}$	Cross-covariance in the statistical linearization approximation

$\mathbf{C}_{\mathrm{U}}$	Cross-covariance in the unscented approximation
d	Positive integer, usually dimensionality of the parameters
d_i	Order of a monomial
$\mathrm{d}t$	Differential of time variable t
$\mathrm{d}\mathbf{x}$	Differential of vector $\mathbf{x}$
$\mathbf{D}$	Derivative of the Cholesky factor, an auxiliary matrix needed in the EM-algorithm, or an arbitrary matrix
$\mathbf{D}_k$	Cross-covariance matrix in a non-linear RTS smoother or an auxiliary matrix used in derivations
$\mathbf{e}_i$	Unit vector in the direction of the coordinate axis i
$\tilde{\mathbf{e}}$	Noise term in statistical linear regression
$\tilde{\mathbf{e}}_k$	Noise term in an enabling Gaussian approximation of a dynamic model at time step k
$\mathbf{f}(\cdot)$	Dynamic transition function in a state space model
$F[\cdot]$	An auxiliary functional needed in the derivation of the EM-algorithm
$\mathbf{F}_{\mathbf{x}}(\cdot)$	Jacobian matrix of the function $\mathbf{x} \mapsto \mathbf{f}(\mathbf{x})$
$\mathbf{F}$	Feedback matrix of a continuous-time linear state space model
$\mathbf{F}^{(i)}_{\mathbf{xx}}(\cdot)$	Hessian matrix of $\mathbf{x} \mapsto f_i(\mathbf{x})$
g	Gravitation acceleration
$g(\cdot)$	An arbitrary function
$g_i(\cdot)$	An arbitrary function
$\mathbf{g}(\cdot)$	An arbitrary vector-valued function
$\mathbf{g}(t)$	Vector of forces
$\mathbf{g}^{-1}(\cdot)$	Inverse function of $\mathbf{g}(\cdot)$
$\tilde{\mathbf{g}}(\cdot)$	Augmented function with elements $(\mathbf{x}, \mathbf{g}(\cdot))$
$\mathbf{G}_k$	Gain matrix in an RTS smoother
$\mathbf{G}_{\mathbf{x}}(\cdot)$	Jacobian matrix of the function $\mathbf{x} \mapsto \mathbf{g}(\mathbf{x})$
$\mathbf{G}^{(i)}_{\mathbf{xx}}(\cdot)$	Hessian matrix of $\mathbf{x} \mapsto g_i(\mathbf{x})$
$\mathbf{h}(\cdot)$	Measurement model function in a state space model
$H_p(\cdot)$	pth order Hermite polynomial
$\mathbf{H}$	Measurement model matrix in a linear Gaussian model, or a Hessian matrix
$\mathbf{H}_k$	Measurement model matrix at time step k in a linear Gaussian or affine model, or approximate measurement model matrix in a statistical linear regression approximation
$\mathbf{H}_k^{(i)}$	Measurement model matrix at the ith iteration of posterior linearization filter

$\mathbf{H}_{\mathbf{x}}(\cdot)$	Jacobian matrix of the function $\mathbf{x} \mapsto \mathbf{h}(\mathbf{x})$
$\mathbf{H}^{(i)}_{\mathbf{xx}}(\cdot)$	Hessian matrix of $\mathbf{x} \mapsto h_i(\mathbf{x})$
i	Integer-valued index variable
$I_{\max}$	Number of iterations in iterated methods
$\mathbf{I}$	Identity matrix
$I_i(\boldsymbol{\theta}, \boldsymbol{\theta}^{(n)})$	An integral term needed in the EM-algorithm
j	Integer-valued index variable
$\mathbf{J}(\cdot)$	Jacobian matrix
k	Time step number
$\mathbf{K}^{(i)}$	Gain at iteration i of iterated posterior linearization
$\mathbf{K}_k$	Gain matrix of a Kalman/Gaussian filter
$\mathbf{K}_k^{(i)}$	Gain matrix at the ith iteration of an iterated filter at time step k
L	Positive constant
$L(\cdot)$	Negative logarithm of distribution
$L_{\mathrm{GN}}(\cdot)$	Gauss–Newton objective function
$\mathbf{L}$	Noise coefficient (i.e., dispersion) matrix of a continuous-time linear state space model
$\mathcal{L}(\cdot)$	Likelihood function
m	Dimensionality of a measurement, mean of the univariate Gaussian distribution, a mass, number of sigma points, or loop counter
$\mathbf{m}$	Mean of a Gaussian distribution
$\tilde{\mathbf{m}}$	Mean of an augmented random variable
$\mathbf{m}^{(i)}$	Mean at iteration i of iterated posterior linearization
$\mathbf{m}_k$	Mean of a Kalman/Gaussian filter at the time step k
$\mathbf{m}_k^{(i)}$	Mean at the ith iteration of a posterior linearization filter, mean of the Kalman filter in the particle i of RBPF at time step k
$\mathbf{m}_{0:T}^{(i)}$	History of means of the Kalman filter in the particle i of RBPF
$\tilde{\mathbf{m}}_k$	Augmented mean at time step k, an auxiliary variable used in derivations, or linearization point
$\mathbf{m}_k^-$	Predicted mean of a Kalman/Gaussian filter at time step k, just before the measurement $\mathbf{y}_k$
$\mathbf{m}_k^{-(i)}$	Predicted mean of the Kalman filter in the particle i of RBPF at time step k
$\tilde{\mathbf{m}}_k^-$	Augmented predicted mean at the time step k

$\mathbf{m}_k^{\mathrm{s}}$	Mean computed by a Gaussian (RTS) smoother for the time step k
$\mathbf{m}_{0:T}^{\mathrm{s}}$	Trajectory of the smoother means from a Gaussian (RTS) smoother
$\mathbf{m}_{0:T}^{\mathrm{s},(i)}$	Trajectory of means at smoother iteration i or history of means of the RTS smoother in the particle i of RBPS
$\mathbf{m}_\pi$	Expected value of $\mathbf{x} \sim \pi(\mathbf{x})$
$\mathbf{m}_k^{\mathbf{f}}$	Mean of $\pi_k^{\mathbf{f}}$
$\mathbf{m}_k^{\mathbf{h}}$	Mean of $\pi_k^{\mathbf{h}}$
M	Constant in rejection sampling
n	Positive integer, usually the dimensionality of the state
n'	Augmented state dimensionality in a non-linear transform
n''	Augmented state dimensionality in a non-linear transform
N	Positive integer, usually the number of Monte Carlo samples
$\mathrm{N}(\cdot)$	Gaussian distribution (i.e., normal distribution)
$O_{i,j}$	Element (i, j) of emission matrix $\mathbf{O}$
$\mathbf{O}$	Emission matrix of a hidden Markov model
p	Order of a Hermite polynomial
p_0	State-switching probability in the Gilbert–Elliot channel model
p_1	State-switching probability in the Gilbert–Elliot channel model
p_2	State-switching probability in the Gilbert–Elliot channel model
$p_{j,k}^-$	Predictive distribution for the discrete state $x_k = j$
$p_{j,k}$	Filtering distribution for the discrete state $x_k = j$
$p_{j,k}^{\mathrm{s}}$	Smoothing distribution for the discrete state $x_k = j$
P	Variance of the univariate Gaussian distribution
$\mathrm{Po}(\cdot)$	Poisson distribution
$\mathbf{P}$	Covariance of the Gaussian distribution
$\tilde{\mathbf{P}}$	Covariance of an augmented random variable
$\mathbf{P}^{\mathrm{xy},(i-1)}$	Predicted cross-covariance from iteration $i-1$ in iterated posterior linearization
$\mathbf{P}^{\mathrm{y},(i-1)}$	Predicted covariance from iteration $i-1$ in iterated posterior linearization
$\mathbf{P}^{(i)}$	Covariance at iteration i of iterated posterior linearization
$\mathbf{P}_k$	Covariance of a Kalman/Gaussian filter at time step k

$\mathbf{P}_k^{(i)}$	Covariance at iteration i of iterated posterior linearization filter, covariance of the Kalman filter in the particle i of RBPF at time step k
$\mathbf{P}_{0:T}^{(i)}$	History of covariances of the Kalman filter in the particle i of RBPF
$\tilde{\mathbf{P}}_k$	Augmented covariance at time step k or an auxiliary variable used in derivations
$\mathbf{P}_k^-$	Predicted covariance of a Kalman/Gaussian filter at the time step k just before the measurement $\mathbf{y}_k$
$\tilde{\mathbf{P}}_k^-$	Augmented predicted covariance at time step k
$\mathbf{P}_k^{-(i)}$	Predicted covariance of the Kalman filter in the particle i of RBPF at time step k
$\mathbf{P}_k^{\mathrm{s}}$	Covariance computed by a Gaussian (RTS) smoother for the time step k
$\mathbf{P}_{0:T}^{\mathrm{s}}$	Trajectory of smoother means from a Gaussian (RTS) smoother
$\mathbf{P}_{0:T}^{\mathrm{s},(i)}$	Trajectory of smoother covariances from iteration i of an iterated smoother or history of covariances of the RTS smoother in the particle i of RBPS
$\mathbf{P}_k^{\mathrm{x}}$	Predicted dynamic model covariance in a statistical linear regression filter
$\mathbf{P}_k^{\mathrm{xx}}$	Predicted dynamic model cross-covariance in a statistical linear regression filter
$\mathbf{P}_k^{\mathrm{xy}}$	Predicted measurement model cross-covariance in a statistical linear regression filter
$\mathbf{P}_k^{\mathrm{y}}$	Predicted measurement model covariance in a statistical linear regression filter
$\mathbf{P}_k^{\mathrm{x}}(\mathbf{x}_{k-1})$	Conditional covariance moment for a dynamic model
$\mathbf{P}_k^{\mathrm{y}}(\mathbf{x}_k)$	Conditional covariance moment for a measurement model
$\mathbf{P}_k^{x,(i)}$	Predicted dynamic model covariance in SPCMKF for sigma point i on time step k
$\mathbf{P}_k^{y,(i)}$	Predicted measurement model covariance in SPCMKF for sigma point i on time step k
$\mathbf{P}_k^{\mathrm{xy},(i-1)}$	Predicted cross-covariance from iteration $i-1$ in the iterated posterior linearization filter
$\mathbf{P}_k^{\mathrm{y},(i-1)}$	Predicted covariance from iteration $i-1$ in the iterated posterior linearization filter
$\mathbf{P}_\pi$	Covariance of $\mathbf{x} \sim \pi(\mathbf{x})$
$\mathbf{P}_k^{\mathbf{f}}$	Covariance of $\pi_k^{\mathbf{f}}$

$\mathbf{P}_k^{\mathbf{h}}$	Covariance of $\pi_k^{\mathbf{h}}$
q_0	Smaller probability of error in the Gilbert–Elliot channel model
q_1	Larger probability of error in the Gilbert–Elliot channel model
q^c	Spectral density of a white noise process
q_i^c	Spectral density of component i of a white noise process
$q(\cdot)$	Proposal distribution in the MCMC algorithm or an arbitrary distribution in the derivation of the EM-algorithm
$q^{(n)}$	Distribution approximation on the nth step of the EM-algorithm
$\mathbf{q}$	Gaussian random vector
$\mathbf{q}_k$	Gaussian process noise
$\tilde{\mathbf{q}}_k$	Euler–Maruyama approximation-based Gaussian process noise
Q	Variance of scalar process noise
$\mathcal{Q}_k^{(\cdot)}$	Sigma point of the process noise $\mathbf{q}_k$
$\mathcal{Q}(\boldsymbol{\theta}, \boldsymbol{\theta}^{(n)})$	An auxiliary function needed in the EM-algorithm
$\mathbf{Q}$	Covariance of the process noise in a time-invariant model
$\mathbf{Q}_k$	Covariance of the process noise at the jump from step k to $k+1$
$\tilde{\mathbf{Q}}_k$	Euler–Maruyama approximation-based covariance of the process noise
$\mathbf{Q}^c$	Spectral density matrix of (vector-valued) white noise
r	Distance to the center of rotation
r_k	Scalar Gaussian measurement noise
$\mathbf{r}_k$	Vector of Gaussian measurement noises
$\mathbf{r}_j(\cdot)$	Residual term in the Gauss–Newton objective function
R	Variance of scalar measurement noise
$\mathcal{R}_k^{(\cdot)}$	Sigma point of the measurement noise $\mathbf{r}_k$
$\mathbf{R}$	Covariance matrix of the measurement in a time-invariant model or the covariance-related parameter in Student's t-distribution
$\mathbf{R}_k$	Covariance matrix of the measurement at the time step k
$\mathbb{R}$	Space of real numbers
$\mathbb{R}^n$	n-dimensional space of real numbers
$\mathbb{R}^{n \times m}$	Space of real $n \times m$ matrices
s	Speed, generic integration variable, or temporary variable
s_k	Regime signal in the Gilbert–Elliot channel model

$s_{i,x}$	x-coordinate of radar i
$s_{i,y}$	y-coordinate of radar i
S	Number of backward-simulation draws
$\mathrm{St}(\cdot)$	Student's t-distribution
$\mathbf{S}^{(i)}$	Covariance at iteration i of iterated posterior linearization
$\mathbf{S}_k$	Innovation covariance of a Kalman/Gaussian filter at time step k
$\mathbf{S}_k^{(i)}$	Innovation covariance at the ith iteration of an iterated filter at time step k
$\mathbf{S}_{\mathrm{G}}$	Covariance in the generalized statistical linear regression
$\mathbf{S}_{\mathrm{L}}$	Covariance in the linear (Taylor series-based) approximation
$\mathbf{S}_{\mathrm{M}}$	Covariance in the Gaussian moment matching approximation
$\mathbf{S}_{\mathrm{Q}}$	Covariance in the quadratic approximation
$\mathbf{S}_{\mathrm{R}}$	Covariance in the statistical linear regression approximation
$\mathbf{S}_{\mathrm{R}}(\mathbf{x})$	Conditional covariance moment in statistical linear regression
$\mathbf{S}_{\mathrm{S}}$	Covariance in the statistical linearization approximation
$\mathbf{S}_{\mathrm{U}}$	Covariance in the unscented approximation
t	Time variable $t \in [0, \infty)$
t'	Another time variable $t' \in [0, \infty)$
$t^{(i)}$	Cumulative sum in resampling
t_k	Time of the step k (usually time of the measurement y_k)
T	Index of the last time step or the final time of a time interval
$\mathcal{T}_k$	Sufficient statistics
u	Scalar (random) variable
$\mathbf{u}_k$	Latent (non-linear) variable in a Rao–Blackwellized particle filter or smoother, or deterministic input to a dynamic model
$\mathbf{u}_k^{(i)}$	Latent variable value in particle i
$\mathbf{u}_{0:k}^{(i)}$	History of latent variable values in particle i
$U(\cdot)$	Utility function
$\mathrm{U}(\cdot)$	Uniform distribution
$\upsilon^{(i)}$	Random variable
υ_k	Bernoulli sequence in the Gilbert–Elliot channel model
$\upsilon_k^{(i)}$	Unnormalized weight in a SIR particle filter-based likelihood evaluation
$\mathbf{v}_k$	Innovation vector of a Kalman/Gaussian filter at time step k
$\tilde{\mathbf{v}}_k$	Noise term in an enabling Gaussian approximation of a measurement model on time step k
$\mathbf{v}_k^{(i)}$	Innovation vector at ith iteration of iterated extended Kalman filter at time step k

V	Volume of space
$V_k(x_k)$	Value function at time step k of the Viterbi algorithm
$\mathbb{V}$	Region in space (e.g., $\mathbb{V} = [-1, 1]$)
$w^{(i)}$	Normalized weight of the particle i in importance sampling
$\tilde{w}^{(i)}$	Weight of the particle i in importance sampling
$w^{*(i)}$	Unnormalized weight of the particle i in importance sampling
$w_k^{(i)}$	Normalized weight of the particle i at time step k of a particle filter
$w_{k\|n}^{(i)}$	Normalized weight of a particle smoother
w_i	Weight i in a regression model
$\mathbf{w}_k$	Vector of weights at time step k in a regression model
$\mathbf{w}(t)$	Gaussian white noise process
W	Weight in the cubature or unscented approximation
W_i	ith weight in a sigma point approximation
$W_i^{(\mathrm{m})}$	Mean weight of the unscented transform
$W_i^{(\mathrm{c})}$	Covariance weight of the unscented transform
$W_{i_1,\ldots,i_n}$	Weight in multivariate Gauss–Hermite cubature
x	Scalar random variable or state, sometimes regressor variable, or a generic scalar variable
$\mathbf{x}$	Random variable or state
$\hat{\mathbf{x}}$	Estimate of $\mathbf{x}$ or nominal $\mathbf{x}$
$\mathbf{x}^{(i)}$	ith Monte Carlo draw from the distribution of $\mathbf{x}$
$\mathbf{x}_k$	State at time step k
$\mathbf{x}_k^*$	Optimal state at time step k
$\mathbf{x}_k^{(i)}$	ith iterate of state estimate for time step k in iterated extended Kalman filter or smoother, or ith Monte Carlo sample of state in MCKF
$\hat{\mathbf{x}}_k^{(i)}$	Predicted ith Monte Carlo sample of the state in MCKF (before prediction)
$\mathbf{x}_k^{-(i)}$	Predicted ith Monte Carlo sample of the state in MCKF (after prediction)
$\mathbf{x}(t)$	State at (continuous) time t
$\tilde{\mathbf{x}}_k$	Augmented state at time step k
$\mathbf{x}_{0:k}$	Set containing the state vectors $\{\mathbf{x}_0, \ldots, \mathbf{x}_k\}$
$\mathbf{x}_{0:k}^{(i)}$	The history of the states in the particle i
$\mathbf{x}_{0:T}^*$	Optimal state trajectory

$\mathbf{x}_{0:T}^{(i)}$	ith iterate of state trajectory estimate in an iterated extended Kalman smoother
$\tilde{\mathbf{x}}_{0:T}^{(j)}$	State trajectory simulated by a backward-simulation particle smoother
$\mathbf{X}$	Matrix of regressors
$\mathbf{X}_k$	Matrix of regressors up to the time step k
$\mathcal{X}^{(\cdot)}$	Sigma point of $\mathbf{x}$
$\tilde{\mathcal{X}}^{(\cdot)}$	Augmented sigma point of $\mathbf{x}$
$\mathcal{X}_k^{(\cdot)}$	Sigma point of the state $\mathbf{x}_k$
$\tilde{\mathcal{X}}_k^{(\cdot)}$	Augmented sigma point of the state $\mathbf{x}_k$
$\hat{\mathcal{X}}_k^{(\cdot)}$	Predicted sigma point of the state $\mathbf{x}_k$
$\mathcal{X}_k^{-(\cdot)}$	Sigma point of the predicted state $\mathbf{x}_k$
$\tilde{\mathcal{X}}_k^{-(\cdot)}$	Augmented sigma point of the predicted state $\mathbf{x}_k$
$\mathbb{X}$	Set of possible values for a discrete-valued state
y	Scalar variable, random variable, or measurement
$\mathbf{y}$	Random variable or measurement
$\mathbf{y}_k$	Measurement at the time step k
$\mathbf{y}_{1:k}$	Set containing the measurement vectors $\{\mathbf{y}_1, \ldots, \mathbf{y}_k\}$
$\mathbf{y}_k^{(i)}$	ith Monte Carlo sample of measurement in MCKF
$\mathcal{Y}^{(\cdot)}$	Sigma point of $\mathbf{y}$
$\tilde{\mathcal{Y}}^{(\cdot)}$	Augmented sigma point of $\mathbf{y}$
$\hat{\mathcal{Y}}_k^{(\cdot)}$	ith predicted sigma point of the measurement $\mathbf{y}_k$ at step k
$\mathbb{Y}$	Set of possible values for a discrete-valued measurement
z	Scalar variable
$\mathbf{z}$	Augmented random variable
Z	Normalization constant
Z_k	Normalization constant at the time step k
Z_p	Normalization constant for distribution p
∞	Infinity

Abbreviations

2D	Two-dimensional
AD	Automatic differentiation
ADF	Assumed density filter
AM	Adaptive Metropolis (algorithm)
AMCMC	Adaptive Markov chain Monte Carlo
APF	Auxiliary particle filter

AR	Autoregressive (model)
ARMA	Autoregressive moving average (model)
ASIR	Auxiliary sequential importance resampling
BS-PS	Backward-simulation particle smoother
CA	Constant acceleration (model)
CDKF	Central differences Kalman filter
CKF	Cubature Kalman filter
CLT	Central limit theorem
CPF	Cubature particle filter
CPU	Central processing unit
CRLB	Cramér–Rao lower bound
CRTSS	Cubature RTS smoother
CT	Coordinated turn (model)
CV	Constant velocity (model)
DLM	Dynamic linear model
DOT	Diffuse optical tomography
DSP	Digital signal processing
EC	Expectation correction
EEG	Electroencephalography
EKF	Extended Kalman filter
EM	Expectation-maximization
EP	Expectation propagation
ERTSS	Extended Rauch–Tung–Striebel smoother
FFBS	Forward-filtering backward-sampling (smoother)
FHKF	Fourier–Hermite Kalman filter
fMRI	Functional magnetic resonance imaging
GHKF	Gauss–Hermite Kalman filter
GHPF	Gauss–Hermite particle filter
GHRTSS	Gauss–Hermite Rauch–Tung–Striebel smoother
GNSS	Global navigation satellite system
GPQF	Gaussian process quadrature filter
GPS	Global positioning system
GPU	Graphics processing unit
GSLR	Generalized statistical linear regression
HMC	Hamiltonian (or hybrid) Monte Carlo
HMM	Hidden Markov model
IEKF	Iterated extended Kalman filter
IEKS	Iterated extended Kalman smoother
IERTSS	Iterated extended RTS smoother
IKS	Iterated Kalman smoother

IMM	Interacting multiple model (algorithm)
INS	Inertial navigation system
IPL	Iterated posterior linearization
IPLF	Iterated posterior linearization filter
IPLS	Iterated posterior linearization smoother
IS	Importance sampling
IURTSS	Iterated unscented RTS smoother
InI	Inverse imaging
KF	Kalman filter
LIE	Law of iterated expectations
LLN	Law of large numbers
LMS	Least mean squares
LQG	Linear quadratic Gaussian (regulator)
MA	Moving average (model)
MAP	Maximum a posteriori
MC	Monte Carlo
MCKF	Monte Carlo Kalman filter
MCMC	Markov chain Monte Carlo
MCRTSS	Monte Carlo RTS smoother
MEG	Magnetoencephalography
MH	Metropolis–Hastings
MKF	Mixture Kalman filter
ML	Maximum likelihood
MLP	Multi-layer perceptron
MMSE	Minimum mean squared error
MNE	Minimum norm estimate
MSE	Mean squared error
PF	Particle filter
PL	Posterior linearization
PLF	Posterior linearization filter
PMCMC	Particle Markov chain Monte Carlo
PMMH	Particle marginal Metropolis–Hastings
QKF	Quadrature Kalman filter
RAM	Robust adaptive Metropolis (algorithm)
RBPF	Rao–Blackwellized particle filter
RBPS	Rao–Blackwellized particle smoother
RMSE	Root mean squared error
RTS	Rauch–Tung–Striebel
RTSS	Rauch–Tung–Striebel smoother
RW	Random walk (model)

SDE	Stochastic differential equation
SIR	Sequential importance resampling
SIR-PS	Sequential importance resampling particle smoother
SIS	Sequential importance sampling
SL	Statistical linearization
SLAM	Simultaneous localization and mapping
SLDS	Switching linear dynamic system
SLF	Statistically linearized filter
SLR	Statistical linear regression
SLRF	Statistical linear regression filter
SLRRTSS	Statistical linear regression RTS smoother
SLRTSS	Statistically linearized RTS smoother
SMC	Sequential Monte Carlo
SPCMKF	Sigma-point conditional moment Kalman filter
TV	Time varying
TVAR	Time-varying autoregressive (model)
UKF	Unscented Kalman filter
UPF	Unscented particle filter
URTSS	Unscented Rauch–Tung–Striebel smoother
UT	Unscented transform

1

What Are Bayesian Filtering and Smoothing?

The term *optimal filtering* traditionally refers to a class of methods that can be used for estimating the state of a time-varying system that is indirectly observed through noisy measurements. The term *optimal* in this context refers to statistical optimality. *Bayesian filtering* refers to the Bayesian way of formulating optimal filtering. In this book we use these terms interchangeably and always mean Bayesian filtering.

In optimal, Bayesian, and Bayesian optimal filtering, the *state* of the system refers to the collection of dynamic variables, such as position, velocity, orientation, and angular velocity, which fully describe the system. The *noise* in the measurements means that they are uncertain; even if we knew the true system state, the measurements would not be deterministic functions of the state but would have a distribution of possible values. The time evolution of the state is modeled as a dynamic system that is perturbed by a certain *process noise*. This noise is used for modeling the uncertainties in the system dynamics. In most cases the system is not truly stochastic, but stochasticity is used to represent the model uncertainties.

Bayesian smoothing (or optimal smoothing) is often considered to be a class of methods within the field of Bayesian filtering. While Bayesian filters in their basic form only compute estimates of the current state of the system given the history of measurements, Bayesian smoothers can be used to reconstruct states that happened before the current time. Although the term *smoothing* is sometimes used in a more general sense for methods that generate a smooth (as opposed to rough) representation of data, in the context of Bayesian filtering, the term (Bayesian) smoothing has this more definite meaning.

1.1 Applications of Bayesian Filtering and Smoothing

Phenomena that can be modeled as time-varying systems of the above type are very common in engineering applications. This kind of model can be found, for example, in navigation, aerospace engineering, space engineering, remote surveillance, telecommunications, physics, audio signal processing, control engineering, finance, and many other fields. Examples of such applications are the following.

- *Global positioning system (GPS)* (Kaplan, 1996) is a widely used satellite navigation system, where the GPS receiver unit measures arrival times of signals from several GPS satellites and computes its position based on these measurements (see Figure 1.1). The GPS receiver typically uses an extended Kalman filter (EKF) or some other optimal filtering algorithm[1] for computing the current position and velocity such that the measurements and the assumed dynamics (laws of physics) are taken into account. Also, the ephemeris information, which is the satellite reference information transmitted from the satellites to the GPS receivers, is typically generated using optimal filters.

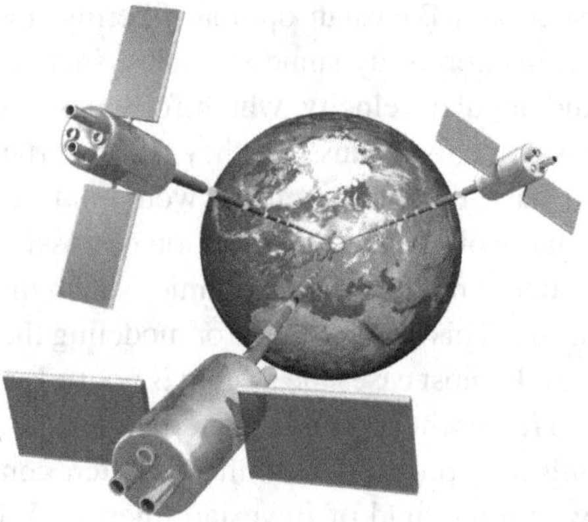

Figure 1.1 In the GPS system, the measurements are time delays of satellite signals, and the optimal filter (e.g., extended Kalman filter, EKF) computes the position and the accurate time.

- *Target tracking* (Bar-Shalom et al., 2001; Crassidis and Junkins, 2004; Challa et al., 2011) refers to the methodology where a set of sensors,

[1] Strictly speaking, the EKF is only an approximate optimal filtering algorithm because it uses a Taylor series-based Gaussian approximation to the non-Gaussian optimal filtering solution.

such as active or passive radars, radio frequency sensors, acoustic arrays, infrared sensors, or other types of sensors, are used for determining the position and velocity of a remote target (see Figure 1.2). When this tracking is done continuously in time, the dynamics of the target and measurements from the different sensors are most naturally combined using an optimal filter or smoother. The target in this (single) target tracking case can be, for example, a robot, a satellite, a car, or an airplane.

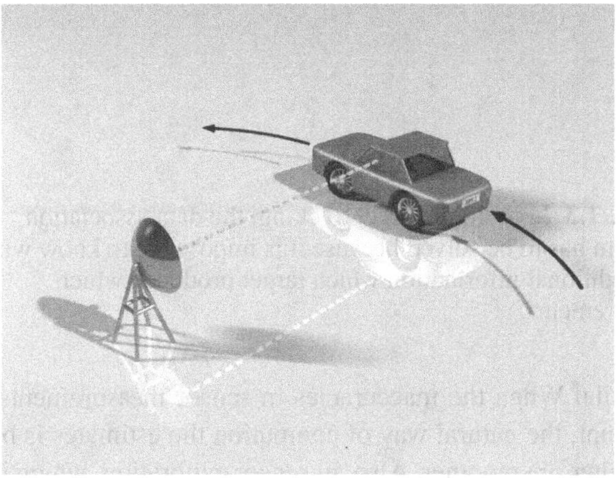

Figure 1.2 In target tracking, a sensor (e.g., radar) generates measurements (e.g., angle and distance measurements) of the target, and the purpose is to determine the target trajectory.

- *Multiple target tracking* (Bar-Shalom and Li, 1995; Blackman and Popoli, 1999; Mahler, 2014; Stone et al., 2014) systems are used for remote surveillance in cases where there are multiple targets moving at the same time in the same geographical area (see Figure 1.3). This introduces the concept of data association (which measurement was from which target?) and the problem of estimating the number of targets. Multiple target tracking systems are typically used in remote surveillance for military purposes, but their civil applications are, for example, monitoring of car tunnels, automatic alarm systems, and people tracking in buildings.
- *Inertial navigation* (Titterton and Weston, 1997; Grewal et al., 2001) uses inertial sensors, such as accelerometers and gyroscopes, for computing the position and velocity of a device, such as a car, an airplane,

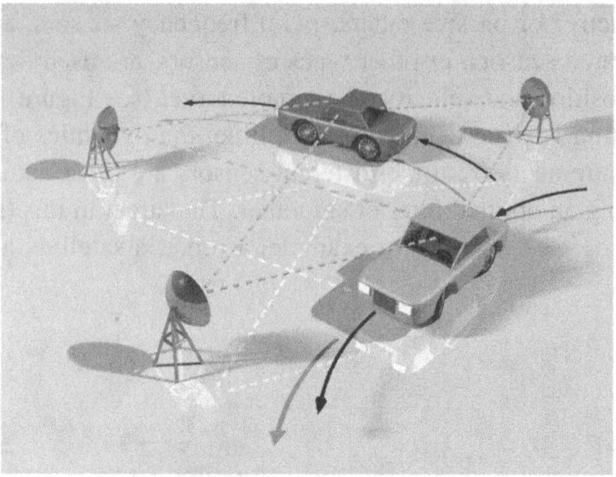

Figure 1.3 In multiple target tracking, the data association problem has to be solved because it is impossible to know without any additional information which target produced which measurement.

or a missile. When the inaccuracies in sensor measurements are taken into account, the natural way of computing the estimates is by using an optimal filter or smoother. Also, in sensor calibration, which is typically done in a time-varying environment, optimal filters and smoothers can be applied.

- *Integrated inertial navigation* (Bar-Shalom et al., 2001; Grewal et al., 2001) combines the good sides of unbiased but inaccurate sensors, such as altimeters and landmark trackers, and biased but locally accurate inertial sensors. Combination of these different sources of information is most naturally performed using an optimal filter, such as the extended Kalman filter. This kind of approach was used, for example, in the guidance system of the Apollo 11 lunar module (Eagle), which landed on the moon in 1969.
- *GPS/INS navigation* (Bar-Shalom et al., 2001; Grewal et al., 2001) is a form of integrated inertial navigation where the inertial navigation system (INS) is combined with a GPS receiver unit. In a GPS/INS navigation system, the short-term fluctuations of the GPS can be compensated by the inertial sensors, and the inertial sensor biases can be compensated by the GPS receiver. An additional advantage of this approach is that it is possible to temporarily switch to pure inertial navigation when the GPS

receiver is unable to compute its position (i.e., has no fix) for some reason. This happens, for example, indoors, in tunnels, and in other cases when there is no direct line-of-sight between the GPS receiver and the satellites.

- *Robotics and autonomous systems* (Thrun et al., 2005; Barfoot, 2017) typically use combinations of tracking and inertial navigation methods, along with sensors that measure the characteristics of the environment in one way or another. Examples of characteristics of the environment are radio signals or the locations of obstacles or landmarks detected from camera images. As the environment of the robot or autonomous system is typically unknown, the map of the environment also needs to be generated during the localization process. This concept is called simultaneous localization and mapping (SLAM), and the methodology for this purpose includes, for example, extended Kalman filters and particle filters.
- *Brain imaging* methods, such as electroencephalography (EEG), magnetoencephalography (MEG), parallel functional magnetic resonance imaging (fMRI), and diffuse optical tomography (DOT) (see Figure 1.4), are based on reconstruction of the source field in the brain from noisy sensor data by using the minimum norm estimates (MNE) technique and its variants (Hauk, 2004; Tarantola, 2004; Kaipio and Somersalo, 2005; Lin et al., 2006). The minimum norm solution can also be interpreted in the Bayesian sense as a problem of estimating the field with certain prior structure from Gaussian observations. With that interpretation, the estimation problem becomes equivalent to a *statistical inversion* or generalized *Gaussian process regression problem* (Tarantola, 2004; Kaipio and Somersalo, 2005; Rasmussen and Williams, 2006; Särkkä, 2011). Including dynamical priors then leads to a linear or non-linear spatio-temporal estimation problem, which can be solved with Kalman filters and smoothers (cf. Hiltunen et al., 2011; Särkkä et al., 2012b). The same can be done in inversion-based approaches to parallel fMRI, such as inverse imaging (InI) (Lin et al., 2006).
- *Spread of infectious diseases* (Keeling and Rohani, 2007) can often be modeled as differential equations for the number of susceptible, infected, recovered, and dead individuals. When uncertainties are introduced into the dynamic equations, and when the measurements are not perfect, the estimation of the spread of the disease can be formulated as an optimal filtering problem (see, e.g., Särkkä and Sottinen, 2008).
- *Biological processes* (Murray, 1993), such as population growth, predator–prey models, and several other dynamic processes in biology,

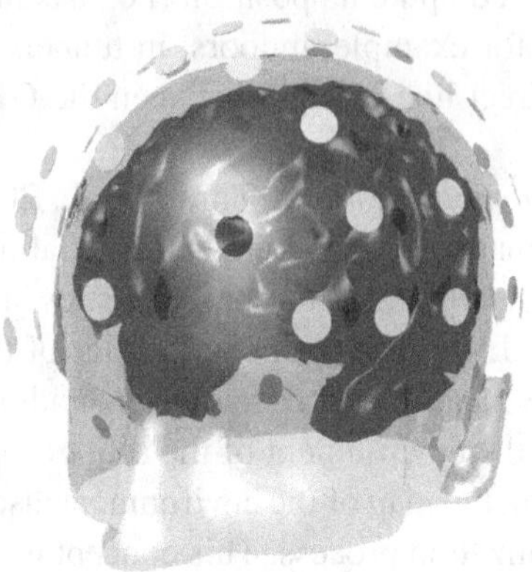

Figure 1.4 Brain imaging methods such as EEG and MEG are based on estimating the state of the brain from sensor readings. In the dynamic case, the related inversion problem can be solved with an optimal filter or smoother.

can also be modeled as (stochastic) differential equations. Estimation of the states of these processes from inaccurate measurements can be formulated as an optimal filtering and smoothing problem.

- *Telecommunications* is also a field where optimal filters are traditionally used. For example, optimal receivers, signal detectors, and phase locked loops can be interpreted to contain optimal filters (Van Trees, 1968, 1971; Proakis, 2001) as components. Also, the celebrated Viterbi algorithm (Viterbi, 1967) can be seen as a method for computing the maximum a posteriori (MAP) Bayesian smoothing solution for the underlying hidden Markov model (HMM).
- *Audio signal processing* applications, such as audio restoration (Godsill and Rayner, 1998) and audio signal enhancement (Fong et al., 2002), often use time-varying autoregressive (TVAR) models as the underlying audio signal models. These kinds of models can be efficiently estimated using optimal filters and smoothers.
- *Stochastic optimal control* (Aoki, 1967; Maybeck, 1982a; Stengel, 1994) considers control of time-varying stochastic systems. Stochastic controllers can typically be found in, for example, airplanes, cars, and rockets. Optimal, in addition to statistical optimality, means that the control

signal is constructed to minimize a performance cost, such as the expected time to reach a predefined state, the amount of fuel consumed, or the average distance from a desired position trajectory. When the state of the system is observed through a set of sensors, as it usually is, optimal filters are needed for reconstructing the state from them.

- *Learning systems* or adaptive systems can often be mathematically formulated in terms of optimal filters and smoothers (Haykin, 2001), and they have a close relationship to Bayesian non-parametric modeling, machine learning, and neural network modeling (Bishop, 2006). Methods similar to the data association methods in multiple target tracking are also applicable to on-line adaptive classification (Andrieu et al., 2002). The connection between Gaussian process regression (Rasmussen and Williams, 2006) and optimal filtering has also been discussed, for example, in Särkkä et al. (2007a), Hartikainen and Särkkä (2010), Särkkä et al. (2013), and Särkkä and Solin (2019).
- *Physical systems* that are time-varying and measured through non-ideal sensors can sometimes be formulated as stochastic state space models, and the time evolution of the system can be estimated using optimal filters (Kaipio and Somersalo, 2005). These kinds of problem are often called *inverse problems* (Tarantola, 2004), and optimal filters and smoothers can be seen as the Bayesian solutions to time-varying inverse problems.

1.2 Origins of Bayesian Filtering and Smoothing

The roots of Bayesian analysis of time-dependent behavior are in the field of optimal linear filtering. The idea of constructing mathematically optimal recursive estimators was first presented for linear systems due to their mathematical simplicity, and the most natural optimality criterion from both the mathematical and modeling points of view was least squares optimality. For linear systems, the optimal Bayesian solution (with minimum mean squared error, MMSE, loss) coincides with the least squares solution, that is, the optimal least squares solution is exactly the posterior mean.

The history of optimal filtering starts from the *Wiener filter* (Wiener, 1950), which is a frequency-domain solution to the problem of least squares optimal filtering of stationary Gaussian signals. The Wiener filter is still important in communication applications (Proakis, 2001), digital signal processing (Hayes, 1996), and image processing (Gonzalez and Woods, 2008). The disadvantage of the Wiener filter is that it can only be applied to stationary signals.

The success of optimal linear filtering in engineering applications is mostly due to the seminal article of Kalman (1960b), which describes the recursive solution to the optimal discrete-time (sampled) linear filtering problem. One reason for its success is that the *Kalman filter* can be understood and applied with very much lighter mathematical machinery than the Wiener filter. Also, despite its mathematical simplicity and generality, the Kalman filter (or actually the Kalman–Bucy filter (Kalman and Bucy, 1961)) contains the Wiener filter as its limiting special case.

In the early stages of its history, the Kalman filter was soon discovered to belong to the class of Bayesian filters (Ho and Lee, 1964; Lee, 1964; Jazwinski, 1966, 1970). The corresponding Bayesian smoothers (Rauch, 1963; Rauch et al., 1965; Leondes et al., 1970) were also developed soon after the invention of the Kalman filter. An interesting historical detail is that while Kalman and Bucy were formulating the linear theory in the United States, Stratonovich was doing the pioneering work on the probabilistic (Bayesian) approach in Russia (Stratonovich, 1968; Jazwinski, 1970).

As discussed in the book of West and Harrison (1997), in the 1960s, Kalman filter-like recursive estimators were also used in the Bayesian community, and it is not clear whether the theory of Kalman filtering or the theory of *dynamic linear models* (DLM) came first. Although these theories were originally derived from slightly different starting points, they are equivalent. Because of the Kalman filter's useful connection to the theory and history of stochastic optimal control, this book approaches the Bayesian filtering problem from the Kalman filtering point of view.

Although the original derivation of the *Kalman filter* was based on the least squares approach, the same equations can be derived from pure probabilistic Bayesian analysis. The Bayesian analysis of Kalman filtering is well covered in the classical book of Jazwinski (1970) and a bit more recently in the book of Bar-Shalom et al. (2001). Kalman filtering, mostly because of its least squares interpretation, has been widely used in stochastic optimal control. A practical reason for this is that the inventor of the Kalman filter, Rudolph E. Kalman, has also made several contributions to the theory of *linear quadratic Gaussian* (LQG) regulators (Kalman, 1960a), which are fundamental tools of stochastic optimal control (Stengel, 1994; Maybeck, 1982a).

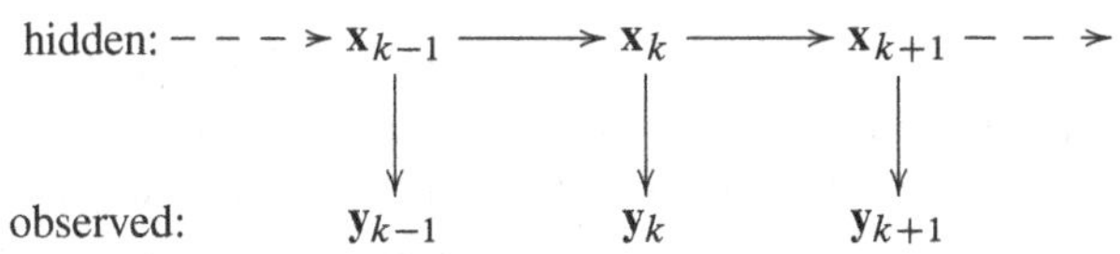

Figure 1.5 In optimal filtering and smoothing problems a sequence of hidden states $\mathbf{x}_k$ is indirectly observed through noisy measurements $\mathbf{y}_k$.

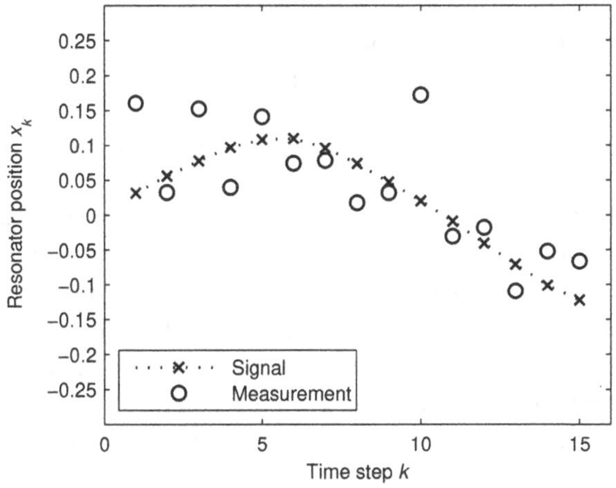

Figure 1.6 An example of a time series, which models a discrete-time resonator. The actual resonator state (signal) is hidden and only observed through the noisy measurements.

1.3 Optimal Filtering and Smoothing as Bayesian Inference

In mathematical terms, optimal filtering and smoothing are considered to be statistical inversion problems, where the unknown quantity is a vector-valued time series $\{\mathbf{x}_0, \mathbf{x}_1, \mathbf{x}_2, \ldots\}$ that is observed through a set of noisy measurements $\{\mathbf{y}_1, \mathbf{y}_2, \ldots\}$, as illustrated in Figure 1.5. An example of this kind of time series is shown in Figure 1.6. The process shown is a noisy resonator with a known angular velocity. The state $\mathbf{x}_k = (x_k \; \dot{x}_k)^\mathsf{T}$ is two dimensional (2D) and consists of the position of the resonator x_k and its time derivative $\dot{x}_k$. The measurements y_k are scalar observations of the resonator position (signal), and they are corrupted by measurement noise.

The purpose of the *statistical inversion* at hand is to estimate the hidden states $\mathbf{x}_{0:T} = \{\mathbf{x}_0, \ldots, \mathbf{x}_T\}$ from the observed measurements $\mathbf{y}_{1:T} = \{\mathbf{y}_1, \ldots, \mathbf{y}_T\}$, which means that in the Bayesian sense we want to compute the joint *posterior distribution* of all the states given all the measurements. In principle, this can be done by a straightforward application of Bayes' rule

$$p(\mathbf{x}_{0:T} \mid \mathbf{y}_{1:T}) = \frac{p(\mathbf{y}_{1:T} \mid \mathbf{x}_{0:T})\, p(\mathbf{x}_{0:T})}{p(\mathbf{y}_{1:T})}, \tag{1.1}$$

where

- $p(\mathbf{x}_{0:T})$ is the *prior distribution* defined by the dynamic model,
- $p(\mathbf{y}_{1:T} \mid \mathbf{x}_{0:T})$ is the likelihood model for the measurements,
- $p(\mathbf{y}_{1:T})$ is the normalization constant defined as

$$p(\mathbf{y}_{1:T}) = \int p(\mathbf{y}_{1:T} \mid \mathbf{x}_{0:T})\, p(\mathbf{x}_{0:T})\, \mathrm{d}\mathbf{x}_{0:T}. \tag{1.2}$$

Unfortunately, this full posterior formulation has the serious disadvantage that each time we obtain a new measurement, the full posterior distribution would have to be recomputed. This is particularly a problem in dynamic estimation (which is exactly the problem we are solving here!), where measurements are typically obtained one at a time, and we would want to compute the best possible estimate after each measurement. When the number of time steps increases, the dimensionality of the full posterior distribution also increases, which means that the computational complexity of a single time step increases. Thus eventually the computations will become intractable, no matter how much computational power is available. Without additional information or restrictive approximations, there is no way of getting over this problem in the full posterior computation.

However, the above problem only arises when we want to compute the *full* posterior distribution of the states at each time step. If we are willing to relax this a bit and be satisfied with selected marginal distributions of the states, the computations become an order of magnitude lighter. To achieve this, we also need to restrict the class of dynamic models to probabilistic Markov sequences, which is not as restrictive as it may at first seem. The model for the states and measurements will be assumed to be of the following type.

- **An initial distribution** specifies the *prior probability distribution* $p(\mathbf{x}_0)$ of the hidden state $\mathbf{x}_0$ at the initial time step $k = 0$.

- **A dynamic model** describes the system dynamics and its uncertainties as a *Markov sequence*, defined in terms of the transition probability distribution $p(\mathbf{x}_k \mid \mathbf{x}_{k-1})$.
- **A measurement model** describes how the measurement $\mathbf{y}_k$ depends on the current state $\mathbf{x}_k$. This dependence is modeled by specifying the conditional probability distribution of the measurement given the state, which is denoted as $p(\mathbf{y}_k \mid \mathbf{x}_k)$.

Thus a general probabilistic *state space model* is usually written in the following form:

$$\begin{aligned}\mathbf{x}_0 &\sim p(\mathbf{x}_0),\\ \mathbf{x}_k &\sim p(\mathbf{x}_k \mid \mathbf{x}_{k-1}),\\ \mathbf{y}_k &\sim p(\mathbf{y}_k \mid \mathbf{x}_k).\end{aligned} \tag{1.3}$$

Because computing the full joint distribution of the states at all time steps is computationally very inefficient and unnecessary in real-time applications, in *Bayesian filtering and smoothing* the following marginal distributions are considered instead (see Figure 1.7).

- *Filtering distributions* computed by the *Bayesian filter* are the marginal distributions of *the current state* $\mathbf{x}_k$ given *the current and previous measurements* $\mathbf{y}_{1:k} = \{\mathbf{y}_1, \ldots, \mathbf{y}_k\}$:

  $$p(\mathbf{x}_k \mid \mathbf{y}_{1:k}), \qquad k = 1, \ldots, T. \tag{1.4}$$

  The result of applying the Bayesian filter to the resonator time series in Figure 1.6 is shown in Figure 1.8.
- *Prediction distributions*, which can be computed with the *prediction step of the Bayesian filter*, are the marginal distributions of the *future state* $\mathbf{x}_{k+n}$, n steps after the current time step:

  $$p(\mathbf{x}_{k+n} \mid \mathbf{y}_{1:k}), \qquad k = 1, \ldots, T, \quad n = 1, 2, \ldots. \tag{1.5}$$

- *Smoothing distributions* computed by the *Bayesian smoother* are the marginal distributions of the state $\mathbf{x}_k$ given a certain interval $\mathbf{y}_{1:T} = \{\mathbf{y}_1, \ldots, \mathbf{y}_T\}$ of measurements with $T > k$:

  $$p(\mathbf{x}_k \mid \mathbf{y}_{1:T}), \qquad k = 1, \ldots, T. \tag{1.6}$$

  The result of applying the Bayesian smoother to the resonator time series is shown in Figure 1.9.

Computing filtering, prediction, and smoothing distributions require only a constant number of computations per time step, and thus the problem of processing arbitrarily long time series is solved.

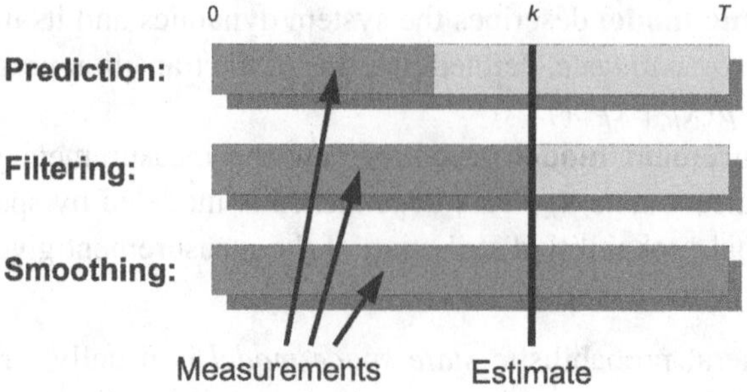

Figure 1.7 State estimation problems can be divided into optimal prediction, filtering, and smoothing, depending on the time span of the measurements available with respect to the time of the estimated state.

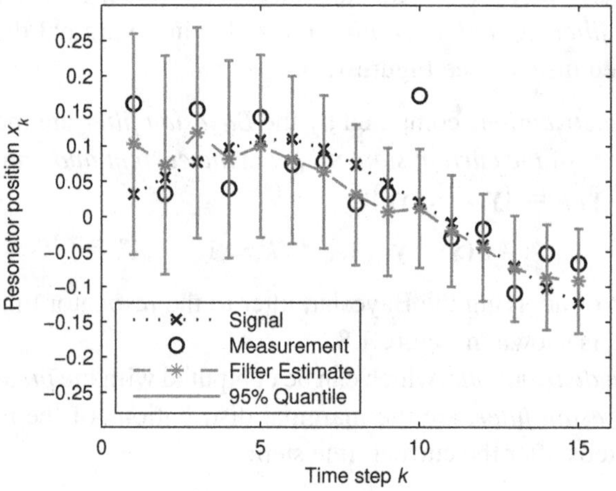

Figure 1.8 The result of computing the filtering distributions for the discrete-time resonator model. The *estimates* are the means of the filtering distributions, and the quantiles are the 95% quantiles of the filtering distributions.

1.4 Algorithms for Bayesian Filtering and Smoothing

There exist a few classes of filtering and smoothing problems that have closed form solutions.

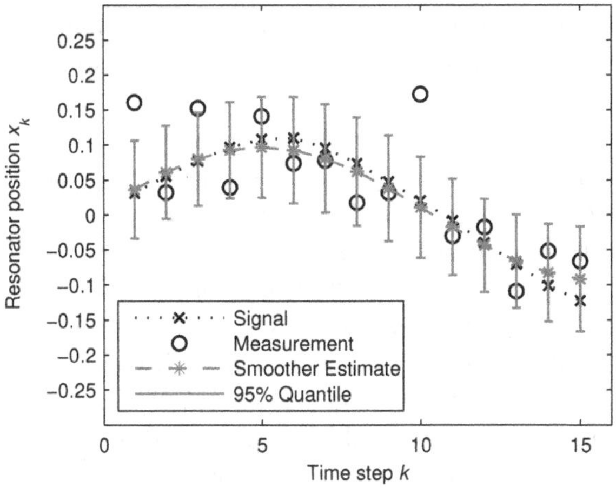

Figure 1.9 The result of computing the smoothing distributions for the discrete-time resonator model. The *estimates* are the means of the smoothing distributions, and the quantiles are the 95% quantiles of the smoothing distributions.

- *The Kalman filter* (KF) is a closed form solution to the linear Gaussian filtering problem. Due to linear Gaussian model assumptions, the posterior distribution is exactly Gaussian, and no numerical approximations are needed.
- *The Rauch–Tung–Striebel smoother* (RTSS) is the corresponding closed form smoother for linear Gaussian state space models.
- *Finite-state filters and smoothers* are solutions for hidden Markov models (HMMs) with finite state spaces.

But because the Bayesian optimal filtering and smoothing equations are generally computationally intractable, many kinds of numerical approximation methods have been developed, for example:

- *The extended Kalman filter* (EKF) approximates the non-linear and non-Gaussian measurement and dynamic models by linearization, that is, by forming a Taylor series expansion at the nominal (or maximum a posteriori, MAP) solution. This results in a Gaussian approximation to the filtering distribution.
- *The extended Rauch–Tung–Striebel smoother* (ERTSS) is the approximate non-linear smoothing algorithm corresponding to the EKF.

- *The unscented Kalman filter* (UKF) approximates the propagation of densities through the non-linearities of measurement and noise processes using the *unscented transform*. This also results in a Gaussian approximation.
- *The unscented Rauch–Tung–Striebel smoother* (URTSS) is the approximate non-linear smoothing algorithm corresponding to the UKF.
- *Sequential Monte Carlo methods* or *particle filters and smoothers* represent the posterior distribution as a weighted set of Monte Carlo samples.
- *The unscented particle filter* (UPF) and *local linearization*-based particle filtering methods use UKFs and EKFs, respectively, for approximating the optimal importance distributions in particle filters.
- *Rao–Blackwellized particle filters and smoothers* use closed form integration (e.g., Kalman filters and RTS smoothers) for some of the state variables and Monte Carlo integration for others.
- *Grid-based approximation methods* approximate the filtering and smoothing distributions as discrete distributions on a finite grid.
- *Other methods* also exist, for example, based on Gaussian mixtures, series expansions, describing functions, basis function expansions, exponential family of distributions, variational Bayesian methods, and batch Monte Carlo (e.g., Markov chain Monte Carlo, MCMC, methods).

1.5 Parameter Estimation

In state space models of dynamic systems, there are often *unknown or uncertain parameters* $\boldsymbol{\theta}$, which should be estimated along with the state itself. For example, in a stochastic resonator model, the frequency of the resonator might be unknown. Also, the noise variances might be only known approximately, or they can be completely unknown. Although, formally, we can always augment unknown parameters as part of the state, in practice it is often useful to consider parameter estimation separately.

In a Bayesian setting, the proper way to estimate the parameters is by setting a prior distribution on the parameters $p(\boldsymbol{\theta})$ and treating them as additional random variables in the model. When unknown parameters are present, the state space model in Equation (1.3) becomes

$$\begin{aligned}
\boldsymbol{\theta} &\sim p(\boldsymbol{\theta}), \\
\mathbf{x}_0 &\sim p(\mathbf{x}_0 \mid \boldsymbol{\theta}), \\
\mathbf{x}_k &\sim p(\mathbf{x}_k \mid \mathbf{x}_{k-1}, \boldsymbol{\theta}), \\
\mathbf{y}_k &\sim p(\mathbf{y}_k \mid \mathbf{x}_k, \boldsymbol{\theta}).
\end{aligned} \tag{1.7}$$

The full Bayesian solution to this problem would require the computation of the full *joint posterior distribution of states and parameters* $p(\mathbf{x}_{0:T}, \boldsymbol{\theta} \mid \mathbf{y}_{1:T})$. Unfortunately, computing this joint posterior of the states and parameters is even harder than computation of the joint distribution of states alone, and thus this task is intractable.

Fortunately, when run with fixed parameters $\boldsymbol{\theta}$, the Bayesian filter algorithm produces the sequence of distributions $p(\mathbf{y}_k \mid \mathbf{y}_{1:k-1}, \boldsymbol{\theta})$ for $k = 1, \ldots, T$ as side products. Once we have these, we can form the *marginal posterior distribution of parameters* as follows:

$$p(\boldsymbol{\theta} \mid \mathbf{y}_{1:T}) \propto p(\boldsymbol{\theta}) \prod_{k=1}^{T} p(\mathbf{y}_k \mid \mathbf{y}_{1:k-1}, \boldsymbol{\theta}), \tag{1.8}$$

where we have denoted $p(\mathbf{y}_1 \mid \mathbf{y}_{1:0}, \boldsymbol{\theta}) \triangleq p(\mathbf{y}_1 \mid \boldsymbol{\theta})$ for notational convenience. When combined with the smoothing distributions, we can form all the marginal joint distributions of states and parameters as follows:

$$p(\mathbf{x}_k, \boldsymbol{\theta} \mid \mathbf{y}_{1:T}) = p(\mathbf{x}_k \mid \mathbf{y}_{1:T}, \boldsymbol{\theta})\, p(\boldsymbol{\theta} \mid \mathbf{y}_{1:T}) \tag{1.9}$$

for $k = 1, \ldots, T$, where $p(\mathbf{x}_k \mid \mathbf{y}_{1:T}, \boldsymbol{\theta})$ is the smoothing distribution of the states with fixed model parameters $\boldsymbol{\theta}$. However, we cannot compute the full joint posterior distribution of states and parameters, which is the price of only using a constant number of computations per time step.

Although here we use the term *parameter estimation*, it might sometimes be the case that we are not actually interested in the values of the parameters as such, but we just do not know their values. In that case the proper Bayesian approach is to *integrate out* the parameters. For example, to compute the smoothing distributions in the presence of unknown parameters we can integrate out the parameters from the joint distribution in Equation (1.9):

$$\begin{aligned} p(\mathbf{x}_k \mid \mathbf{y}_{1:T}) &= \int p(\mathbf{x}_k, \boldsymbol{\theta} \mid \mathbf{y}_{1:T})\, \mathrm{d}\boldsymbol{\theta} \\ &= \int p(\mathbf{x}_k \mid \mathbf{y}_{1:T}, \boldsymbol{\theta})\, p(\boldsymbol{\theta} \mid \mathbf{y}_{1:T})\, \mathrm{d}\boldsymbol{\theta}. \end{aligned} \tag{1.10}$$

Many of the Bayesian methods for parameter estimation indeed allow this to be done (approximately). For example, by using the parameter samples produced by a Markov chain Monte Carlo (MCMC) method, it is possible to form a Monte Carlo approximation to the above integral.

1.6 Exercises

1.1 Find the seminal article of Kalman (1960b) from the internet (or from a library) and investigate the orthogonal projections approach that is taken in the article. How would you generalize the approach to the non-linear/non-Gaussian case? Is it possible?

1.2 An alternative to Bayesian estimation would be to formulate the state estimation problem as maximum likelihood (ML) estimation. This would amount to estimating the state sequence as the ML-estimate

$$\hat{\mathbf{x}}_{0:T} = \arg\max_{\mathbf{x}_{0:T}} p(\mathbf{y}_{1:T} \mid \mathbf{x}_{0:T}). \tag{1.11}$$

Do you see any problem with this approach? *Hint:* Where is the dynamic model?

1.3 Assume that in an electronics shop, the salesperson decides to give you a chance to win a brand new GPS receiver. He lets you choose one of three packages of which one contains the GPS receiver and two others are empty. After you have chosen the package, the salesperson opens one of the packages that you have not chosen – and that package turns out to be empty. He gives you a chance to switch to the other yet unopened package. Is it advantageous for you to do that?

2

Bayesian Inference

This chapter provides a brief presentation of the philosophical and mathematical foundations of Bayesian inference. The connections to classical statistical inference are also briefly discussed.

2.1 Philosophy of Bayesian Inference

The purpose of *Bayesian inference* (Bernardo and Smith, 1994; Gelman et al., 2013) is to provide a mathematical machinery that can be used for modeling systems, where the uncertainties of the system are taken into account, and the decisions are made according to rational principles. The tools of this machinery are probability distributions and the rules of probability calculus.

If we compare the so-called frequentist philosophy of statistical analysis to Bayesian inference, the difference is that in Bayesian inference, the probability of an event does not mean the proportion of the event in an infinite number of trials but the certainty of the event in a single trial. Because models in Bayesian inference are formulated in terms of probability distributions, the probability axioms and computation rules of probability theory (see, e.g., Shiryaev, 1996) also apply in Bayesian inference.

2.2 Connection to Maximum Likelihood Estimation

Consider a situation where we know the conditional distribution $p(\mathbf{y}_k \mid \boldsymbol{\theta})$ of conditionally independent random variables (measurements) $\mathbf{y}_{1:T} = \{\mathbf{y}_1, \ldots, \mathbf{y}_T\}$, but the parameter $\boldsymbol{\theta} \in \mathbb{R}^d$ is unknown. The classical statistical method for estimating the parameter is the *maximum likelihood method* (Milton and Arnold, 1995), where we maximize the joint probability of the measurements, also called the likelihood function,

$$\mathcal{L}(\boldsymbol{\theta}) = \prod_{k=1}^{T} p(\mathbf{y}_k \mid \boldsymbol{\theta}). \tag{2.1}$$

The maximum of the likelihood function with respect to $\boldsymbol{\theta}$ gives the *maximum likelihood estimate* (ML-estimate)

$$\hat{\boldsymbol{\theta}} = \arg\max_{\boldsymbol{\theta}} \mathcal{L}(\boldsymbol{\theta}). \tag{2.2}$$

The difference between Bayesian inference and the maximum likelihood method is that the starting point of Bayesian inference is to formally consider the parameter $\boldsymbol{\theta}$ as a random variable. Then the posterior distribution of the parameter $\boldsymbol{\theta}$ can be computed by using *Bayes' rule*,

$$p(\boldsymbol{\theta} \mid \mathbf{y}_{1:T}) = \frac{p(\mathbf{y}_{1:T} \mid \boldsymbol{\theta})\, p(\boldsymbol{\theta})}{p(\mathbf{y}_{1:T})}, \tag{2.3}$$

where $p(\boldsymbol{\theta})$ is the prior distribution, which models the prior beliefs on the parameter before we have seen any data, and $p(\mathbf{y}_{1:T})$ is a normalization term, which is independent of the parameter $\boldsymbol{\theta}$. This normalization constant is often left out, and if the measurements $\mathbf{y}_{1:T}$ are conditionally independent given $\boldsymbol{\theta}$, the posterior distribution of the parameter can be written as

$$p(\boldsymbol{\theta} \mid \mathbf{y}_{1:T}) \propto p(\boldsymbol{\theta}) \prod_{k=1}^{T} p(\mathbf{y}_k \mid \boldsymbol{\theta}). \tag{2.4}$$

Because we are dealing with a distribution, we might now choose the most probable value of the random variable, the *maximum a posteriori* (MAP) estimate, which is given by the maximum of the posterior distribution. The optimal estimate in the mean squared sense is the posterior mean of the parameter (MMSE-estimate). There are an infinite number of other ways of choosing the point estimate from the distribution, and the best way depends on the assumed loss or cost function (or utility function). The ML-estimate can be seen as a MAP-estimate with uniform prior $p(\boldsymbol{\theta}) \propto 1$ on the parameter $\boldsymbol{\theta}$.

We can also interpret Bayesian inference as a convenient method for including regularization terms into maximum likelihood estimation. The basic ML-framework does not have a self-consistent method for including regularization terms or prior information into statistical models. However, this regularization interpretation of Bayesian inference is quite limited because Bayesian inference is much more than this.

2.3 The Building Blocks of Bayesian Models

The basic blocks of a Bayesian model are the *prior model* containing the preliminary information on the parameter and the *measurement model* determining the stochastic mapping from the parameter to the measurements. Using combination rules, namely Bayes' rule, it is possible to infer an estimate of the parameters from the measurements. The probability distribution of the parameters, conditional on the observed measurements, is called the *posterior distribution*, and it is the distribution representing the state of knowledge about the parameters when all the information in the observed measurements and the model is used. The *predictive posterior distribution* is the distribution of new (not yet observed) measurements when all the information in the observed measurements and the model is used.

- **Prior model**
  The prior information consists of subjective experience-based beliefs about the possible and impossible parameter values and their relative likelihoods before anything has been observed. The prior distribution is a mathematical representation of this information:
  $$p(\boldsymbol{\theta}) = \text{information on parameter } \boldsymbol{\theta} \text{ before seeing any observations.} \tag{2.5}$$
  The lack of prior information can be expressed by using a non-informative prior. The non-informative prior distribution can be selected in various different ways (Gelman et al., 2013).
- **Measurement model**
  Between the true parameters and the measurements, there often is a causal, but inaccurate or noisy relationship. This relationship is mathematically modeled using the measurement model:
  $$p(\mathbf{y} \mid \boldsymbol{\theta}) = \text{distribution of observation } \mathbf{y} \text{ given the parameters } \boldsymbol{\theta}. \tag{2.6}$$
- **Posterior distribution**
  The posterior distribution is the conditional distribution of the parameters given the observations. It represents the information we have after the measurement $\mathbf{y}$ has been obtained. It can be computed by using Bayes' rule,
  $$p(\boldsymbol{\theta} \mid \mathbf{y}) = \frac{p(\mathbf{y} \mid \boldsymbol{\theta})\, p(\boldsymbol{\theta})}{p(\mathbf{y})} \propto p(\mathbf{y} \mid \boldsymbol{\theta})\, p(\boldsymbol{\theta}), \tag{2.7}$$

where the normalization constant is given as

$$p(\mathbf{y}) = \int p(\mathbf{y} \mid \boldsymbol{\theta})\, p(\boldsymbol{\theta})\, \mathrm{d}\boldsymbol{\theta}. \tag{2.8}$$

In the case of multiple measurements $\mathbf{y}_{1:T}$, if the measurements are conditionally independent, the joint likelihood of all measurements is the product of distributions of the individual measurements, and the posterior distribution is

$$p(\boldsymbol{\theta} \mid \mathbf{y}_{1:T}) \propto p(\boldsymbol{\theta}) \prod_{k=1}^{T} p(\mathbf{y}_k \mid \boldsymbol{\theta}), \tag{2.9}$$

where the normalization term can be computed by integrating the right-hand side over $\boldsymbol{\theta}$. If the random variable is discrete, the integration is replaced by summation.

- **Predictive posterior distribution**
  The predictive posterior distribution is the distribution of new measurements $\mathbf{y}_{T+1}$ given the observed measurements

$$p(\mathbf{y}_{T+1} \mid \mathbf{y}_{1:T}) = \int p(\mathbf{y}_{T+1} \mid \boldsymbol{\theta})\, p(\boldsymbol{\theta} \mid \mathbf{y}_{1:T})\, \mathrm{d}\boldsymbol{\theta}. \tag{2.10}$$

  Thus, after obtaining the measurements $\mathbf{y}_{1:T}$, the predictive posterior distribution can be used to compute the probability distribution for measurement index $T + 1$, which has not been observed yet.

In the case of tracking, we could imagine that the parameter is the sequence of dynamic states of a target, where the state contains the position and velocity. The measurements could be, for example, noisy distance and direction measurements produced by a radar. In this book we will divide the parameters into two classes: the dynamic state of the system and the static parameters of the model. But from the Bayesian estimation point of view, both the states and static parameters are unknown (random) parameters of the system.

2.4 Bayesian Point Estimates

In many practical applications, distributions alone have no use; we need finite dimensional summaries (point estimates). This selection of a point based on observed values of random variables is a statistical decision, and therefore this selection procedure is most naturally formulated in terms of *statistical decision theory* (Berger, 1985; Bernardo and Smith, 1994; Raiffa and Schlaifer, 2000).

Definition 2.1 (Loss function). *A loss function or cost function* $C(\boldsymbol{\theta}, \mathbf{a})$ *is a scalar-valued function that determines the loss of taking the* action $\mathbf{a}$ *when the true parameter value is* $\boldsymbol{\theta}$. *The action (or control) is the statistical decision to be made based on the currently available information.*

Instead of loss functions, it is also possible to work with utility functions $U(\boldsymbol{\theta}, \mathbf{a})$, which determine the reward from taking the action $\mathbf{a}$ with parameter values $\boldsymbol{\theta}$. Loss functions can be converted to utility functions and vice versa by defining $U(\boldsymbol{\theta}, \mathbf{a}) = -C(\boldsymbol{\theta}, \mathbf{a})$.

If the value of the parameter $\boldsymbol{\theta}$ is not known, but the knowledge of the parameter can be expressed in terms of the posterior distribution $p(\boldsymbol{\theta} \mid \mathbf{y}_{1:T})$, then the natural choice is the action that gives the *minimum (maximum) of the expected loss (utility)* (Berger, 1985):

$$\mathrm{E}[C(\boldsymbol{\theta}, \mathbf{a}) \mid \mathbf{y}_{1:T}] = \int C(\boldsymbol{\theta}, \mathbf{a}) \, p(\boldsymbol{\theta} \mid \mathbf{y}_{1:T}) \, \mathrm{d}\boldsymbol{\theta}. \tag{2.11}$$

Commonly used loss functions are the following.

- *Quadratic error loss.* If the loss function is quadratic,

  $$C(\boldsymbol{\theta}, \mathbf{a}) = (\boldsymbol{\theta} - \mathbf{a})^{\mathsf{T}} (\boldsymbol{\theta} - \mathbf{a}), \tag{2.12}$$

  then the optimal choice $\mathbf{a}_{\mathrm{o}}$ is the *mean* of the posterior distribution of $\boldsymbol{\theta}$,

  $$\mathbf{a}_{\mathrm{o}} = \int \boldsymbol{\theta} \, p(\boldsymbol{\theta} \mid \mathbf{y}_{1:T}) \, \mathrm{d}\boldsymbol{\theta}. \tag{2.13}$$

  This posterior mean-based estimate is often called the *minimum mean squared error (MMSE)* estimate of the parameter $\boldsymbol{\theta}$. The quadratic loss is the most commonly used loss function for regression problems because it is easy to handle mathematically and because in the case of a Gaussian posterior distribution, the MAP estimate and the median coincide with the posterior mean.
- *Absolute error loss.* The loss function of the form

  $$C(\boldsymbol{\theta}, \mathbf{a}) = \sum_i |\theta_i - a_i| \tag{2.14}$$

  is called an absolute error loss, and in this case the optimal choice is the *median* of the distribution (the medians of the marginal distributions in the multi-dimensional case).
- *0–1 loss.* If the loss function is of the form

  $$C(\boldsymbol{\theta}, \mathbf{a}) = -\delta(\mathbf{a} - \boldsymbol{\theta}), \tag{2.15}$$

where $\delta(\cdot)$ is the Dirac delta function, then the optimal choice is the maximum (mode) of the posterior distribution, that is, the *maximum a posteriori (MAP)* estimate of the parameter. If the random variable $\boldsymbol{\theta}$ is discrete, the corresponding loss function can be defined as

$$C(\boldsymbol{\theta}, \mathbf{a}) = \begin{cases} 0, & \text{if } \boldsymbol{\theta} = \mathbf{a}, \\ 1, & \text{if } \boldsymbol{\theta} \neq \mathbf{a}. \end{cases} \tag{2.16}$$

2.5 Numerical Methods

In principle, Bayesian inference provides the equations for computing the posterior distributions and point estimates for any model once the model specification has been set up. However, the practical difficulty is that computation of the integrals involved in the equations can rarely be performed analytically, and numerical methods are needed. Here we briefly describe numerical methods that are also applicable in higher-dimensional problems: Gaussian approximations, multi-dimensional quadratures, Monte Carlo methods, and importance sampling.

- *Gaussian approximations* (Gelman et al., 2013) are very common, and in them the posterior distribution is approximated by a Gaussian distribution (see Section A.1)

$$p(\boldsymbol{\theta} \mid \mathbf{y}_{1:T}) \simeq \mathrm{N}(\boldsymbol{\theta} \mid \mathbf{m}, \mathbf{P}). \tag{2.17}$$

  The mean $\mathbf{m}$ and covariance $\mathbf{P}$ of the Gaussian approximation can be computed either by matching the first two moments of the posterior distribution or by using the mode of the distribution as the approximation of $\mathbf{m}$ and by approximating $\mathbf{P}$ using the curvature of the posterior at the mode. Note that above we have introduced the notation $\simeq$, which here means that the left-hand side is *assumed* to be approximately equal to the right-hand side, even though we know that this will not be true in most practical situations nor can we control the approximation error in any practical way.
- *Multi-dimensional quadrature or cubature integration methods*, such as Gauss–Hermite quadrature, can also be used if the dimensionality of the integral is moderate. The idea is to deterministically form a representative set of sample points $\{\boldsymbol{\theta}^{(i)} : i = 1, \ldots, N\}$ (sometimes called *sigma*

points) and form the approximation of the integral as the weighted average

$$\mathrm{E}[\mathbf{g}(\boldsymbol{\theta}) \mid \mathbf{y}_{1:T}] \approx \sum_{i=1}^{N} W_i \, \mathbf{g}(\boldsymbol{\theta}^{(i)}), \tag{2.18}$$

where the numerical values of the weights W_i are determined by the algorithm. The sample points and weights can be selected, for example, to give exact answers for polynomials up to certain degree or to account for the moments up to certain degree. Above we have used the notation $\approx$ to mean that the expressions are approximately equal in some suitable limit (here $N \to \infty$) or in some verifiable conditions.

- In direct *Monte Carlo methods*, a set of N samples from the posterior distribution is randomly drawn,

$$\boldsymbol{\theta}^{(i)} \sim p(\boldsymbol{\theta} \mid \mathbf{y}_{1:T}), \qquad i = 1, \ldots, N, \tag{2.19}$$

and the expectation of any function $\mathbf{g}(\cdot)$ can be then approximated as the sample average

$$\mathrm{E}[\mathbf{g}(\boldsymbol{\theta}) \mid \mathbf{y}_{1:T}] \approx \frac{1}{N} \sum_{i} \mathbf{g}(\boldsymbol{\theta}^{(i)}). \tag{2.20}$$

Another interpretation of this is that Monte Carlo methods form an approximation of the posterior density of the form

$$p(\boldsymbol{\theta} \mid \mathbf{y}_{1:T}) \approx \frac{1}{N} \sum_{i=1}^{N} \delta(\boldsymbol{\theta} - \boldsymbol{\theta}^{(i)}), \tag{2.21}$$

where $\delta(\cdot)$ is the Dirac delta function. The convergence of Monte Carlo approximation is guaranteed by the *central limit theorem (CLT)* (see, e.g., Liu, 2001), and the error term is, at least in theory, under certain ideal conditions, independent of the dimensionality of $\boldsymbol{\theta}$. The rule of thumb is that the error should decrease as the square root of the number of samples, regardless of the dimensions.

- Efficient methods for generating Monte Carlo samples are the *Markov chain Monte Carlo* (MCMC) methods (see, e.g., Gilks et al., 1996; Liu, 2001; Brooks et al., 2011). In MCMC methods, a Markov chain is constructed such that it has the target distribution as its stationary distribution. By simulating the Markov chain, samples from the target distribution can be generated.
- *Importance sampling* (see, e.g., Liu, 2001) is a simple algorithm for generating *weighted* samples from the target distribution. The difference

between this and direct Monte Carlo sampling and MCMC is that each of the particles has an associated weight, which corrects for the difference between the actual target distribution and the approximate importance distribution $\pi(\cdot)$ from which the sample was drawn.

An importance sampling estimate can be formed by drawing N samples from the *importance distribution*

$$\boldsymbol{\theta}^{(i)} \sim \pi(\boldsymbol{\theta} \mid \mathbf{y}_{1:T}), \qquad i = 1, \ldots, N. \tag{2.22}$$

The *importance weights* are then computed as

$$\tilde{w}^{(i)} = \frac{1}{N} \frac{p(\boldsymbol{\theta}^{(i)} \mid \mathbf{y}_{1:T})}{\pi(\boldsymbol{\theta}^{(i)} \mid \mathbf{y}_{1:T})}, \tag{2.23}$$

and the expectation of any function $\mathbf{g}(\cdot)$ can be then approximated as

$$\mathrm{E}[\mathbf{g}(\boldsymbol{\theta}) \mid \mathbf{y}_{1:T}] \approx \sum_{i=1}^{N} \tilde{w}^{(i)} \, \mathbf{g}(\boldsymbol{\theta}^{(i)}), \tag{2.24}$$

or alternatively as

$$\mathrm{E}[\mathbf{g}(\boldsymbol{\theta}) \mid \mathbf{y}_{1:T}] \approx \frac{\sum_{i=1}^{N} \tilde{w}^{(i)} \, \mathbf{g}(\boldsymbol{\theta}^{(i)})}{\sum_{i=1}^{N} \tilde{w}^{(i)}}. \tag{2.25}$$

2.6 Exercises

2.1 Prove that the median of distribution $p(\theta)$ minimizes the expected value of the absolute error loss function

$$\mathrm{E}[|\theta - a|] = \int |\theta - a| \, p(\theta) \, \mathrm{d}\theta. \tag{2.26}$$

2.2 Find the optimal point estimate $\mathbf{a}$ that minimizes the expected value of the loss function

$$C(\boldsymbol{\theta}, \mathbf{a}) = (\boldsymbol{\theta} - \mathbf{a})^{\mathsf{T}} \, \mathbf{R} \, (\boldsymbol{\theta} - \mathbf{a}), \tag{2.27}$$

where $\mathbf{R}$ is a positive definite matrix, and the distribution of the parameter is $\boldsymbol{\theta} \sim p(\boldsymbol{\theta} \mid \mathbf{y}_{1:T})$.

2.3 Assume that we have obtained T measurement pairs (x_k, y_k) from the linear regression model

$$y_k = \theta_1 \, x_k + \theta_2, \qquad k = 1, 2, \ldots, T. \tag{2.28}$$

The purpose is now to derive estimates of the parameters θ_1 and θ_2 such that the following error is minimized (least squares estimate):

$$E(\theta_1, \theta_2) = \sum_{k=1}^{T} (y_k - \theta_1 \, x_k - \theta_2)^2. \tag{2.29}$$

(a) Define $\mathbf{y} = (y_1 \; \ldots \; y_T)^\mathsf{T}$ and $\boldsymbol{\theta} = (\theta_1 \; \theta_2)^\mathsf{T}$. Show that the set of Equations (2.28) can be written in matrix form as

$$\mathbf{y} = \mathbf{X} \, \boldsymbol{\theta},$$

with a suitably defined matrix $\mathbf{X}$.

(b) Write the error function in Equation (2.29) in matrix form in terms of $\mathbf{y}$, $\mathbf{X}$, and $\boldsymbol{\theta}$.

(c) Compute the gradient of the matrix form error function, and solve the least squares estimate of the parameter $\boldsymbol{\theta}$ by finding the point where the gradient is zero.

2.4 Assume that in the linear regression model above (Equation (2.28)), we set independent Gaussian priors for the parameters θ_1 and θ_2 as follows:

$$\theta_1 \sim \mathrm{N}(0, \sigma^2),$$
$$\theta_2 \sim \mathrm{N}(0, \sigma^2),$$

where the variance σ^2 is known. The measurements y_k are modeled as

$$y_k = \theta_1 \, x_k + \theta_2 + \varepsilon_k, \qquad k = 1, 2, \ldots, T,$$

where the terms ε_ks are independent Gaussian errors with mean 0 and variance 1, that is, $\varepsilon_k \sim \mathrm{N}(0, 1)$. The values x_k are fixed and known. The posterior distribution can be now written as

$$p(\boldsymbol{\theta} \mid y_1, \ldots, y_T)$$
$$\propto \exp\left(-\frac{1}{2}\sum_{k=1}^{T}(y_k - \theta_1 \, x_k - \theta_2)^2\right) \exp\left(-\frac{1}{2\sigma^2}\theta_1^2\right) \exp\left(-\frac{1}{2\sigma^2}\theta_2^2\right).$$

The posterior distribution can be seen to be Gaussian, and your task is to derive its mean and covariance.

(a) Write the exponent of the posterior distribution in matrix form as in Exercise 2.3 (in terms of $\mathbf{y}$, $\mathbf{X}$, $\boldsymbol{\theta}$, and σ^2).

(b) Because a Gaussian distribution is always symmetric, its mean $\mathbf{m}$ is at the maximum of the distribution. Find the posterior mean by computing the gradient of the exponent and finding where it vanishes.

(c) Find the covariance of the distribution by computing the second derivative matrix (Hessian matrix) $\mathbf{H}$ of the exponent. The posterior covariance is then $\mathbf{P} = -\mathbf{H}^{-1}$ (why?).

(d) What is the resulting posterior distribution? What is the relationship with the least squares estimate in Exercise 2.3?

2.5 Implement an importance sampling-based approximation for the Bayesian linear regression problem in the above Exercise 2.4. Use a suitable Gaussian distribution as the importance distribution for the parameters $\boldsymbol{\theta}$. Check that the posterior mean and covariance (approximately) coincide with the exact values computed in Exercise 2.4.

3

Batch and Recursive Bayesian Estimation

In order to understand the meaning and applicability of Bayesian filtering and its relationship to recursive estimation, it is useful to go through an example where we solve a simple and familiar linear regression problem in a recursive manner. After that we generalize this concept to include a dynamic model in order to illustrate the differences in dynamic and batch estimation.

3.1 Batch Linear Regression

Consider the *linear regression model*

$$y_k = \theta_1 + \theta_2 \, t_k + \varepsilon_k, \tag{3.1}$$

where we assume that the measurement noise is a zero mean Gaussian with a given variance $\varepsilon_k \sim \mathrm{N}(0, \sigma^2)$, and the prior distribution of the parameters $\boldsymbol{\theta} = (\theta_1 \;\; \theta_2)^\mathsf{T}$ is Gaussian with known mean and covariance $\boldsymbol{\theta} \sim \mathrm{N}(\mathbf{m}_0, \mathbf{P}_0)$. In the classical linear regression problem, we want to estimate the parameters $\boldsymbol{\theta}$ from a set of measurement data $\mathcal{D} = \{(t_1, y_1), \ldots, (t_T, y_T)\}$. The measurement data and the true linear function used in the simulation are illustrated in Figure 3.1.

In compact *probabilistic notation*, the linear regression model can be written as

$$\begin{aligned} p(y_k \mid \boldsymbol{\theta}) &= \mathrm{N}(y_k \mid \mathbf{H}_k \, \boldsymbol{\theta}, \sigma^2), \\ p(\boldsymbol{\theta}) &= \mathrm{N}(\boldsymbol{\theta} \mid \mathbf{m}_0, \mathbf{P}_0), \end{aligned} \tag{3.2}$$

where we have introduced the row vector $\mathbf{H}_k = (1 \;\; t_k)$, and $\mathrm{N}(\cdot)$ denotes the Gaussian probability density function (see Section A.1). Note that we denote the row vector $\mathbf{H}_k$ in matrix notation because it generally is a matrix (when the measurements are vector valued), and we want to avoid using different notations for scalar and vector measurements. The likelihood of y_k is also conditional on the regressors t_k (or equivalently $\mathbf{H}_k$), but because

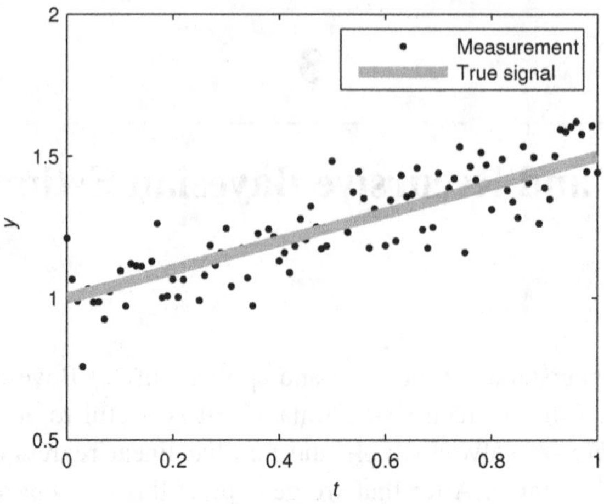

Figure 3.1 The underlying truth and the measurement data in the simple linear regression problem.

the regressors are assumed to be known, to simplify the notation we will not denote this dependence explicitly, and from now on this dependence is assumed to be understood from the context.

The *batch solution* to the linear regression problem in Equation (3.2) can be obtained by a straightforward application of Bayes' rule:

$$
\begin{aligned}
p(\boldsymbol{\theta} \mid y_{1:T}) &\propto p(\boldsymbol{\theta}) \prod_{k=1}^{T} p(y_k \mid \boldsymbol{\theta}) \\
&= \mathrm{N}(\boldsymbol{\theta} \mid \mathbf{m}_0, \mathbf{P}_0) \prod_{k=1}^{T} \mathrm{N}(y_k \mid \mathbf{H}_k \, \boldsymbol{\theta}, \sigma^2).
\end{aligned}
$$

In the *posterior distribution* above, we assume the conditioning on t_k and $\mathbf{H}_k$ but will not denote it explicitly. Thus the posterior distribution is denoted to be conditional on $y_{1:T}$, and not on the data set $\mathcal{D}$ also containing the regressor values t_k. The reason for this simplification is that the simplified notation will also work in more general filtering problems, where there is no natural way of defining the associated regressor variables.

Because the prior and likelihood are Gaussian, the *posterior distribution* will also be Gaussian:

$$
p(\boldsymbol{\theta} \mid y_{1:T}) = \mathrm{N}(\boldsymbol{\theta} \mid \mathbf{m}_T, \mathbf{P}_T). \tag{3.3}
$$

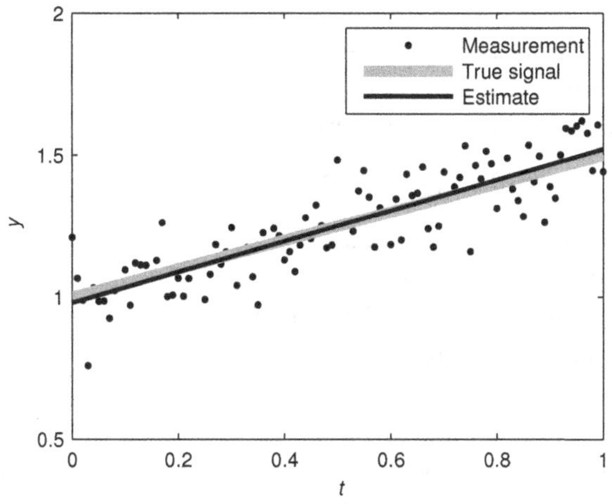

Figure 3.2 The result of simple linear regression with a slight regularization prior used for the regression parameters. For simplicity, the variance was assumed to be known.

The mean and covariance can be obtained by completing the quadratic form in the exponent, which gives:

$$\begin{aligned} \mathbf{m}_T &= \left[\mathbf{P}_0^{-1} + \frac{1}{\sigma^2}\,\mathbf{H}^\mathsf{T}\,\mathbf{H}\right]^{-1} \left[\frac{1}{\sigma^2}\,\mathbf{H}^\mathsf{T}\,\mathbf{y} + \mathbf{P}_0^{-1}\,\mathbf{m}_0\right], \\ \mathbf{P}_T &= \left[\mathbf{P}_0^{-1} + \frac{1}{\sigma^2}\,\mathbf{H}^\mathsf{T}\,\mathbf{H}\right]^{-1}, \end{aligned} \tag{3.4}$$

where $\mathbf{H}_k = (1\ t_k)$, and

$$\mathbf{H} = \begin{pmatrix} \mathbf{H}_1 \\ \vdots \\ \mathbf{H}_T \end{pmatrix} = \begin{pmatrix} 1 & t_1 \\ \vdots & \vdots \\ 1 & t_T \end{pmatrix}, \qquad \mathbf{y} = \begin{pmatrix} y_1 \\ \vdots \\ y_T \end{pmatrix}. \tag{3.5}$$

Figure 3.2 shows the result of batch linear regression, where the posterior mean parameter values are used as the linear regression parameters.

3.2 Recursive Linear Regression

A *recursive solution* to the regression problem (3.2) can be obtained by assuming that we have already obtained the *posterior distribution* conditioned on the previous measurements $1, \dots, k-1$, as follows:

$$p(\boldsymbol{\theta} \mid y_{1:k-1}) = \mathrm{N}(\boldsymbol{\theta} \mid \mathbf{m}_{k-1}, \mathbf{P}_{k-1}).$$

Now assume that we have obtained a new measurement y_k, and we want to compute the posterior distribution of $\boldsymbol{\theta}$ given the old measurements $y_{1:k-1}$ *and* the new measurement y_k. According to the model specification, the new measurement has the likelihood

$$p(y_k \mid \boldsymbol{\theta}) = \mathrm{N}(y_k \mid \mathbf{H}_k \, \boldsymbol{\theta}, \sigma^2).$$

Using the batch version equations such that we interpret the *previous posterior* as the *prior*, we can calculate the distribution

$$\begin{aligned} p(\boldsymbol{\theta} \mid y_{1:k}) &\propto p(y_k \mid \boldsymbol{\theta}) \, p(\boldsymbol{\theta} \mid y_{1:k-1}) \\ &\propto \mathrm{N}(\boldsymbol{\theta} \mid \mathbf{m}_k, \mathbf{P}_k), \end{aligned} \tag{3.6}$$

where the Gaussian distribution parameters are

$$\begin{aligned} \mathbf{m}_k &= \left[\mathbf{P}_{k-1}^{-1} + \frac{1}{\sigma^2} \mathbf{H}_k^\mathsf{T} \, \mathbf{H}_k \right]^{-1} \left[\frac{1}{\sigma^2} \mathbf{H}_k^\mathsf{T} \, y_k + \mathbf{P}_{k-1}^{-1} \, \mathbf{m}_{k-1} \right], \\ \mathbf{P}_k &= \left[\mathbf{P}_{k-1}^{-1} + \frac{1}{\sigma^2} \mathbf{H}_k^\mathsf{T} \, \mathbf{H}_k \right]^{-1}. \end{aligned} \tag{3.7}$$

By using the *matrix inversion lemma* (see Corollary A.5), the covariance calculation can be written as

$$\mathbf{P}_k = \mathbf{P}_{k-1} - \mathbf{P}_{k-1} \, \mathbf{H}_k^\mathsf{T} \left[\mathbf{H}_k \, \mathbf{P}_{k-1} \, \mathbf{H}_k^\mathsf{T} + \sigma^2 \right]^{-1} \mathbf{H}_k \, \mathbf{P}_{k-1}.$$

By introducing temporary variables S_k and $\mathbf{K}_k$, the calculation of the mean and covariance can then be written in the form

$$\begin{aligned} S_k &= \mathbf{H}_k \, \mathbf{P}_{k-1} \, \mathbf{H}_k^\mathsf{T} + \sigma^2, \\ \mathbf{K}_k &= \mathbf{P}_{k-1} \, \mathbf{H}_k^\mathsf{T} \, S_k^{-1}, \\ \mathbf{m}_k &= \mathbf{m}_{k-1} + \mathbf{K}_k \, [y_k - \mathbf{H}_k \, \mathbf{m}_{k-1}], \\ \mathbf{P}_k &= \mathbf{P}_{k-1} - \mathbf{K}_k \, S_k \, \mathbf{K}_k^\mathsf{T}. \end{aligned} \tag{3.8}$$

Note that $S_k = \mathbf{H}_k \, \mathbf{P}_{k-1} \, \mathbf{H}_k^\mathsf{T} + \sigma^2$ is scalar because the measurements are scalar and thus no matrix inversion is required.

The equations above actually are special cases of the Kalman filter update equations. Only the update part of the equations (as opposed to the

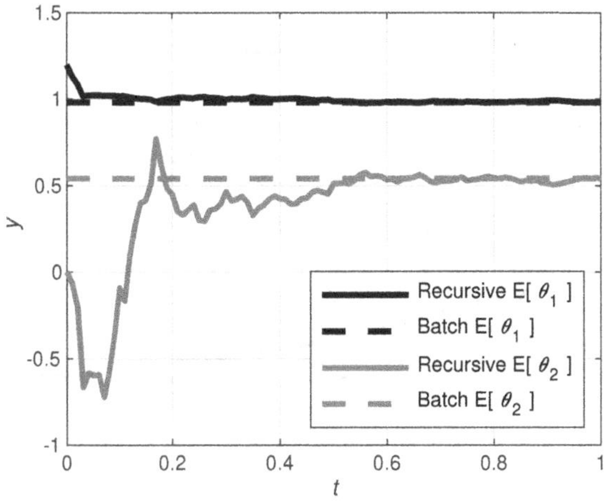

Figure 3.3 Convergence of the means of recursive linear regression parameters. The final value is exactly the same as that obtained with batch linear regression.

prediction and update) is required because the estimated parameters are assumed to be constant, that is, there is no stochastic dynamics model for the parameters $\boldsymbol{\theta}$. Figures 3.3 and 3.4 illustrate the convergence of the means and variances of the parameters during the recursive estimation.

3.3 Batch versus Recursive Estimation

In this section we generalize the recursion idea used in the previous section to general probabilistic models. The underlying idea is simply that at each measurement, we treat the *posterior distribution of the previous time step* as the *prior for the current time step*. This way we can compute the same solution in a recursive manner that we would obtain by direct application of Bayes' rule to the whole (batch) data set.

The *batch Bayesian solution* to a statistical estimation problem can be formulated as follows.

1. Specify the likelihood model of measurements $p(\mathbf{y}_k \mid \boldsymbol{\theta})$ given the parameter $\boldsymbol{\theta}$. Typically the measurements $\mathbf{y}_k$ are assumed to be conditionally independent such that

$$p(\mathbf{y}_{1:T} \mid \boldsymbol{\theta}) = \prod_{k=1}^{T} p(\mathbf{y}_k \mid \boldsymbol{\theta}).$$

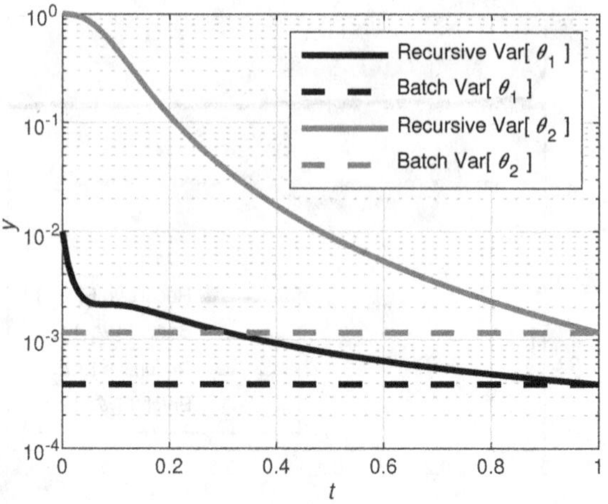

Figure 3.4 Convergence of the variances of linear regression parameters plotted on logarithmic scale. As can be seen, every measurement brings more information, and the uncertainty decreases monotonically. The final values are the same as the variances obtained from the batch solution.

2. The prior information about the parameter $\boldsymbol{\theta}$ is encoded into the prior distribution $p(\boldsymbol{\theta})$.
3. The observed data set is $\mathcal{D} = \{(t_1, \mathbf{y}_1), \ldots, (t_T, \mathbf{y}_T)\}$, or if we drop the explicit conditioning on t_k, the data is $\mathcal{D} = \mathbf{y}_{1:T}$.
4. The batch Bayesian solution to the statistical estimation problem can be computed by applying Bayes' rule:

$$p(\boldsymbol{\theta} \mid \mathbf{y}_{1:T}) = \frac{1}{Z} p(\boldsymbol{\theta}) \prod_{k=1}^{T} p(\mathbf{y}_k \mid \boldsymbol{\theta}),$$

where Z is the *normalization constant*

$$Z = \int p(\boldsymbol{\theta}) \prod_{k=1}^{T} p(\mathbf{y}_k \mid \boldsymbol{\theta}) \, \mathrm{d}\boldsymbol{\theta}.$$

For example, the batch solution of the above kind to the linear regression problem (3.2) was given by Equations (3.3) and (3.4).

The *recursive Bayesian solution* to the above statistical estimation problem can be formulated as follows.

1. The distribution of measurements is again modeled by the likelihood function $p(\mathbf{y}_k \mid \boldsymbol{\theta})$, and the measurements are assumed to be conditionally independent.
2. In the beginning of estimation (i.e., at step 0), all the information about the parameter $\boldsymbol{\theta}$ we have is contained in the prior distribution $p(\boldsymbol{\theta})$.
3. The measurements are assumed to be obtained one at a time – first $\mathbf{y}_1$, then $\mathbf{y}_2$, and so on. At each step we use the posterior distribution from the previous time step as the current prior distribution:

$$
\begin{aligned}
p(\boldsymbol{\theta} \mid \mathbf{y}_1) &= \frac{1}{Z_1} p(\mathbf{y}_1 \mid \boldsymbol{\theta}) p(\boldsymbol{\theta}), \\
p(\boldsymbol{\theta} \mid \mathbf{y}_{1:2}) &= \frac{1}{Z_2} p(\mathbf{y}_2 \mid \boldsymbol{\theta}) p(\boldsymbol{\theta} \mid \mathbf{y}_1), \\
p(\boldsymbol{\theta} \mid \mathbf{y}_{1:3}) &= \frac{1}{Z_3} p(\mathbf{y}_3 \mid \boldsymbol{\theta}) p(\boldsymbol{\theta} \mid \mathbf{y}_{1:2}), \\
&\vdots \\
p(\boldsymbol{\theta} \mid \mathbf{y}_{1:T}) &= \frac{1}{Z_T} p(\mathbf{y}_T \mid \boldsymbol{\theta}) p(\boldsymbol{\theta} \mid \mathbf{y}_{1:T-1}).
\end{aligned}
$$

   It is easy to show that the posterior distribution at the final step above is exactly the posterior distribution obtained by the batch solution. Also, reordering of measurements does not change the final solution.

For example, Equations (3.6) and (3.7) give the one step update rule for the linear regression problem in Equation (3.2).

The recursive formulation of Bayesian estimation has many useful properties.

- The recursive solution can be considered as the *on-line learning* solution to the Bayesian learning problem. That is, the information on the parameters is updated in an on-line manner using new pieces of information as they arrive.
- Because each step in the recursive estimation is a full Bayesian update step, *batch* Bayesian inference is a *special case of recursive* Bayesian inference.
- Due to the sequential nature of estimation, we can also model the effect of time on the parameters. That is, we can model what happens to the parameter $\boldsymbol{\theta}$ between the measurements – this is actually the *basis of filtering theory*, where time behavior is modeled by assuming the parameter to be a time-dependent stochastic process $\boldsymbol{\theta}(t)$.

3.4 Drift Model for Linear Regression

Assume that we have a similar linear regression model as in Equation (3.2), but the parameter $\boldsymbol{\theta}$ is allowed to perform a *Gaussian random walk* between the measurements:

$$\begin{aligned} p(y_k \mid \boldsymbol{\theta}_k) &= \mathrm{N}(y_k \mid \mathbf{H}_k \, \boldsymbol{\theta}_k, \sigma^2), \\ p(\boldsymbol{\theta}_k \mid \boldsymbol{\theta}_{k-1}) &= \mathrm{N}(\boldsymbol{\theta}_k \mid \boldsymbol{\theta}_{k-1}, \mathbf{Q}), \\ p(\boldsymbol{\theta}_0) &= \mathrm{N}(\boldsymbol{\theta}_0 \mid \mathbf{m}_0, \mathbf{P}_0), \end{aligned} \tag{3.9}$$

where $\mathbf{Q}$ is the covariance of the random walk. Now, given the distribution

$$p(\boldsymbol{\theta}_{k-1} \mid y_{1:k-1}) = \mathrm{N}(\boldsymbol{\theta}_{k-1} \mid \mathbf{m}_{k-1}, \mathbf{P}_{k-1}),$$

the joint distribution of $\boldsymbol{\theta}_k$ and $\boldsymbol{\theta}_{k-1}$ is[1]

$$p(\boldsymbol{\theta}_k, \boldsymbol{\theta}_{k-1} \mid y_{1:k-1}) = p(\boldsymbol{\theta}_k \mid \boldsymbol{\theta}_{k-1}) \, p(\boldsymbol{\theta}_{k-1} \mid y_{1:k-1}).$$

The distribution of $\boldsymbol{\theta}_k$ given the measurement history up to time step $k-1$ can be calculated by integrating over $\boldsymbol{\theta}_{k-1}$:

$$p(\boldsymbol{\theta}_k \mid y_{1:k-1}) = \int p(\boldsymbol{\theta}_k \mid \boldsymbol{\theta}_{k-1}) \, p(\boldsymbol{\theta}_{k-1} \mid y_{1:k-1}) \, \mathrm{d}\boldsymbol{\theta}_{k-1}.$$

This relationship is sometimes called the *Chapman–Kolmogorov equation*. Because $p(\boldsymbol{\theta}_k \mid \boldsymbol{\theta}_{k-1})$ and $p(\boldsymbol{\theta}_{k-1} \mid y_{1:k-1})$ are Gaussian, the result of the marginalization is Gaussian,

$$p(\boldsymbol{\theta}_k \mid y_{1:k-1}) = \mathrm{N}(\boldsymbol{\theta}_k \mid \mathbf{m}_k^-, \mathbf{P}_k^-),$$

where

$$\begin{aligned} \mathbf{m}_k^- &= \mathbf{m}_{k-1}, \\ \mathbf{P}_k^- &= \mathbf{P}_{k-1} + \mathbf{Q}. \end{aligned}$$

By using this as the prior distribution for the measurement likelihood $p(y_k \mid \boldsymbol{\theta}_k)$, we get the parameters of the posterior distribution

$$p(\boldsymbol{\theta}_k \mid y_{1:k}) = \mathrm{N}(\boldsymbol{\theta}_k \mid \mathbf{m}_k, \mathbf{P}_k),$$

[1] Note that this formula is correct only for Markovian dynamic models, where $p(\boldsymbol{\theta}_k \mid \boldsymbol{\theta}_{k-1}, y_{1:k-1}) = p(\boldsymbol{\theta}_k \mid \boldsymbol{\theta}_{k-1})$.

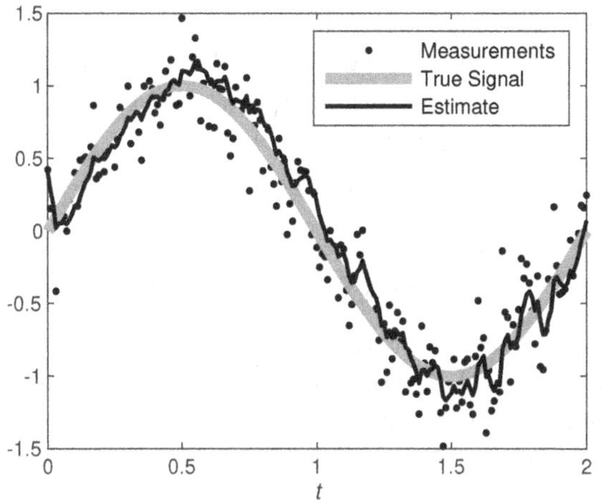

Figure 3.5 Example of tracking a sine signal with a linear model with drift, where the parameters are allowed to vary according to the Gaussian random walk model.

which are given by Equations (3.8), when $\mathbf{m}_{k-1}$ and $\mathbf{P}_{k-1}$ are replaced by $\mathbf{m}_k^-$ and $\mathbf{P}_k^-$:

$$\begin{aligned} S_k &= \mathbf{H}_k \, \mathbf{P}_k^- \, \mathbf{H}_k^\mathsf{T} + \sigma^2, \\ \mathbf{K}_k &= \mathbf{P}_k^- \, \mathbf{H}_k^\mathsf{T} \, S_k^{-1}, \\ \mathbf{m}_k &= \mathbf{m}_k^- + \mathbf{K}_k \, [y_k - \mathbf{H}_k \, \mathbf{m}_k^-], \\ \mathbf{P}_k &= \mathbf{P}_k^- - \mathbf{K}_k \, S_k \, \mathbf{K}_k^\mathsf{T}. \end{aligned} \tag{3.10}$$

This recursive computational algorithm for the time-varying linear regression weights is again a special case of the Kalman filter algorithm. Figure 3.5 shows the result of recursive estimation of a sine signal assuming a small diagonal Gaussian drift model for the parameters.

At this point we change from the *regression notation* used so far into *state space model notation*, which is commonly used in Kalman filtering and related dynamic estimation literature. Because this notation easily causes confusion to people who have got used to regression notation, this point is emphasized.

- In *state space notation*, $\mathbf{x}$ means the unknown state of the system, that is, the vector of *unknown parameters in the system*. It is *not* the regressor, covariate, or input variable of the system.

- For example, the time-varying linear regression model with drift presented in this section can be transformed into the more standard *state space model notation* by replacing the variable $\boldsymbol{\theta}_k = (\theta_{1,k}\ \theta_{2,k})^\mathsf{T}$ with the variable $\mathbf{x}_k = (x_{1,k}\ x_{2,k})^\mathsf{T}$:

$$\begin{aligned} p(y_k \mid \mathbf{x}_k) &= \mathrm{N}(y_k \mid \mathbf{H}_k\, \mathbf{x}_k, \sigma^2), \\ p(\mathbf{x}_k \mid \mathbf{x}_{k-1}) &= \mathrm{N}(\mathbf{x}_k \mid \mathbf{x}_{k-1}, \mathbf{Q}), \\ p(\mathbf{x}_0) &= \mathrm{N}(\mathbf{x}_0 \mid \mathbf{m}_0, \mathbf{P}_0). \end{aligned} \tag{3.11}$$

From now on, the symbol $\boldsymbol{\theta}$ is reserved for denoting the static parameters of the state space model. Although there is no fundamental difference between states and the static parameters of the model (we can always augment the parameters as part of the state), it is useful to treat them separately.

3.5 State Space Model for Linear Regression with Drift

The linear regression model with drift in the previous section had the disadvantage that the covariates t_k occurred explicitly in the model specification. The problem with this is that when we get more and more measurements, the parameter t_k grows without bound. Thus the conditioning of the problem also gets worse in time. For practical reasons it would also be desirable to have a *time-invariant model*, that is, a model that is not dependent on the absolute time but only on the relative positions of states and measurements in time.

The alternative state space formulation of the linear regression model with drift, without using explicit covariates, can be done as follows. Let us denote the time difference between consecutive times as $\Delta t_{k-1} = t_k - t_{k-1}$. The idea is that if the underlying phenomenon (signal, state, parameter) x_k was exactly linear, the difference between adjacent time points could be written exactly as

$$x_k - x_{k-1} = \dot{x}\, \Delta t_{k-1}, \tag{3.12}$$

where $\dot{x}$ is the derivative, which is constant in the exactly linear case. The divergence from the exactly linear function can be modeled by assuming that the above equation does not hold exactly, but there is a small *noise term* on the right-hand side. The derivative can also be assumed to perform a small *random walk* and thus not be exactly constant. This model can be written as follows:

$$\begin{aligned} x_{1,k} &= x_{1,k-1} + \Delta t_{k-1} x_{2,k-1} + q_{1,k-1}, \\ x_{2,k} &= x_{2,k-1} + q_{2,k-1}, \\ y_k &= x_{1,k} + r_k, \end{aligned} \tag{3.13}$$

where the signal is the first components of the state, $x_{1,k} \triangleq x_k$, and the derivative is the second, $x_{2,k} \triangleq \dot{x}_k$. The noises are $r_k \sim \mathrm{N}(0, \sigma^2)$ and $(q_{1,k-1}, q_{2,k-1}) \sim \mathrm{N}(\mathbf{0}, \mathbf{Q})$. The model can also be written in the form

$$\begin{aligned} p(y_k \mid \mathbf{x}_k) &= \mathrm{N}(y_k \mid \mathbf{H}\, \mathbf{x}_k, \sigma^2), \\ p(\mathbf{x}_k \mid \mathbf{x}_{k-1}) &= \mathrm{N}(\mathbf{x}_k \mid \mathbf{A}_{k-1}\, \mathbf{x}_{k-1}, \mathbf{Q}), \end{aligned} \tag{3.14}$$

where

$$\mathbf{A}_{k-1} = \begin{pmatrix} 1 & \Delta t_{k-1} \\ 0 & 1 \end{pmatrix}, \qquad \mathbf{H} = \begin{pmatrix} 1 & 0 \end{pmatrix}.$$

With a suitable $\mathbf{Q}$, this model is actually equivalent to model (3.9), but in this formulation we explicitly estimate the state of the signal (point on the regression line) instead of the linear regression parameters.

We could now explicitly derive the recursion equations in the same manner as we did in the previous sections. However, we can also use the *Kalman filter*, which is a readily derived recursive solution to generic linear Gaussian models of the form

$$\begin{aligned} p(\mathbf{y}_k \mid \mathbf{x}_k) &= \mathrm{N}(\mathbf{y}_k \mid \mathbf{H}_k\, \mathbf{x}_k, \mathbf{R}_k), \\ p(\mathbf{x}_k \mid \mathbf{x}_{k-1}) &= \mathrm{N}(\mathbf{x}_k \mid \mathbf{A}_{k-1}\, \mathbf{x}_{k-1}, \mathbf{Q}_{k-1}). \end{aligned}$$

Our alternative linear regression model in Equation (3.13) can be seen to be a special case of these models. The *Kalman filter equations* are often expressed as *prediction and update steps* as follows.

1. *Prediction step:*

$$\begin{aligned} \mathbf{m}_k^- &= \mathbf{A}_{k-1}\, \mathbf{m}_{k-1}, \\ \mathbf{P}_k^- &= \mathbf{A}_{k-1}\, \mathbf{P}_{k-1}\, \mathbf{A}_{k-1}^\mathsf{T} + \mathbf{Q}_{k-1}. \end{aligned}$$

2. *Update step:*

$$\begin{aligned} \mathbf{S}_k &= \mathbf{H}_k\, \mathbf{P}_k^-\, \mathbf{H}_k^\mathsf{T} + \mathbf{R}_k, \\ \mathbf{K}_k &= \mathbf{P}_k^-\, \mathbf{H}_k^\mathsf{T}\, \mathbf{S}_k^{-1}, \\ \mathbf{m}_k &= \mathbf{m}_k^- + \mathbf{K}_k\, [\mathbf{y}_k - \mathbf{H}_k\, \mathbf{m}_k^-], \\ \mathbf{P}_k &= \mathbf{P}_k^- - \mathbf{K}_k\, \mathbf{S}_k\, \mathbf{K}_k^\mathsf{T}. \end{aligned}$$

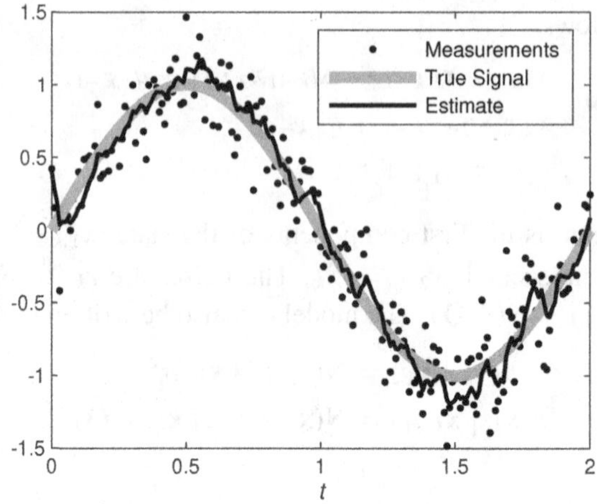

Figure 3.6 Example of tracking a sine signal with a Kalman filter using the locally linear state space model. The result differs slightly from the random walk parameter model because of a slightly different choice of process noise. It could be made equivalent if desired.

The result of tracking the sine signal with a Kalman filter is shown in Figure 3.6. All the mean and covariance calculation equations given in this book so far have been special cases of the above equations, including the batch solution to the scalar measurement case (which is a one-step solution). The Kalman filter recursively computes the mean and covariance of the posterior distributions of the form

$$p(\mathbf{x}_k \mid \mathbf{y}_{1:k}) = \mathrm{N}(\mathbf{x}_k \mid \mathbf{m}_k, \mathbf{P}_k).$$

Note that the estimates of $\mathbf{x}_k$ derived from this distribution are non-anticipative in the sense that they are only conditional on the measurements obtained before and at the time step k. However, after we have obtained the measurements $\mathbf{y}_1, \dots, \mathbf{y}_k$, we could compute estimates of $\mathbf{x}_{k-1}, \mathbf{x}_{k-2}, \dots$, which are also conditional to the measurements after the corresponding state time steps. Because more measurements and more information are available for the estimator, these estimates can be expected to be more accurate than the non-anticipative measurements computed by the filter.

The abovementioned problem of computing estimates of the state by conditioning not only on previous measurements but also on future measurements is called *Bayesian smoothing*, as already mentioned in

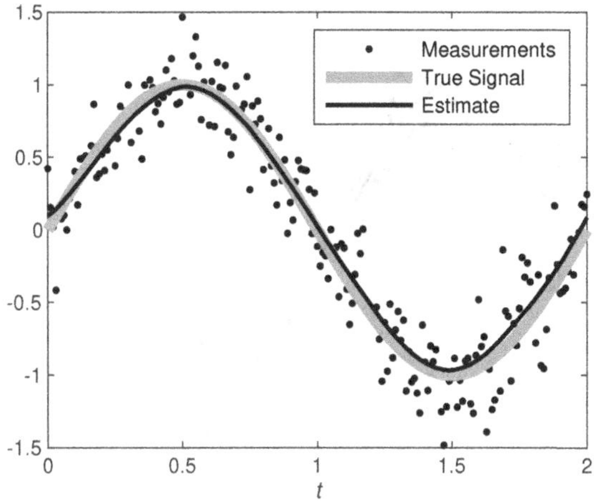

Figure 3.7 Example of tracking a sine signal with the Rauch–Tung–Striebel (RTS) smoother using the locally linear state space model. The result is much "smoother" and more accurate than the result of the Kalman filter.

Section 1.3. The Bayesian smoothing solution to linear Gaussian state space models is given by the *Rauch–Tung–Striebel (RTS) smoother*. The full Bayesian theory of smoothing will be presented in Chapter 12. The result of tracking the sine signal with the RTS smoother is shown in Figure 3.7.

It is also possible to predict the time behavior of the state in the future that we have not yet measured. This procedure is called *optimal prediction*. Because optimal prediction can always be done by iterating the prediction step of the optimal filter, no specialized algorithms are needed for this. The result of prediction of the future values of the sine signal is shown in Figure 3.8.

3.6 Toward Bayesian Filtering and Smoothing

The models that we have seen in this chapter can be seen as special cases of probabilistic state models having the following form:

$$\begin{aligned} \text{dynamics:} \quad \mathbf{x}_k &\sim p(\mathbf{x}_k \mid \mathbf{x}_{k-1}), \\ \text{measurements:} \quad \mathbf{y}_k &\sim p(\mathbf{y}_k \mid \mathbf{x}_k), \end{aligned} \tag{3.15}$$

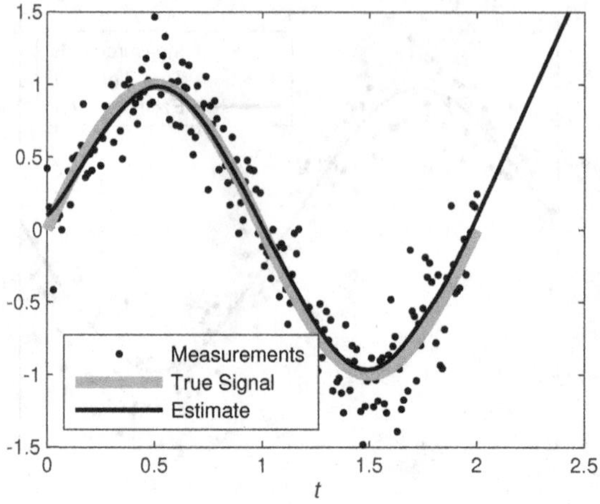

Figure 3.8 Example of prediction of a sine signal with an optimal linear predictor (the Kalman filter prediction step) using the locally linear state space model. The prediction is a straight line extending to infinity, as the model states.

although so far we have only looked at Gaussian examples of dynamic and measurement models. However, non-linear generalizations of Bayesian filtering, smoothing, and prediction problems and solutions can be obtained by replacing the Gaussian distributions and linear functions in the model with non-Gaussian and non-linear ones. The Bayesian filtering and smoothing theory described in this book can be applied to generic non-linear filtering models of the above form.

To understand the generality of this model, it is useful to note that if we dropped the time-dependence from the state, we would get the model

$$\begin{aligned} \text{dynamics:} \quad & \mathbf{x} \sim p(\mathbf{x}), \\ \text{measurements:} \quad & \mathbf{y}_k \sim p(\mathbf{y}_k \mid \mathbf{x}), \end{aligned} \tag{3.16}$$

which can be seen as a general Bayesian statistical model (Gelman et al., 2013) where $\mathbf{y}_k$ contains the observed quantities (data) and $\mathbf{x}$ is the vector of all unknown parameters. Because $\mathbf{x}$ can be an arbitrary set of parameters or hyper-parameters of the system, all static Bayesian models are special cases of this model. Thus in the dynamic estimation context, we extend the static models by allowing for a Markov model for the time-behavior of the (hyper-)parameters.

The Markovianity is also less of a restriction than it sounds because what we have is a vector-valued Markov process, not a scalar one. Similar to the fact that differential equations of an arbitrary order can always be transformed into vector-valued differential equations of the first order, Markov processes of an arbitrary order can be transformed into vector-valued first-order Markov processes.

3.7 Exercises

3.1 Recall that the batch solution to the linear regression problem in Equations (3.2) is given by the mean and covariance in Equations (3.4). The implementation of these equations can be found in the companion codes of the book[2].

(a) Use the matrix inversion lemmas in Corollary A.5 to convert the batch linear regression solution in Equations (3.4) to a similar form as Equations (3.8). *Hint:* $\mathbf{R} = \sigma^2 \mathbf{I}$.

(b) Check numerically that Equations (3.4) and the equations that you derived give exactly the same mean and covariance.

3.2 Note that the model in Exercise 2.4 can be rewritten as a linear state space model

$$\begin{aligned} \mathbf{w}_k &= \mathbf{w}_{k-1}, \\ y_k &= \mathbf{H}_k \, \mathbf{w}_k + \varepsilon_k, \end{aligned}$$

where $\mathbf{H}_k = (x_k \;\; 1)$, $\mathbf{w}_0 \sim \mathrm{N}(0, \sigma^2 \, \mathbf{I})$, and $\varepsilon_k \sim \mathrm{N}(0, 1)$. The state in the model is now $\mathbf{w}_k = (\theta_1 \;\; \theta_2)^\mathsf{T}$, and the measurements are y_k for $k = 1, \ldots, T$. Assume that the Kalman filter is used for processing the measurements $y_1, \ldots, y_T$. Your task is to prove that at time step T, the mean and covariance of $\mathbf{w}_T$ computed by the Kalman filter are the same as the mean and covariance of the posterior distribution computed in Exercise 2.4.
The Kalman filter equations for the above model can be written as:

$$\begin{aligned} S_k &= \mathbf{H}_k \, \mathbf{P}_{k-1} \, \mathbf{H}_k^\mathsf{T} + 1, \\ \mathbf{K}_k &= \mathbf{P}_{k-1} \, \mathbf{H}_k^\mathsf{T} \, S_k^{-1}, \\ \mathbf{m}_k &= \mathbf{m}_{k-1} + \mathbf{K}_k \, (y_k - \mathbf{H}_k \, \mathbf{m}_{k-1}), \\ \mathbf{P}_k &= \mathbf{P}_{k-1} - \mathbf{K}_k \, S_k \, \mathbf{K}_k^\mathsf{T}. \end{aligned}$$

(a) Write formulas for the posterior mean $\mathbf{m}_{k-1}$ and covariance $\mathbf{P}_{k-1}$ assuming that they are the same as those that would be obtained if the pairs $\{(x_i, y_i) \, : \, i = 1, \ldots, k-1\}$ were (batch) processed as in Exercise 2.4.

[2] https://github.com/EEA-sensors/Bayesian-Filtering-and-Smoothing

Write similar equations for the mean $\mathbf{m}_k$ and covariance $\mathbf{P}_k$. Show that the posterior means can be expressed in the form

$$\mathbf{m}_{k-1} = \mathbf{P}_{k-1}\,\mathbf{X}_{k-1}^\mathsf{T}\,\mathbf{y}_{k-1},$$
$$\mathbf{m}_k = \mathbf{P}_k\,\mathbf{X}_k^\mathsf{T}\,\mathbf{y}_k,$$

where $\mathbf{X}_{k-1}$ and $\mathbf{y}_{k-1}$ have been constructed as $\mathbf{X}$ and $\mathbf{y}$ in Exercise 2.4, except that only the pairs $\{(x_i, y_i) : i = 1, \ldots, k-1\}$ have been used, and $\mathbf{X}_k$ and $\mathbf{y}_k$ have been constructed similarly from pairs up to the step k.

(b) Rewrite the expressions $\mathbf{X}_k^\mathsf{T}\,\mathbf{X}_k$ and $\mathbf{X}_k^\mathsf{T}\,\mathbf{y}_k$ in terms of $\mathbf{X}_{k-1}$, $\mathbf{y}_{k-1}$, $\mathbf{H}_k$ and y_k. Substitute these into the expressions of $\mathbf{m}_k$ and $\mathbf{P}_k$ obtained in (a).

(c) Expand the expression of the covariance $\mathbf{P}_k = \mathbf{P}_{k-1} - \mathbf{K}_k\,S_k\,\mathbf{K}_k^\mathsf{T}$ by substituting the expressions for $\mathbf{K}_k$ and S_k. Convert it to a simpler form by applying the matrix inversion lemma (see Corollary A.5)

$$\mathbf{P}_{k-1} - \mathbf{P}_{k-1}\,\mathbf{H}_k^\mathsf{T}\,(\mathbf{H}_k\,\mathbf{P}_{k-1}\,\mathbf{H}_k^\mathsf{T} + 1)^{-1}\,\mathbf{H}_k\,\mathbf{P}_{k-1} = (\mathbf{P}_{k-1}^{-1} + \mathbf{H}_k^\mathsf{T}\,\mathbf{H}_k)^{-1}.$$

Show that this expression for $\mathbf{P}_k$ is equivalent to the expression in (a).

(d) Expand the expression of the mean $\mathbf{m}_k = \mathbf{m}_{k-1} + \mathbf{K}_k\,(y_k - \mathbf{H}_k\,\mathbf{m}_{k-1})$, and show that the result is equivalent to the expression obtained in (a). *Hint:* The Kalman gain can also be written as $\mathbf{K}_k = \mathbf{P}_k\,\mathbf{H}_k^\mathsf{T}$.

(e) Prove by an induction argument that the mean and covariance computed by the Kalman filter at step T is the same as the posterior mean and covariance obtained in Exercise 2.4.

3.3 Recall that the Gaussian probability density is defined as

$$\mathrm{N}(\mathbf{x} \mid \mathbf{m}, \mathbf{P}) = \frac{1}{(2\,\pi)^{n/2}\,|\mathbf{P}|^{1/2}} \exp\left(-\frac{1}{2}(\mathbf{x}-\mathbf{m})^\mathsf{T}\,\mathbf{P}^{-1}\,(\mathbf{x}-\mathbf{m})\right).$$

Derive the following Gaussian identities.

(a) Let $\mathbf{x}$ and $\mathbf{y}$ have the Gaussian densities

$$p(\mathbf{x}) = \mathrm{N}(\mathbf{x} \mid \mathbf{m}, \mathbf{P}), \qquad p(\mathbf{y} \mid \mathbf{x}) = \mathrm{N}(\mathbf{y} \mid \mathbf{H}\,\mathbf{x}, \mathbf{R}),$$

then the joint distribution of $\mathbf{x}$ and $\mathbf{y}$ is

$$\begin{pmatrix}\mathbf{x}\\ \mathbf{y}\end{pmatrix} \sim \mathrm{N}\left(\begin{pmatrix}\mathbf{m}\\ \mathbf{H}\,\mathbf{m}\end{pmatrix}, \begin{pmatrix}\mathbf{P} & \mathbf{P}\,\mathbf{H}^\mathsf{T}\\ \mathbf{H}\,\mathbf{P} & \mathbf{H}\,\mathbf{P}\,\mathbf{H}^\mathsf{T} + \mathbf{R}\end{pmatrix}\right)$$

and the marginal distribution of $\mathbf{y}$ is

$$\mathbf{y} \sim \mathrm{N}(\mathbf{H}\,\mathbf{m}, \mathbf{H}\,\mathbf{P}\,\mathbf{H}^\mathsf{T} + \mathbf{R}).$$

Hint: Use the properties of expectation $\mathrm{E}[\mathbf{H}\,\mathbf{x} + \mathbf{r}] = \mathbf{H}\,\mathrm{E}[\mathbf{x}] + \mathrm{E}[\mathbf{r}]$ and $\mathrm{Cov}[\mathbf{H}\,\mathbf{x} + \mathbf{r}] = \mathbf{H}\,\mathrm{Cov}[\mathbf{x}]\,\mathbf{H}^\mathsf{T} + \mathrm{Cov}[\mathbf{r}]$ (if $\mathbf{x}$ and $\mathbf{r}$ are independent).

(b) Write down the explicit expression for the joint and marginal probability densities above:

$$p(\mathbf{x}, \mathbf{y}) = p(\mathbf{y} \mid \mathbf{x})\, p(\mathbf{x}) = ?$$
$$p(\mathbf{y}) = \int p(\mathbf{y} \mid \mathbf{x})\, p(\mathbf{x})\, d\mathbf{x} = ?$$

(c) If the random variables $\mathbf{x}$ and $\mathbf{y}$ have the joint Gaussian probability density

$$\begin{pmatrix} \mathbf{x} \\ \mathbf{y} \end{pmatrix} \sim \mathrm{N}\left(\begin{pmatrix} \mathbf{a} \\ \mathbf{b} \end{pmatrix}, \begin{pmatrix} \mathbf{A} & \mathbf{C} \\ \mathbf{C}^\mathsf{T} & \mathbf{B} \end{pmatrix} \right),$$

then the conditional density of $\mathbf{x}$ given $\mathbf{y}$ is

$$\mathbf{x} \mid \mathbf{y} \sim \mathrm{N}(\mathbf{a} + \mathbf{C}\, \mathbf{B}^{-1}\, (\mathbf{y} - \mathbf{b}), \mathbf{A} - \mathbf{C}\, \mathbf{B}^{-1} \mathbf{C}^\mathsf{T}).$$

Hints:

- Denote inverse covariance as $\mathbf{D} = \begin{pmatrix} \mathbf{D}_{11} & \mathbf{D}_{12} \\ \mathbf{D}_{12}^\mathsf{T} & \mathbf{D}_{22} \end{pmatrix}$, and expand the quadratic form in the Gaussian exponent.
- Compute the derivative with respect to $\mathbf{x}$ and set it to zero. Conclude that due to symmetry, the point where the derivative vanishes is the mean.
- From the block matrix inverse formulas given in Theorem A.4, we get that the inverse of $\mathbf{D}_{11}$ is

$$\mathbf{D}_{11}^{-1} = \mathbf{A} - \mathbf{C}\, \mathbf{B}^{-1}\, \mathbf{C}^\mathsf{T}$$

  and that $\mathbf{D}_{12}$ can be then written as

$$\mathbf{D}_{12} = -\mathbf{D}_{11}\, \mathbf{C}\, \mathbf{B}^{-1}.$$

- Find the simplified expression for the mean by applying the identities above.
- Find the second derivative of the negative Gaussian exponent with respect to $\mathbf{x}$. Conclude that it must be the inverse conditional covariance of $\mathbf{x}$.
- Use the Schur complement expression above for computing the conditional covariance.

4

Discretization of Continuous-Time Dynamic Models

A probabilistic state space model requires both a dynamic model and a measurement model. The measurement model relates states to measurements via the conditional distribution $p(\mathbf{y}_k \mid \mathbf{x}_k)$ and enables us to use measurements to learn about states. The dynamic model instead uses the transition distribution $p(\mathbf{x}_k \mid \mathbf{x}_{k-1})$ to describe how the state evolves over time, and therefore indirectly enables us to make use of measurements collected at other time steps to learn about $\mathbf{x}_k$ at the current time step k. This chapter aims to describe how to construct dynamic models for tracking. We will complete these state space models with measurement models in the next chapter. Although the focus is on tracking, the same ideas also work for building models of, for example, electromechanical systems or biological systems.

Dynamic models that arise from target tracking can often be described in terms of differential equations, or their random counterparts stochastic differential equations (SDEs), which evolve continuously in time (as opposed to jumping from one time step to another). Therefore, in this chapter, we first construct models in continuous time and then discretize them to obtain models that operate in discrete time steps. This conversion to discrete time is needed because our filters and smoothers in the next chapters all operate in discrete time. For more details on the mathematics behind continuous-time models, and in particular SDEs, the reader is referred to Särkkä and Solin (2019).

It is worth noting that in this chapter (and book), we have purposely chosen the approach of first forming a continuous-time white-noise-driven model and then discretizing it. This approach has the advantage of having an inherent mathematical connection with the theory of SDEs. Another approach to form the model would be to drive the stochastic differential equation with a piecewise constant or impulse noise. The relationships and differences between these modeling approaches are discussed, for example, in Bar-Shalom et al. (2001).

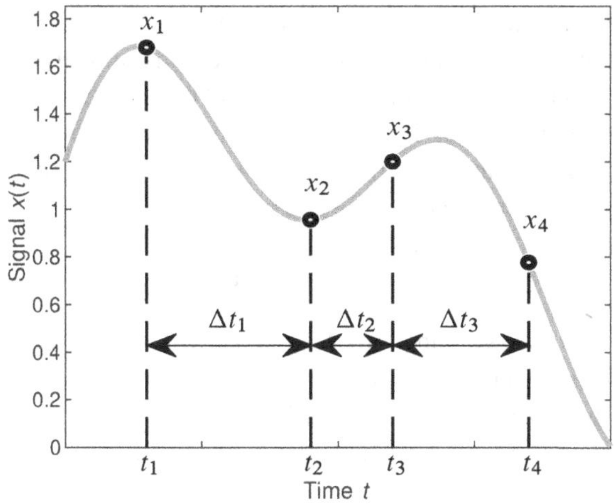

Figure 4.1 A scalar continuous time signal $x(t)$ sampled at four different times: $x_1 = x(t_1), x_2 = x(t_2), x_3 = x(t_3)$, and $x_4 = x(t_4)$.

4.1 Discrete-Time and Continuous-Time Dynamic Models

A discrete-time dynamic model refers to a model of $\mathbf{x}_k$ given $\mathbf{x}_{k-1}$, where k is an integer-valued time step index. We can often express this model as a functional relationship

$$\mathbf{x}_k = \mathbf{f}_{k-1}(\mathbf{x}_{k-1}) + \mathbf{q}_{k-1}, \tag{4.1}$$

where $\mathbf{q}_{k-1} \sim \mathrm{N}(\mathbf{0}, \mathbf{Q}_{k-1})$ is a Gaussian process noise. This can also be expressed as a conditional distribution,

$$p(\mathbf{x}_k \mid \mathbf{x}_{k-1}) = \mathrm{N}(\mathbf{x}_k \mid \mathbf{f}_{k-1}(\mathbf{x}_{k-1}), \mathbf{Q}_{k-1}). \tag{4.2}$$

By accurately modeling the dynamics of the state, we can better predict the distribution of future states and thereby obtain better prediction, filtering, and smoothing performance.

In many cases, we can view the discrete-time state sequence $\mathbf{x}_k$, where $k = 1, 2, 3, \ldots$, as samples from a continuous-time function $\mathbf{x}(t)$ where $t \in [0, \infty)$ such that $\mathbf{x}_k = \mathbf{x}(t_k)$. We use t_k to denote the time of the discrete time step k, which is usually the time when we observe the measurement $\mathbf{y}_k$, and $\Delta t_{k-1} = t_k - t_{k-1}$ to denote the sampling period. This is illustrated in Figure 4.1.

The connection to the continous-time function $\mathbf{x}(t)$ is useful because instead of directly modeling the discrete-time dynamics of $\mathbf{x}_k$, it is often easier to formulate a model for the continuous-time state sequence, for instance, by using underlying laws of physics to describe how objects move. The continuous-time analog of the discrete-time model in Equation (4.1) is the stochastic differential equation (SDE)

$$\frac{\mathrm{d}\mathbf{x}(t)}{\mathrm{d}t} = \mathbf{a}(\mathbf{x}(t)) + \mathbf{L}\,\mathbf{w}(t), \tag{4.3}$$

where $\mathbf{a}(\cdot)$ is a known function, $\mathbf{L}$ is a known matrix, and $\mathbf{w}(t)$ is a continuous-time Gaussian process noise. The process noise $\mathbf{w}(t)$ is assumed to be a zero mean white noise process, which means that the process has the following mean and covariance functions:

$$\begin{aligned} \mathbb{E}\left[\mathbf{w}(t)\right] &= \mathbf{0}, \\ \operatorname{Cov}\left[\mathbf{w}(\tau_1), \mathbf{w}(\tau_2)\right] &= \delta(\tau_1 - \tau_2)\, \mathbf{Q}^{\mathrm{c}}. \end{aligned} \tag{4.4}$$

Here $\delta(\cdot)$ is the Dirac delta function, and $\mathbf{Q}^{\mathrm{c}}$ is referred to as the spectral density matrix of the white noise, which can be seen as a continuous-time analog of a covariance matrix. Furthermore, the white noise enters the SDE causally in the sense that $\mathbf{w}(t')$ and $\mathbf{x}(t)$ are uncorrelated whenever $t' \geq t$. Considering that the noise is also white, see Equation (4.4), this implies that $\mathbf{x}(t)$ is Markovian. In engineering literature, it is common to express the models in terms of white noise like this because that enables us to develop the models without introducing new mathematical concepts – rigorous formulations of SDEs often rely on Itô calculus (see, e.g., Särkkä and Solin, 2019); this requires a redefinition of the concept of an integral, which adds mathematical complexity.

In the following sections, we explain how to construct a dynamic model for $\mathbf{x}_k$. Specifically, we describe general techniques to express the time behaviour of $\mathbf{x}(t)$ as an SDE of the form of Equation (4.3). The discrete-time dynamic model such as Equation (4.1) is then obtained as a model for the transition from $\mathbf{x}_{k-1} = \mathbf{x}(t_{k-1})$ to $\mathbf{x}_k = \mathbf{x}(t_k)$. The process of obtaining a discrete-time model from the continuous-time counterpart is referred to as discretization. Importantly, since $\mathbf{x}(t)$ is Markovian, so is the discrete-time sequence $\mathbf{x}_k$. It also turns out that it is always possible to find a transition density $p(\mathbf{x}_k \mid \mathbf{x}_{k-1}) = p(\mathbf{x}(t_k) \mid \mathbf{x}(t_{k-1}))$ that is an exact probabilistic description of the transition from $\mathbf{x}_{k-1}$ to $\mathbf{x}_k$ (Särkkä and Solin, 2019), but unfortunately, this density is often intractable and cannot be written in a simple form as Equation (4.1). Therefore, for non-linear models, we need to use approximate discretization methods.

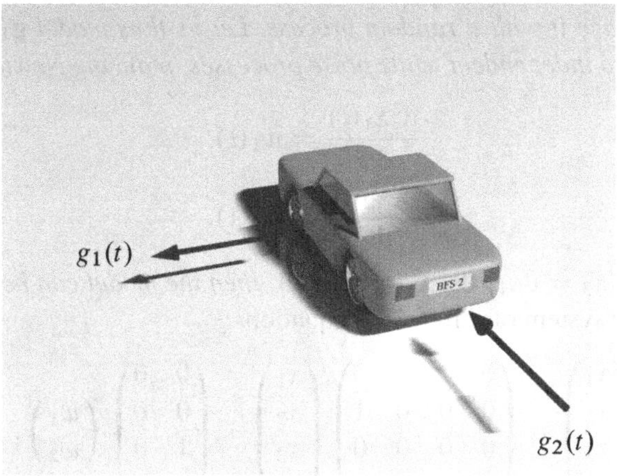

Figure 4.2 Illustration of car's dynamic model in Example 4.1. In the dynamic model, the unknown forces $g_1(t)$ and $g_2(t)$ are modeled as white noise processes.

4.2 Discretizing Linear Dynamic Models

Let us start with a time-invariant continuous-time linear model, expressed as an SDE,

$$\frac{\mathrm{d}\mathbf{x}(t)}{\mathrm{d}t} = \mathbf{F}\,\mathbf{x}(t) + \mathbf{L}\,\mathbf{w}(t), \tag{4.5}$$

where $\mathbf{w}(t)$ is white Gaussian noise with moments (i.e., mean and covariance) as given in Equation (4.4). The key feature in this model is that the right-hand side is a linear function of $\mathbf{x}(t)$. This kind of model is very useful in practical applications because simple physics-based dynamic models often have this form, as seen from the following example.

Example 4.1 (Dynamic model of a car). *Let us consider a car moving in 2D as shown in Figure 4.2. We can model the car as a point mass located at its center point* (x_1, x_2). *A simple way to model its movement is to assume that it is pushed around by certain unknown, time-dependent forces* $g_1(t)$ *and* $g_2(t)$, *which are related to the dynamical behavior by* Newton's second law

$$\mathbf{g}(t) = m\,\mathbf{a}(t), \tag{4.6}$$

which thus relates the vector of (unknown) forces $\mathbf{g} = (g_1, g_2)$ *to the accelerations* $\mathbf{a} = (\mathrm{d}^2x_1/\mathrm{d}t^2\,,\, \mathrm{d}^2x_2/\mathrm{d}t^2)$ *and the mass of the car* m.

Because the function $\mathbf{g}(t)$ *is unknown, a sensible (Bayesian) way to model it is to replace it with a random process. Let us thus model* $\mathbf{g}(t)/m$ *as a vector of two independent white noise processes, reducing Newton's law to*

$$\begin{aligned}\frac{\mathrm{d}^2 x_1(t)}{\mathrm{d}t^2} &= w_1(t),\\ \frac{\mathrm{d}^2 x_2(t)}{\mathrm{d}t^2} &= w_2(t).\end{aligned} \tag{4.7}$$

If we define $x_3 = \mathrm{d}x_1/\mathrm{d}t$, $x_4 = \mathrm{d}x_2/\mathrm{d}t$, *then the model can be written as a first order* system of differential equations*:*

$$\frac{\mathrm{d}}{\mathrm{d}t}\begin{pmatrix} x_1 \\ x_2 \\ x_3 \\ x_4 \end{pmatrix} = \underbrace{\begin{pmatrix} 0 & 0 & 1 & 0 \\ 0 & 0 & 0 & 1 \\ 0 & 0 & 0 & 0 \\ 0 & 0 & 0 & 0 \end{pmatrix}}_{\mathbf{F}} \begin{pmatrix} x_1 \\ x_2 \\ x_3 \\ x_4 \end{pmatrix} + \underbrace{\begin{pmatrix} 0 & 0 \\ 0 & 0 \\ 1 & 0 \\ 0 & 1 \end{pmatrix}}_{\mathbf{L}} \begin{pmatrix} w_1 \\ w_2 \end{pmatrix}. \tag{4.8}$$

In shorter matrix form, *this can be written as a continuous-time linear dynamic model of the form*

$$\frac{\mathrm{d}\mathbf{x}}{\mathrm{d}t} = \mathbf{F}\,\mathbf{x} + \mathbf{L}\,\mathbf{w},$$

where the spectral density of white noise $\mathbf{w}(t)$ *has the form* $\mathbf{Q}^{\mathrm{c}} = \mathrm{diag}(q_1^{\mathrm{c}}, q_2^{\mathrm{c}})$, *where* q_1^{c} *and* q_2^{c} *are the spectral densities of* $w_1(t)$ *and* $w_2(t)$, *respectively.*

The model obtained above is a special case of the continuous-time constant velocity models, also called Wiener velocity models. More generally, we have the following class of models.

Example 4.2 (Continuous-time constant velocity model). *A* d*-dimensional continuous-time constant velocity (CV) model, also called the Wiener velocity model, can be written as*

$$\frac{\mathrm{d}\mathbf{x}(t)}{\mathrm{d}t} = \begin{pmatrix} \mathbf{0} & \mathbf{I} \\ \mathbf{0} & \mathbf{0} \end{pmatrix} \mathbf{x}(t) + \begin{pmatrix} \mathbf{0} \\ \mathbf{I} \end{pmatrix} \mathbf{w}(t), \tag{4.9}$$

where the first d *elements in* $\mathbf{x}$ *are the position of an object (or the value of a variable), and the remaining* d *elements represent the velocity. The model specifies that the time derivative of the position is the velocity, whereas the time derivative of the velocity is the process noise* $\mathbf{w}(t)$.

The solution $\mathbf{x}(t)$ to the linear SDE in Equation (4.5) is Gaussian because it is produced by a linear system driven by Gaussian noise. This also implies that the transition density $p(\mathbf{x}(t_k) \mid \mathbf{x}(t_{k-1}))$, and thus the corresponding discrete-time model, is a discrete-time linear Gaussian model. The following theorem provides a closed-form expression for computing the transition density and hence the closed-form discretization of a time-invariant linear SDE.

Theorem 4.3 (Discretization of linear dynamic models). *The transition density for the linear time-invariant SDE in Equation* (4.5), *where the white noise process* $\mathbf{w}(t)$ *has the spectral density* $\mathbf{Q}^c$, *is given as*

$$p(\mathbf{x}(t_k) \mid \mathbf{x}(t_{k-1})) = \mathrm{N}(\mathbf{x}(t_k) \mid \exp(\mathbf{F}\,\Delta t_{k-1})\,\mathbf{x}(t_{k-1}), \mathbf{Q}_{k-1}), \tag{4.10}$$

where $\Delta t_{k-1} = t_k - t_{k-1}$ *and*

$$\mathbf{Q}_{k-1} = \int_0^{\Delta t_{k-1}} \exp(\mathbf{F}\,s)\,\mathbf{L}\,\mathbf{Q}^c\,\mathbf{L}^\mathsf{T} \exp(\mathbf{F}\,s)^\mathsf{T}\,\mathrm{d}s. \tag{4.11}$$

Please note that above, $\exp(\cdot)$ *is the matrix exponential function, not an element-wise exponential. Consequently, discretization gives a linear and Gaussian model*

$$\mathbf{x}_k = \mathbf{A}_{k-1}\,\mathbf{x}_{k-1} + \mathbf{q}_{k-1}, \tag{4.12}$$

where $\mathbf{A}_{k-1} = \exp(\mathbf{F}\,\Delta t_{k-1})$ *and* $\mathbf{q}_{k-1} \sim \mathrm{N}(\mathbf{0}, \mathbf{Q}_{k-1})$.

Proof This theorem follows from Lemma A.9, by setting $\mathbf{b} = \mathbf{0}$ and $\tau = \Delta t_{k-1}$ in the lemma. □

Example 4.4 (Discretization of the Wiener process). *Let us then consider a* d*-dimensional Wiener process, which can be considered as a time-integral of a* d*-dimensional white noise process and hence has the representation*

$$\frac{\mathrm{d}\mathbf{x}(t)}{\mathrm{d}t} = \mathbf{w}(t), \tag{4.13}$$

for which $\mathbf{F} = \mathbf{0}$. *Assume that the spectral density of the white noise is* $\mathbf{Q}^c$. *Because* $\exp(\mathbf{0}) = \mathbf{I}$, *it follows from Theorem 4.3 that*

$$p(\mathbf{x}(t_k) \mid \mathbf{x}(t_{k-1})) = \mathrm{N}(\mathbf{x}(t_k) \mid \mathbf{x}(t_{k-1}), \mathbf{Q}_{k-1}), \tag{4.14}$$

where $\mathbf{Q}_{k-1} = \int_0^{\Delta t_{k-1}} \mathbf{Q}^c\,\mathrm{d}s = \mathbf{Q}^c\,\Delta t_{k-1}$. *The discretized version of the Wiener process is thus a random walk*

$$\mathbf{x}_k = \mathbf{x}_{k-1} + \mathbf{q}_{k-1}, \tag{4.15}$$

where $\mathbf{q}_{k-1} \sim \mathrm{N}(\mathbf{0}, \mathbf{Q}_{k-1})$ *and* $\mathbf{Q}_{k-1} = \mathbf{Q}^c\,\Delta t_{k-1}$. *See Figure 4.3 for an illustration.*

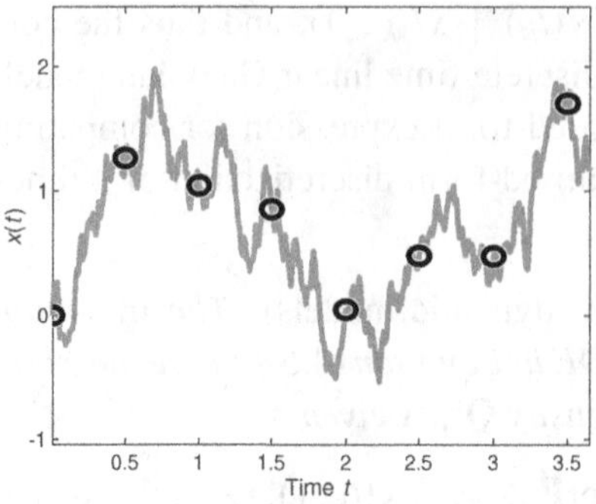

(a)

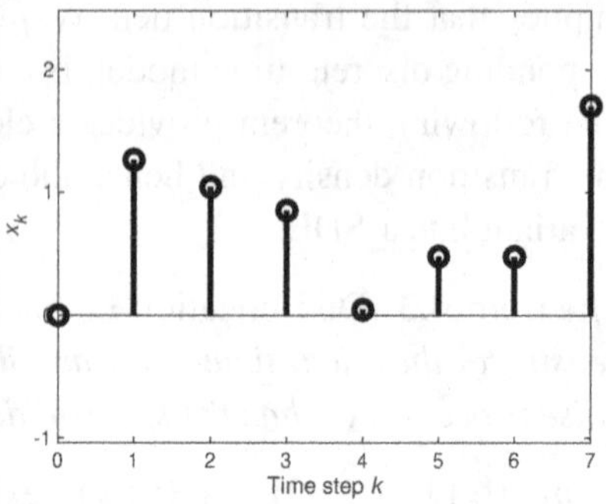

(b)

Figure 4.3 (a) One realization of a scalar Wiener process $x(t)$ with $Q^c = 2$. (b) A random walk, obtained by sampling $x(t)$ with sampling period $\Delta t = 0.5$, which means that we have $Q_k = 1$ for all $k = 0, 1, 2, \ldots$.

Example 4.5 (Discrete-time constant/Wiener velocity model). *We can use Theorem 4.3 to discretize the constant/Wiener velocity model in Example 4.2. From Equation* (4.9), *we conclude that*

$$\mathbf{F} = \begin{pmatrix} \mathbf{0} & \mathbf{I} \\ \mathbf{0} & \mathbf{0} \end{pmatrix}, \quad \mathbf{L} = \begin{pmatrix} \mathbf{0} \\ \mathbf{I} \end{pmatrix}. \tag{4.16}$$

It follows from Theorem 4.3, and the definition of the matrix exponential, that

$$\begin{aligned} \mathbf{A}_{k-1} &= \exp(\mathbf{F}\, \Delta t_{k-1}) = \sum_{n=0}^{\infty} \frac{(\mathbf{F}\, \Delta t_{k-1})^n}{n!} \\ &= \begin{pmatrix} \mathbf{I} & \mathbf{0} \\ 0 & \mathbf{I} \end{pmatrix} + \mathbf{F}\, \Delta t_{k-1} = \begin{pmatrix} \mathbf{I} & \mathbf{I}\, \Delta t_{k-1} \\ 0 & \mathbf{I} \end{pmatrix}, \end{aligned} \tag{4.17}$$

becase $\mathbf{F}^n = \mathbf{0}$ *except for* $n = 0$ *and* $n = 1$, *that is,* $\mathbf{F}$ *is a nilpotent matrix with index 2. Ignoring the process noise, this transition matrix says that the velocity is constant, whereas the new position is the previous position plus* Δt_{k-1} *times the velocity.*

If we assume that the spectral density of the white noise is $\mathbf{Q}^c$, *then the discrete-time covariance matrix is*

$$\begin{aligned}
\mathbf{Q}_{k-1} &= \int_0^{\Delta t_{k-1}} \exp(\mathbf{F}\, s)\, \mathbf{L}\, \mathbf{Q}^c\, \mathbf{L}^\mathsf{T} \exp(\mathbf{F}\, s)^\mathsf{T} \, \mathrm{d}s \\
&= \int_0^{\Delta t_{k-1}} \begin{pmatrix} \mathbf{I} & s\,\mathbf{I} \\ \mathbf{0} & \mathbf{I} \end{pmatrix} \begin{pmatrix} \mathbf{0} \\ \mathbf{I} \end{pmatrix} \mathbf{Q}^c \begin{pmatrix} \mathbf{0} & \mathbf{I} \end{pmatrix} \begin{pmatrix} \mathbf{I} & \mathbf{0} \\ s\,\mathbf{I} & \mathbf{I} \end{pmatrix} \mathrm{d}s \\
&= \int_0^{\Delta t_{k-1}} \begin{pmatrix} s^2\,\mathbf{Q}^c & s\,\mathbf{Q}^c \\ s\,\mathbf{Q}^c & \mathbf{Q}^c \end{pmatrix} \mathrm{d}s \\
&= \begin{pmatrix} \frac{\Delta t_{k-1}^3}{3}\,\mathbf{Q}^c & \frac{\Delta t_{k-1}^2}{2}\,\mathbf{Q}^c \\ \frac{\Delta t_{k-1}^2}{2}\,\mathbf{Q}^c & \Delta t_{k-1}\,\mathbf{Q}^c \end{pmatrix},
\end{aligned} \tag{4.18}$$

where we used (4.17) *to rewrite* $\exp(\mathbf{F}\, s)$. *If* $\Delta t_{k-1} \ll 1$, *the overall process noise is small and dominated by the noise on the velocity state. If* $\Delta t_{k-1} \gg 1$, *the process noise is large and dominated by the noise on the position state.*

Example 4.6 (Discretized dynamic model of a car). *Using the result in Example 4.5 with* $\mathbf{Q}^c = \mathrm{diag}(q_1^c, q_2^c)$, *we can discretize the dynamic model of the car in Example 4.1 with discretization step* Δt *to get*

$$\begin{pmatrix} x_{1,k} \\ x_{2,k} \\ x_{3,k} \\ x_{4,k} \end{pmatrix} = \underbrace{\begin{pmatrix} 1 & 0 & \Delta t & 0 \\ 0 & 1 & 0 & \Delta t \\ 0 & 0 & 1 & 0 \\ 0 & 0 & 0 & 1 \end{pmatrix}}_{\mathbf{A}} \begin{pmatrix} x_{1,k-1} \\ x_{2,k-1} \\ x_{3,k-1} \\ x_{4,k-1} \end{pmatrix} + \mathbf{q}_{k-1}, \tag{4.19}$$

where $\mathbf{q}_{k-1}$ *is a discrete-time Gaussian noise process with zero mean and covariance:*

$$\mathbf{Q} = \begin{pmatrix} \frac{q_1^c\,\Delta t^3}{3} & 0 & \frac{q_1^c\,\Delta t^2}{2} & 0 \\ 0 & \frac{q_2^c\,\Delta t^3}{3} & 0 & \frac{q_2^c\,\Delta t^2}{2} \\ \frac{q_1^c\,\Delta t^2}{2} & 0 & q_1^c\,\Delta t & 0 \\ 0 & \frac{q_2^c\,\Delta t^2}{2} & 0 & q_2^c\,\Delta t \end{pmatrix}. \tag{4.20}$$

This can be seen to be a (discrete-time) linear dynamic model of the form

$$\mathbf{x}_k = \mathbf{A}\,\mathbf{x}_{k-1} + \mathbf{q}_{k-1},$$

where $\mathbf{x}_k = \mathbf{x}(t_k)$, *and* $\mathbf{A}$ *is the transition matrix given in Equation* (4.19).

For models where the integral expression for the covariance in Equation (4.11) does not have a convenient analytical expression, it can still be numerically evaluated efficiently using matrix fraction decomposition (see, e.g., Axelsson and Gustafsson, 2015; Särkkä and Solin, 2019).

4.3 The Euler–Maruyama Method

In Example 4.1 we derived a dynamic model of a car, and the model had a closed-form discretization given in Example 4.6 because the model was linear. However, often the dynamics need to be modeled with non-linear (stochastic) differential equations, in which case the discretization cannot be done in closed form. One method to approximately discretize a non-linear SDE model is the Euler–Maruyama method (see, e.g., Särkkä and Solin, 2019).

Let us now assume that we can formulate the dynamics of our modeled phenomenon as a non-linear SDE of the form

$$\frac{\mathrm{d}\mathbf{x}(t)}{\mathrm{d}t} = \mathbf{a}(\mathbf{x}(t)) + \mathbf{L}\,\mathbf{w}(t), \tag{4.21}$$

where $\mathbf{L}$ is a known matrix and $\mathbf{w}(t)$ is a white noise process with spectral density $\mathbf{Q}^{\mathrm{c}}$ as before, but now $\mathbf{a}(\mathbf{x})$ is some non-linear function. In this case, the transition density $p(\mathbf{x}(t_k) \mid \mathbf{x}(t_{k-1}))$ is no longer Gaussian in general and generally lacks a closed form expression. In spite of this, we generally approximate it as Gaussian and seek a non-linear model with additive Gaussian noise

$$\mathbf{x}_k = \mathbf{f}_{k-1}(\mathbf{x}_{k-1}) + \mathbf{q}_{k-1}, \quad \mathbf{q}_{k-1} \sim \mathrm{N}(\mathbf{0}, \mathbf{Q}_{k-1}). \tag{4.22}$$

An example of a non-linear model is given in the following.

Example 4.7 (Noisy pendulum model)**.** *The differential equation for a simple pendulum (see Figure 4.4) with unit length and mass can be written as*

$$\frac{\mathrm{d}^2\alpha}{\mathrm{d}t^2} = -g\,\sin(\alpha) + w(t), \tag{4.23}$$

where α is the angle, g is the gravitational acceleration, and $w(t)$ is a white noise process with a spectral density q^{c} modeling random forces acting on the pendulum. This model can be converted into the following state space model:

$$\frac{\mathrm{d}}{\mathrm{d}t}\begin{pmatrix} x_1 \\ x_2 \end{pmatrix} = \begin{pmatrix} x_2 \\ -g\,\sin(x_1) \end{pmatrix} + \begin{pmatrix} 0 \\ 1 \end{pmatrix} w(t), \tag{4.24}$$

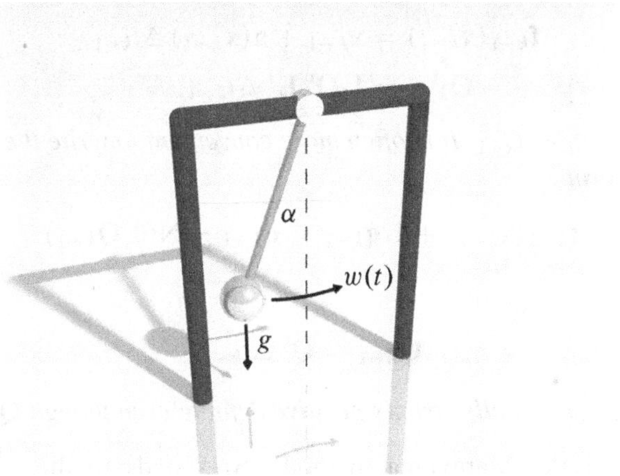

Figure 4.4 Illustration of pendulum example. In addition to the gravitation g, there is an additional unknown force component $w(t)$, which is modeled as white noise.

where $x_1 = \alpha$ and $x_2 = \mathrm{d}\alpha / \mathrm{d}t$. This can be seen as a particular case of a continuous-time non-linear dynamic model of the form

$$\frac{\mathrm{d}\mathbf{x}}{\mathrm{d}t} = \mathbf{a}(\mathbf{x}) + \mathbf{L}\,\mathbf{w}, \tag{4.25}$$

where $\mathbf{a}(\mathbf{x})$ is a non-linear function.

The Euler–Maruyama method (see, e.g., Särkkä and Solin, 2019) is an extension of the Euler method to SDEs. The method corresponds to the approximation

$$\frac{\mathrm{d}\mathbf{x}(t)}{\mathrm{d}t} \approx \mathbf{a}(\mathbf{x}(t_{k-1})) + \mathbf{L}\,\mathbf{w}(t), \quad \forall t \in [t_{k-1}, t_k), \tag{4.26}$$

that is, it uses the approximation $\mathbf{a}(\mathbf{x}(t)) \approx \mathbf{a}(\mathbf{x}(t_{k-1}))$ on a short interval $[t_{k-1}, t_k)$ to simplify the solution.

Theorem 4.8 (The Euler–Maruyama method)**.** *For*

$$\frac{\mathrm{d}\mathbf{x}(t)}{\mathrm{d}t} = \mathbf{a}(\mathbf{x}(t)) + \mathbf{L}\,\mathbf{w}(t), \tag{4.27}$$

the Euler–Maruyama method leads to the discrete-time model

$$\mathbf{x}_k = \mathbf{f}_{k-1}(\mathbf{x}_{k-1}) + \mathbf{q}_{k-1}, \quad \mathbf{q}_{k-1} \sim \mathrm{N}(\mathbf{0}, \mathbf{Q}_{k-1}), \tag{4.28}$$

where

$$\begin{aligned}\mathbf{f}_{k-1}(\mathbf{x}_{k-1}) &= \mathbf{x}_{k-1} + \mathbf{a}(\mathbf{x}_{k-1})\,\Delta t_{k-1},\\ \mathbf{Q}_{k-1} &= \mathbf{L}\,\mathbf{Q}^{\mathrm{c}}\,\mathbf{L}^{\mathsf{T}}\,\Delta t_{k-1},\end{aligned} \tag{4.29}$$

and $\Delta t_{k-1} = t_k - t_{k-1}$. *It is often more convenient to write the discretization of the form*

$$\mathbf{x}_k = \mathbf{f}_{k-1}(\mathbf{x}_{k-1}) + \mathbf{L}\,\tilde{\mathbf{q}}_{k-1}, \quad \tilde{\mathbf{q}}_{k-1} \sim \mathrm{N}(\mathbf{0}, \tilde{\mathbf{Q}}_{k-1}) \tag{4.30}$$

where

$$\tilde{\mathbf{Q}}_{k-1} = \mathbf{Q}^{\mathrm{c}}\,\Delta t_{k-1} \tag{4.31}$$

because $\tilde{\mathbf{Q}}_{k-1}$ *is usually strictly positive definite even though* $\mathbf{Q}_{k-1}$ *is not.*

Proof The Euler–Maruyama method corresponds to the approximate model

$$\frac{\mathrm{d}\mathbf{x}(t)}{\mathrm{d}t} \approx \mathbf{a}(\mathbf{x}(t_{k-1})) + \mathbf{L}\,\mathbf{w}(t). \tag{4.32}$$

By integrating both sides from t_{k-1} to t_k, we get

$$\mathbf{x}(t_k) - \mathbf{x}(t_{k-1}) = \mathbf{a}(\mathbf{x}(t_{k-1}))\,\Delta t_{k-1} + \mathbf{L}\int_{t_{k-1}}^{t_k} \mathbf{w}(t)\,\mathrm{d}t. \tag{4.33}$$

We know from Example 4.4 that

$$\int_{t_{k-1}}^{t_k} \mathbf{w}(t)\,\mathrm{d}t \sim \mathrm{N}(\mathbf{0}, \mathbf{Q}^{\mathrm{c}}\,\Delta t_{k-1}), \tag{4.34}$$

and hence $\mathbf{L}\int_{t_{k-1}}^{t_k} \mathbf{w}(t)\,\mathrm{d}t \sim \mathrm{N}(\mathbf{0}, \mathbf{L}\,\mathbf{Q}^{\mathrm{c}}\,\mathbf{L}^{\mathsf{T}}\,\Delta t_{k-1})$, which gives the results by rearranging the terms. □

Let us now apply the Euler–Maruyama method to the noisy pendulum model that we introduced above.

Example 4.9 (Euler–Maruyama discretization of pendulum model). *In the case of the noisy pendulum model in Example 4.7, we have*

$$\mathbf{a}(\mathbf{x}) = \begin{pmatrix} x_2 \\ -g\,\sin(x_1) \end{pmatrix}, \qquad \mathbf{L} = \begin{pmatrix} 0 \\ 1 \end{pmatrix}. \tag{4.35}$$

If the spectral density of the white noise process $w(t)$ *is* q^{c}, *then the first version of the Euler–Maruyama method in Equation* (4.28) *gives*

$$\begin{pmatrix} x_{1,k} \\ x_{2,k} \end{pmatrix} = \begin{pmatrix} x_{1,k-1} \\ x_{2,k-1} \end{pmatrix} + \begin{pmatrix} x_{2,k-1} \\ -g\,\sin(x_{1,k-1}) \end{pmatrix} \Delta t_{k-1} + \begin{pmatrix} q_{1,k-1} \\ q_{2,k-1} \end{pmatrix}, \tag{4.36}$$

which implies that the discrete-time dynamic model function is

$$\mathbf{f}_{k-1}(\mathbf{x}_{k-1}) = \begin{pmatrix} x_{1,k-1} + x_{2,k-1}\,\Delta t_{k-1} \\ x_{2,k-1} - g\,\sin(x_{1,k-1})\,\Delta t_{k-1} \end{pmatrix}, \tag{4.37}$$

and the joint covariance of $\mathbf{q}_{k-1} = (q_{1,k-1}, q_{2,k-1})^\mathsf{T}$ *is*

$$\mathbf{Q}_{k-1} = \mathbf{L}\,q^c\,\mathbf{L}^\mathsf{T}\,\Delta t_{k-1} = \begin{pmatrix} 0 & 0 \\ 0 & q^c\,\Delta t_{k-1} \end{pmatrix}, \tag{4.38}$$

which is a singular matrix. The second version of the Euler–Maruyama method in Equation (4.30) *gives*

$$\begin{pmatrix} x_{1,k} \\ x_{2,k} \end{pmatrix} = \begin{pmatrix} x_{1,k-1} \\ x_{2,k-1} \end{pmatrix} + \begin{pmatrix} x_{2,k-1} \\ -g\,\sin(x_{1,k-1}) \end{pmatrix} \Delta t_{k-1} + \begin{pmatrix} 0 \\ 1 \end{pmatrix} \tilde{q}_{k-1}, \tag{4.39}$$

where $\tilde{q}_{k-1} \sim \mathrm{N}(0, q^c\,\Delta t_{k-1})$. *This discretization has the same* $\mathbf{f}_{k-1}(\mathbf{x}_{k-1})$ *defined in Equation* (4.37), *but now the noise is non-singular and only enters into the second component of the state.*

The challenge with the Euler–Maruyama discretization in the example above is that the overall process noise $\mathbf{q}_{k-1}$ is singular in the sense that its covariance matrix $\mathbf{L}\,q^c\,\mathbf{L}^\mathsf{T}\,\Delta t_{k-1}$ is singular. Although this problem can sometimes be avoided by using the second form of the discretization, it does not resolve the general issue that the transition density approximation implied by the Euler–Maruyama method is singular whenever $\mathbf{L}\,\mathbf{Q}^c\,\mathbf{L}^\mathsf{T}$ is not invertible. This singularity is not a problem for most of the filtering methods that we see later in this book, but in iterated filters and smoothers this singularity causes numerical problems.

Let us now look at a more complicated example of a coordinated turn model, which is useful in target tracking applications.

Example 4.10 (The polar coordinated turn (CT) model). *Consider the problem of modeling the motion of a car as shown in Figure 4.5. Instead of assuming random forces pushing the car as in Example 4.1, let us assume that the car has a direction* $\varphi(t)$, *which defines the front-direction of the car as the counterclockwise angle from the east direction. If we assume that the car moves forward, then the velocity of the car is given as*

$$\begin{aligned} \frac{\mathrm{d}x_1(t)}{\mathrm{d}t} &= s(t)\,\cos(\varphi(t)), \\ \frac{\mathrm{d}x_2(t)}{\mathrm{d}t} &= s(t)\,\sin(\varphi(t)), \end{aligned} \tag{4.40}$$

where x_1, x_2 *are the position coordinates, and* $s(t)$ *is the speed of the car.*

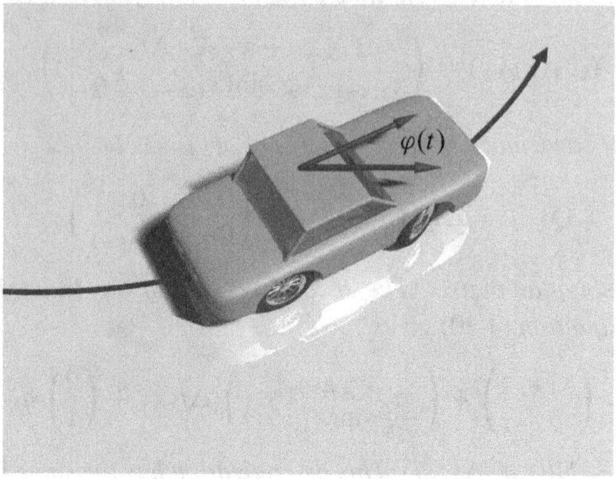

Figure 4.5 Illustration of polar coordinated turn (CT) model in Example 4.10. The car moves to a direction defined by the angle $\varphi(t)$, and the speed and angular velocity are perturbed by white noises.

Furthermore, let us assume that the car is turning with an angular velocity $\omega(t)$, which thus gives the following differential equation for the direction:

$$\frac{d\varphi(t)}{dt} = \omega(t). \tag{4.41}$$

We can now form a stochastic model by assuming that $s(t)$ and $\omega(t)$ are perturbed by white noises via the differential equations

$$\begin{aligned} \frac{ds(t)}{dt} &= w_1(t), \\ \frac{d\omega(t)}{dt} &= w_2(t). \end{aligned} \tag{4.42}$$

We can then write the resulting model in the form

$$\frac{d}{dt}\begin{pmatrix} x_1(t) \\ x_2(t) \\ s(t) \\ \varphi(t) \\ \omega(t) \end{pmatrix} = \begin{pmatrix} s(t)\cos(\varphi(t)) \\ s(t)\sin(\varphi(t)) \\ 0 \\ \omega(t) \\ 0 \end{pmatrix} + \begin{pmatrix} 0 & 0 \\ 0 & 0 \\ 1 & 0 \\ 0 & 0 \\ 0 & 1 \end{pmatrix} \begin{pmatrix} w_1(t) \\ w_2(t) \end{pmatrix}, \tag{4.43}$$

which is a model of the form

$$\frac{\mathrm{d}\mathbf{x}}{\mathrm{d}t} = \mathbf{a}(\mathbf{x}) + \mathbf{L}\,\mathbf{w}, \tag{4.44}$$

with the state $\mathbf{x}(t) = \big(x_1(t), x_2(t), s(t), \varphi(t), \omega(t)\big)$.

We can now apply the Euler–Maruyama method to the model above.

Example 4.11 (Euler–Maruyama discretization of polar CT model)**.** *If we use the Euler–Maruyama method to discretize the polar CT model in Example 4.10, we get (by the second form of discretization)*

$$\mathbf{x}_k = \mathbf{x}_{k-1} + \begin{pmatrix} s_{k-1}\,\cos(\varphi_{k-1}) \\ s_{k-1}\,\sin(\varphi_{k-1}) \\ 0 \\ \omega_{k-1} \\ 0 \end{pmatrix} \Delta t_{k-1} + \begin{pmatrix} 0 & 0 \\ 0 & 0 \\ 1 & 0 \\ 0 & 0 \\ 0 & 1 \end{pmatrix} \tilde{\mathbf{q}}_{k-1}, \tag{4.45}$$

where $\tilde{\mathbf{q}}_{k-1} = (\tilde{q}_{1,k-1}, \tilde{q}_{2,k-1})^{\mathsf{T}}$ *with*

$$\tilde{\mathbf{q}}_{k-1} \sim \mathrm{N}(\mathbf{0}, \tilde{\mathbf{Q}}_{k-1}), \quad \tilde{\mathbf{Q}}_{k-1} = \begin{pmatrix} q_1^{\mathrm{c}} & 0 \\ 0 & q_2^{\mathrm{c}} \end{pmatrix} \Delta t_{k-1}, \tag{4.46}$$

and where q_1^{c} *and* q_2^{c} *are the spectral densities of the white noise processes* $w_1(t)$ *and* $w_2(t)$*. Thus the discretized model has the dynamic model function*

$$\mathbf{f}_{k-1}(\mathbf{x}_{k-1}) = \begin{pmatrix} x_{1,k-1} + s_{k-1}\,\cos(\varphi_{k-1})\,\Delta t_{k-1} \\ x_{2,k-1} + s_{k-1}\,\sin(\varphi_{k-1})\,\Delta t_{k-1} \\ s_{k-1} \\ \varphi_{k-1} + \omega_{k-1}\,\Delta t_{k-1} \\ \omega_{k-1} \end{pmatrix}, \tag{4.47}$$

with the state $\mathbf{x} = \big(x_1, x_2, s, \varphi, \omega\big)^{\mathsf{T}}$*. However,* $p(\mathbf{x}_k \mid \mathbf{x}_{k-1}) = \mathrm{N}(\mathbf{x}_k \mid \mathbf{f}(\mathbf{x}_{k-1}), \mathbf{Q}_{k-1})$ *is not a proper density since* $\mathbf{Q}_{k-1} = \mathbf{L}\,\mathbf{Q}^{\mathrm{c}}\,\mathbf{L}^{\mathsf{T}}\,\Delta t_{k-1}$ *is not invertible.*

Originally, the Euler–Maruyama method was developed to simulate solutions of SDEs (see, e.g., Kloeden and Platen, 1999; Särkkä and Solin, 2019) instead of forming discretizations of dynamic models for filtering and smoothing problems as we do here. To simulate a solution of the SDE in Equation (4.21) between t_{k-1} and t_k by using the Euler–Maruyama method, we can do the following.

Algorithm 4.12 (Simulating SDE solution with Euler–Maruyama)**.** *Split interval $[t_{k-1}, t_k]$ into n steps of length $\Delta\tau = (t_k - t_{k-1})/n$; then do the following:*

- *Start simulation from the state $\hat{\mathbf{x}}_0 = \mathbf{x}(t_{k-1})$.*
- *For $j = 1, \ldots, n$, simulate solution using Euler–Maruyama:*

$$\hat{\mathbf{x}}_j = \hat{\mathbf{x}}_{j-1} + \mathbf{a}(\hat{\mathbf{x}}_{j-1})\,\Delta\tau + \mathbf{L}\,\hat{\mathbf{q}}_{j-1}, \tag{4.48}$$

  where $\hat{\mathbf{q}}_{j-1}$ is a random draw from $\mathrm{N}(\mathbf{0}, \mathbf{Q}^{\mathrm{c}}\,\Delta\tau)$.
- *The approximation of the trajectory $\mathbf{x}(t_{k-1}) \to \mathbf{x}(t_k)$ is then given by the piece-wise constant trajectory consisting of the points $\{\hat{\mathbf{x}}_0, \ldots, \hat{\mathbf{x}}_n\}$.*

It can be shown (Kloeden and Platen, 1999) that the trajectories simulated using the Euler–Maruyama method above converge to true trajectory samples of the SDE when $\Delta\tau \to 0$. When using multiple steps in the Euler–Maryuama method, we also avoid the variance singularity problem arising in the method. As discussed in Section 4.6, this can be used to create multi-step discretizations on the basis of simple approximations such as Euler–Maruyama. Furthermore, in Monte Carlo-based particle filtering methods, Euler–Maruyama simulation can be used as such to implement the prediction step (or actually the importance sampling step) as it only needs us to be able to simulate trajectories from the SDE. Simulating solutions of SDEs is also useful in generating simulated data from dynamic models, such as the pendulum model and coordinate turn model that we have seen in this section. Because we are using more than one step in the simulation, the singularity of the single step vanishes as the effective multi-step transition density is no longer singular.

There are a wide range of other simulation methods for SDEs (see Kloeden and Platen, 1999; Särkkä and Solin, 2019), which can often also be used as a basis for discretization of continuous-time models in the sense that we do here. However, simulation often requires slightly different approximations than discretization. For example, linearization, which we discuss in the next sections, and which is very useful in discretization of dynamic models, is not as useful in simulation as the rate of convergence, in a theoretical sense, is no better than that of Euler–Maruyama.

4.4 Discretization via Continuous-Time Linearization

The Euler–Maruyama method, which uses a zeroth order approximation of $\mathbf{a}(\mathbf{x})$, is a general and simple discretization method, but it often yields rough approximations when the sampling period is long. Another problem with the Euler–Maruyama method is that it often leads to singular process noise covariances. One way to cope with these problems is to use a first order Taylor series approximation

$$\mathbf{a}(\mathbf{x}(t)) \approx \mathbf{a}(\bar{\mathbf{x}}(t)) + \mathbf{A}_{\mathbf{x}}(\bar{\mathbf{x}}(t))\,(\mathbf{x}(t) - \bar{\mathbf{x}}(t)), \tag{4.49}$$

where $\bar{\mathbf{x}}(t)$ is some nominal or linearization trajectory. Above, $\mathbf{A}_{\mathbf{x}}(\bar{\mathbf{x}}(t))$ denotes the Jacobian of $\mathbf{a}(\mathbf{x})$ evaluated at $\bar{\mathbf{x}}(t)$, whose ith row is the gradient of $a_i(\mathbf{x})$.

There are different ways to form the nominal trajectory. One way is to use the expected value $\mathbf{m}(t) = \mathrm{E}[\mathbf{x}(t)]$ of the continuous-time model solution as the linearization trajectory:

$$\mathbf{a}(\mathbf{x}(t)) \approx \mathbf{a}(\mathbf{m}(t)) + \mathbf{A}_{\mathbf{x}}(\mathbf{m}(t))\,(\mathbf{x}(t) - \mathbf{m}(t)). \tag{4.50}$$

It may seem counterintuitive to use an unknown quantity $\mathbf{m}(t)$ to approximate $\mathbf{a}(\mathbf{x}(t))$, but this approximation actually helps us find $\mathbf{m}(t)$ and to form the discretized model. The same approximation is also used in the continuous-discrete time extended Kalman filter (e.g., Särkkä and Solin, 2019) to approximate a non-linear dynamic model with an affine one.

Theorem 4.13 (Linearized continuous-time dynamic model). *Suppose that we approximate the SDE as*

$$\frac{\mathrm{d}\mathbf{x}(t)}{\mathrm{d}t} \approx \mathbf{a}(\mathbf{m}(t)) + \mathbf{A}_{\mathbf{x}}(\mathbf{m}(t))\,(\mathbf{x}(t) - \mathbf{m}(t)) + \mathbf{L}\,\mathbf{w}(t). \tag{4.51}$$

Given $\mathbf{x}(t_{k-1}) = \mathbf{x}_{k-1}$, $\mathbf{x}(t)$ *is then Gaussian for* $t \geq t_{k-1}$ *with moments that satisfy*

$$\frac{\mathrm{d}\mathbf{m}(t)}{\mathrm{d}t} = \mathbf{a}(\mathbf{m}(t)), \tag{4.52}$$

$$\frac{\mathrm{d}\mathbf{P}(t)}{\mathrm{d}t} = \mathbf{A}_{\mathbf{x}}(\mathbf{m}(t))\,\mathbf{P}(t) + \mathbf{P}(t)\,\mathbf{A}_{\mathbf{x}}^{\mathsf{T}}(\mathbf{m}(t)) + \mathbf{L}\,\mathbf{Q}^{\mathrm{c}}\,\mathbf{L}^{\mathsf{T}}, \tag{4.53}$$

where $\mathbf{m}(t_{k-1}) = \mathbf{x}_{k-1}$ *and* $\mathbf{P}(t_{k-1}) = \mathbf{0}$.

Proof Taking expectations of both sides of Equation (4.51) and using $\mathrm{E}\,[\mathbf{A}_{\mathbf{x}}(\mathbf{m}(t))(\mathbf{x}(t) - \mathbf{m}(t))] = 0$ and $\mathrm{E}\,[\mathbf{L}\,\mathbf{w}(t)] = 0$ gives Equation (4.52). The derivation of (4.53) is more involved and is given in Appendix A.5. □

The above approximation has the useful feature that the solution to the equation $\mathrm{d}\mathbf{m}(t)/\mathrm{d}t = \mathbf{a}(\mathbf{m}(t))$, with initial condition $\mathbf{m}(t_{k-1}) = \mathbf{x}_{k-1}$, is just the solution to the deterministic dynamic equation without noise.

It turns out that this equation can be solved in closed form in several important examples. However, the expression for $\mathbf{P}(t)$ is known as the Lyapunov differential equation and generally lacks an analytical solution unless $\mathbf{A_x}(\mathbf{m}(t))$ is constant. In any case, it is possible to solve the equation for $\mathbf{P}(t)$, or both of the equations, using differential equation solvers such as Runge–Kutta methods (Hairer et al., 2008). Regardless of the solution method, we get the following conceptual discretization method.

Algorithm 4.14 (Discretization by mean linearization). *Let us denote the solution to Equation* (4.52) *at time t_k as $\mathbf{f}_{k-1}(\mathbf{x}_{k-1}) = \mathbf{m}(t_k)$ and the corresponding solution to Equation* (4.53) *as $\mathbf{Q}_{k-1}(\mathbf{x}_{k-1}) = \mathbf{P}(t_k)$. Note that the dependence on $\mathbf{x}_{k-1}$ comes from the initial condition $\mathbf{m}(t_{k-1}) = \mathbf{x}_{k-1}$. Then we can form a discretization of the continuous-time model*

$$\frac{\mathrm{d}\mathbf{x}(t)}{\mathrm{d}t} = \mathbf{a}(\mathbf{x}(t)) + \mathbf{L}\,\mathbf{w}(t), \tag{4.54}$$

as

$$\mathbf{x}_k = \mathbf{f}_{k-1}(\mathbf{x}_{k-1}) + \mathbf{q}_{k-1}, \qquad \mathbf{q}_{k-1} \sim \mathrm{N}(\mathbf{0}, \mathbf{Q}_{k-1}(\mathbf{x}_{k-1})). \tag{4.55}$$

The covariance $\mathbf{Q}_{k-1}(\mathbf{x}_{k-1})$ in the above discretization is, in general, a function of $\mathbf{x}_{k-1}$ and has a more complicated form than the additive noise discretization that we obtained from the Euler–Maruyama method. Because of this, and due to the difficulty of solving the Lyapunov equation, the covariance expression is often further approximated, as we will see in the next section. However, before that, let us take a look at an example.

Example 4.15 (Analytical mean for the polar CT model). *The deterministic part of the polar CT model considered in Example 4.10 has the form*

$$\frac{\mathrm{d}}{\mathrm{d}t}\begin{pmatrix} x_1(t) \\ x_2(t) \\ s(t) \\ \varphi(t) \\ \omega(t) \end{pmatrix} = \begin{pmatrix} s(t)\,\cos(\varphi(t)) \\ s(t)\,\sin(\varphi(t)) \\ 0 \\ \omega(t) \\ 0 \end{pmatrix}, \tag{4.56}$$

which is thus also the differential equation for the mean. To determine the equation for $\mathbf{f}_{k-1}(\mathbf{x}_{k-1})$ for the discretization with Algorithm 4.14, we need to solve this differential equation, with initial condition

$\mathbf{x}_{k-1} = (x_{1,k-1}, x_{2,k-1}, s_{k-1}, \varphi_{k-1}, \omega_{k-1})^{\mathsf{T}}$, *which can be done analytically. As shown in Appendix A.7, this gives*

$$\mathbf{f}_{k-1}(\mathbf{x}_{k-1}) = \begin{pmatrix} x_{1,k-1} + \frac{2s_{k-1}}{\omega_{k-1}} \sin(\frac{\Delta t_{k-1}\,\omega_{k-1}}{2}) \cos(\varphi_{k-1} + \frac{\Delta t_{k-1}\,\omega_{k-1}}{2}) \\ x_{2,k-1} + \frac{2s_{k-1}}{\omega_{k-1}} \sin(\frac{\Delta t_{k-1}\,\omega_{k-1}}{2}) \sin(\varphi_{k-1} + \frac{\Delta t_{k-1}\,\omega_{k-1}}{2}) \\ s_{k-1} \\ \varphi_{k-1} + \omega_{k-1}\,\Delta t_{k-1} \\ \omega_{k-1} \end{pmatrix}. \tag{4.57}$$

Note that when the angular rate is $\omega_{k-1} \approx 0$, *we get the approximation* $\frac{2s_{k-1}}{\omega_{k-1}} \sin(\frac{\Delta t_{k-1}\,\omega_{k-1}}{2}) \approx s_{k-1}\Delta t_{k-1}$ *and the relations*

$$\begin{aligned} f_1(\mathbf{x}_{k-1}) &\approx x_{1,k-1} + s_{k-1}\,\Delta t_{k-1}\,\cos(\varphi_{k-1}), \\ f_2(\mathbf{x}_{k-1}) &\approx x_{2,k-1} + s_{k-1}\,\Delta t_{k-1}\,\sin(\varphi_{k-1}), \end{aligned} \tag{4.58}$$

which can be used as numerically stable approximations when ω_{k-1} *is small.*

Unfortunately, the solution to the covariance differential equation (4.53) for the polar CT model is intractable. It is possible to approximate the covariance of $\mathbf{q}_{k-1}$ using, for example, the Euler–Maruyama approximation given in Example 4.11; however, unfortunately this is singular and thus unsuited for many cases. We will come back to approximating the covariance in the next section, but before that, let us take a look at another example.

Example 4.16 (Analytical mean for the Cartesian CT model)**.** *It is also possible to express the polar coordinated turn (CT) model in Example 4.10 in Cartesian coordinates such that the angle and speed are replaced by a velocity vector. If we take time derivatives of Equation* (4.40) *and use Equations* (4.41) *and* (4.42)*, we get*

$$\begin{aligned} \frac{\mathrm{d}^2 x_1(t)}{\mathrm{d}t^2} &= \frac{\mathrm{d}s(t)}{\mathrm{d}t} \cos(\varphi(t)) - s(t)\,\frac{\mathrm{d}\varphi(t)}{\mathrm{d}t} \sin(\varphi(t)) \\ &= w_1(t)\cos(\varphi(t)) - \omega(t)\,\frac{\mathrm{d}x_2(t)}{\mathrm{d}t}, \\ \frac{\mathrm{d}^2 x_2(t)}{\mathrm{d}t^2} &= \frac{\mathrm{d}s(t)}{\mathrm{d}t} \sin(\varphi(t)) + s(t)\,\frac{\mathrm{d}\varphi(t)}{\mathrm{d}t} \cos(\varphi(t)) \\ &= w_1(t)\sin(\varphi(t)) + \omega(t)\,\frac{\mathrm{d}x_1(t)}{\mathrm{d}t}. \end{aligned} \tag{4.59}$$

Note that the derivation is somewhat heuristic and that a more rigorous derivation would involve the Itô formula (Särkkä and Solin, 2019). If we *now rename the angular velocity noise to $w_3(t)$ and approximate the noise terms above with independent noises $w_1(t) \to w_1(t)\cos(\varphi(t))$ and $w_2(t) \to w_1(t)\sin(\varphi(t))$, we end up with the following Cartesian coordinated turn (CT) model:*

$$\frac{\mathrm{d}}{\mathrm{d}t}\begin{pmatrix} x_1(t) \\ x_2(t) \\ \dot{x}_1(t) \\ \dot{x}_2(t) \\ \omega(t) \end{pmatrix} = \begin{pmatrix} \dot{x}_1(t) \\ \dot{x}_2(t) \\ -\omega(t)\,\dot{x}_2(t) \\ \omega(t)\,\dot{x}_1(t) \\ 0 \end{pmatrix} + \begin{pmatrix} 0 & 0 & 0 \\ 0 & 0 & 0 \\ 1 & 0 & 0 \\ 0 & 1 & 0 \\ 0 & 0 & 1 \end{pmatrix} \begin{pmatrix} w_1(t) \\ w_2(t) \\ w_3(t) \end{pmatrix}, \tag{4.60}$$

where $(\dot{x}_1, \dot{x}_2)$ is the velocity vector. Although this model is conceptually simpler than the polar CT model, the disadvantage is that we no longer have a noise process that directly affects the tangential (speed) direction, and we are instead limited to additive noise to the velocity vector.

The analytical solution to the deterministic part of the aforementioned model now results in

$$\begin{aligned}
&\mathbf{f}_{k-1}(\mathbf{x}_{k-1}) \\
&= \begin{pmatrix} x_{1,k-1} + \frac{\dot{x}_{1,k-1}}{\omega_{k-1}}\sin(\Delta t_{k-1}\,\omega_{k-1}) - \frac{\dot{x}_{2,k-1}}{\omega_{k-1}}(1-\cos(\Delta t_{k-1}\,\omega_{k-1})) \\ x_{2,k-1} + \frac{\dot{x}_{1,k-1}}{\omega_{k-1}}(1-\cos(\Delta t_{k-1}\,\omega_{k-1})) + \frac{\dot{x}_{2,k-1}}{\omega_{k-1}}\sin(\Delta t_{k-1}\,\omega_{k-1}) \\ \dot{x}_{1,k-1}\cos(\Delta t_{k-1}\,\omega_{k-1}) - \dot{x}_{2,k-1}\sin(\Delta t_{k-1}\,\omega_{k-1}) \\ \dot{x}_{1,k-1}\sin(\Delta t_{k-1}\,\omega_{k-1}) + \dot{x}_{2,k-1}\cos(\Delta t_{k-1}\,\omega_{k-1}) \\ \omega_{k-1} \end{pmatrix} \\
&= \begin{pmatrix} 1 & 0 & \frac{\sin(\Delta t_{k-1}\,\omega_{k-1})}{\omega_{k-1}} & -\frac{(1-\cos(\Delta t_{k-1}\,\omega_{k-1}))}{\omega_{k-1}} & 0 \\ 0 & 1 & \frac{(1-\cos(\Delta t_{k-1}\,\omega_{k-1}))}{\omega_{k-1}} & \frac{\sin(\Delta t_{k-1}\,\omega_{k-1})}{\omega_{k-1}} & 0 \\ 0 & 0 & \cos(\Delta t_{k-1}\,\omega_{k-1}) & -\sin(\Delta t_{k-1}\,\omega_{k-1}) & 0 \\ 0 & 0 & \sin(\Delta t_{k-1}\,\omega_{k-1}) & \cos(\Delta t_{k-1}\,\omega_{k-1}) & 0 \\ 0 & 0 & 0 & 0 & 1 \end{pmatrix} \begin{pmatrix} x_{1,k-1} \\ x_{2,k-1} \\ \dot{x}_{1,k-1} \\ \dot{x}_{2,k-1} \\ \omega_{k-1} \end{pmatrix},
\end{aligned} \tag{4.61}$$

as we will see in Exercise 4.5. When $\omega_{k-1} \approx 0$, for numerical stability, it is useful to replace this expression with its first order limit when $\omega_{k-1} \to 0$, which can be written as

$$\lim_{\omega_{k-1}\to 0} \mathbf{f}_{k-1}(\mathbf{x}_{k-1}) = \begin{pmatrix} 1 & 0 & \Delta t_{k-1} & 0 & 0 \\ 0 & 1 & 0 & \Delta t_{k-1} & 0 \\ 0 & 0 & 1 & 0 & 0 \\ 0 & 0 & 0 & 1 & 0 \\ 0 & 0 & 0 & 0 & 1 \end{pmatrix} \begin{pmatrix} x_{1,k-1} \\ x_{2,k-1} \\ \dot{x}_{1,k-1} \\ \dot{x}_{2,k-1} \\ \omega_{k-1} \end{pmatrix}. \tag{4.62}$$

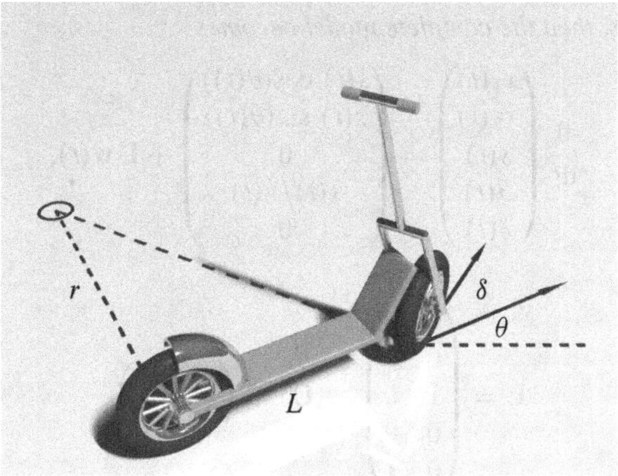

Figure 4.6 An illustration of the parameters in the bicycle model described in Example 4.17.

The solution to the covariance is again intractable, but we come back to finding suitable approximations for the covariance in the next sections.

Another model, which is useful for modeling vehicle dynamics, is the bicycle model. Like the above CT models, the bicycle model also yields closed-form solutions to the predicted mean in Equation (4.52).

Example 4.17 (Bicycle model)**.** *The bicycle model is a simple model for cars and vehicles, which models the vehicle as having the shape of a bicycle, or an electric scooter, as shown in Figure 4.6.*

Let $(x_1(t), x_2(t))$ *refer to the position of the rear wheel,* $\theta(t)$ *the heading,* $s(t)$ *the speed,* $\delta(t)$ *the steering angle for the front wheel, and let* L *be the length of the wheelbase. Apart from the state variables* $\mathbf{x}(t) = (x_1(t), x_2(t), s(t), \theta(t), \delta(t))$*, we use* $r(t)$ *to denote the (signed) distance to the current center of rotation.*

From Figure 4.6 we conclude that $\tan(\delta(t)) = L/r(t)$*. We can now describe the turn rate as*

$$\frac{\mathrm{d}\theta(t)}{\mathrm{d}t} = \frac{s(t)}{r(t)} = \frac{s(t)\tan(\delta(t))}{L}. \tag{4.63}$$

If we assume that the speed and steering angle are driven by independent white noises, then the complete model becomes

$$\frac{\mathrm{d}}{\mathrm{d}t}\begin{pmatrix} x_1(t) \\ x_2(t) \\ s(t) \\ \theta(t) \\ \delta(t) \end{pmatrix} = \begin{pmatrix} s(t)\,\cos(\theta(t)) \\ s(t)\,\sin(\theta(t)) \\ 0 \\ s(t)/r(t) \\ 0 \end{pmatrix} + \mathbf{L}\,\mathbf{w}(t), \tag{4.64}$$

with

$$\mathbf{L} = \begin{pmatrix} 0 & 0 \\ 0 & 0 \\ 1 & 0 \\ 0 & 0 \\ 0 & 1 \end{pmatrix}, \quad \mathbf{Q}^{\mathrm{c}} = \begin{pmatrix} q_s^{\mathrm{c}} & 0 \\ 0 & q_\delta^{\mathrm{c}} \end{pmatrix}. \tag{4.65}$$

The mean in Equation (4.52) *again has a closed-form solution and, as shown in Appendix A.6, this leads to the discretization*

$$\mathbf{f}(\mathbf{x}_{k-1}) = \begin{pmatrix} x_{1,k-1} + r_{k-1}\,(\sin(\theta_{k-1} + \beta_{k-1}) - \sin(\theta_{k-1})) \\ x_{2,k-1} + r_{k-1}\,(\cos(\theta_{k-1}) - \cos(\theta_{k-1} + \beta_{k-1})) \\ s_{k-1} \\ \theta_{k-1} + \beta_{k-1} \\ \delta_{k-1} \end{pmatrix}, \tag{4.66}$$

where $r_{k-1} = L/\tan(\delta_{k-1})$ *and* $\beta_{k-1} = \Delta t_{k-1} s_{k-1}/r_{k-1}$. *Similarly to the CT model, the solution to the covariance using Equation* (4.53) *is intractable, and we can either resort to the Euler–Maryuama approximation for it or use the approximations discussed in the next section.*

As discussed at the beginning of this section, we have many different choices for the linearization trajectory $\bar{\mathbf{x}}(t)$. The linearization with respect to the mean $\mathbf{m}(t)$, which we have discussed so far, is useful when we can solve the deterministic dynamic model analytically or otherwise quickly. However, if that is not the case, then another possible approximation is to linearize with respect to the initial point, that is,

$$\mathbf{a}(\mathbf{x}(t)) \approx \mathbf{a}(\mathbf{x}(t_{k-1})) + \mathbf{A}_{\mathbf{x}}(\mathbf{x}(t_{k-1}))\,(\mathbf{x}(t) - \mathbf{x}(t_{k-1})). \tag{4.67}$$

This leads to the following kind of approximation.

Theorem 4.18 (Linearization with respect to initial condition)**.** *Suppose that we approximate*

$$\frac{\mathrm{d}\mathbf{x}(t)}{\mathrm{d}t} = \mathbf{a}(\mathbf{x}(t)) + \mathbf{L}\,\mathbf{w}(t) \tag{4.68}$$

as

$$\frac{\mathrm{d}\mathbf{x}(t)}{\mathrm{d}t} \approx \mathbf{a}(\mathbf{x}(t_{k-1})) + \mathbf{A}_{\mathbf{x}}(\mathbf{x}(t_{k-1}))\,(\mathbf{x}(t) - \mathbf{x}(t_{k-1})) + \mathbf{L}\,\mathbf{w}(t). \quad (4.69)$$

The discretized model is then

$$\begin{aligned}\mathbf{x}_k &= \mathbf{x}_{k-1} + \left(\int_0^{\Delta t_{k-1}} \exp(\mathbf{A}_{\mathbf{x}}(\mathbf{x}_{k-1})\,s)\,\mathrm{d}s\right)\mathbf{a}(\mathbf{x}_{k-1}) + \mathbf{q}_{k-1} \\ &= \underbrace{\mathbf{x}_{k-1} + \left(\sum_{i=1}^{\infty} \frac{\Delta t_{k-1}^i\,\mathbf{A}_{\mathbf{x}}(\mathbf{x}_{k-1})^{i-1}}{i!}\right)\mathbf{a}(\mathbf{x}_{k-1})}_{\mathbf{f}_{k-1}(\mathbf{x}_{k-1})} + \mathbf{q}_{k-1},\end{aligned} \quad (4.70)$$

where $\mathbf{q}_{k-1} \sim \mathrm{N}(\mathbf{0}, \mathbf{Q}_{k-1}(\mathbf{x}_{k-1}))$ *and*

$$\mathbf{Q}_{k-1}(\mathbf{x}_{k-1}) = \int_0^{\Delta t_{k-1}} \exp(\mathbf{A}_{\mathbf{x}}(\mathbf{x}_{k-1})\,s)\,\mathbf{L}\,\mathbf{Q}^{\mathrm{c}}\,\mathbf{L}^{\mathsf{T}} \exp(\mathbf{A}_{\mathbf{x}}(\mathbf{x}_{k-1})\,s)^{\mathsf{T}}\,\mathrm{d}s. \quad (4.71)$$

Proof See Exercise 4.7. □

As a sanity check, we note that Theorem 4.18 yields the same discretized model as Theorem 4.3 when $\mathbf{a}(\mathbf{x}(t_{k-1})) = \mathbf{F}\,\mathbf{x}(t_{k-1})$, since that implies that $\mathbf{A}_{\mathbf{x}}(\mathbf{x}_{k-1}) = \mathbf{F}$, for which the deterministic part of the discretization simplifies to

$$\mathbf{x}_{k-1} + \left(\sum_{i=1}^{\infty} \frac{\Delta t_{k-1}^i \mathbf{A}_{\mathbf{x}}(\mathbf{x}_{k-1})^{i-1}}{i!}\right)\mathbf{a}(\mathbf{x}_{k-1}) = \exp(\mathbf{F}\,\Delta t_{k-1})\,\mathbf{x}_{k-1}. \quad (4.72)$$

The method above has the advantage that it provides a tractable approximation for the covariance as well. However, the approximation of the deterministic part can be quite rough as it is based on linearization at the initial point only. It can also be cumbersome to compute. Now a promising idea is to combine the approximations, which is discussed in the next section.

4.5 Covariance Approximation via Constant Gradients

Recall that we are aiming at finding discretizations of the form

$$\mathbf{x}_k = \mathbf{f}_{k-1}(\mathbf{x}_{k-1}) + \mathbf{q}_{k-1}, \quad \mathbf{q}_{k-1} \sim \mathrm{N}(\mathbf{0}, \mathbf{Q}_{k-1}), \quad (4.73)$$

where both the deterministic part $\mathbf{f}_{k-1}(\mathbf{x}_{k-1})$ and the noise covariance $\mathbf{Q}_{k-1}$ should be as accurate as possible. The covariance should preferably also be non-singular and independent of $\mathbf{x}_{k-1}$. In the previous section we saw that by using a continuous-time linearization with respect to the mean,

$$\mathbf{a}(\mathbf{x}(t)) \approx \mathbf{a}(\mathbf{m}(t)) + \mathbf{A}_{\mathbf{x}}(\mathbf{m}(t))\,(\mathbf{x}(t) - \mathbf{m}(t)), \tag{4.74}$$

we can often obtain good approximations to the deterministic part $\mathbf{f}_{k-1}(\mathbf{x}_{k-1})$ by solving the mean differential equation in closed form. On the other hand, by linearizing with respect to the initial point,

$$\mathbf{a}(\mathbf{x}(t)) \approx \mathbf{a}(\mathbf{x}_{k-1}) + \mathbf{A}_{\mathbf{x}}(\mathbf{x}_{k-1})\,(\mathbf{x}(t) - \mathbf{x}_{k-1}), \tag{4.75}$$

we obtained a slightly inferior approximation for the deterministic part but with the advantage that the covariance approximation was tractable. To get the benefits of both of the approaches, it is possible to use a linearization of the form

$$\mathbf{a}(\mathbf{x}(t)) \approx \mathbf{a}(\mathbf{m}(t)) + \mathbf{A}_{\mathbf{x}}(\mathbf{x}_{k-1})\,(\mathbf{x}(t) - \mathbf{m}(t)), \tag{4.76}$$

where $\mathbf{A}_{\mathbf{x}}(\mathbf{x}_{k-1})$ is now a constant matrix. This linearization yields the same expression for $\mathbf{m}(t)$ as the linearization with respect to the mean, and thus $\mathbf{f}_{k-1}(\cdot)$, as the linearization in (4.50).

Theorem 4.19 (Discretization with a constant gradient). *Suppose that we approximate the dynamic model SDE by*

$$\frac{\mathrm{d}\mathbf{x}(t)}{\mathrm{d}t} \approx \mathbf{a}(\mathbf{m}(t)) + \mathbf{A}_{\mathbf{x}}(\mathbf{x}_{k-1})\,(\mathbf{x}(t) - \mathbf{m}(t)) + \mathbf{L}\,\mathbf{w}(t), \tag{4.77}$$

with the initial condition $\mathbf{x}(t_{k-1}) = \mathbf{x}_{k-1}$. *The mean* $\mathbf{m}(t)$ *is still the solution to*

$$\frac{\mathrm{d}\mathbf{m}(t)}{\mathrm{d}t} = \mathbf{a}(\mathbf{m}(t)) \tag{4.78}$$

with $\mathbf{m}(t_{k-1}) = \mathbf{x}_{k-1}$, *and the covariance is given by*

$$\mathbf{P}(t_k) = \int_0^{\Delta t_{k-1}} \exp(\mathbf{A}_{\mathbf{x}}(\mathbf{x}_{k-1})\,s)\,\mathbf{L}\,\mathbf{Q}^{\mathrm{c}}\,\mathbf{L}^{\mathsf{T}} \exp(\mathbf{A}_{\mathbf{x}}(\mathbf{x}_{k-1})\,s)^{\mathsf{T}}\,\mathrm{d}s. \tag{4.79}$$

The discretization can then be formed by putting $\mathbf{f}_{k-1}(\mathbf{x}_{k-1}) = \mathbf{m}(t_k)$ *and* $\mathbf{Q}_{k-1}(\mathbf{x}_{k-1}) = \mathbf{P}(t_k)$.

Proof See Exercise 4.9. □

Although the above theorem provides approximations for the mean and covariance in a form that can be easily numerically approximated, one practical challenge is that the covariance expressions explicitly depend on the value of $\mathbf{x}_{k-1}$. That is, the dynamic model takes the form $p(\mathbf{x}_k \mid \mathbf{x}_{k-1}) = \mathrm{N}(\mathbf{x}_k \mid \mathbf{f}_{k-1}(\mathbf{x}_{k-1}), \mathbf{Q}_{k-1}(\mathbf{x}_{k-1}))$. Some of the filters (and smoothers) presented later can easily handle models of this type, such as Gaussian filters that use the conditional moments formulation (see, e.g., Algorithm 9.22) and the particle filters (see, e.g., Algorithm 11.8). Still, to obtain simpler filters and smoothers, we would prefer the covariance not to depend on the state. One simple solution is to introduce additional approximations and evaluate the Jacobian $\mathbf{A}_{\mathbf{x}}(\cdot)$ at some fixed point, other than $\mathbf{x}_{k-1}$.

In fact, the aforementioned procedure can be combined with an Euler–Maruyama approximation for $\mathbf{f}_{k-1}(\mathbf{x}_{k-1})$. Thus we get the following procedure.

Algorithm 4.20 (Additive noise approximation). *An additive noise approximation to the SDE in Equation* (4.21) *can be obtained by first forming, for example, the Euler–Maruyama approximation or continuous-time linearization approximation for* $\mathbf{f}_{k-1}(\mathbf{x}_{k-1})$. *Then the noise can be approximated by linearizing the model at some fixed point* $\hat{\mathbf{x}}$ *and by approximating the covariance by the covariance of*

$$\frac{\mathrm{d}\mathbf{x}(t)}{\mathrm{d}t} = \mathbf{A}_{\mathbf{x}}(\hat{\mathbf{x}})\,\mathbf{x}(t) + \mathbf{L}\,\mathbf{w}(t), \tag{4.80}$$

which leads to the approximation

$$\mathbf{Q}_{k-1} \approx \int_0^{\Delta t_{k-1}} \exp(\mathbf{A}_{\mathbf{x}}(\hat{\mathbf{x}})\,s)\,\mathbf{L}\,\mathbf{Q}^{\mathrm{c}}\,\mathbf{L}^{\mathsf{T}} \exp(\mathbf{A}_{\mathbf{x}}(\hat{\mathbf{x}})\,s)^{\mathsf{T}}\,\mathrm{d}s. \tag{4.81}$$

The covariance resulting from the above is usually non-singular even in the case when the Euler–Maruyama method gives a singular covariance. Furthermore, it is independent of $\mathbf{x}_{k-1}$ since we use a fixed linearization point $\hat{\mathbf{x}}$. We are free to select any linearization point, for example, the origin. However, in practice, it is often desirable to linearize $\mathbf{a}(\mathbf{x}_k)$ close to the true value of the state, and it may therefore be better to use, for instance, the posterior mean computed by a filter as linearization point, $\hat{\mathbf{x}} = \mathbf{m}_k$.

The approximation in Algorithm 4.20 can often be made even simpler by only linearizing with respect to some variables and leaving others constant. This is illustrated in the following example.

Example 4.21 (Additive noise approximation for pendulum model). *The noisy pendulum model in Example 4.7 can be linearized around* $\hat{\mathbf{x}} = (0, 0)$

to give

$$\mathbf{a}(\mathbf{x}) = \begin{pmatrix} x_2 \\ -g\,\sin(x_1) \end{pmatrix} \approx \underbrace{\begin{pmatrix} 0 & 1 \\ -g & 0 \end{pmatrix}}_{\mathbf{A}_\mathbf{x}(\hat{\mathbf{x}})} \begin{pmatrix} x_1 \\ x_2 \end{pmatrix}, \tag{4.82}$$

for which the approximate process noise given by Equation (4.81) *is constant. Another way is just to linearize with respect to x_2 and leave x_1 fixed to the origin, which gives an even simpler approximation:*

$$\mathbf{a}(\mathbf{x}) = \begin{pmatrix} x_2 \\ -g\,\sin(x_1) \end{pmatrix} \approx \begin{pmatrix} 0 & 1 \\ 0 & 0 \end{pmatrix} \begin{pmatrix} x_1 \\ x_2 \end{pmatrix}. \tag{4.83}$$

With this approximation, we get a closed form covariance approximation via Equation (4.18)*:*

$$\mathbf{Q}_{k-1} = q^{\mathrm{c}} \begin{pmatrix} \frac{\Delta t_{k-1}^3}{3} & \frac{\Delta t_{k-1}^2}{2} \\ \frac{\Delta t_{k-1}^2}{2} & \Delta t_{k-1} \end{pmatrix}. \tag{4.84}$$

Sometimes it is not possible to form a simple approximation by linearization at a single point. In that case it is sometimes possible to first form an approximation via linearization at $\mathbf{x}_{k-1}$ and then approximating it. This is demonstrated in the following example.

Example 4.22 (Constant gradient covariance for polar CT model)**.** *The expression for the covariance for the polar coordinated turn model in Example 4.10, by a constant gradient approximation at $\mathbf{x}_{k-1}$, is given in Appendix A.8. By setting the speed to zero, approximating the cross-terms involving sines and cosines to zero, and by dominating squares of sines and cosines by 1, we obtain the following approximation to the covariance:*

$$\mathbf{Q}_{k-1} = \begin{pmatrix} q_s^{\mathrm{c}}\,\frac{\Delta t_{k-1}^3}{3} & 0 & 0 & 0 & 0 \\ 0 & q_s^{\mathrm{c}}\,\frac{\Delta t_{k-1}^3}{3} & 0 & 0 & 0 \\ 0 & 0 & q_s^{\mathrm{c}}\,\Delta t_{k-1} & 0 & 0 \\ 0 & 0 & 0 & q_\omega^{\mathrm{c}}\,\frac{\Delta t_{k-1}^3}{3} & q_\omega^{\mathrm{c}}\,\frac{\Delta t_{k-1}^2}{2} \\ 0 & 0 & 0 & q_\omega^{\mathrm{c}}\,\frac{\Delta t_{k-1}^2}{2} & q_\omega^{\mathrm{c}}\,\Delta t_{k-1} \end{pmatrix}, \tag{4.85}$$

where q_s^{c} and q_ω^{c} are the spectral densities of the white noises driving the speed and angular velocity, respectively.

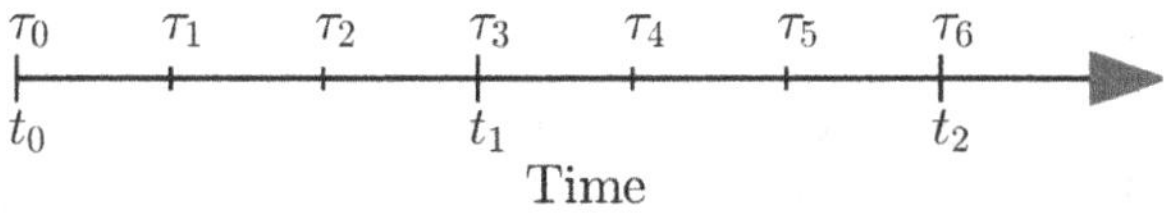

Figure 4.7 The figure illustrates a possible relation between the sequences $(t_0, t_1, \ldots)$ and $(\tau_0, \tau_1, \ldots)$ when $n_1 = n_2 = 3$. We normally collect measurements at $\tau_0 = t_0$, $\tau_3 = t_1$ and $\tau_6 = t_2$, but not at the intermediate times τ_1, τ_2, τ_4, and τ_5.

4.6 Fast Sampling

As long as the sampling period Δt_{k-1} is small, even the Euler–Maruyama method in Theorem 4.8 yields an accurate discrete-time dynamic model. We have presented several techniques that aim to also perform accurate discretization for larger sampling periods. However, we limited our discussion to models that approximate $p(\mathbf{x}(t_k) \mid \mathbf{x}(t_{k-1}))$ as Gaussian in a single step, which may be a poor approximation for large sampling periods. One way to improve the discretization is to use high order SDE approximations, such as Itô–Taylor expansions (see Kloeden and Platen, 1999; Särkkä and Solin, 2019), which unfortunately often leads to overly complicated solutions. An alternative strategy is to reformulate the problem such that we sample the continuous-time state sequence more frequently, to reduce the sampling period. We here refer to this as a fast sampling approach. In this formulation, it is easy to obtain an accurate (Gaussian) dynamic model, but the reduced sampling period may instead result in increased complexity for the discrete-time filtering and smoothing algorithms.

One way to formalize this strategy is by introducing additional $n_k - 1$ sampling times τ_i between t_{k-1} and t_k such that $\tau_i = t_k$ when $i = \sum_{j=1}^{k} n_k$ and defining the discrete time sequence as $\mathbf{x}_k = \mathbf{x}(\tau_k)$. This is illustrated in Figure 4.7. This approach yields a different state sequence because we sample the continuous time state sequence more frequently. Since we are still only collecting measurements at $t_1, t_2, \ldots$, fast sampling means that there are n_k discrete time steps between every measurement that we collect. In spite of this, the filters, predictors, and smoothers presented in this book can still be applied directly, with the minor difference that we skip the update step when we do not have a measurement. By computing the posterior distribution of $\mathbf{x}_0 = \mathbf{x}(\tau_0), \mathbf{x}_1 = \mathbf{x}(\tau_1), \ldots$, we also obtain the posterior distribution at the original sampling times $\mathbf{x}(t_0), \mathbf{x}(t_1), \ldots$

since we have merely introduced new sampling times between the original ones.

Fast sampling is closely related to the numerical solution of the SDE in Algorithm 4.12 since we again split the time interval $[t_{k-1}, t_k]$ into subintervals and approximate $\mathbf{a}(\mathbf{x}(t))$ differently in each interval. If we use the Euler–Maruyama method with the faster sampling rate, we implicitly assume that

$$\mathbf{a}(\mathbf{x}(t)) \approx \mathbf{a}(\mathbf{x}(\tau_{k-1})), \quad t \in [\tau_{k-1}, \tau_k), \tag{4.86}$$

but we now only use this approximation in a time interval of length $\Delta\tau_{k-1} = \tau_k - \tau_{k-1}$. An important special case is when the sampling time is constant, $\Delta t_k = \Delta t$, and we further split Δt_k into n equally long subintervals such that also $\Delta\tau_k = \Delta\tau$ is the same for all k and $\Delta\tau = \Delta t/n$. By sampling faster, the Euler–Maruyama method uses (4.86) in an interval that is $1/n$ of the original sampling period, which leads to more accurate approximations.

4.7 Exercises

4.1 Use Algorithm 4.12 to sample from a scalar Wiener process

$$\frac{dx(t)}{dt} = w(t), \tag{4.87}$$

where $w(t)$ has the spectral density $q^c = 1$. Use $\Delta\tau = 0.01$ to generate and illustrate $N = 3$ samples in the interval $t \in [0, 4]$ when $x(0) = 0$.

4.2 In continuous time, a d-dimensional constant acceleration (CA) model can be described as

$$\frac{d\mathbf{x}(t)}{dt} = \begin{pmatrix} \mathbf{0} & \mathbf{I} & \mathbf{0} \\ \mathbf{0} & \mathbf{0} & \mathbf{I} \\ \mathbf{0} & \mathbf{0} & \mathbf{0} \end{pmatrix} \mathbf{x}(t) + \begin{pmatrix} \mathbf{0} \\ \mathbf{0} \\ \mathbf{I} \end{pmatrix} \mathbf{w}(t), \tag{4.88}$$

where the first d elements in $\mathbf{x}(t)$ represent the position of an object, the next d elements represent the velocity, and the final d elements are the acceleration. Use Theorem 4.3 to obtain the discrete-time CA model. For simplicity, you can assume that the spectral density of the noise is $\mathbf{Q}^c = \sigma^2\mathbf{I}$.

4.3 Discretize the CA model in Exercise 4.2 using the Euler–Maruyama method in Theorem 4.8, and compare the result to the exact discretization obtained in Exercise 4.2.

4.4 In order to build intuition for when different models are useful, it is helpful to observe sequences sampled from candidate models. Generate sequences from the scalar random walk (RW) model given in Equation (4.15), the constant velocity (CV) model described in Example 4.5, and the CA model described in Exercise 4.2. Plot the positions for sequences generated using the RW and CV models, and the velocities for sequences generated using the CV and CA models. To facilitate the comparison, adjust the parameters such that the sequences that you plot jointly take values in similar ranges. Compare and comment on the roughness of the different sequences.

4.5 Derive the expressions for $\mathbf{f}_{k-1}(\mathbf{x}_{k-1})$ for the Cartesian coordinated turn model given in Example 4.16. Note that $\mathbf{f}_{k-1}(\mathbf{x}_{k-1}) = \mathbf{m}(t_k)$, where $\mathbf{m}(t_k)$ is the solution to the dynamic equation without noise.

4.6 Derive the limit for $\mathbf{f}_{k-1}(\mathbf{x}_{k-1})$ when $\omega_{k-1} \to 0$ for the Cartesian coordinated turn model given in Example 4.16.

4.7 Prove Theorem 4.18. *Hint:* You can obtain the discrete-time model using Lemma A.9 and simplify the expression by recalling the definition of the matrix exponential.

4.8 An interesting special case of Theorem 4.13 is when $\mathbf{a}(\mathbf{x}) = \mathbf{F}\,\mathbf{x}$, which implies that $\mathbf{A}_{\mathbf{x}}(\mathbf{x}) = \mathbf{F}$. It then holds that $\mathbf{a}(\mathbf{m}(t)) = \mathbf{A}_{\mathbf{x}}(\mathbf{m}(t))\,\mathbf{m}(t)$, and (4.51) then simplifies to (4.5). For such linear and time-invariant systems, Theorem 4.3 specifies that

$$\begin{aligned} \mathbf{m}(t) &= \exp(\mathbf{F}\,(t - t_{k-1}))\,\mathbf{x}(t_{k-1}), \\ \mathbf{P}(t) &= \int_0^{t-t_{k-1}} \exp(\mathbf{F}\,s)\;\mathbf{L}\,\mathbf{Q}^{\mathrm{c}}\,\mathbf{L}^{\mathsf{T}}\,\exp(\mathbf{F}\,s)^{\mathsf{T}}\;\mathrm{d}s. \end{aligned} \tag{4.89}$$

Verify that the moments in Equation (4.89) satisfy the description in Theorem 4.13 when we have a scalar state variable $x(t)$. Note that $P(t)$ in (4.89) has a simple closed form solution when all variables are scalar.

4.9 Prove Theorem 4.19.

4.10 Write down the expression for the process noise covariance $\mathbf{Q}_{k-1}(\mathbf{x}_{k-1})$ resulting from the method in Theorem 4.19 for the bicycle model considered in Example 4.17. Also use the discretized model to simulate trajectories from the dynamics.

4.11 Show that using the additive noise approximation in Algorithm 4.20 on the Cartesian coordinated turn model in Equation (4.60) at origin leads to the following process noise covariance:

$$\mathbf{Q}_{k-1} = \begin{pmatrix} q_1^{\mathrm{c}}\,\frac{\Delta t_{k-1}^3}{3} & 0 & q_1^{\mathrm{c}}\,\frac{\Delta t_{k-1}^2}{2} & 0 & 0 \\ 0 & q_2^{\mathrm{c}}\,\frac{\Delta t_{k-1}^3}{3} & 0 & q_2^{\mathrm{c}}\,\frac{\Delta t_{k-1}^2}{2} & 0 \\ q_1^{\mathrm{c}}\,\frac{\Delta t_{k-1}^2}{2} & 0 & q_1^{\mathrm{c}}\,\Delta t_{k-1} & 00 & \\ 0 & q_2^{\mathrm{c}}\,\frac{\Delta t_{k-1}^2}{2} & 0 & q_2^{\mathrm{c}}\,\Delta t_{k-1} & 0 \\ 0 & 0 & 0 & 0 & q_3^{\mathrm{c}}\,\Delta t_{k-1} \end{pmatrix}, \tag{4.90}$$

where the spectral densities of white noises $w_1(t)$, $w_2(t)$, and $w_3(t)$ are q_1^c, q_2^c, and q_3^c, respectively.

5

Modeling with State Space Models

In the previous chapter, we discussed the discretization of continuous-time dynamic models as a strategy to construct discrete-time dynamic models that describe the transition density, $p(\mathbf{x}_k \mid \mathbf{x}_{k-1})$, for a probabilistic state space model. However, a probabilistic state space model also requires a measurement model, $p(\mathbf{y}_k \mid \mathbf{x}_k)$, which describes the conditional distribution of a measurement $\mathbf{y}_k$ given a state $\mathbf{x}_k$ at time step k. In this chapter, we discuss the construction of measurement models in general and provide specific examples of models that we can combine with the dynamic models introduced in the previous chapter. Finally, we present two more model families that we can represent as state space models.

5.1 Linear and Non-Linear Gaussian Models

In Chapter 4 we presented various ways to construct dynamic models of the form

$$\mathbf{x}_k = \mathbf{f}(\mathbf{x}_{k-1}) + \mathbf{q}_{k-1}, \tag{5.1}$$

where $\mathbf{q}_{k-1} \sim \mathrm{N}(\mathbf{0}, \mathbf{Q}_{k-1})$ is Gaussian process noise. In general, the dynamic function $\mathbf{f}(\cdot)$ may depend on the time step $k-1$, and the covariance $\mathbf{Q}_{k-1}$ may depend on the state $\mathbf{x}_{k-1}$, but here we omit both of these dependencies from our notation.

The model in Equation (5.1) thus corresponds to a conditional distribution model

$$p(\mathbf{x}_k \mid \mathbf{x}_{k-1}) = \mathrm{N}(\mathbf{x}_k \mid \mathbf{f}(\mathbf{x}_{k-1}), \mathbf{Q}_{k-1}). \tag{5.2}$$

In reality, we often cannot observe the states $\mathbf{x}_k$ directly, but we only get some indirect, noisy observations $\mathbf{y}_k$ of them. The aim is then to estimate the states from the measurements, which is a task that can be accomplished by using Bayesian filters and smoothers (such as Kalman filters and Rauch–Tung–Striebel smoothers), which we will encounter in the subsequent chapters.

We can often express the measurement model as a functional relation

$$\mathbf{y}_k = \mathbf{h}(\mathbf{x}_k) + \mathbf{r}_k, \quad \mathbf{r}_k \sim \mathrm{N}(\mathbf{0}, \mathbf{R}_k), \tag{5.3}$$

which implies that we have

$$p(\mathbf{y}_k \mid \mathbf{x}_k) = \mathrm{N}(\mathbf{y}_k \mid \mathbf{h}(\mathbf{x}_k), \mathbf{R}_k). \tag{5.4}$$

Even more generally, we may have $\mathbf{y}_k = \mathbf{h}(\mathbf{x}_k, \mathbf{r}_k)$, and the noise $\mathbf{r}_k$ may also be non-Gaussian, but the special case in Equation (5.3) is more common in the literature. We will, however, come back to these extensions in the next section.

There is a wide range of sensors, including radars, speedometers, gyroscopes, global navigation satellite systems (GNSSs), radars, accelerometers, thermometers, and barometers. The functional form of $\mathbf{h}(\cdot)$, and whether the function is linear or non-linear, depend on what the sensor measures and how the measurement is related to the chosen state representation. Linear models are common when the measurements are noisy observations of some state variables. Non-linear models may appear, for example, if the state variables need to be converted into a different coordinate system to enable a simple relation to the measurements.

Let us start with a linear measurement model for a car.

Example 5.1 (Position measurement model for a car)**.** *Recall that in Example 4.6 we derived a discrete-time linear dynamic model for a car of the form $\mathbf{x}_k = \mathbf{A}\,\mathbf{x}_{k-1} + \mathbf{q}_{k-1}$, where the state $\mathbf{x} = (x_1, x_2, x_3, x_4)$ contains the position variables (x_1, x_2) and the velocity vector (x_3, x_4). Assume that the position of the car (x_1, x_2) is measured and that the measurements are corrupted by independent Gaussian measurement noises $r_{1,k} \sim \mathrm{N}(0, \sigma_1^2)$ and $r_{2,k} \sim \mathrm{N}(0, \sigma_2^2)$ (see Figure 5.1):*

$$\begin{aligned} y_{1,k} &= x_{1,k} + r_{1,k}, \\ y_{2,k} &= x_{2,k} + r_{2,k}. \end{aligned} \tag{5.5}$$

The measurement model can now be written as

$$\mathbf{y}_k = \mathbf{H}\,\mathbf{x}_k + \mathbf{r}_k, \qquad \mathbf{H} = \begin{pmatrix} 1 & 0 & 0 & 0 \\ 0 & 1 & 0 & 0 \end{pmatrix}. \tag{5.6}$$

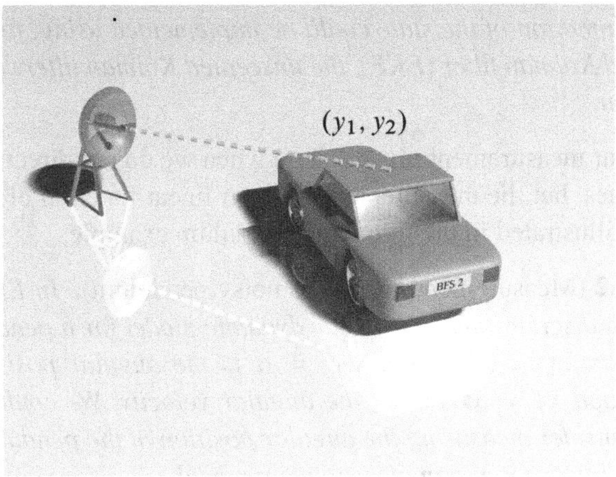

Figure 5.1 Illustration of car's measurement model. The measurements (y_1, y_2) are modeled as noise-corrupted observations of the car's position.

The dynamic and measurement models of the car now form a linear Gaussian state space model

$$\begin{aligned} \mathbf{x}_k &= \mathbf{A}\,\mathbf{x}_{k-1} + \mathbf{q}_{k-1}, \\ \mathbf{y}_k &= \mathbf{H}\,\mathbf{x}_k + \mathbf{r}_k, \end{aligned}$$

where $\mathbf{q}_{k-1} \sim \mathrm{N}(\mathbf{0}, \mathbf{Q})$ *and* $\mathbf{r}_k \sim \mathrm{N}(\mathbf{0}, \mathbf{R})$ *with* $\mathbf{R} = \mathrm{diag}(\sigma_1^2, \sigma_2^2)$. *Because this model is linear and Gaussian, the state of the car can be estimated from the noisy measurements using the Kalman filter.*

Suppose that we instead use the non-linear polar coordinated turn model in Example 4.10 as the dynamic model, whose discrete-time versions were considered in Examples 4.11 and 4.15. We would then get a measurement model

$$\mathbf{y}_k = \begin{pmatrix} 1 & 0 & 0 & 0 & 0 \\ 0 & 1 & 0 & 0 & 0 \end{pmatrix} \mathbf{x}_k + \mathbf{r}_k, \tag{5.7}$$

as the state in that model is defined as $\mathbf{x} = \big(x_1, x_2, s, \varphi, \omega\big)$, *where* (x_1, x_2) *is the position,* s *is the speed,* φ *is the direction, and* ω *is the angular velocity. This then leads to a model that has non-linear dynamics but a linear measurement model:*

$$\begin{aligned} \mathbf{x}_k &= \mathbf{f}\,(\mathbf{x}_{k-1}) + \mathbf{q}_{k-1}, \\ \mathbf{y}_k &= \mathbf{H}\,\mathbf{x}_k + \mathbf{r}_k, \end{aligned}$$

where it is common to assume that $\mathbf{q}_{k-1}$ and $\mathbf{r}_k$ are again Gaussian. In this case, the estimation of the state could be implemented using, for example, the extended Kalman filter (EKF), the unscented Kalman filter (UKF), or a particle filter.

Non-linear measurement models arise when we do not directly measure state variables, but the measurement is a non-linear function of state components, as illustrated in the following pendulum example.

Example 5.2 (Measurement model for noisy pendulum). *In Example 4.9 we formed a discrete-time non-linear dynamic model for a pendulum with the state $\mathbf{x} = (x_1, x_2)$, where $x_1 = \alpha$ is the angular position of the pendulum, and $x_2 = \mathrm{d}\alpha/\mathrm{d}t$ is the angular velocity. We could then, for example, consider measuring the angular position of the pendulum, which leads to the linear Gaussian measurement model*

$$y_k = \alpha(t_k) + \mathit{noise}. \tag{5.8}$$

However, if we measure only the horizontal position of the pendulum, we get a non-linear measurement model

$$y_k = \sin(\alpha(t_k)) + r_k, \tag{5.9}$$

where r_k is the measurement noise, which we can model, for example, as a Gaussian $r_k \sim \mathrm{N}(0, R_k)$. If we combine the dynamic and measurement models together, we get a model of the general form

$$\begin{aligned} \mathbf{x}_k &= \mathbf{f}(\mathbf{x}_{k-1}) + \mathbf{q}_{k-1}, \\ \mathbf{y}_k &= \mathbf{h}(\mathbf{x}_k) + \mathbf{r}_k, \end{aligned} \tag{5.10}$$

where $\mathbf{y}_k$ is the vector of measurements, $\mathbf{q}_{k-1} \sim \mathrm{N}(\mathbf{0}, \mathbf{Q}_{k-1})$, and $\mathbf{r}_k \sim \mathrm{N}(\mathbf{0}, \mathbf{R}_k)$. Estimation of the pendulum state can be now implemented using, for example, the extended Kalman filter (EKF), unscented Kalman filter (UKF), or particle filter.

Non-linear measurement models also arise, for example, when we measure range and bearing, which is a type of measurement that a radar produces. This is illustrated in the following example.

Example 5.3 (Radar measurements). *Let $\mathbf{y}_k$ be an observation from a radar sensor, which provides a noisy observation of the range (distance) and bearing (angle) to the position of an object in 2D. The sensor is located at the origin, and we model the state dynamics using a coordinated turn model in Example 4.10 (with discretizations in Examples 4.11 and 4.15) such that the state is represented by $\mathbf{x} = (x_1, x_2, s, \varphi, \omega)$, where*

(x_1, x_2) is the position, s is the speed, φ is the direction, and ω is the angular velocity. To model the measurement, we can express range and bearing as a function of $\mathbf{x}_k$ plus noise as follows:

$$\mathbf{y}_k = \underbrace{\begin{pmatrix} \sqrt{x_{1,k}^2 + x_{2,k}^2} \\ \arctan\left(\frac{x_{2,k}}{x_{1,k}}\right) \end{pmatrix}}_{\mathbf{h}(\mathbf{x}_k)} + \underbrace{\begin{pmatrix} r_{1,k} \\ r_{2,k} \end{pmatrix}}_{\mathbf{r}_k}. \tag{5.11}$$

The measurement model above would be identical if we modeled the dynamics using the constant velocity model in Example 4.6.

In target tracking scenarios, we often have multiple sensors (such as radars) instead of a single one. In this case, the measurements from the multiple sensors can be simply stacked into a larger-dimensional measurement vector as illustrated in the following.

Example 5.4 (Multiple range measurements)**.** *Let us assume that we have four radars located in positions $(s_{1,x}, s_{1,y}), \ldots, (s_{4,x}, s_{4,y})$, and they measure range to the target whose position coordinates are given by the first two elements (x_1, x_2) of the state $\mathbf{x}$. Then the measurement model can be written as*

$$\mathbf{y}_k = \underbrace{\begin{pmatrix} \sqrt{(x_{1,k} - s_{1,x})^2 + (x_{2,k} - s_{1,y})^2} \\ \sqrt{(x_{1,k} - s_{2,x})^2 + (x_{2,k} - s_{2,y})^2} \\ \sqrt{(x_{1,k} - s_{3,x})^2 + (x_{2,k} - s_{3,y})^2} \\ \sqrt{(x_{1,k} - s_{4,x})^2 + (x_{2,k} - s_{4,y})^2} \end{pmatrix}}_{\mathbf{h}(\mathbf{x}_k)} + \underbrace{\begin{pmatrix} r_{1,k} \\ r_{2,k} \\ r_{3,k} \\ r_{4,k} \end{pmatrix}}_{\mathbf{r}_k}. \tag{5.12}$$

Similarly we could stack a combination of range and bearings measurements into a single measurement vector.

When the radar measurement model in Example 5.3 or range measurement model in Example 5.4 is combined with a dynamic model, this leads to a model of the form

$$\begin{aligned} \mathbf{x}_k &= \mathbf{f}(\mathbf{x}_{k-1}) + \mathbf{q}_{k-1}, \\ \mathbf{y}_k &= \mathbf{h}(\mathbf{x}_k) + \mathbf{r}_k, \end{aligned} \tag{5.13}$$

where we can process the measurements $\mathbf{y}_k$ to obtain estimates of the state by using, for example, the extended Kalman filter (EKF) or one of the other non-linear filters or smoothers that we encounter later in this book.

5.2 Non-Gaussian Measurement Models

In the previous section we discussed models of the form

$$p(\mathbf{y}_k \mid \mathbf{x}_k) = \mathrm{N}(\mathbf{y}_k \mid \mathbf{h}(\mathbf{x}_k), \mathbf{R}_k), \tag{5.14}$$

which can also be expressed as $\mathbf{y}_k = \mathbf{h}(\mathbf{x}_k) + \mathbf{r}_k$ where $\mathbf{r}_k \sim \mathrm{N}(\mathbf{0}, \mathbf{R}_k)$. In other words, we have a possibly non-linear measurement function $\mathbf{h}(\mathbf{x}_k)$ that describes the conditional mean $\mathrm{E}[\mathbf{y}_k \mid \mathbf{x}_k]$, and we have additive Gaussian noise $\mathbf{r}_k$. Even though this is a commonly used model, there are many situations where the noise cannot be accurately modeled as Gaussian.

One distribution that is often used to model the noise is the Student's t-distribution. The n-dimensional Student's t-distribution with ν degrees of freedom has the form (Kotz and Nadarajah, 2004; Roth, 2012)

$$\begin{aligned}\mathrm{St}(\mathbf{y} \mid \boldsymbol{\mu}, \mathbf{R}, \nu) \\ = \frac{\Gamma((\nu+n)/2)}{\Gamma(\nu/2)\,\nu^{n/2}\,\pi^{n/2}\,|\mathbf{R}|^{1/2}} \left(1 + \frac{1}{\nu}\,(\mathbf{y}-\boldsymbol{\mu})^{\mathsf{T}}\,\mathbf{R}^{-1}\,(\mathbf{y}-\boldsymbol{\mu})\right)^{-(\nu+n)/2},\end{aligned} \tag{5.15}$$

where $\Gamma(z) = \int_0^\infty x^{z-1} \exp(-x)\,\mathrm{d}x$ is the gamma function. The Student's t-distribution has heavier tails than the Gaussian distribution, that is, more of its probability is far away from the mean (see Figure 5.2). Due to the heavy tails, the model expects some measurements with larger errors than the typical ones; such measurements are often referred to as outliers, and the model is therefore more robust to outliers than a Gaussian model.

Using this distribution for the additive noise, we can now form a more robust measurement model as

$$p(\mathbf{y}_k \mid \mathbf{x}_k) = \mathrm{St}(\mathbf{y}_k \mid \mathbf{h}(\mathbf{x}_k), \mathbf{R}, \nu), \tag{5.16}$$

which thus sets the mean to be equal to the sensor model $\mathbf{h}(\mathbf{x}_k)$. The other parameters $\mathbf{R}$ and ν control the covariance and heaviness of the tails of the distribution. For example, the covariance is given as $\frac{\nu}{\nu-2}\mathbf{R}$, which is only finite when $\nu > 2$.

Example 5.5 (Robust measurement model for a car). *A more robust measurement model for the car considered in Example 5.1 can be constructed by replacing the Gaussian noise model with a Student's t-distribution model:*

$$p(\mathbf{y}_k \mid \mathbf{x}_k) = \mathrm{St}(\mathbf{y}_k \mid \mathbf{H}\,\mathbf{x}_k, \mathbf{R}, \nu). \tag{5.17}$$

In this case the Kalman filter is no longer applicable, though it is possible to use Student's t extensions of the Kalman filter on it (see, e.g., Roth et al.,

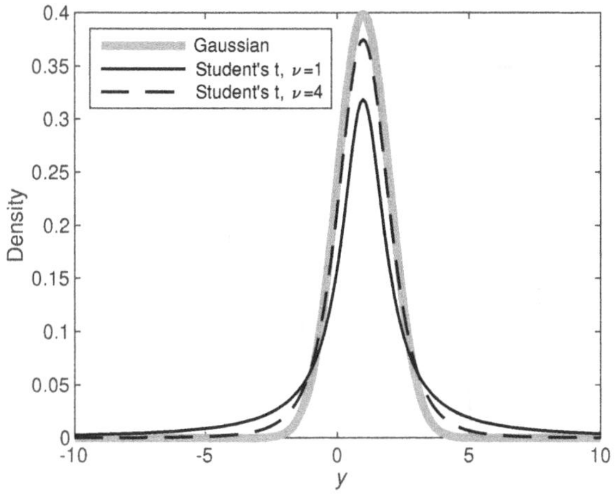

Figure 5.2 An illustration of three scalar densities with mean $\mu = 1$ and parameter $R = 1$: one Gaussian density $\mathrm{N}(y \mid \mu, R)$ and two Student's t-distributions, $\mathrm{St}(y \mid \mu, R, \nu)$, with $\nu = 1$ and $\nu = 4$, respectively.

2013), and both the generalized Gaussian filters discussed in Chapters 9 and 10, and the particle filters presented in Chapter 11, are also applicable to this kind of model.

The Student's t-distribution is also closely related to Gaussian distributions, because if we have

$$\begin{aligned} u &\sim \mathrm{Ga}(\nu/2, \nu/2), \\ \mathbf{y} &\sim \mathrm{N}(\boldsymbol{\mu}, \mathbf{R}/u), \end{aligned} \tag{5.18}$$

where $\mathrm{Ga}(\cdot, \cdot)$ is the gamma distribution, then the marginal distribution of $\mathbf{y}$ is $\mathrm{St}(\boldsymbol{\mu}, \mathbf{R}, \nu)$ (see, e.g., Roth, 2012). Thus in this sense a Student's t-distribution can be seen as a Gaussian distribution with uncertainty in the noise covariance modeled by scaling $\mathbf{R} \to \mathbf{R}/u$ with an unobserved gamma-distributed random variable u. This in turn makes it closely connected with Kalman filtering with unknown variance (see, e.g., Särkkä and Nummenmaa, 2009).

Another way to model outliers is to state that we obtain a regular measurement with some probability $\rho \in [0, 1]$ and an outlier with probability $1 - \rho$. We can then obtain a measurement model by introducing a separate model for the outliers. That model could be a Gaussian density with a

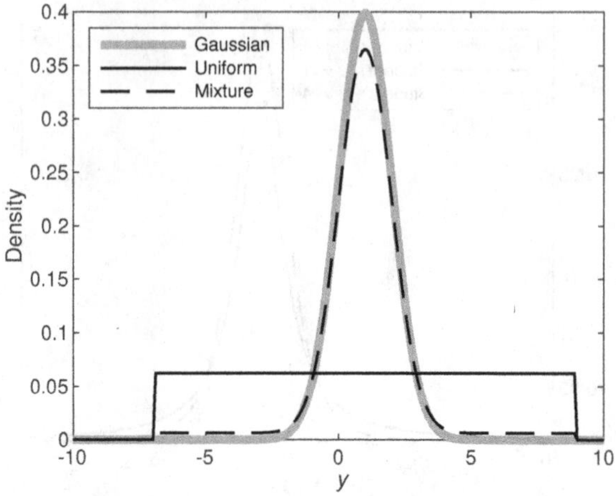

Figure 5.3 Here we show a Gaussian density with mean $\mu = 1$ and covariance $R = 1$, a uniform density $\mathrm{U}(y \mid \mathbb{V})$ where $\mathbb{V} = [-7, 9]$, and a mixture of densities, as in Equation 5.21, with $\rho = 0.9$.

larger covariance, or it could be a uniform density over some region $\mathbb{V}$:

$$\mathrm{U}(\mathbf{y}_k \mid \mathbb{V}) = \begin{cases} \frac{1}{V}, & \text{if } \mathbf{y}_k \in \mathbb{V}, \\ 0, & \text{otherwise,} \end{cases} \tag{5.19}$$

where V is the volume of $\mathbb{V}$. The complete measurement model then takes the form

$$p(\mathbf{y}_k \mid \mathbf{x}_k) = \begin{cases} \mathrm{N}(\mathbf{y}_k \mid \mathbf{h}(\mathbf{x}_k), \mathbf{R}_k), & \text{with probability } \rho, \\ \mathrm{U}(\mathbf{y}_k \mid \mathbb{V}), & \text{with probability } 1 - \rho, \end{cases} \tag{5.20}$$

where $\rho \in [0, 1]$ is the detection probability. The density in Equation 5.20 can also be expressed as a mixture model

$$p(\mathbf{y}_k \mid \mathbf{x}_k) = (1 - \rho)\, \mathrm{U}(\mathbf{y}_k \mid \mathbb{V}) + \rho\, \mathrm{N}(\mathbf{y}_k \mid \mathbf{h}(\mathbf{x}_k), \mathbf{R}_k). \tag{5.21}$$

We illustrate this density in Figure 5.3 in a scalar setting that resembles the example shown in Figure 5.2.

Alternatively, this mixture model can be seen as a model with a Bernoulli-distributed binary indicator variable $c_k \sim \mathrm{Be}(\rho)$ taking value 1 with probability ρ and value 0 with probability $1 - \rho$ (see Equation (5.26) below). In this formulation, the model takes the form

$$\begin{aligned} p(c_k) &= \mathrm{Be}(c_k \mid \rho), \\ p(\mathbf{y}_k \mid \mathbf{x}_k, c_k) &= \begin{cases} \mathrm{N}(\mathbf{y}_k \mid \mathbf{h}(\mathbf{x}_k), \mathbf{R}_k), & \text{if } c_k = 1, \\ \mathrm{U}(\mathbf{y}_k \mid \mathbb{V}), & \text{if } c_k = 0, \end{cases} \end{aligned} \tag{5.22}$$

where c_k is unobserved.

Example 5.6 (Clutter model for pendulum). *In Example 5.2 we constructed a non-linear Gaussian measurement model for a pendulum. We can now allow for outlier measurements in the model by replacing it with*

$$p(y_k \mid \mathbf{x}_k) = \begin{cases} \mathrm{N}(y_k \mid \sin(\alpha_k), R_k), & \text{with probability } \rho, \\ \mathrm{U}(y_k \mid [-L, L]), & \text{with probability } 1 - \rho \end{cases} \tag{5.23}$$

for some suitable constant L. The state estimation problem in this kind of model can no longer be done with a method based on an extended Kalman filter, but we can still use a general Gaussian filter or a particle filter.

Yet another useful model is a Poisson distribution

$$\mathrm{Po}(y \mid \lambda) = \exp(-\lambda) \frac{\lambda^y}{y!}, \tag{5.24}$$

which can be used to model counts of events $y = 0, 1, 2, \ldots$. Because the Poisson parameter λ must be positive, we typically form the model via exp-transformation as $\lambda = \lambda_0 \exp(u)$, where u is a variable that can take positive and negative values, and λ_0 is a constant. This leads to measurement models of the form

$$p(y_k \mid \mathbf{x}_k) = \mathrm{Po}(y_k \mid \lambda_0 \exp(h(\mathbf{x}_k))), \tag{5.25}$$

where $h(\mathbf{x}_k)$ is the sensor model that maps the state to the log-intensity relative to λ_0. State-estimation problems with this kind of measurement model can be tackled with either particle filters or generalized Gaussian filters.

In classification problems arising in machine learning (see, e.g., Rasmussen and Williams, 2006; Bishop, 2006), the Bernoulli distribution can be used as a measurement model:

$$\mathrm{Be}(y \mid \rho) = \rho^y (1 - \rho)^{1-y} = \begin{cases} \rho, & \text{if } y = 1, \\ 1 - \rho, & \text{if } y = 0, \end{cases} \tag{5.26}$$

where $y \in \{0, 1\}$. Because we need to have $\rho \in [0, 1]$, the model is typically expressed in terms of a transformation, such as a standard cumulative distribution function or a standard logistic transformation, of a real-valued variable:

$$\begin{aligned} p(y_k = 1 \mid x_k) &= \Phi(h(\mathbf{x}_k)), \text{ or} \\ p(y_k = 1 \mid x_k) &= \frac{1}{1 + \exp(-h(\mathbf{x}_k))}, \end{aligned} \tag{5.27}$$

where $\Phi(z) = \int_{-\infty}^{z} \mathrm{N}(x \mid 0, 1)\, \mathrm{d}x$. The corresponding state-estimation problem can then be approached by using particle filters or generalized Gaussian filters (cf. García-Fernández et al., 2019).

5.3 Measurements as Inputs

For practical reasons, it is sometimes useful to view some of the observations as inputs to the dynamic model. The objective is generally to simplify the problem and to reduce the number of state variables, and, when applicable, this technique often yields good performance as long as the measurement noise is small. For notation, we write the dynamic model as

$$\mathbf{x}_k = \mathbf{f}(\mathbf{x}_{k-1}, \mathbf{u}_{k-1}) + \mathbf{q}_{k-1}, \tag{5.28}$$

where $\mathbf{u}_{k-1}$ is a measured signal that we describe as a known input to the system.

Example 5.7 (Gyroscope measurement as input)**.** *Let us recall the dynamic model function corresponding to the polar coordinated turn model that we considered in Example 4.15:*

$$\mathbf{f}(\mathbf{x}_{k-1}) = \begin{pmatrix} x_{1,k-1} + \frac{2 s_{k-1}}{\omega_{k-1}} \sin(\frac{\Delta t_{k-1}\, \omega_{k-1}}{2}) \cos(\varphi_{k-1} + \frac{\Delta t_{k-1}\, \omega_{k-1}}{2}) \\ x_{2,k-1} + \frac{2 s_{k-1}}{\omega_{k-1}} \sin(\frac{\Delta t_{k-1}\, \omega_{k-1}}{2}) \sin(\varphi_{k-1} + \frac{\Delta t_{k-1}\, \omega_{k-1}}{2}) \\ s_{k-1} \\ \varphi_{k-1} + \omega_{k-1}\, \Delta t_{k-1} \\ \omega_{k-1} \end{pmatrix}, \tag{5.29}$$

where the state is $\mathbf{x}_{k-1} = \begin{pmatrix} x_{1,k-1}, x_{2,k-1}, s_{k-1}, \varphi_{k-1}, \omega_{k-1} \end{pmatrix}$. *If we now have a gyroscope sensor available, then it gives a direct measurement* ω_k^{gyro} *of the angular velocity* ω_k. *It is possible to model this as a noisy measurement of* ω_k*:*

$$\omega_k^{\text{gyro}} = \omega_k + r_k = \begin{pmatrix} 0 & 0 & 0 & 0 & 1 \end{pmatrix} \mathbf{x}_k + r_k, \tag{5.30}$$

where r_k is measurement noise. However, we can also put $u_{k-1} = \omega_{k-1}^{\text{gyro}}$ and redefine the dynamic model function as

$$\mathbf{f}(\mathbf{x}_{k-1}, u_{k-1}) = \begin{pmatrix} x_{1,k-1} + \frac{2 s_{k-1}}{u_{k-1}} \sin(\frac{\Delta t_{k-1} u_{k-1}}{2}) \cos(\varphi_{k-1} + \frac{\Delta t_{k-1} u_{k-1}}{2}) \\ x_{2,k-1} + \frac{2 s_{k-1}}{u_{k-1}} \sin(\frac{\Delta t_{k-1} u_{k-1}}{2}) \sin(\varphi_{k-1} + \frac{\Delta t_{k-1} u_{k-1}}{2}) \\ s_{k-1} \\ \varphi_{k-1} + u_{k-1}\, \Delta t_{k-1} \end{pmatrix}, \tag{5.31}$$

where the state is now $\mathbf{x}_{k-1} = \left(x_{1,k-1}, x_{2,k-1}, s_{k-1}, \varphi_{k-1}\right)$, and $\omega_{k-1}^{\text{gyro}}$ is fed in as the input signal u_{k-1}.

Viewing the measurements as input can be used in many contexts and for many sensors. For instance, for the above polar coordinated turn model, we could also view measurements from a speedometer as inputs instead of including s_k as part of the state (see Exercise 5.4). However, it should be noted that the simplifications that this gives rise to may come at the price of reduced performance.

5.4 Linear-in-Parameters Models in State Space Form

In Chapter 3 we saw that a simple one-dimensional linear regression problem can be converted into a state space model that can be solved using the Kalman filter. In an similar manner, more general linear-in-parameters models can be converted into state space models. This is demonstrated in the following example.

Example 5.8 (Linear-in-parameters regression model I). *Consider the following parametric model:*

$$y_k = w_0 + w_1\, g_1(t_k) + \cdots + w_d\, g_d(t_k) + \varepsilon_k, \tag{5.32}$$

where $g_1(t_k), \ldots, g_d(t_k)$ are some given functions of time t_k, and ε_k is Gaussian measurement noise. The problem of determining the weights $w_0, \ldots, w_d$ from a set of measurements $\{(t_1, y_1), \ldots, (t_T, y_T)\}$ can be converted into a Kalman filtering problem as follows.

Introducing the variables $\mathbf{H}_k = (1\ g_1(t_k)\ \cdots\ g_d(t_k))$ and $\mathbf{x}_k = \mathbf{x} = (w_0, w_1, \ldots, w_d)$, we can rewrite the model as a linear Gaussian state space model:

$$\begin{aligned} \mathbf{x}_k &= \mathbf{x}_{k-1}, \\ y_k &= \mathbf{H}_k\, \mathbf{x}_k + \varepsilon_k. \end{aligned} \tag{5.33}$$

Because the model is a linear Gaussian state space model, we can now use a linear Kalman filter to estimate the parameters.

The presented strategy is not limited to functions $g_1, \ldots, g_d$ that are explicit functions of time. Instead, they can be functions of arbitrary regressors, as illustrated in the following example.

Example 5.9 (Linear-in-parameters regression model II). *Consider the following parametric model:*

$$y_k = w_0 + w_1\, g_1(\mathbf{s}_k) + \cdots + w_d\, g_d(\mathbf{s}_k) + \varepsilon_k, \tag{5.34}$$

where $g_1(\mathbf{s}), \ldots, g_d(\mathbf{s})$ are some given functions of a the variable $\mathbf{s} \in \mathbb{R}^n$, and where the weights $w_0, \ldots, w_d$ are to be estimated from a sequence of measurements $((\mathbf{s}_1, y_1), \ldots, (\mathbf{s}_T, y_T))$. Similarly to the previous example, we can convert the problem into a linear Gaussian state space model by defining $\mathbf{H}_k = (1 \; g_1(\mathbf{s}_k) \; \cdots \; g_d(\mathbf{s}_k))$ and $\mathbf{x}_k = \mathbf{x} = (w_0, w_1, \ldots, w_d)$, and then solve for $\mathbf{x}_k$ using a Kalman filter.

Linearity of the state space models in the above examples resulted from the property that the models are linear in their parameters. Generalized linear models involving non-linear link functions will lead to non-linear state space models, as is illustrated in the following example.

Example 5.10 (Generalized linear model). *An example of a generalized linear model is*

$$y_k = g(w_0 + \mathbf{w}^\mathsf{T}\, \mathbf{s}_k) + \varepsilon_k, \tag{5.35}$$

where $g(\cdot)$ is some given non-linear link function, and $\mathbf{s}_k$ is a vector of regressors. If we define the state as $\mathbf{x} = (w_0, \mathbf{w})$ and $h_k(\mathbf{x}) \triangleq g(w_0 + \mathbf{w}^\mathsf{T}\, \mathbf{s}_k)$, we get the following non-linear Gaussian state space model:

$$\begin{aligned} \mathbf{x}_k &= \mathbf{x}_{k-1}, \\ y_k &= h_k(\mathbf{x}_k) + \varepsilon_k. \end{aligned} \tag{5.36}$$

Because the state space model is non-linear, instead of the linear Kalman filter, we need to use non-linear Kalman filters, such as the extended Kalman filter (EKF) or the unscented Kalman filter (UKF), to cope with the non-linearity.

One general class of non-linear regression models, which can be converted into state space models using an analogous approach to the above, is the multi-layer perceptron (MLP) (see, e.g., Bishop, 2006), also known as a neural network. Using a non-linear Kalman filter is indeed one way to train (to estimate the parameters of) such models (Haykin, 2001), even though it remains an unconventional technique to train neural networks. However, non-linear regression models of this kind arise in various others contexts as well.

Many sensors have an unknown offset $\mathbf{b}$, also known as a bias, such that we observe, say,

$$\mathbf{y}_k = \mathbf{h}(\mathbf{x}_k) + \mathbf{b} + \mathbf{r}_k, \quad \mathbf{r}_k \sim \mathrm{N}(\mathbf{0}, \mathbf{R}_k), \tag{5.37}$$

instead of $\mathbf{y}_k = \mathbf{h}_k(\mathbf{x}_k) + \mathbf{r}_k$. One way to handle an unknown bias is to include it in the state vector. That is, we can introduce a new state vector $\tilde{\mathbf{x}}_k = (\mathbf{x}_k, \mathbf{b}_k)$ where the original state is augmented with the bias $\mathbf{b}_k = \mathbf{b}_{k-1}$. The model then becomes a standard state estimation problem with the redefined measurement model function

$$\tilde{\mathbf{h}}(\tilde{\mathbf{x}}_k) = \mathbf{h}(\mathbf{x}_k) + \mathbf{b}_k. \tag{5.38}$$

Instead of the singular dynamics, $\mathbf{b}_k = \mathbf{b}_{k-1}$, we can include a small noise in the dynamics of the bias $\mathbf{b}_k = \mathbf{b}_{k-1} + \mathbf{q}_{k-1}^{\mathbf{b}}$ to allow for slow drifting of the bias.

Example 5.11 (Including bias in state vector). *Consider a constant velocity model (see Example 4.5) with $\mathbf{x}_k = \left(x_{1,k}, x_{2,k}\right)$, where $x_{1,k}$ is the position and $x_{2,k}$ is the velocity. Suppose that we observe*

$$\begin{pmatrix} y_{1,k} \\ y_{2,k} \end{pmatrix} = \begin{pmatrix} x_{1,k} \\ x_{2,k} \end{pmatrix} + \begin{pmatrix} 0 \\ b_k \end{pmatrix} + \begin{pmatrix} r_{1,k} \\ r_{2,k} \end{pmatrix}, \tag{5.39}$$

where b_k is a bias in the velocity measurement that drifts slowly with time. We can then redefine $\tilde{\mathbf{x}}_k = \left(x_{1,k}, x_{2,k}, b_k\right)$ as a state for which (5.39) *is a measurement model without any unknown parameters (see Exercise 5.6). We can then introduce the dynamic model*

$$\begin{pmatrix} x_{1,k} \\ x_{2,k} \\ b_k \end{pmatrix} = \begin{pmatrix} 1 & \Delta t_{k-1} & 0 \\ 0 & 1 & 0 \\ 0 & 0 & 1 \end{pmatrix} \begin{pmatrix} x_{1,k-1} \\ x_{2,k-1} \\ b_{k-1} \end{pmatrix} + \begin{pmatrix} q_{1,k-1} \\ q_{2,k-1} \\ q_{3,k-1} \end{pmatrix}, \tag{5.40}$$

where $\mathbf{q}_{k-1} = \left(q_{1,k-1}, q_{2,k-1}, q_{3,k-1}\right) \sim \mathrm{N}(\mathbf{0}, \mathbf{Q}_{k-1})$, which leads to a complete probabilistic state space model that enables us to estimate b_k jointly with $x_{1,k}$ and $x_{2,k}$. We note that (5.40) *is a constant velocity model for $(x_{1,k}, x_{2,k})$ and a random walk model for b_k, and we can use Examples 4.4 and 4.5 to find a suitable structure for $\mathbf{Q}_{k-1}$.*

5.5 Autoregressive Models

In digital signal processing (DSP), an important class of models is linear signal models, such as autoregressive (AR) models, moving average (MA) models, autoregressive moving average models (ARMA), and their generalizations (see, e.g., Hayes, 1996). In those models one is often interested in performing adaptive filtering, which refers to the methodology where the parameters of the signal model are estimated from data. These kinds of adaptive filtering problems can often be formulated using a linear Gaussian state space model (and solved using a Kalman filter), as illustrated in the following example.

Example 5.12 (Autoregressive (AR) model). *An autoregressive (AR) model of order d has the form*

$$y_k = w_1 \, y_{k-1} + \cdots + w_d \, y_{k-d} + \varepsilon_k, \tag{5.41}$$

where ε_k is a white Gaussian noise process. The problem of adaptive filtering is to estimate the weights $w_1, \ldots, w_d$ given the observed signal $y_1, y_2, y_3, \ldots$. If we let $\mathbf{H}_k = (y_{k-1} \;\; \cdots \;\; y_{k-d})$ and define the state as $\mathbf{x}_k = (w_1, \ldots, w_d)$, we get a linear Gaussian state space model

$$\begin{aligned} \mathbf{x}_k &= \mathbf{x}_{k-1}, \\ y_k &= \mathbf{H}_k \, \mathbf{x}_k + \varepsilon_k. \end{aligned} \tag{5.42}$$

Thus the adaptive filtering problem can be solved with a linear Kalman filter.

The classical algorithm for adaptive filtering is called the least mean squares (LMS) algorithm, and it can be interpreted as an approximate version of the above Kalman filter. However, in LMS algorithms it is common to allow the model to change over time, which in the state space context corresponds to setting up a dynamic model for the model parameters. This kind of model is illustrated in the next example.

Example 5.13 (Time-varying autoregressive (TVAR) model). *In a time-varying autoregressive (TVAR) model, the weights are assumed to depend on the time step k as*

$$y_k = w_{1,k} \, y_{k-1} + \cdots + w_{d,k} \, y_{k-d} + \varepsilon_k. \tag{5.43}$$

A typical model for the time dependence of weights is the random walk model

$$w_{i,k} = w_{i,k-1} + q_{i,k-1}, \quad q_{i,k-1} \sim \mathrm{N}(0, Q_i), \quad i = 1, \ldots, d. \tag{5.44}$$

If we define the state as $\mathbf{x}_k = (w_{1,k}, \ldots, w_{d,k})$*, this model can be written as a linear Gaussian state space model with process noise* $\mathbf{q}_{k-1} = (q_{1,k-1}, \ldots, q_{d,k-1})$*:*

$$\begin{aligned} \mathbf{x}_k &= \mathbf{x}_{k-1} + \mathbf{q}_{k-1}, \\ y_k &= \mathbf{H}_k \, \mathbf{x}_k + \varepsilon_k. \end{aligned} \tag{5.45}$$

More general (TV)ARMA models can be handled similarly.

5.6 Discrete-State Hidden Markov Models (HMMs)

One useful class of probabilistic state space models that is often used in speech processing and telecommunications applications is that of finite-state hidden Markov models (HMMs, Rabiner, 1989). In these models, the state is assumed to take values in some finite set of possible values $\mathbf{x}_k \in \mathbb{X}$. Often (but not always), the measurements can also be assumed to take values in another finite set $\mathbf{y}_k \in \mathbb{Y}$.

When the sets $\mathbb{X}$ and $\mathbb{Y}$ are both finite, we can without a loss of generality assume scalar states x_k and measurements y_k taking values in sets $\mathbb{X} = \{1, 2, \ldots, X\}$ and $\mathbb{Y} = \{1, 2, \ldots, Y\}$, respectively. The states x_k now form a Markov chain whose transition probabilities can be conveniently collected into a state transition matrix $\mathbf{\Pi}$ with elements

$$\Pi_{i,j} = p(x_k = j \mid x_{k-1} = i). \tag{5.46}$$

The element $\Pi_{i,j}$ thus gives the probability of jumping from the state $x_{k-1} = i$ to the state $x_k = j$. In general, this matrix can also depend on the time step k.

The measurements can be conveniently described in terms of an emission matrix $\mathbf{O}$, which can also generally depend on the time step k and has the elements

$$O_{i,j} = p(y_k = j \mid x_k = i), \tag{5.47}$$

that is, the element $O_{i,j}$ is the probability of observing the measurement $y_k = j$ given that we are on the state $x_k = i$.

Example 5.14 (Gilbert–Elliot channel model)**.** *In this example we consider the Gilbert–Elliot model (see, e.g., Cappé et al. (2005), Sec. 1.3.1), which is an example of a finite-state hidden Markov model (HMM). The model consists of a binary input sequence* $\{b_k : k = 1, \ldots, T\}$ *to be transmitted and a binary channel regime signal* $\{s_k : k = 1, \ldots, T\}$*. The binary measurements are given as* $b_k \oplus v_k$*, where* v_k *is a Bernoulli sequence,*

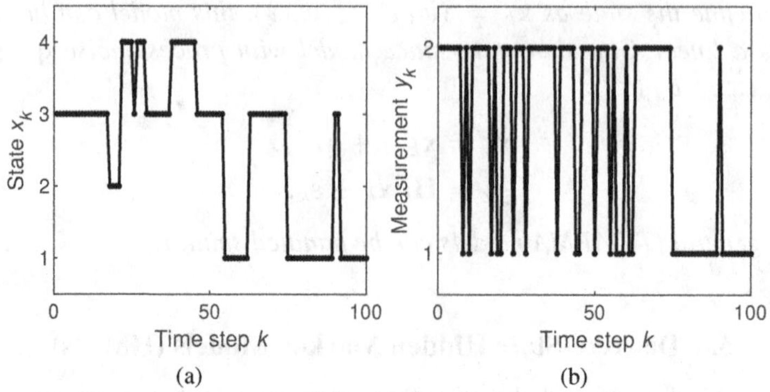

Figure 5.4 Simulated state (a) and measurement (b) sequences from Gilbert–Elliot channel model in Example 5.14.

and $\oplus$ denotes the exclusive or (i.e., xor) operation. These measurements can be encoded as values $y_k = 1$ for $b_k \oplus v_k = 0$ and $y_k = 2$ when $b_k \oplus v_k = 1$.

The regime signal s_k defines the probability of an error (i.e., $p(v_k = 1)$) such that when $s_k = 0$, then the probability of error has a small value q_0 and when $s_k = 1$, the probability of error has a larger value q_1. The sequence $\{s_k : k = 1, \ldots, T\}$ is modeled as a first order Markov chain, and the probability of moving to state $s_k = 1$ from state $s_{k-1} = 0$ is given by p_0 and the probability of moving to state $s_k = 0$ from state $s_{k-1} = 1$ is p_1. We model the input signal $\{b_k : k = 1, \ldots, T\}$ as a first order Markov chain with state-switching probability p_2. The joint model for the state (b_k, s_k) is a four-state Markov chain for the pair consisting of the states $\{(0,0), (0,1), (1,0), (1,1)\}$ which are encoded as states $x_k \in \{1, \ldots, 4\}$ in the Markov chain, respectively.

The state transition and emission matrices are given as

$$\mathbf{\Pi} = \begin{pmatrix} (1-p_0)(1-p_2) & p_0(1-p_2) & (1-p_0)\,p_2 & p_0\,p_2 \\ p_1(1-p_2) & (1-p_1)(1-p_2) & p_1\,p_2 & (1-p_1)\,p_2 \\ (1-p_0)\,p_2 & p_0\,p_2 & (1-p_0)(1-p_2) & p_0(1-p_2) \\ p_1\,p_2 & (1-p_1)\,p_2 & p_1(1-p_2) & (1-p_1)(1-p_2) \end{pmatrix},$$

$$\mathbf{O} = \begin{pmatrix} (1-q_0) & q_0 \\ (1-q_1) & q_1 \\ q_0 & (1-q_0) \\ q_1 & (1-q_1) \end{pmatrix}.$$

Figure 5.4 shows examples of simulated state and measurement trajectories from this model.

HMMs can also be often used as part of conventional continuous-state tracking models to model the dynamics of indicator variables, as is illustrated in Exercise 5.8.

5.7 Exercises

5.1 Modify Example 5.4 to correspond to bearings (i.e., direction measurements) from the four radars located at positions $(s_{1,x}, s_{1,y}), \ldots, (s_{4,x}, s_{4,y})$.

5.2 Show that Student's t-distribution in (5.15) converges to a Gaussian as the degrees of freedom $\nu \to \infty$.

5.3 Suppose we have a constant velocity model in 2D and that a speedometer observes the speed in additive Gaussian noise. Write down the measurement model in the functional relation form (see Equation (5.3)) and in conditional density form (see Equation (5.4)).

5.4 Modify the gyroscope measurement and input models in Example 5.7 to correspond to modeling speedometer readings instead. How would you combine the two types of measurements (gyroscope and speedometer) in the same model?

5.5 Consider the regression problem

$$\begin{aligned} y_k &= a_1 \, s_k + a_2 \, \sin(s_k) + b + \varepsilon_k, \qquad k = 1, \ldots, T, \\ \varepsilon_k &\sim \mathrm{N}(0, R), \\ a_1 &\sim \mathrm{N}(0, \sigma_1^2), \\ a_2 &\sim \mathrm{N}(0, \sigma_2^2), \\ b &\sim \mathrm{N}(0, \sigma_b^2), \end{aligned} \tag{5.48}$$

where $s_k \in \mathbb{R}$ are the known regressor values, $R, \sigma_1^2, \sigma_2^2, \sigma_b^2$ are given positive constants, $y_k \in \mathbb{R}$ are the observed output variables, and ε_k are independent Gaussian measurement errors. The scalars a_1, a_2, and b are the unknown parameters to be estimated. Formulate the estimation problem as a linear Gaussian state space model.

5.6 Determine a matrix $\mathbf{H}$ such that the biased measurement model in Equation (5.39) can be rewritten in linear form $\mathbf{y}_k = \mathbf{H}\,\tilde{\mathbf{x}}_k + \mathbf{r}_k$, where $\tilde{\mathbf{x}}_k = (x_{1,k}, x_{2,k}, b_k)$.

5.7 Consider the model

$$\begin{aligned} x_k &= \sum_{i=1}^{d} a_i \, x_{k-i} + q_{k-1}, \\ y_k &= x_k + \varepsilon_k, \end{aligned} \tag{5.49}$$

where the values $\{y_k\}$ are observed, q_{k-1} and ε_k are independent Gaussian noises, and the weights $a_1, \ldots, a_d$ are known. The aim is to estimate the sequence $\{x_k\}$. Rewrite the problem as an estimation problem in a linear Gaussian state space model.

5.8 Let us now consider the clutter measurement model from Equation (5.22) formulated for the car model.

(a) Write down the dynamic and measurement models with a clutter model of the form of Equation (5.22) for the linear Gaussian car model considered in Example 5.1. You can assume that the clutter measurements are uniformly distributed in a box $[-L, L] \times [-L, L]$.

(b) Instead of assuming that each measurement is independently clutter or non-clutter, write down a Markov model (in terms of the state-transition matrix) for the clutter indicator c_k that assigns a probability of α for having another non-clutter measurement after a non-clutter measurement, and β for having another clutter measurement after a clutter measurement.

(c) Formulate a joint probabilistic state space model for the car and the clutter state by using the augmented state $\tilde{\mathbf{x}}_k = (\mathbf{x}_k, c_k)$, where $\mathbf{x}_k$ contains the linear dynamics, and c_k is the clutter indicator.

6

Bayesian Filtering Equations and Exact Solutions

In this chapter, we derive the Bayesian filtering equations, which are the general equations for computing Bayesian filtering solutions to both linear Gaussian and non-linear/non-Gaussian state space models. We also derive the Kalman filtering equations that give the closed-form solution to the linear Gaussian Bayesian filtering problem. Additionally, we specialize the Bayesian filtering equations to discrete-state hidden Markov models (HMMs).

6.1 Probabilistic State Space Models

Bayesian filtering is considered with state estimation in general probabilistic state space models, which have the following form.

Definition 6.1 (Probabilistic state space model). A probabilistic state space model *or non-linear filtering model consists of a sequence of conditional probability distributions:*

$$\begin{aligned} \mathbf{x}_k &\sim p(\mathbf{x}_k \mid \mathbf{x}_{k-1}), \\ \mathbf{y}_k &\sim p(\mathbf{y}_k \mid \mathbf{x}_k) \end{aligned} \tag{6.1}$$

for $k = 1, 2, \ldots,$ *where*

- $\mathbf{x}_k \in \mathbb{R}^n$ *is the* state *of the system at time step* k*,*
- $\mathbf{y}_k \in \mathbb{R}^m$ *is the measurement at time step* k*,*
- $p(\mathbf{x}_k \mid \mathbf{x}_{k-1})$ *is the* dynamic model *that describes the stochastic dynamics of the system. The dynamic model can be a probability density, a counting measure, or a combination of these, depending on whether the state* $\mathbf{x}_k$ *is continuous, discrete, or hybrid.*
- $p(\mathbf{y}_k \mid \mathbf{x}_k)$ *is the* measurement model, *which is the distribution of measurements given the state.*

The model is assumed to be Markovian, which means that it has the following two properties.

Property 6.2 (Markov property of states).
The states $\{\mathbf{x}_k : k = 0, 1, 2, \dots\}$ *form a Markov sequence (or Markov chain if the state is discrete). This* Markov property *means that* $\mathbf{x}_k$ *(and actually the whole future* $\mathbf{x}_{k+1}, \mathbf{x}_{k+2}, \dots$*) given* $\mathbf{x}_{k-1}$ *is independent of anything that has happened before the time step* $k - 1$*:*

$$p(\mathbf{x}_k \mid \mathbf{x}_{0:k-1}, \mathbf{y}_{1:k-1}) = p(\mathbf{x}_k \mid \mathbf{x}_{k-1}). \tag{6.2}$$

Also, the past is independent of the future given the present:

$$p(\mathbf{x}_{k-1} \mid \mathbf{x}_{k:T}, \mathbf{y}_{k:T}) = p(\mathbf{x}_{k-1} \mid \mathbf{x}_k). \tag{6.3}$$

Property 6.3 (Conditional independence of measurements).
The current measurement $\mathbf{y}_k$ *given the current state* $\mathbf{x}_k$ *is conditionally independent of the measurement and state histories:*

$$p(\mathbf{y}_k \mid \mathbf{x}_{0:k}, \mathbf{y}_{1:k-1}) = p(\mathbf{y}_k \mid \mathbf{x}_k). \tag{6.4}$$

A simple example of a Markovian sequence is the Gaussian random walk. When this is combined with noisy measurements, we obtain the following example of a probabilistic state space model.

Example 6.4 (Gaussian random walk). *A Gaussian random walk model can be written as*

$$\begin{aligned} x_k &= x_{k-1} + q_{k-1}, & q_{k-1} &\sim \mathrm{N}(0, Q), \\ y_k &= x_k + r_k, & r_k &\sim \mathrm{N}(0, R), \end{aligned} \tag{6.5}$$

where x_k *is the hidden state (or signal), and* y_k *is the measurement. In terms of probability densities, the model can be written as*

$$\begin{aligned} p(x_k \mid x_{k-1}) &= \mathrm{N}(x_k \mid x_{k-1}, Q) \\ &= \frac{1}{\sqrt{2\pi Q}} \exp\left(-\frac{1}{2Q}(x_k - x_{k-1})^2\right), \\ p(y_k \mid x_k) &= \mathrm{N}(y_k \mid x_k, R) \\ &= \frac{1}{\sqrt{2\pi R}} \exp\left(-\frac{1}{2R}(y_k - x_k)^2\right), \end{aligned} \tag{6.6}$$

which is a probabilistic state space model. Example realizations of the signal x_k *and measurements* y_k *are shown in Figure 6.1. The parameter values in the simulation were* $Q = R = 1$.

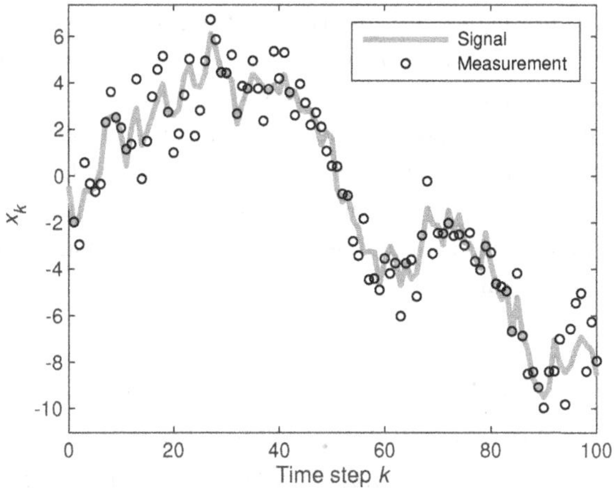

Figure 6.1 Simulated signal and measurements from the Gaussian random walk model in Example 6.4.

With the Markovian assumption and the filtering model (6.1) the *joint prior distribution of the states* $\mathbf{x}_{0:T} = \{\mathbf{x}_0, \ldots, \mathbf{x}_T\}$, and the *joint likelihood of the measurements* $\mathbf{y}_{1:T} = \{\mathbf{y}_1, \ldots, \mathbf{y}_T\}$ are, respectively,

$$p(\mathbf{x}_{0:T}) = p(\mathbf{x}_0) \prod_{k=1}^{T} p(\mathbf{x}_k \mid \mathbf{x}_{k-1}), \tag{6.7}$$

$$p(\mathbf{y}_{1:T} \mid \mathbf{x}_{0:T}) = \prod_{k=1}^{T} p(\mathbf{y}_k \mid \mathbf{x}_k). \tag{6.8}$$

In principle, for a given T we could simply compute the posterior distribution of the states by Bayes' rule:

$$\begin{aligned} p(\mathbf{x}_{0:T} \mid \mathbf{y}_{1:T}) &= \frac{p(\mathbf{y}_{1:T} \mid \mathbf{x}_{0:T})\, p(\mathbf{x}_{0:T})}{p(\mathbf{y}_{1:T})} \\ &\propto p(\mathbf{y}_{1:T} \mid \mathbf{x}_{0:T})\, p(\mathbf{x}_{0:T}). \end{aligned} \tag{6.9}$$

However, this kind of explicit usage of the full Bayes' rule is not feasible in real-time applications, because the number of computations per time step increases as new observations arrive. Thus, this way we could only work with small data sets, because if the amount of data is unbounded (as in real-time sensing applications), then at some point of time the computations will become intractable. To cope with real-time data, we need to have an algorithm that performs a constant number of computations per time step.

As discussed in Section 1.3, *filtering distributions*, *prediction distributions*, and *smoothing distributions* can be computed recursively such that only a constant number of computations is done on each time step. For this reason we shall not consider the full posterior computation at all, but concentrate on the abovementioned distributions instead. In this and the next few chapters, we consider computation of the filtering and prediction distributions; algorithms for computing the smoothing distributions will be considered in later chapters.

6.2 Bayesian Filtering Equations

The purpose of *Bayesian filtering* is to compute the *marginal posterior distribution* or *filtering distribution* of the state $\mathbf{x}_k$ at each time step k given the history of the measurements up to the time step k:

$$p(\mathbf{x}_k \mid \mathbf{y}_{1:k}). \tag{6.10}$$

The fundamental equations of Bayesian filtering theory are given by the following theorem.

Theorem 6.5 (Bayesian filtering equations). *The recursive equations (the* Bayesian filter*) for computing the* predicted distribution $p(\mathbf{x}_k \mid \mathbf{y}_{1:k-1})$ *and the* filtering distribution $p(\mathbf{x}_k \mid \mathbf{y}_{1:k})$ *at time step k are given by the following* Bayesian filtering equations.

- Initialization. *The recursion starts from the prior distribution* $p(\mathbf{x}_0)$.
- Prediction step. *The predictive distribution of the state* $\mathbf{x}_k$ *at time step k, given the dynamic model, can be computed by the Chapman–Kolmogorov equation*

$$p(\mathbf{x}_k \mid \mathbf{y}_{1:k-1}) = \int p(\mathbf{x}_k \mid \mathbf{x}_{k-1})\, p(\mathbf{x}_{k-1} \mid \mathbf{y}_{1:k-1})\, \mathrm{d}\mathbf{x}_{k-1}. \tag{6.11}$$

- Update step. *Given the measurement* $\mathbf{y}_k$ *at time step k, the posterior distribution of the state* $\mathbf{x}_k$ *can be computed by Bayes' rule*

$$p(\mathbf{x}_k \mid \mathbf{y}_{1:k}) = \frac{1}{Z_k} p(\mathbf{y}_k \mid \mathbf{x}_k)\, p(\mathbf{x}_k \mid \mathbf{y}_{1:k-1}), \tag{6.12}$$

  where the normalization constant Z_k *is given as*

$$Z_k = \int p(\mathbf{y}_k \mid \mathbf{x}_k)\, p(\mathbf{x}_k \mid \mathbf{y}_{1:k-1})\, \mathrm{d}\mathbf{x}_k. \tag{6.13}$$

If some of the components of the state are discrete, the corresponding integrals are replaced with summations. This is discussed more in Section 6.5. The prediction and update steps are illustrated in Figures 6.2 and 6.3, respectively.

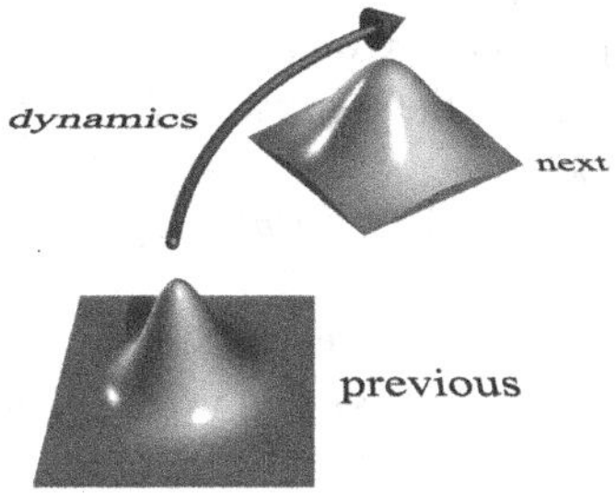

Figure 6.2 Visualization of the prediction step: Prediction propagates the state distribution of the previous measurement step through the dynamic model such that the uncertainties (stochastics) in the dynamic model are taken into account.

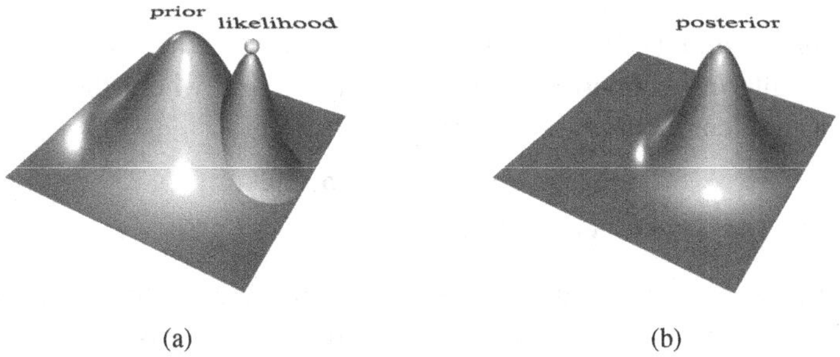

Figure 6.3 Visualization of the update step: (a) prior distribution from prediction and the likelihood of measurement just before the update step; (b) the posterior distribution after combining the prior and likelihood by Bayes' rule.

Proof The joint distribution of $\mathbf{x}_k$ and $\mathbf{x}_{k-1}$ given $\mathbf{y}_{1:k-1}$ can be computed as

$$\begin{aligned} p(\mathbf{x}_k, \mathbf{x}_{k-1} \mid \mathbf{y}_{1:k-1}) &= p(\mathbf{x}_k \mid \mathbf{x}_{k-1}, \mathbf{y}_{1:k-1})\, p(\mathbf{x}_{k-1} \mid \mathbf{y}_{1:k-1}) \\ &= p(\mathbf{x}_k \mid \mathbf{x}_{k-1})\, p(\mathbf{x}_{k-1} \mid \mathbf{y}_{1:k-1}), \end{aligned} \tag{6.14}$$

where the disappearance of the measurement history $\mathbf{y}_{1:k-1}$ is due to the Markov property of the sequence $\{\mathbf{x}_k : k = 1, 2, \dots\}$. The marginal distribution of $\mathbf{x}_k$ given $\mathbf{y}_{1:k-1}$ can be obtained by integrating the distribution (6.14) over $\mathbf{x}_{k-1}$, which gives the *Chapman–Kolmogorov equation*

$$p(\mathbf{x}_k \mid \mathbf{y}_{1:k-1}) = \int p(\mathbf{x}_k \mid \mathbf{x}_{k-1})\, p(\mathbf{x}_{k-1} \mid \mathbf{y}_{1:k-1})\, \mathrm{d}\mathbf{x}_{k-1}. \tag{6.15}$$

If $\mathbf{x}_{k-1}$ is discrete, then the above integral is replaced with summation over $\mathbf{x}_{k-1}$. The distribution of $\mathbf{x}_k$ given $\mathbf{y}_k$ and $\mathbf{y}_{1:k-1}$, that is, given $\mathbf{y}_{1:k}$, can be computed by *Bayes' rule*

$$\begin{aligned} p(\mathbf{x}_k \mid \mathbf{y}_{1:k}) &= \frac{1}{Z_k} p(\mathbf{y}_k \mid \mathbf{x}_k, \mathbf{y}_{1:k-1})\, p(\mathbf{x}_k \mid \mathbf{y}_{1:k-1}) \\ &= \frac{1}{Z_k} p(\mathbf{y}_k \mid \mathbf{x}_k)\, p(\mathbf{x}_k \mid \mathbf{y}_{1:k-1}), \end{aligned} \tag{6.16}$$

where the normalization constant is given by Equation (6.13). The disappearance of the measurement history $\mathbf{y}_{1:k-1}$ in Equation (6.16) is due to the conditional independence of $\mathbf{y}_k$ of the measurement history, given $\mathbf{x}_k$. □

6.3 Kalman Filter

The Kalman filter (Kalman, 1960b) is the closed form solution to the Bayesian filtering equations for the filtering model, where the dynamic and measurement models are linear Gaussian:

$$\begin{aligned} \mathbf{x}_k &= \mathbf{A}_{k-1}\, \mathbf{x}_{k-1} + \mathbf{q}_{k-1}, \\ \mathbf{y}_k &= \mathbf{H}_k\, \mathbf{x}_k + \mathbf{r}_k, \end{aligned} \tag{6.17}$$

where $\mathbf{x}_k \in \mathbb{R}^n$ is the state, $\mathbf{y}_k \in \mathbb{R}^m$ is the measurement, $\mathbf{q}_{k-1} \sim \mathrm{N}(\mathbf{0}, \mathbf{Q}_{k-1})$ is the process noise, $\mathbf{r}_k \sim \mathrm{N}(\mathbf{0}, \mathbf{R}_k)$ is the measurement noise, and the prior distribution is Gaussian $\mathbf{x}_0 \sim \mathrm{N}(\mathbf{m}_0, \mathbf{P}_0)$. The matrix $\mathbf{A}_{k-1}$ is the transition matrix of the dynamic model, and $\mathbf{H}_k$ is the measurement model matrix. In probabilistic terms the model is

$$\begin{aligned} p(\mathbf{x}_k \mid \mathbf{x}_{k-1}) &= \mathrm{N}(\mathbf{x}_k \mid \mathbf{A}_{k-1}\, \mathbf{x}_{k-1}, \mathbf{Q}_{k-1}), \\ p(\mathbf{y}_k \mid \mathbf{x}_k) &= \mathrm{N}(\mathbf{y}_k \mid \mathbf{H}_k\, \mathbf{x}_k, \mathbf{R}_k). \end{aligned} \tag{6.18}$$

Theorem 6.6 (Kalman filter). *The Bayesian filtering equations for the linear filtering model* (6.17) *can be evaluated in closed form, and the resulting distributions are Gaussian:*

$$\begin{aligned} p(\mathbf{x}_k \mid \mathbf{y}_{1:k-1}) &= \mathrm{N}(\mathbf{x}_k \mid \mathbf{m}_k^-, \mathbf{P}_k^-), \\ p(\mathbf{x}_k \mid \mathbf{y}_{1:k}) &= \mathrm{N}(\mathbf{x}_k \mid \mathbf{m}_k, \mathbf{P}_k), \\ p(\mathbf{y}_k \mid \mathbf{y}_{1:k-1}) &= \mathrm{N}(\mathbf{y}_k \mid \mathbf{H}_k \, \mathbf{m}_k^-, \mathbf{S}_k). \end{aligned} \tag{6.19}$$

The parameters of the distributions above can be computed with the following Kalman filter prediction *and* update steps.

- *The* prediction step *is*

$$\begin{aligned} \mathbf{m}_k^- &= \mathbf{A}_{k-1} \, \mathbf{m}_{k-1}, \\ \mathbf{P}_k^- &= \mathbf{A}_{k-1} \, \mathbf{P}_{k-1} \, \mathbf{A}_{k-1}^\mathsf{T} + \mathbf{Q}_{k-1}. \end{aligned} \tag{6.20}$$

- *The* update step *is*

$$\begin{aligned} \mathbf{v}_k &= \mathbf{y}_k - \mathbf{H}_k \, \mathbf{m}_k^-, \\ \mathbf{S}_k &= \mathbf{H}_k \, \mathbf{P}_k^- \, \mathbf{H}_k^\mathsf{T} + \mathbf{R}_k, \\ \mathbf{K}_k &= \mathbf{P}_k^- \, \mathbf{H}_k^\mathsf{T} \, \mathbf{S}_k^{-1}, \\ \mathbf{m}_k &= \mathbf{m}_k^- + \mathbf{K}_k \, \mathbf{v}_k, \\ \mathbf{P}_k &= \mathbf{P}_k^- - \mathbf{K}_k \, \mathbf{S}_k \, \mathbf{K}_k^\mathsf{T}. \end{aligned} \tag{6.21}$$

The recursion is started from the prior mean $\mathbf{m}_0$ *and covariance* $\mathbf{P}_0$.

Proof The Kalman filter equations can be derived as follows.

1. By Lemma A.2, the joint distribution of $\mathbf{x}_k$ and $\mathbf{x}_{k-1}$ given $\mathbf{y}_{1:k-1}$ is

$$\begin{aligned} &p(\mathbf{x}_{k-1}, \mathbf{x}_k \mid \mathbf{y}_{1:k-1}) \\ &= p(\mathbf{x}_k \mid \mathbf{x}_{k-1}) \, p(\mathbf{x}_{k-1} \mid \mathbf{y}_{1:k-1}) \\ &= \mathrm{N}(\mathbf{x}_k \mid \mathbf{A}_{k-1} \, \mathbf{x}_{k-1}, \mathbf{Q}_{k-1}) \, \mathrm{N}(\mathbf{x}_{k-1} \mid \mathbf{m}_{k-1}, \mathbf{P}_{k-1}) \\ &= \mathrm{N}\left(\begin{pmatrix} \mathbf{x}_{k-1} \\ \mathbf{x}_k \end{pmatrix} \middle| \, \mathbf{m}', \mathbf{P}' \right), \end{aligned} \tag{6.22}$$

where

$$\begin{aligned} \mathbf{m}' &= \begin{pmatrix} \mathbf{m}_{k-1} \\ \mathbf{A}_{k-1} \, \mathbf{m}_{k-1} \end{pmatrix}, \\ \mathbf{P}' &= \begin{pmatrix} \mathbf{P}_{k-1} & \mathbf{P}_{k-1} \, \mathbf{A}_{k-1}^\mathsf{T} \\ \mathbf{A}_{k-1} \, \mathbf{P}_{k-1} & \mathbf{A}_{k-1} \, \mathbf{P}_{k-1} \, \mathbf{A}_{k-1}^\mathsf{T} + \mathbf{Q}_{k-1} \end{pmatrix}, \end{aligned} \tag{6.23}$$

and the marginal distribution of $\mathbf{x}_k$ is by Lemma A.3

$$p(\mathbf{x}_k \mid \mathbf{y}_{1:k-1}) = \mathrm{N}(\mathbf{x}_k \mid \mathbf{m}_k^-, \mathbf{P}_k^-), \tag{6.24}$$

where

$$\mathbf{m}_k^- = \mathbf{A}_{k-1}\, \mathbf{m}_{k-1}, \quad \mathbf{P}_k^- = \mathbf{A}_{k-1}\, \mathbf{P}_{k-1}\, \mathbf{A}_{k-1}^\mathsf{T} + \mathbf{Q}_{k-1}. \tag{6.25}$$

2. By Lemma A.2, the joint distribution of $\mathbf{y}_k$ and $\mathbf{x}_k$ is

$$\begin{aligned} p(\mathbf{x}_k, \mathbf{y}_k \mid \mathbf{y}_{1:k-1}) &= p(\mathbf{y}_k \mid \mathbf{x}_k)\, p(\mathbf{x}_k \mid \mathbf{y}_{1:k-1}) \\ &= \mathrm{N}(\mathbf{y}_k \mid \mathbf{H}_k\, \mathbf{x}_k, \mathbf{R}_k)\, \mathrm{N}(\mathbf{x}_k \mid \mathbf{m}_k^-, \mathbf{P}_k^-) \\ &= \mathrm{N}\left(\begin{pmatrix} \mathbf{x}_k \\ \mathbf{y}_k \end{pmatrix} \,\middle|\, \mathbf{m}'', \mathbf{P}'' \right), \end{aligned} \tag{6.26}$$

where

$$\mathbf{m}'' = \begin{pmatrix} \mathbf{m}_k^- \\ \mathbf{H}_k\, \mathbf{m}_k^- \end{pmatrix}, \qquad \mathbf{P}'' = \begin{pmatrix} \mathbf{P}_k^- & \mathbf{P}_k^-\, \mathbf{H}_k^\mathsf{T} \\ \mathbf{H}_k\, \mathbf{P}_k^- & \mathbf{H}_k\, \mathbf{P}_k^-\, \mathbf{H}_k^\mathsf{T} + \mathbf{R}_k \end{pmatrix}. \tag{6.27}$$

3. By Lemma A.3, the conditional distribution of $\mathbf{x}_k$ is

$$\begin{aligned} p(\mathbf{x}_k \mid \mathbf{y}_k, \mathbf{y}_{1:k-1}) &= p(\mathbf{x}_k \mid \mathbf{y}_{1:k}) \\ &= \mathrm{N}(\mathbf{x}_k \mid \mathbf{m}_k, \mathbf{P}_k), \end{aligned} \tag{6.28}$$

where

$$\begin{aligned} \mathbf{m}_k &= \mathbf{m}_k^- + \mathbf{P}_k^-\, \mathbf{H}_k^\mathsf{T} (\mathbf{H}_k\, \mathbf{P}_k^-\, \mathbf{H}_k^\mathsf{T} + \mathbf{R}_k)^{-1} [\mathbf{y}_k - \mathbf{H}_k\, \mathbf{m}_k^-], \\ \mathbf{P}_k &= \mathbf{P}_k^- - \mathbf{P}_k^-\, \mathbf{H}_k^\mathsf{T}\, (\mathbf{H}_k\, \mathbf{P}_k^-\, \mathbf{H}_k^\mathsf{T} + \mathbf{R}_k)^{-1}\, \mathbf{H}_k\, \mathbf{P}_k^-, \end{aligned} \tag{6.29}$$

which can be also written in the form (6.21).

□

The functional form of the Kalman filter equations given here is not the only possible one. From a numerical stability point of view it would be better to work with matrix square roots of covariances instead of plain covariance matrices. The theory and details of implementation of this kind of method is well covered, for example, in the book of Grewal and Andrews (2015).

Example 6.7 (Kalman filter for a Gaussian random walk). *Assume that we are observing measurements y_k of the Gaussian random walk model given in Example 6.4, and we want to estimate the state x_k at each time step. The information obtained up to time step k is summarized by the Gaussian filtering density*

$$p(x_k \mid y_{1:k}) = \mathrm{N}(x_k \mid m_k, P_k). \tag{6.30}$$

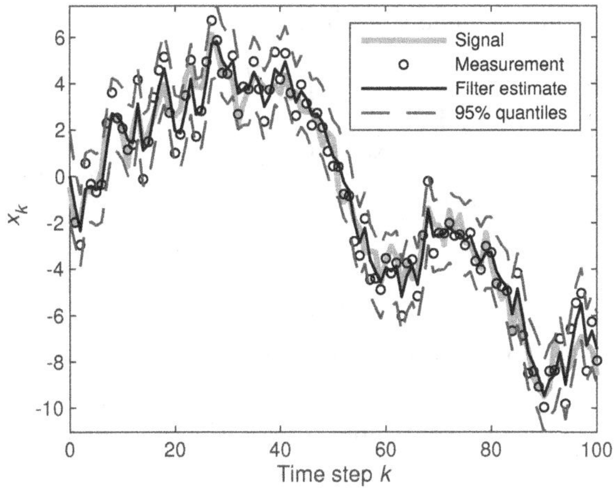

Figure 6.4 Signal, measurements, and result of Kalman filtering of the Gaussian random walk in Example 6.7.

The Kalman filter prediction and update equations are now given as

$$\begin{aligned}
m_k^- &= m_{k-1}, \\
P_k^- &= P_{k-1} + Q, \\
m_k &= m_k^- + \frac{P_k^-}{P_k^- + R}\,(y_k - m_k^-), \\
P_k &= P_k^- - \frac{(P_k^-)^2}{P_k^- + R}.
\end{aligned} \tag{6.31}$$

The result of applying this Kalman filter to the data in Figure 6.1 is shown in Figure 6.4.

Example 6.8 (Kalman filter for car tracking)**.** *By discretizing the state space model for the car in Example 4.6 and by forming a position measurement model for it in Example 5.1, we got the following linear state space model:*

$$\mathbf{x}_k = \mathbf{A}\,\mathbf{x}_{k-1} + \mathbf{q}_{k-1}, \qquad \mathbf{q}_{k-1} \sim \mathrm{N}(\mathbf{0}, \mathbf{Q}), \tag{6.32}$$

$$\mathbf{y}_k = \mathbf{H}\,\mathbf{x}_k + \mathbf{r}_k, \qquad \mathbf{r}_k \sim \mathrm{N}(\mathbf{0}, \mathbf{R}), \tag{6.33}$$

where the state is four dimensional, $\mathbf{x} = (x_1, x_2, x_3, x_4)$*, such that the position of the car is* (x_1, x_2) *and the corresponding velocities are* (x_3, x_4)*.*

The matrices in the dynamic model are

$$\mathbf{A} = \begin{pmatrix} 1 & 0 & \Delta t & 0 \\ 0 & 1 & 0 & \Delta t \\ 0 & 0 & 1 & 0 \\ 0 & 0 & 0 & 1 \end{pmatrix}, \quad \mathbf{Q} = \begin{pmatrix} \frac{q_1^c \, \Delta t^3}{3} & 0 & \frac{q_1^c \, \Delta t^2}{2} & 0 \\ 0 & \frac{q_2^c \, \Delta t^3}{3} & 0 & \frac{q_2^c \, \Delta t^2}{2} \\ \frac{q_1^c \, \Delta t^2}{2} & 0 & q_1^c \, \Delta t & 0 \\ 0 & \frac{q_2^c \, \Delta t^2}{2} & 0 & q_2^c \, \Delta t \end{pmatrix},$$

where q_1^c and q_2^c are the spectral densities (continuous time variances) of the process noises in each direction. The matrices in the measurement model are

$$\mathbf{H} = \begin{pmatrix} 1 & 0 & 0 & 0 \\ 0 & 1 & 0 & 0 \end{pmatrix}, \quad \mathbf{R} = \begin{pmatrix} \sigma_1^2 & 0 \\ 0 & \sigma_2^2 \end{pmatrix},$$

where σ_1^2 and σ_2^2 are the measurement noise variances in each position coordinate.

- *The Kalman filter prediction step now becomes the following:*

$$\mathbf{m}_k^- = \begin{pmatrix} 1 & 0 & \Delta t & 0 \\ 0 & 1 & 0 & \Delta t \\ 0 & 0 & 1 & 0 \\ 0 & 0 & 0 & 1 \end{pmatrix} \mathbf{m}_{k-1},$$

$$\mathbf{P}_k^- = \begin{pmatrix} 1 & 0 & \Delta t & 0 \\ 0 & 1 & 0 & \Delta t \\ 0 & 0 & 1 & 0 \\ 0 & 0 & 0 & 1 \end{pmatrix} \mathbf{P}_{k-1} \begin{pmatrix} 1 & 0 & \Delta t & 0 \\ 0 & 1 & 0 & \Delta t \\ 0 & 0 & 1 & 0 \\ 0 & 0 & 0 & 1 \end{pmatrix}^{\mathsf{T}}$$
$$+ \begin{pmatrix} \frac{q_1^c \, \Delta t^3}{3} & 0 & \frac{q_1^c \, \Delta t^2}{2} & 0 \\ 0 & \frac{q_2^c \, \Delta t^3}{3} & 0 & \frac{q_2^c \, \Delta t^2}{2} \\ \frac{q_1^c \, \Delta t^2}{2} & 0 & q_1^c \, \Delta t & 0 \\ 0 & \frac{q_2^c \, \Delta t^2}{2} & 0 & q_2^c \, \Delta t \end{pmatrix}.$$

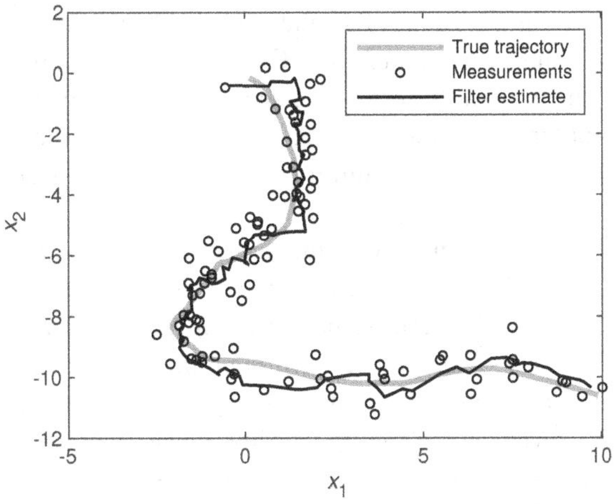

Figure 6.5 Simulated trajectory, measurements, and result of the Kalman filter-based car tracking in Example 6.8. The starting point is at the top of the trajectory. The RMSE position error based on the measurements only is 0.77, whereas the position RMSE of the Kalman filter estimate is 0.43.

- *The corresponding Kalman filter update step is*

$$\begin{aligned}
\mathbf{S}_k &= \begin{pmatrix} 1 & 0 & 0 & 0 \\ 0 & 1 & 0 & 0 \end{pmatrix} \mathbf{P}_k^- \begin{pmatrix} 1 & 0 & 0 & 0 \\ 0 & 1 & 0 & 0 \end{pmatrix}^\mathsf{T} + \begin{pmatrix} \sigma_1^2 & 0 \\ 0 & \sigma_2^2 \end{pmatrix}, \\
\mathbf{K}_k &= \mathbf{P}_k^- \begin{pmatrix} 1 & 0 & 0 & 0 \\ 0 & 1 & 0 & 0 \end{pmatrix}^\mathsf{T} \mathbf{S}_k^{-1}, \\
\mathbf{m}_k &= \mathbf{m}_k^- + \mathbf{K}_k \left(\mathbf{y}_k - \begin{pmatrix} 1 & 0 & 0 & 0 \\ 0 & 1 & 0 & 0 \end{pmatrix} \mathbf{m}_k^- \right), \\
\mathbf{P}_k &= \mathbf{P}_k^- - \mathbf{K}_k \, \mathbf{S}_k \, \mathbf{K}_k^\mathsf{T}.
\end{aligned}$$

The result of applying this filter to simulated data is shown in Figure 6.5. The parameter values used in the simulation were $\sigma_1 = \sigma_2 = 1/2$, $q_1^c = q_2^c = 1$, *and* $\Delta t = 1/10$.

Although the Kalman filter can be seen as the closed form Bayesian solution to the linear Gaussian filtering problem, the original derivation of Kalman (1960b) is based on a different principle. In his seminal article, Kalman (1960b) derived the filter by considering orthogonal projections on the linear manifold spanned by the measurements. A similar approach is also employed in many other works concentrating on estimation in linear systems, such as in the book of Kailath et al. (2000). The advantage of the approach is that it avoids explicit Gaussian assumptions in the noise processes, as the Kalman filter can be seen as a more general best linear unbiased estimator of the state. The disadvantage is that the route to non-linear models is much less straightforward than in the present Bayesian approach.

When the model is time invariant, that is, when the model matrices in Equation (6.17) do not explicitly depend on the time step k, then as $k \to \infty$, the Kalman filter approaches a time-invariant filter called the *stationary Kalman filter* or *steady-state Kalman filter* (Anderson and Moore, 1979; Kailath et al., 2000). What happens in the Kalman filter is that the prediction step variance and the update step covariance approach steady-state values $\mathbf{P}_k \to \mathbf{P}$ and $\mathbf{P}_k^- \to \mathbf{P}^-$ when $k \to \infty$. Then the gain also becomes constant $\mathbf{K}$, and the Kalman filter becomes a time-invariant linear filter.

6.4 Affine Kalman Filter

An often needed simple extension of the Kalman filter is the *affine Kalman filter*, which provides the closed-form filtering solution to models of the affine form

$$\begin{aligned} \mathbf{x}_k &= \mathbf{A}_{k-1}\,\mathbf{x}_{k-1} + \mathbf{a}_{k-1} + \mathbf{q}_{k-1}, \\ \mathbf{y}_k &= \mathbf{H}_k\,\mathbf{x}_k + \mathbf{b}_k + \mathbf{r}_k, \end{aligned} \tag{6.34}$$

where the difference to the linear Gaussian model in Equation (6.17) is that we have additional (known and deterministic) offsets $\mathbf{a}_{k-1}$ and $\mathbf{b}_k$ in the dynamic and measurement models.

Theorem 6.9 (Affine Kalman filter)**.** *The Bayesian filtering equations for the affine filtering model in Equation* (6.34) *can be evaluated in closed form, and the resulting distributions are Gaussian:*

$$\begin{aligned} p(\mathbf{x}_k \mid \mathbf{y}_{1:k-1}) &= \mathrm{N}(\mathbf{x}_k \mid \mathbf{m}_k^-, \mathbf{P}_k^-), \\ p(\mathbf{x}_k \mid \mathbf{y}_{1:k}) &= \mathrm{N}(\mathbf{x}_k \mid \mathbf{m}_k, \mathbf{P}_k), \\ p(\mathbf{y}_k \mid \mathbf{y}_{1:k-1}) &= \mathrm{N}(\mathbf{y}_k \mid \mathbf{H}_k\,\mathbf{m}_k^- + \mathbf{b}_k, \mathbf{S}_k). \end{aligned} \tag{6.35}$$

The parameters of these distributions can be computed recursively with the following affine Kalman filter prediction *and* update steps.

- *The* prediction step *is*

$$\begin{aligned} \mathbf{m}_k^- &= \mathbf{A}_{k-1}\,\mathbf{m}_{k-1} + \mathbf{a}_{k-1}, \\ \mathbf{P}_k^- &= \mathbf{A}_{k-1}\,\mathbf{P}_{k-1}\,\mathbf{A}_{k-1}^\mathsf{T} + \mathbf{Q}_{k-1}. \end{aligned} \tag{6.36}$$

- *The* update step *is*

$$\begin{aligned} \mathbf{v}_k &= \mathbf{y}_k - \mathbf{H}_k\,\mathbf{m}_k^- - \mathbf{b}_k, \\ \mathbf{S}_k &= \mathbf{H}_k\,\mathbf{P}_k^-\,\mathbf{H}_k^\mathsf{T} + \mathbf{R}_k, \\ \mathbf{K}_k &= \mathbf{P}_k^-\,\mathbf{H}_k^\mathsf{T}\,\mathbf{S}_k^{-1}, \\ \mathbf{m}_k &= \mathbf{m}_k^- + \mathbf{K}_k\,\mathbf{v}_k, \\ \mathbf{P}_k &= \mathbf{P}_k^- - \mathbf{K}_k\,\mathbf{S}_k\,\mathbf{K}_k^\mathsf{T}. \end{aligned} \tag{6.37}$$

The recursion is started from the prior mean $\mathbf{m}_0$ *and covariance* $\mathbf{P}_0$.

Proof The derivation is analogous to the derivation of the Kalman filter in Theorem 6.6. □

Note that the expressions for the distributions in Equation (6.35) are identical to Equation (6.19) when $\mathbf{b}_k = 0$. Similarly, the prediction and update steps in Equations (6.36) and (6.37) are identical to the original Kalman filter, except for a minor adjustment in the expressions for $\mathbf{m}_k^-$ and $\mathbf{v}_k$.

6.5 Bayesian Filter for Discrete State Space

In Section 6.2, the Bayesian filtering equations were formulated for the case of continuous states $\mathbf{x}_k \in \mathbb{R}^n$, but it turns out that the equations also work for discrete state spaces $\mathbf{x}_k \in \mathbb{X}$, where $\mathbb{X}$ is some discrete space such as $\mathbb{X} = \{1, 2, 3, \ldots\}$. The only difference is that the integrals in Equations (6.11) and (6.13) are replaced with summations. Discrete models are common in applications such as speech recognition and telecommunications, and they are often referred to as hidden Markov models (HMMs, Rabiner, 1989).

Corollary 6.10 (Bayesian filtering equations for discrete state spaces). *When* $\mathbf{x}_k$ *takes values in a discrete state-space* $\mathbb{X}$, *the Bayesian filter equations in Theorem 6.5 take the following form.*

- Initialization. *The recursion starts from the prior distribution* $p(\mathbf{x}_0)$.

- Prediction step. *The predictive distribution of the state* $\mathbf{x}_k$ *can be computed by a discrete version of the Chapman–Kolmogorov equation:*

$$p(\mathbf{x}_k \mid \mathbf{y}_{1:k-1}) = \sum_{\mathbf{x}_{k-1} \in \mathbb{X}} p(\mathbf{x}_k \mid \mathbf{x}_{k-1})\, p(\mathbf{x}_{k-1} \mid \mathbf{y}_{1:k-1}). \tag{6.38}$$

- Update step. *The update step takes the form*

$$p(\mathbf{x}_k \mid \mathbf{y}_{1:k}) = \frac{1}{Z_k} p(\mathbf{y}_k \mid \mathbf{x}_k)\, p(\mathbf{x}_k \mid \mathbf{y}_{1:k-1}), \tag{6.39}$$

where

$$Z_k = \sum_{\mathbf{x}_k \in \mathbb{X}} p(\mathbf{y}_k \mid \mathbf{x}_k)\, p(\mathbf{x}_k \mid \mathbf{y}_{1:k-1}). \tag{6.40}$$

As discussed in Section 5.6, when the sets of possible states and measurements are finite, and of the form $x_k \in \mathbb{X}$ with $\mathbb{X} = \{1, 2, \ldots, X\}$ and $y_k \in \mathbb{Y}$ with $\mathbb{Y} = \{1, 2, \ldots, Y\}$, then it is convenient to define a state transition matrix $\mathbf{\Pi}$ and an emission matrix $\mathbf{O}$, which have elements

$$\begin{aligned} \Pi_{i,j} &= p(x_k = j \mid x_{k-1} = i), \\ O_{i,j} &= p(y_k = j \mid x_k = i). \end{aligned} \tag{6.41}$$

These matrices can also depend on the time step k, but the dependence is omitted for notational convenience. If we further denote

$$\begin{aligned} p_{j,k}^- &= p(x_k = j \mid y_{1:k-1}), \\ p_{j,k} &= p(x_k = j \mid y_{1:k}), \end{aligned} \tag{6.42}$$

then the Bayesian filter for the resulting hidden Markov model (HMM) can be written as follows.

Corollary 6.11 (Bayesian filtering equations for discrete HMMs). *The Bayesian filtering equations for hidden Markov models (HMMs) with finite state and measurement spaces can be written as*

$$\begin{aligned} p_{j,k}^- &= \sum_i \Pi_{i,j}\, p_{i,k-1}, \\ p_{i,k} &= \frac{O_{i,y_k}\, p_{i,k}^-}{\sum_j O_{j,y_k}\, p_{j,k}^-}, \end{aligned} \tag{6.43}$$

where the dynamic and measurement model matrices (i.e., transition and emission matrices) are defined in Equations (6.41)*, and the predictive and posterior probabilities* $p_{j,k}^-$ *and* $p_{j,k}$ *in Equations* (6.42).

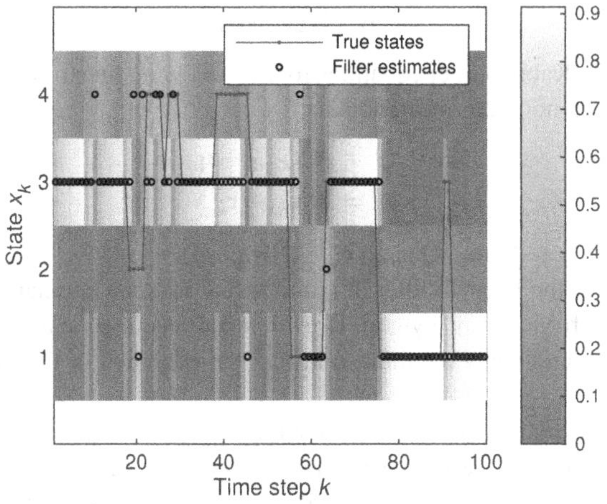

Figure 6.6 Filtering result for the Gilbert–Elliot channel model in Example 6.12. The probabilities of the states are denoted by shades of gray, and the filter estimates are the most probable states computed as maxima of the filtering distributions. The error rate of the filter when the most probable state is guessed is 11%, whereas guessing by no errors, that is, $v_k = 1$, gives an error rate of 14%.

The above algorithm is closely related to the forward portion of the forward-backward algorithm for HMMs (see, e.g., Rabiner, 1989). The smoother version of the above algorithm along with the Viterbi algorithm that targets the MAP estimation of the trajectory are presented in Sections 12.4 and 12.5, respectively.

Example 6.12 (Gilbert–Elliot channel filter). *In this example we consider the filtering problem on the Gilbert–Elliot model considered in Example 5.14. The aim is to estimate the hidden state sequence on the left in Figure 5.4 from the measurements shown on the right in the same figure. Figure 6.6 shows the result of applying the Bayesian filter in Corollary 6.11 to the measurements shown in Figure 5.4.*

6.6 Exercises

6.1 Derive the Kalman filter equations for the following linear-Gaussian filtering model with non-zero-mean noises:

$$\begin{aligned} \mathbf{x}_k &= \mathbf{A}\,\mathbf{x}_{k-1} + \mathbf{q}_{k-1}, \\ \mathbf{y}_k &= \mathbf{H}\,\mathbf{x}_k + \mathbf{r}_k, \end{aligned} \tag{6.44}$$

where $\mathbf{q}_{k-1} \sim \mathrm{N}(\mathbf{m}_q, \mathbf{Q})$ and $\mathbf{r}_k \sim \mathrm{N}(\mathbf{m}_r, \mathbf{R})$.

6.2 Implement the Kalman filter in Example 6.7 for the Gaussian random walk model with Matlab or Python. Draw realizations of state and measurement sequences and apply the filter to it. Plot the evolution of the filtering distribution.

6.3 Derive the stationary Kalman filter for the Gaussian random walk model. That is, compute the limiting Kalman filter gain when $k \to \infty$, and write down the mean equation of the resulting constant-gain Kalman filter. Plot the frequency response of the resulting time-invariant filter. What type of digital filter is it?

6.4 Consider the following dynamic model:

$$\begin{aligned} \mathbf{x}_k &= \begin{pmatrix} \cos(\omega) & \frac{\sin(\omega)}{\omega} \\ -\omega\,\sin(\omega) & \cos(\omega) \end{pmatrix} \mathbf{x}_{k-1} + \mathbf{q}_{k-1}, \\ y_k &= \begin{pmatrix} 1 & 0 \end{pmatrix} \mathbf{x}_k + r_k, \end{aligned}$$

where $\mathbf{x}_k \in \mathbb{R}^2$ is the state, y_k is the measurement, $r_k \sim \mathrm{N}(0, 0.1)$ is a white Gaussian measurement noise, and $\mathbf{q}_k \sim \mathrm{N}(\mathbf{0}, \mathbf{Q})$, where

$$\mathbf{Q} = \begin{pmatrix} \frac{q^c\,\omega - q^c\,\cos(\omega)\,\sin(\omega)}{2\omega^3} & \frac{q^c\,\sin^2(\omega)}{2\omega^2} \\ \frac{q^c\,\sin^2(\omega)}{2\omega^2} & \frac{q^c\,\omega + q^c\,\cos(\omega)\,\sin(\omega)}{2\omega} \end{pmatrix}. \tag{6.45}$$

The angular velocity is $\omega = 1/2$ and the spectral density is $q^c = 0.01$. The model is a discretized version of the noisy resonator model with a given angular velocity ω.

In the file `exercise6_4.m` (MATLAB) or `Exercise6_4.ipynb` (Python) of the book's companion code repository[1] there is simulation of the dynamic model together with a baseline solution, where the measurement is directly used as the estimate of the state component x_1, and the second component x_2 is computed as a weighted average of the measurement differences. Implement the Kalman filter for the model and compare its performance (in RMSE sense) to the baseline solution. Plot figures of the solutions.

6.5 Let us consider a binary HMM with two states $x_k \in \{1, 2\}$ where the probability of jumping from state $x_{k-1} = 1$ to $x_k = 2$ is given by $\rho_{1,2} = 0.2$,

[1] https://github.com/EEA-sensors/Bayesian-Filtering-and-Smoothing

and the probability of jumping from $x_{k-1} = 2$ to $x_k = 1$ is $\rho_{2,1} = 0.1$. Furthermore, assume that $y_k \in \{1, 2\}$ as well and that the probability of getting $y_k = x_k$ as the measurement is 0.3 for both of the states.

(a) Verify that the state transition and emission matrices are given by

$$\boldsymbol{\Pi} = \begin{pmatrix} 0.8 & 0.2 \\ 0.1 & 0.9 \end{pmatrix}, \qquad \mathbf{O} = \begin{pmatrix} 0.7 & 0.3 \\ 0.3 & 0.7 \end{pmatrix}. \tag{6.46}$$

(b) Starting from prior $\mathbf{p}_0 = (0.5, 0.5)$, simulate 100 steps of states and measurements, and compute the error of using y_k as the guess of x_k. What error rate would you expect?

(c) Implement an HMM filter for the problem, and compute estimates of the state by taking the higher probability state at each step based on the filter distribution. Simulate 100 independent realizations of the problem, and check that on average we get a lower error rate than by just guessing y_k as in (b).

6.6 It is also possible to approximate the filtering distributions on a grid and use the HMM filter to solve the resulting finite-state problem. Select a finite interval in the state space, say $x \in [-10, 10]$, and discretize it evenly to N subintervals (e.g., $N = 1000$). Using a suitable numerical approximation to the integrals in the Bayesian filtering equations, implement a finite grid approximation to the Bayesian filter for the Gaussian random walk in Example 6.4. Verify that the result is practically the same as that of the Kalman filter in Exercise 6.2.

7

Extended Kalman Filtering

It often happens in practical applications that the dynamic and measurement models are not linear, and the Kalman filter is not appropriate. However, often the filtering distributions of this kind of model can be approximated by Gaussian distributions. In this chapter, we investigate methods that use Taylor series expansions to perform approximate Bayesian filtering using such Gaussian approximations. The first method is the (first order) extended Kalman filter (EKF), which is the classical and probably the most widely used non-linear filter. We then proceed to discuss higher order EKFs and then present the iterated EKF, which aims to improve the Taylor series approximation by using iteration.

7.1 Taylor Series Approximation of Non-Linear Transform

Consider the following transformation of a Gaussian random variable $\mathbf{x}$ into another random variable $\mathbf{y}$:

$$\begin{aligned} \mathbf{x} &\sim \mathrm{N}(\mathbf{m}, \mathbf{P}), \\ \mathbf{y} &= \mathbf{g}(\mathbf{x}), \end{aligned} \tag{7.1}$$

where $\mathbf{x} \in \mathbb{R}^n$, $\mathbf{y} \in \mathbb{R}^m$, and $\mathbf{g} : \mathbb{R}^n \to \mathbb{R}^m$ is a general non-linear function. Formally, the probability density of the random variable $\mathbf{y}$ is[1] (see, e.g., Gelman et al., 2013)

$$p(\mathbf{y}) = |\mathbf{J}(\mathbf{y})| \; \mathrm{N}(\mathbf{g}^{-1}(\mathbf{y}) \mid \mathbf{m}, \mathbf{P}), \tag{7.2}$$

where $|\mathbf{J}(\mathbf{y})|$ is the determinant of the Jacobian matrix of the inverse transform $\mathbf{g}^{-1}(\mathbf{y})$. However, it is not generally possible to handle this distribution directly, because it is non-Gaussian for all but linear $\mathbf{g}$.

[1] This actually only applies to invertible $\mathbf{g}(\cdot)$, but it can be easily generalized to the non-invertible case.

A first order Taylor series-based Gaussian approximation to the distribution of $\mathbf{y}$ can now be formed as follows. If we let $\mathbf{x} = \mathbf{m} + \delta\mathbf{x}$, where $\delta\mathbf{x} \sim \mathrm{N}(\mathbf{0}, \mathbf{P})$, we can form the Taylor series expansion of the function $\mathbf{g}(\cdot)$ as follows (provided that the function is sufficiently differentiable):

$$\mathbf{g}(\mathbf{x}) = \mathbf{g}(\mathbf{m}+\delta\mathbf{x}) \approx \mathbf{g}(\mathbf{m}) + \mathbf{G}_{\mathbf{x}}(\mathbf{m})\,\delta\mathbf{x} + \sum_i \frac{1}{2}\delta\mathbf{x}^\mathsf{T}\,\mathbf{G}_{\mathbf{xx}}^{(i)}(\mathbf{m})\,\delta\mathbf{x}\,\mathbf{e}_i + \cdots, \tag{7.3}$$

where $\mathbf{G}_{\mathbf{x}}(\mathbf{m})$ is the Jacobian matrix of $\mathbf{g}$ with elements

$$[\mathbf{G}_{\mathbf{x}}(\mathbf{m})]_{j,j'} = \left.\frac{\partial g_j(\mathbf{x})}{\partial x_{j'}}\right|_{\mathbf{x}=\mathbf{m}}, \tag{7.4}$$

and $\mathbf{G}_{\mathbf{xx}}^{(i)}(\mathbf{m})$ is the Hessian matrix of $g_i(\cdot)$ evaluated at $\mathbf{m}$:

$$\left[\mathbf{G}_{\mathbf{xx}}^{(i)}(\mathbf{m})\right]_{j,j'} = \left.\frac{\partial^2 g_i(\mathbf{x})}{\partial x_j\,\partial x_{j'}}\right|_{\mathbf{x}=\mathbf{m}}. \tag{7.5}$$

Furthermore, $\mathbf{e}_i = (0\ \cdots\ 0\ 1\ 0\ \cdots\ 0)^\mathsf{T}$ is a vector with 1 at position i and all other elements zero, that is, it is the unit vector in the direction of the coordinate axis i.

The linear approximation can be obtained by approximating the function by the first two terms in the Taylor series:

$$\mathbf{g}(\mathbf{x}) \simeq \mathbf{g}(\mathbf{m}) + \mathbf{G}_{\mathbf{x}}(\mathbf{m})\,\delta\mathbf{x}. \tag{7.6}$$

Computing the expected value with respect to $\mathbf{x}$ gives:

$$\begin{aligned} \mathrm{E}[\mathbf{g}(\mathbf{x})] &\simeq \mathrm{E}[\mathbf{g}(\mathbf{m}) + \mathbf{G}_{\mathbf{x}}(\mathbf{m})\,\delta\mathbf{x}] \\ &= \mathbf{g}(\mathbf{m}) + \mathbf{G}_{\mathbf{x}}(\mathbf{m})\;\mathrm{E}[\delta\mathbf{x}] \\ &= \mathbf{g}(\mathbf{m}). \end{aligned} \tag{7.7}$$

The covariance can then be approximated as

$$
\begin{aligned}
&\mathrm{E}\left[(\mathbf{g}(\mathbf{x}) - \mathrm{E}[\mathbf{g}(\mathbf{x})])\,(\mathbf{g}(\mathbf{x}) - \mathrm{E}[\mathbf{g}(\mathbf{x})])^{\mathsf{T}}\right] \\
&\quad \simeq \mathrm{E}\left[(\mathbf{g}(\mathbf{x}) - \mathbf{g}(\mathbf{m}))\,(\mathbf{g}(\mathbf{x}) - \mathbf{g}(\mathbf{m}))^{\mathsf{T}}\right] \\
&\quad \simeq \mathrm{E}\left[(\mathbf{g}(\mathbf{m}) + \mathbf{G}_{\mathbf{x}}(\mathbf{m})\,\delta\mathbf{x} - \mathbf{g}(\mathbf{m})])\,(\mathbf{g}(\mathbf{m}) + \mathbf{G}_{\mathbf{x}}(\mathbf{m})\,\delta\mathbf{x} - \mathbf{g}(\mathbf{m}))^{\mathsf{T}}\right] \\
&\quad = \mathrm{E}\left[(\mathbf{G}_{\mathbf{x}}(\mathbf{m})\,\delta\mathbf{x})\,(\mathbf{G}_{\mathbf{x}}(\mathbf{m})\,\delta\mathbf{x})^{\mathsf{T}}\right] \\
&\quad = \mathbf{G}_{\mathbf{x}}(\mathbf{m})\,\mathrm{E}\left[\delta\mathbf{x}\,\delta\mathbf{x}^{\mathsf{T}}\right]\mathbf{G}_{\mathbf{x}}^{\mathsf{T}}(\mathbf{m}) \\
&\quad = \mathbf{G}_{\mathbf{x}}(\mathbf{m})\,\mathbf{P}\,\mathbf{G}_{\mathbf{x}}^{\mathsf{T}}(\mathbf{m}).
\end{aligned}
\tag{7.8}
$$

We often are also interested in the the joint covariance between the variables $\mathbf{x}$ and $\mathbf{y}$. Approximation of the joint covariance can be achieved by considering the augmented transformation

$$
\tilde{\mathbf{g}}(\mathbf{x}) = \begin{pmatrix} \mathbf{x} \\ \mathbf{g}(\mathbf{x}) \end{pmatrix}. \tag{7.9}
$$

The resulting mean and covariance are

$$
\begin{aligned}
\mathrm{E}[\tilde{\mathbf{g}}(\mathbf{x})] &\simeq \begin{pmatrix} \mathbf{m} \\ \mathbf{g}(\mathbf{m}) \end{pmatrix}, \\
\mathrm{Cov}[\tilde{\mathbf{g}}(\mathbf{x})] &\simeq \begin{pmatrix} \mathbf{I} \\ \mathbf{G}_{\mathbf{x}}(\mathbf{m}) \end{pmatrix} \mathbf{P} \begin{pmatrix} \mathbf{I} \\ \mathbf{G}_{\mathbf{x}}(\mathbf{m}) \end{pmatrix}^{\mathsf{T}} \\
&= \begin{pmatrix} \mathbf{P} & \mathbf{P}\,\mathbf{G}_{\mathbf{x}}^{\mathsf{T}}(\mathbf{m}) \\ \mathbf{G}_{\mathbf{x}}(\mathbf{m})\,\mathbf{P} & \mathbf{G}_{\mathbf{x}}(\mathbf{m})\,\mathbf{P}\,\mathbf{G}_{\mathbf{x}}^{\mathsf{T}}(\mathbf{m}) \end{pmatrix}.
\end{aligned}
\tag{7.10}
$$

In the derivation of the extended Kalman filter equations, we need a slightly more general transformation of the form

$$
\begin{aligned}
\mathbf{x} &\sim \mathrm{N}(\mathbf{m}, \mathbf{P}), \\
\mathbf{q} &\sim \mathrm{N}(\mathbf{0}, \mathbf{Q}), \\
\mathbf{y} &= \mathbf{g}(\mathbf{x}) + \mathbf{q},
\end{aligned}
\tag{7.11}
$$

where $\mathbf{q}$ is independent of $\mathbf{x}$. The joint distribution of $\mathbf{x}$ and $\mathbf{y}$, as defined above, is now the same as in Equations (7.10) except that the covariance $\mathbf{Q}$ is added to the lower right block of the covariance matrix of $\tilde{\mathbf{g}}(\cdot)$. Thus we get the following algorithm.

Algorithm 7.1 (Linear approximation of an additive transform)**.** *The linear approximation-based Gaussian approximation to the joint distribution of* $\mathbf{x}$

and the transformed random variable $\mathbf{y} = \mathbf{g}(\mathbf{x}) + \mathbf{q}$, *where* $\mathbf{x} \sim \mathrm{N}(\mathbf{m}, \mathbf{P})$ *and* $\mathbf{q} \sim \mathrm{N}(\mathbf{0}, \mathbf{Q})$ *is given by*

$$\begin{pmatrix} \mathbf{x} \\ \mathbf{y} \end{pmatrix} \sim \mathrm{N}\left(\begin{pmatrix} \mathbf{m} \\ \boldsymbol{\mu}_{\mathrm{L}} \end{pmatrix}, \begin{pmatrix} \mathbf{P} & \mathbf{C}_{\mathrm{L}} \\ \mathbf{C}_{\mathrm{L}}^{\mathsf{T}} & \mathbf{S}_{\mathrm{L}} \end{pmatrix} \right), \tag{7.12}$$

where

$$\begin{aligned} \boldsymbol{\mu}_{\mathrm{L}} &= \mathbf{g}(\mathbf{m}), \\ \mathbf{S}_{\mathrm{L}} &= \mathbf{G}_{\mathbf{x}}(\mathbf{m})\, \mathbf{P}\, \mathbf{G}_{\mathbf{x}}^{\mathsf{T}}(\mathbf{m}) + \mathbf{Q}, \\ \mathbf{C}_{\mathrm{L}} &= \mathbf{P}\, \mathbf{G}_{\mathbf{x}}^{\mathsf{T}}(\mathbf{m}), \end{aligned} \tag{7.13}$$

and $\mathbf{G}_{\mathbf{x}}(\mathbf{m})$ *is the Jacobian matrix of* $\mathbf{g}$ *with respect to* $\mathbf{x}$, *evaluated at* $\mathbf{x} = \mathbf{m}$, *with elements*

$$[\mathbf{G}_{\mathbf{x}}(\mathbf{m})]_{j,j'} = \left. \frac{\partial g_j(\mathbf{x})}{\partial x_{j'}} \right|_{\mathbf{x}=\mathbf{m}}. \tag{7.14}$$

In filtering models where the process noise is not additive, we often need to approximate transformations of the form

$$\begin{aligned} \mathbf{x} &\sim \mathrm{N}(\mathbf{m}, \mathbf{P}), \\ \mathbf{q} &\sim \mathrm{N}(\mathbf{0}, \mathbf{Q}), \\ \mathbf{y} &= \mathbf{g}(\mathbf{x}, \mathbf{q}), \end{aligned} \tag{7.15}$$

where $\mathbf{x}$ and $\mathbf{q}$ are independent random variables. The mean and covariance can now be computed by substituting the augmented vector $(\mathbf{x}, \mathbf{q})$ for the vector $\mathbf{x}$ in Equation (7.10). The joint Jacobian matrix can then be written as $\mathbf{G}_{\mathbf{x},\mathbf{q}} = (\mathbf{G}_{\mathbf{x}} \; \mathbf{G}_{\mathbf{q}})$. Here $\mathbf{G}_{\mathbf{q}}$ is the Jacobian matrix of $\mathbf{g}(\cdot)$ with respect to $\mathbf{q}$, and both Jacobian matrices are evaluated at $\mathbf{x} = \mathbf{m}$, $\mathbf{q} = \mathbf{0}$. The approximations to the mean and covariance of the augmented transform as in Equation (7.10) are then given as

$$\begin{aligned} \mathrm{E}[\tilde{\mathbf{g}}(\mathbf{x}, \mathbf{q})] &\simeq \mathbf{g}(\mathbf{m}, \mathbf{0}), \\ \mathrm{Cov}[\tilde{\mathbf{g}}(\mathbf{x}, \mathbf{q})] &\simeq \begin{pmatrix} \mathbf{I} & \mathbf{0} \\ \mathbf{G}_{\mathbf{x}}(\mathbf{m}) & \mathbf{G}_{\mathbf{q}}(\mathbf{m}) \end{pmatrix} \begin{pmatrix} \mathbf{P} & \mathbf{0} \\ \mathbf{0} & \mathbf{Q} \end{pmatrix} \begin{pmatrix} \mathbf{I} & \mathbf{0} \\ \mathbf{G}_{\mathbf{x}}(\mathbf{m}) & \mathbf{G}_{\mathbf{q}}(\mathbf{m}) \end{pmatrix}^{\mathsf{T}} \\ &= \begin{pmatrix} \mathbf{P} & \mathbf{P}\, \mathbf{G}_{\mathbf{x}}^{\mathsf{T}}(\mathbf{m}) \\ \mathbf{G}_{\mathbf{x}}(\mathbf{m})\, \mathbf{P} & \mathbf{G}_{\mathbf{x}}(\mathbf{m})\, \mathbf{P}\, \mathbf{G}_{\mathbf{x}}^{\mathsf{T}}(\mathbf{m}) + \mathbf{G}_{\mathbf{q}}(\mathbf{m})\, \mathbf{Q}\, \mathbf{G}_{\mathbf{q}}^{\mathsf{T}}(\mathbf{m}) \end{pmatrix}. \end{aligned} \tag{7.16}$$

The approximation above can be formulated as the following algorithm.

Algorithm 7.2 (Linear approximation of a non-additive transform)**.** *The linear approximation-based Gaussian approximation to the joint distribution of* $\mathbf{x}$ *and the transformed random variable* $\mathbf{y} = \mathbf{g}(\mathbf{x}, \mathbf{q})$ *when* $\mathbf{x} \sim \mathrm{N}(\mathbf{m}, \mathbf{P})$ *and* $\mathbf{q} \sim \mathrm{N}(\mathbf{0}, \mathbf{Q})$ *is given by*

$$\begin{pmatrix} \mathbf{x} \\ \mathbf{y} \end{pmatrix} \sim \mathrm{N}\left(\begin{pmatrix} \mathbf{m} \\ \boldsymbol{\mu}_{\mathrm{L}} \end{pmatrix}, \begin{pmatrix} \mathbf{P} & \mathbf{C}_{\mathrm{L}} \\ \mathbf{C}_{\mathrm{L}}^{\mathsf{T}} & \mathbf{S}_{\mathrm{L}} \end{pmatrix} \right), \tag{7.17}$$

where

$$\begin{aligned} \boldsymbol{\mu}_{\mathrm{L}} &= \mathbf{g}(\mathbf{m}, \mathbf{0}), \\ \mathbf{S}_{\mathrm{L}} &= \mathbf{G}_{\mathbf{x}}(\mathbf{m})\, \mathbf{P}\, \mathbf{G}_{\mathbf{x}}^{\mathsf{T}}(\mathbf{m}) + \mathbf{G}_{\mathbf{q}}(\mathbf{m})\, \mathbf{Q}\, \mathbf{G}_{\mathbf{q}}^{\mathsf{T}}(\mathbf{m}), \\ \mathbf{C}_{\mathrm{L}} &= \mathbf{P}\, \mathbf{G}_{\mathbf{x}}^{\mathsf{T}}(\mathbf{m}), \end{aligned} \tag{7.18}$$

$\mathbf{G}_{\mathbf{x}}(\mathbf{m})$ *is the Jacobian matrix of* $\mathbf{g}$ *with respect to* $\mathbf{x}$*, evaluated at* $\mathbf{x} = \mathbf{m}, \mathbf{q} = \mathbf{0}$*, with elements*

$$[\mathbf{G}_{\mathbf{x}}(\mathbf{m})]_{j,j'} = \left. \frac{\partial g_j(\mathbf{x}, \mathbf{q})}{\partial x_{j'}} \right|_{\mathbf{x}=\mathbf{m}, \mathbf{q}=\mathbf{0}}, \tag{7.19}$$

and $\mathbf{G}_{\mathbf{q}}(\mathbf{m})$ *is the corresponding Jacobian matrix with respect to* $\mathbf{q}$*:*

$$\left[\mathbf{G}_{\mathbf{q}}(\mathbf{m})\right]_{j,j'} = \left. \frac{\partial g_j(\mathbf{x}, \mathbf{q})}{\partial q_{j'}} \right|_{\mathbf{x}=\mathbf{m}, \mathbf{q}=\mathbf{0}}. \tag{7.20}$$

In quadratic approximations, in addition to the first order terms, the second order terms in the Taylor series expansion of the non-linear function are also retained.

Algorithm 7.3 (Quadratic approximation of an additive non-linear transform)**.** *The second order approximation is of the form*

$$\begin{pmatrix} \mathbf{x} \\ \mathbf{y} \end{pmatrix} \sim \mathrm{N}\left(\begin{pmatrix} \mathbf{m} \\ \boldsymbol{\mu}_{\mathrm{Q}} \end{pmatrix}, \begin{pmatrix} \mathbf{P} & \mathbf{C}_{\mathrm{Q}} \\ \mathbf{C}_{\mathrm{Q}}^{\mathsf{T}} & \mathbf{S}_{\mathrm{Q}} \end{pmatrix} \right), \tag{7.21}$$

where the parameters are

$$\begin{aligned} \boldsymbol{\mu}_{\mathrm{Q}} &= \mathbf{g}(\mathbf{m}) + \frac{1}{2} \sum_i \mathbf{e}_i \operatorname{tr}\left\{ \mathbf{G}_{\mathbf{xx}}^{(i)}(\mathbf{m})\, \mathbf{P} \right\}, \\ \mathbf{S}_{\mathrm{Q}} &= \mathbf{G}_{\mathbf{x}}(\mathbf{m})\, \mathbf{P}\, \mathbf{G}_{\mathbf{x}}^{\mathsf{T}}(\mathbf{m}) + \frac{1}{2} \sum_{i,i'} \mathbf{e}_i\, \mathbf{e}_{i'}^{\mathsf{T}} \operatorname{tr}\left\{ \mathbf{G}_{\mathbf{xx}}^{(i)}(\mathbf{m})\, \mathbf{P}\, \mathbf{G}_{\mathbf{xx}}^{(i')}(\mathbf{m})\, \mathbf{P} \right\} + \mathbf{Q}, \\ \mathbf{C}_{\mathrm{Q}} &= \mathbf{P}\, \mathbf{G}_{\mathbf{x}}^{\mathsf{T}}(\mathbf{m}), \end{aligned} \tag{7.22}$$

$\mathbf{G_x}(\mathbf{m})$ *is the Jacobian matrix* (7.14)*, and* $\mathbf{G}^{(i)}_{\mathbf{xx}}(\mathbf{m})$ *is the Hessian matrix of* $g_i(\cdot)$ *evaluated at* $\mathbf{m}$*:*

$$\left[\mathbf{G}^{(i)}_{\mathbf{xx}}(\mathbf{m})\right]_{j,j'} = \left.\frac{\partial^2 g_i(\mathbf{x})}{\partial x_j\,\partial x_{j'}}\right|_{\mathbf{x}=\mathbf{m}}, \tag{7.23}$$

where $\mathbf{e}_i = (0 \ \cdots \ 0\ 1\ 0 \ \cdots \ 0)^\mathsf{T}$ *is a vector with 1 at position i and other elements zero, that is, it is the unit vector in the direction of the coordinate axis i.*

7.2 Extended Kalman Filter

The extended Kalman filter (EKF) (see, e.g., Jazwinski, 1970; Maybeck, 1982b; Bar-Shalom et al., 2001; Grewal and Andrews, 2015) is an extension of the Kalman filter to non-linear filtering problems. If the process and measurement noises can be assumed to be additive, the state space model can be written as

$$\begin{aligned} \mathbf{x}_k &= \mathbf{f}(\mathbf{x}_{k-1}) + \mathbf{q}_{k-1}, \\ \mathbf{y}_k &= \mathbf{h}(\mathbf{x}_k) + \mathbf{r}_k, \end{aligned} \tag{7.24}$$

where $\mathbf{x}_k \in \mathbb{R}^n$ is the state, $\mathbf{y}_k \in \mathbb{R}^m$ is the measurement, $\mathbf{q}_{k-1} \sim \mathrm{N}(\mathbf{0}, \mathbf{Q}_{k-1})$ is the Gaussian process noise, $\mathbf{r}_k \sim \mathrm{N}(\mathbf{0}, \mathbf{R}_k)$ is the Gaussian measurement noise, $\mathbf{f}(\cdot)$ is the dynamic model function, and $\mathbf{h}(\cdot)$ is the measurement model function. The functions $\mathbf{f}$ and $\mathbf{h}$ can also depend on the step number k, but for notational convenience, this dependence has not been explicitly denoted.

The idea of the extended Kalman filter is to use (or assume) Gaussian approximations

$$p(\mathbf{x}_k \mid \mathbf{y}_{1:k}) \simeq \mathrm{N}(\mathbf{x}_k \mid \mathbf{m}_k, \mathbf{P}_k) \tag{7.25}$$

to the filtering densities. In the EKF, these approximations are formed by utilizing Taylor series approximations to the non-linearities, and the result is the following algorithm.

Algorithm 7.4 (Extended Kalman filter I)**.** *The prediction and update steps of the first order additive noise extended Kalman filter (EKF) are:*

- *Prediction:*

$$\begin{aligned} \mathbf{m}_k^- &= \mathbf{f}(\mathbf{m}_{k-1}), \\ \mathbf{P}_k^- &= \mathbf{F_x}(\mathbf{m}_{k-1})\,\mathbf{P}_{k-1}\,\mathbf{F}_\mathbf{x}^\mathsf{T}(\mathbf{m}_{k-1}) + \mathbf{Q}_{k-1}, \end{aligned} \tag{7.26}$$

- *Update:*

$$\begin{aligned}
\mathbf{v}_k &= \mathbf{y}_k - \mathbf{h}(\mathbf{m}_k^-), \\
\mathbf{S}_k &= \mathbf{H}_{\mathbf{x}}(\mathbf{m}_k^-)\,\mathbf{P}_k^-\,\mathbf{H}_{\mathbf{x}}^\mathsf{T}(\mathbf{m}_k^-) + \mathbf{R}_k, \\
\mathbf{K}_k &= \mathbf{P}_k^-\,\mathbf{H}_{\mathbf{x}}^\mathsf{T}(\mathbf{m}_k^-)\,\mathbf{S}_k^{-1}, \\
\mathbf{m}_k &= \mathbf{m}_k^- + \mathbf{K}_k\,\mathbf{v}_k, \\
\mathbf{P}_k &= \mathbf{P}_k^- - \mathbf{K}_k\,\mathbf{S}_k\,\mathbf{K}_k^\mathsf{T}.
\end{aligned} \tag{7.27}$$

Above, $\mathbf{F}_{\mathbf{x}}(\mathbf{m})$ *and* $\mathbf{H}_{\mathbf{x}}(\mathbf{m})$ *denote the Jacobian matrices of functions* $\mathbf{f}(\cdot)$ *and* $\mathbf{h}(\cdot)$*, respectively, with elements*

$$[\mathbf{F}_{\mathbf{x}}(\mathbf{m})]_{j,j'} = \left.\frac{\partial f_j(\mathbf{x})}{\partial x_{j'}}\right|_{\mathbf{x}=\mathbf{m}}, \tag{7.28}$$

$$[\mathbf{H}_{\mathbf{x}}(\mathbf{m})]_{j,j'} = \left.\frac{\partial h_j(\mathbf{x})}{\partial x_{j'}}\right|_{\mathbf{x}=\mathbf{m}}. \tag{7.29}$$

Derivation These filtering equations can be derived by repeating the same steps as in the derivation of the Kalman filter in Section 6.3 and by applying Taylor series approximations on the appropriate steps.

1. The joint distribution of $\mathbf{x}_k$ and $\mathbf{x}_{k-1}$ is non-Gaussian, but we can form a Gaussian approximation to it by applying the approximation Algorithm 7.1 to the function

$$\mathbf{f}(\mathbf{x}_{k-1}) + \mathbf{q}_{k-1}, \tag{7.30}$$

which results in the Gaussian approximation

$$p(\mathbf{x}_{k-1}, \mathbf{x}_k \mid \mathbf{y}_{1:k-1}) \simeq \mathrm{N}\left(\begin{pmatrix}\mathbf{x}_{k-1} \\ \mathbf{x}_k\end{pmatrix} \,\middle|\, \mathbf{m}', \mathbf{P}'\right), \tag{7.31}$$

where

$$\begin{aligned}
\mathbf{m}' &= \begin{pmatrix}\mathbf{m}_{k-1} \\ \mathbf{f}(\mathbf{m}_{k-1})\end{pmatrix}, \\
\mathbf{P}' &= \begin{pmatrix}\mathbf{P}_{k-1} & \mathbf{P}_{k-1}\,\mathbf{F}_{\mathbf{x}}^\mathsf{T} \\ \mathbf{F}_{\mathbf{x}}\,\mathbf{P}_{k-1} & \mathbf{F}_{\mathbf{x}}\,\mathbf{P}_{k-1}\,\mathbf{F}_{\mathbf{x}}^\mathsf{T} + \mathbf{Q}_{k-1}\end{pmatrix},
\end{aligned} \tag{7.32}$$

and the Jacobian matrix $\mathbf{F}_{\mathbf{x}}$ of $\mathbf{f}(\mathbf{x})$ is evaluated at $\mathbf{x} = \mathbf{m}_{k-1}$. The approximations of the marginal mean and covariance of $\mathbf{x}_k$ are thus

$$\begin{aligned}
\mathbf{m}_k^- &= \mathbf{f}(\mathbf{m}_{k-1}), \\
\mathbf{P}_k^- &= \mathbf{F}_{\mathbf{x}}\,\mathbf{P}_{k-1}\,\mathbf{F}_{\mathbf{x}}^\mathsf{T} + \mathbf{Q}_{k-1}.
\end{aligned} \tag{7.33}$$

2. The joint distribution of $\mathbf{y}_k$ and $\mathbf{x}_k$ is also non-Gaussian, but we can again approximate it by applying Algorithm 7.1 to the function
$$\mathbf{h}(\mathbf{x}_k) + \mathbf{r}_k. \tag{7.34}$$
We get the approximation
$$p(\mathbf{x}_k, \mathbf{y}_k \mid \mathbf{y}_{1:k-1}) \simeq \mathrm{N}\left(\begin{pmatrix} \mathbf{x}_k \\ \mathbf{y}_k \end{pmatrix} \,\middle|\, \mathbf{m}'', \mathbf{P}'' \right), \tag{7.35}$$
where
$$\mathbf{m}'' = \begin{pmatrix} \mathbf{m}_k^- \\ \mathbf{h}(\mathbf{m}_k^-) \end{pmatrix}, \qquad \mathbf{P}'' = \begin{pmatrix} \mathbf{P}_k^- & \mathbf{P}_k^- \, \mathbf{H}_\mathbf{x}^\mathsf{T} \\ \mathbf{H}_\mathbf{x} \, \mathbf{P}_k^- & \mathbf{H}_\mathbf{x} \, \mathbf{P}_k^- \, \mathbf{H}_\mathbf{x}^\mathsf{T} + \mathbf{R}_k \end{pmatrix}, \tag{7.36}$$
and the Jacobian matrix $\mathbf{H}_\mathbf{x}$ of $\mathbf{h}(\mathbf{x})$ is evaluated at $\mathbf{x} = \mathbf{m}_k^-$.
3. By Lemma A.3, the conditional distribution of $\mathbf{x}_k$ is approximately
$$p(\mathbf{x}_k \mid \mathbf{y}_k, \mathbf{y}_{1:k-1}) \simeq \mathrm{N}(\mathbf{x}_k \mid \mathbf{m}_k, \mathbf{P}_k), \tag{7.37}$$
where
$$\begin{aligned} \mathbf{m}_k &= \mathbf{m}_k^- + \mathbf{P}_k^- \, \mathbf{H}_\mathbf{x}^\mathsf{T} \, (\mathbf{H}_\mathbf{x} \, \mathbf{P}_k^- \, \mathbf{H}_\mathbf{x}^\mathsf{T} + \mathbf{R}_k)^{-1} \, [\mathbf{y}_k - \mathbf{h}(\mathbf{m}_k^-)], \\ \mathbf{P}_k &= \mathbf{P}_k^- - \mathbf{P}_k^- \, \mathbf{H}_\mathbf{x}^\mathsf{T} \, (\mathbf{H}_\mathbf{x} \, \mathbf{P}_k^- \, \mathbf{H}_\mathbf{x}^\mathsf{T} + \mathbf{R}_k)^{-1} \, \mathbf{H}_\mathbf{x} \, \mathbf{P}_k^-. \end{aligned} \tag{7.38}$$

□

A more general state space model, with non-additive noise, can be written as
$$\begin{aligned} \mathbf{x}_k &= \mathbf{f}(\mathbf{x}_{k-1}, \mathbf{q}_{k-1}), \\ \mathbf{y}_k &= \mathbf{h}(\mathbf{x}_k, \mathbf{r}_k), \end{aligned} \tag{7.39}$$
where $\mathbf{q}_{k-1} \sim \mathrm{N}(\mathbf{0}, \mathbf{Q}_{k-1})$ and $\mathbf{r}_k \sim \mathrm{N}(\mathbf{0}, \mathbf{R}_k)$ are the Gaussian process and measurement noises, respectively. Again, the functions $\mathbf{f}$ and $\mathbf{h}$ can also depend on the step number k. The EKF algorithm for the above model is the following.

Algorithm 7.5 (Extended Kalman filter II). *The prediction and update steps of the (first order) extended Kalman filter (EKF) in the non-additive noise case are:*

- *Prediction:*
$$\begin{aligned} \mathbf{m}_k^- &= \mathbf{f}(\mathbf{m}_{k-1}, \mathbf{0}), \\ \mathbf{P}_k^- &= \mathbf{F}_\mathbf{x}(\mathbf{m}_{k-1}) \, \mathbf{P}_{k-1} \, \mathbf{F}_\mathbf{x}^\mathsf{T}(\mathbf{m}_{k-1}) + \mathbf{F}_\mathbf{q}(\mathbf{m}_{k-1}) \, \mathbf{Q}_{k-1} \, \mathbf{F}_\mathbf{q}^\mathsf{T}(\mathbf{m}_{k-1}), \end{aligned} \tag{7.40}$$

- *Update:*

$$
\begin{aligned}
\mathbf{v}_k &= \mathbf{y}_k - \mathbf{h}(\mathbf{m}_k^-, \mathbf{0}), \\
\mathbf{S}_k &= \mathbf{H}_\mathbf{x}(\mathbf{m}_k^-)\,\mathbf{P}_k^-\,\mathbf{H}_\mathbf{x}^\mathsf{T}(\mathbf{m}_k^-) + \mathbf{H}_\mathbf{r}(\mathbf{m}_k^-)\,\mathbf{R}_k\,\mathbf{H}_\mathbf{r}^\mathsf{T}(\mathbf{m}_k^-), \\
\mathbf{K}_k &= \mathbf{P}_k^-\,\mathbf{H}_\mathbf{x}^\mathsf{T}(\mathbf{m}_k^-)\,\mathbf{S}_k^{-1}, \\
\mathbf{m}_k &= \mathbf{m}_k^- + \mathbf{K}_k\,\mathbf{v}_k, \\
\mathbf{P}_k &= \mathbf{P}_k^- - \mathbf{K}_k\,\mathbf{S}_k\,\mathbf{K}_k^\mathsf{T},
\end{aligned}
\tag{7.41}
$$

where the matrices $\mathbf{F}_\mathbf{x}(\mathbf{m})$, $\mathbf{F}_\mathbf{q}(\mathbf{m})$, $\mathbf{H}_\mathbf{x}(\mathbf{m})$, *and* $\mathbf{H}_\mathbf{r}(\mathbf{m})$ *are the Jacobian matrices of* $\mathbf{f}$ *and* $\mathbf{h}$ *with respect to state and noise, with elements*

$$
[\mathbf{F}_\mathbf{x}(\mathbf{m})]_{j,j'} = \left.\frac{\partial f_j(\mathbf{x},\mathbf{q})}{\partial x_{j'}}\right|_{\mathbf{x}=\mathbf{m},\,\mathbf{q}=\mathbf{0}}, \tag{7.42}
$$

$$
\left[\mathbf{F}_\mathbf{q}(\mathbf{m})\right]_{j,j'} = \left.\frac{\partial f_j(\mathbf{x},\mathbf{q})}{\partial q_{j'}}\right|_{\mathbf{x}=\mathbf{m},\,\mathbf{q}=\mathbf{0}}, \tag{7.43}
$$

$$
[\mathbf{H}_\mathbf{x}(\mathbf{m})]_{j,j'} = \left.\frac{\partial h_j(\mathbf{x},\mathbf{r})}{\partial x_{j'}}\right|_{\mathbf{x}=\mathbf{m},\,\mathbf{r}=\mathbf{0}}, \tag{7.44}
$$

$$
[\mathbf{H}_\mathbf{r}(\mathbf{m})]_{j,j'} = \left.\frac{\partial h_j(\mathbf{x},\mathbf{r})}{\partial r_{j'}}\right|_{\mathbf{x}=\mathbf{m},\,\mathbf{r}=\mathbf{0}}. \tag{7.45}
$$

Derivation These filtering equations can be derived by repeating the same steps as in the derivation of the extended Kalman filter above, but instead of using Algorithm 7.1, we use Algorithm 7.2 for computing the approximations. □

The advantage of the EKF over other non-linear filtering methods is its relative simplicity compared to its performance. Linearization is a very common engineering way of constructing approximations to non-linear systems, and thus it is very easy to understand and apply. A disadvantage is that because it is based on a local linear approximation, it will not work in problems with considerable non-linearities. The filtering model is also restricted in the sense that only Gaussian noise processes are allowed, and thus the model cannot contain, for example, discrete-valued random variables. The Gaussian restriction also prevents the handling of hierarchical models or other models where significantly non-Gaussian distribution models would be needed.

The EKF also requires the measurement model and the dynamic model functions to be differentiable, which is a restriction. In some cases

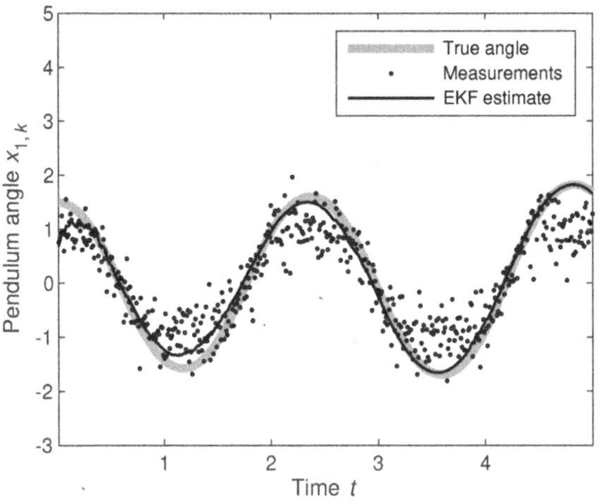

Figure 7.1 Simulated pendulum data and the result of tracking the pendulum angle and angular rate with the EKF in Example 7.6. The resulting RMSE was 0.17.

it might also be simply impossible to compute the required Jacobian matrices, which renders the use of the EKF impossible. Even when the Jacobian matrices exist and could be computed, the actual computation and programming of Jacobian matrices can be quite error prone and hard to debug. However, the manual computation of Jacobians can sometimes be avoided by using automatic differentiation (Griewank and Walther, 2008), which we discuss more in Section 7.6.

Example 7.6 (Pendulum tracking with EKF). *The Euler–Maruyama discretization of the pendulum model given in Example 4.9 in combination with the measurement model given in Example 5.2 leads to the following model:*

$$\begin{aligned}\underbrace{\begin{pmatrix} x_{1,k} \\ x_{2,k} \end{pmatrix}}_{\mathbf{x}_k} &= \underbrace{\begin{pmatrix} x_{1,k-1} + x_{2,k-1}\,\Delta t \\ x_{2,k-1} - g\,\sin(x_{1,k-1})\,\Delta t \end{pmatrix}}_{\mathbf{f}(\mathbf{x}_{k-1})} + \mathbf{q}_{k-1}, \\ y_k &= \underbrace{\sin(x_{1,k})}_{h(\mathbf{x}_k)} + r_k,\end{aligned} \tag{7.46}$$

where $r_k \sim \mathrm{N}(0, R)$. *In order to avoid the singularity of the noise resulting from the Euler–Maruyama discretization, we use the additive noise*

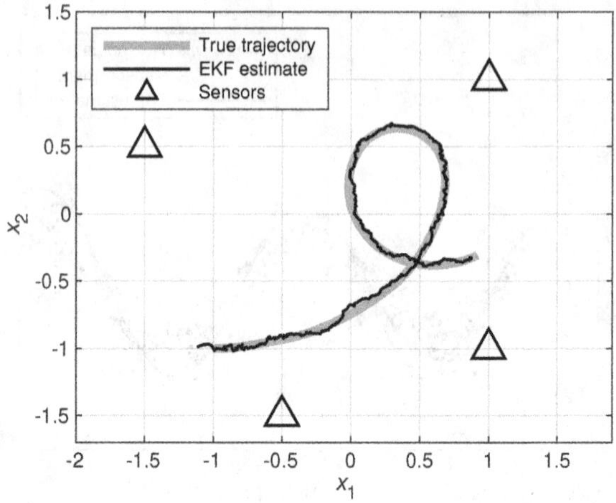

Figure 7.2 The result of applying EKF to simulated data from the (polar) coordinated turn model in Example 7.7. The positions of the radar sensors that measure the range to the target are depicted by triangles. The resulting position RMSE was 0.031.

approximation given in Example 4.21, which puts $\mathbf{q}_{k-1} \sim \mathrm{N}(\mathbf{0}, \mathbf{Q})$ *with*

$$\mathbf{Q} = \begin{pmatrix} \frac{q^c\, \Delta t^3}{3} & \frac{q^c\, \Delta t^2}{2} \\ \frac{q^c\, \Delta t^2}{2} & q^c\, \Delta t \end{pmatrix}, \tag{7.47}$$

where q^c *is the spectral density of the continuous-time process noise. The required Jacobian matrices of* $\mathbf{f}(\mathbf{x})$ *and* $h(\mathbf{x})$ *for the first order EKF are:*

$$\mathbf{F}_{\mathbf{x}}(\mathbf{x}) = \begin{pmatrix} 1 & \Delta t \\ -g\, \cos(x_1)\, \Delta t & 1 \end{pmatrix}, \qquad \mathbf{H}_{\mathbf{x}}(\mathbf{x}) = \begin{pmatrix} \cos(x_1) & 0 \end{pmatrix}. \tag{7.48}$$

An example result of applying the EKF to simulated data from the pendulum model is shown in Figure 7.1. The prior mean and covariance were selected to be $\mathbf{m}_0 = (0, 0)$ *and* $\mathbf{P}_0 = \mathbf{I}$*. After an initial transient, the EKF is indeed able to track the pendulum angle quite well despite the nonlinearity of the model. The RMSE in the angle is* 0.17*, which is much lower than the standard deviation of the measurement noise, which was* 0.32*.*

Example 7.7 (Coordinated turn model with EKF)**.** *As a more complicated example, we consider a polar coordinated turn model, which has a dynamic model of the form*

$$\mathbf{x}_k = \mathbf{f}(\mathbf{x}_{k-1}) + \mathbf{q}_{k-1}, \tag{7.49}$$

with state consisting of 2D-position, speed, direction, and angular velocity: $\mathbf{x}_k = (x_{1,k}, x_{2,k}, s_k, \varphi_k, \omega_k)$. *As the dynamic model function we use the analytical mean discretization given in Example 4.15:*

$$\mathbf{f}(\mathbf{x}_{k-1}) = \begin{pmatrix} x_{1,k-1} + \frac{2 s_{k-1}}{\omega_{k-1}} \sin(\frac{\Delta t\, \omega_{k-1}}{2}) \cos(\varphi_{k-1} + \frac{\Delta t\, \omega_{k-1}}{2}) \\ x_{2,k-1} + \frac{2 s_{k-1}}{\omega_{k-1}} \sin(\frac{\Delta t\, \omega_{k-1}}{2}) \sin(\varphi_{k-1} + \frac{\Delta t\, \omega_{k-1}}{2}) \\ s_{k-1} \\ \varphi_{k-1} + \omega_{k-1}\, \Delta t \\ \omega_{k-1} \end{pmatrix}. \tag{7.50}$$

As the process noise covariance $\mathbf{Q}_{k-1}$ *for* $\mathbf{q}_{k-1} \sim \mathrm{N}(\mathbf{0}, \mathbf{Q}_{k-1})$ *we use the additive noise approximation provided in Example 4.22. The measurements are range measurements from four radars positioned in locations* $(s_{1,x}, s_{1,y}), \ldots, (s_{4,x}, s_{4,y})$ *as described by the model given in Example 5.4:*

$$\mathbf{y}_k = \underbrace{\begin{pmatrix} \sqrt{(x_{1,k} - s_{1,x})^2 + (x_{2,k} - s_{1,y})^2} \\ \sqrt{(x_{1,k} - s_{2,x})^2 + (x_{2,k} - s_{2,y})^2} \\ \sqrt{(x_{1,k} - s_{3,x})^2 + (x_{2,k} - s_{3,y})^2} \\ \sqrt{(x_{1,k} - s_{4,x})^2 + (x_{2,k} - s_{4,y})^2} \end{pmatrix}}_{\mathbf{h}(\mathbf{x}_k)} + \underbrace{\begin{pmatrix} r_{1,k} \\ r_{2,k} \\ r_{3,k} \\ r_{4,k} \end{pmatrix}}_{\mathbf{r}_k}. \tag{7.51}$$

In this case the required Jacobians of $\mathbf{f}(\cdot)$ *and* $\mathbf{h}(\cdot)$ *are more complicated than in the pendulum example above, but they can still be easily derived.*

Figure 7.2 shows the result of applying EKF to simulated measurements from the model. Note that instead of using the approximate additive noise model in the simulation, the trajectory was simulated using 100 steps of Euler–Maruyama between each measurement. The EKF can be seen to recover the trajectory quite well, resulting in a position RMSE of 0.031.

7.3 Higher Order Extended Kalman Filters

In the so-called second order EKF (see, e.g., Gelb, 1974), the non-linearity is approximated by retaining the second order terms in the Taylor series expansion as in Algorithm 7.3. The resulting algorithm is the following.

Algorithm 7.8 (Second order extended Kalman filter)**.** *The prediction and update steps of the second order extended Kalman filter (in the additive noise case) are:*

• *Prediction:*

$$\begin{aligned}
\mathbf{m}_k^- &= \mathbf{f}(\mathbf{m}_{k-1}) + \frac{1}{2}\sum_i \mathbf{e}_i \operatorname{tr}\left\{\mathbf{F}_{\mathbf{xx}}^{(i)}(\mathbf{m}_{k-1})\,\mathbf{P}_{k-1}\right\}, \\
\mathbf{P}_k^- &= \mathbf{F}_{\mathbf{x}}(\mathbf{m}_{k-1})\,\mathbf{P}_{k-1}\,\mathbf{F}_{\mathbf{x}}^\mathsf{T}(\mathbf{m}_{k-1}) \\
&\quad + \frac{1}{2}\sum_{i,i'} \mathbf{e}_i\,\mathbf{e}_{i'}^\mathsf{T} \operatorname{tr}\left\{\mathbf{F}_{\mathbf{xx}}^{(i)}(\mathbf{m}_{k-1})\,\mathbf{P}_{k-1}\,\mathbf{F}_{\mathbf{xx}}^{(i')}(\mathbf{m}_{k-1})\,\mathbf{P}_{k-1}\right\} + \mathbf{Q}_{k-1},
\end{aligned} \tag{7.52}$$

• *Update:*

$$\begin{aligned}
\mathbf{v}_k &= \mathbf{y}_k - \mathbf{h}(\mathbf{m}_k^-) - \frac{1}{2}\sum_i \mathbf{e}_i \operatorname{tr}\left\{\mathbf{H}_{\mathbf{xx}}^{(i)}(\mathbf{m}_k^-)\,\mathbf{P}_k^-\right\}, \\
\mathbf{S}_k &= \mathbf{H}_{\mathbf{x}}(\mathbf{m}_k^-)\,\mathbf{P}_k^-\,\mathbf{H}_{\mathbf{x}}^\mathsf{T}(\mathbf{m}_k^-) \\
&\quad + \frac{1}{2}\sum_{i,i'} \mathbf{e}_i\,\mathbf{e}_{i'}^\mathsf{T} \operatorname{tr}\left\{\mathbf{H}_{\mathbf{xx}}^{(i)}(\mathbf{m}_k^-)\,\mathbf{P}_k^-\,\mathbf{H}_{\mathbf{xx}}^{(i')}(\mathbf{m}_k^-)\,\mathbf{P}_k^-\right\} + \mathbf{R}_k, \qquad (7.53) \\
\mathbf{K}_k &= \mathbf{P}_k^-\,\mathbf{H}_{\mathbf{x}}^\mathsf{T}(\mathbf{m}_k^-)\,\mathbf{S}_k^{-1}, \\
\mathbf{m}_k &= \mathbf{m}_k^- + \mathbf{K}_k\,\mathbf{v}_k, \\
\mathbf{P}_k &= \mathbf{P}_k^- - \mathbf{K}_k\,\mathbf{S}_k\,\mathbf{K}_k^\mathsf{T},
\end{aligned}$$

where the matrices $\mathbf{F}_{\mathbf{x}}(\mathbf{m})$ *and* $\mathbf{H}_{\mathbf{x}}(\mathbf{m})$ *are given by Equations* (7.28) *and* (7.29). *The matrices* $\mathbf{F}_{\mathbf{xx}}^{(i)}(\mathbf{m})$ *and* $\mathbf{H}_{\mathbf{xx}}^{(i)}(\mathbf{m})$ *are the Hessian matrices of* f_i *and* h_i *respectively:*

$$\left[\mathbf{F}_{\mathbf{xx}}^{(i)}(\mathbf{m})\right]_{j,j'} = \left.\frac{\partial^2 f_i(\mathbf{x})}{\partial x_j\,\partial x_{j'}}\right|_{\mathbf{x}=\mathbf{m}}, \tag{7.54}$$

$$\left[\mathbf{H}_{\mathbf{xx}}^{(i)}(\mathbf{m})\right]_{j,j'} = \left.\frac{\partial^2 h_i(\mathbf{x})}{\partial x_j\,\partial x_{j'}}\right|_{\mathbf{x}=\mathbf{m}}. \tag{7.55}$$

The non-additive version can be derived in an analogous manner, but due to its complicated appearance, it is not presented here.

It would also be possible to attempt to use even higher order Taylor series expansions for the dynamic and measurement models in order to get higher order EKFs. However, the computational complexity of these methods increases rapidly with the Taylor series order, and the inclusion of higher order terms can lead to numerically unstable filters. Therefore a better idea is to use iteration as described in the next section.

7.4 Iterated Extended Kalman Filter

The iterated extended Kalman filter (IEKF, Gelb, 1974) is an iterated version of the EKF where the idea is that after performing the EKF update using a linearization at the predicted mean, we have improved our knowledge about the state $\mathbf{x}_k$ and can linearize again at the updated mean. We can then perform another update using the resulting (presumably better) linearization to obtain a new approximation to the posterior mean, using which we can perform another linearization. This iterative procedure is the update step in the IEKF algorihm.

It turns out that the above iterative procedure can be interpreted as a Gauss–Newton method for finding the maximum a posteriori (MAP) estimate of the state at each update step (Bell and Cathey, 1993), which also guarantees the convergence of this procedure in well-defined conditions. Even though the above description may be enough to formulate the algorithm, we now derive the IEKF measurement update for additive noise models from the perspective of a MAP estimator, to provide additional insights.

The prediction step of an IEKF is identical to an EKF and yields a Gaussian approximation:

$$p(\mathbf{x}_k \mid \mathbf{y}_{1:k-1}) \simeq \mathrm{N}(\mathbf{x}_k \mid \mathbf{m}_k^-, \mathbf{P}_k^-). \tag{7.56}$$

The likelihood corresponding to the additive noise measurement model in Equation (7.24) is

$$p(\mathbf{y}_k \mid \mathbf{x}_k) = \mathrm{N}(\mathbf{y}_k \mid \mathbf{h}(\mathbf{x}_k), \mathbf{R}_k). \tag{7.57}$$

By using Bayes' rule, we can then write the posterior distribution of $\mathbf{x}_k$ as

$$\begin{aligned} p(\mathbf{x}_k \mid \mathbf{y}_{1:k}) &= \frac{p(\mathbf{y}_k \mid \mathbf{x}_k)\, p(\mathbf{x}_k \mid \mathbf{y}_{1:k-1})}{\int p(\mathbf{y}_k \mid \mathbf{x}_k)\, p(\mathbf{x}_k \mid \mathbf{y}_{1:k-1})\, \mathrm{d}\mathbf{x}_k} \\ &\propto p(\mathbf{y}_k \mid \mathbf{x}_k)\, p(\mathbf{x}_k \mid \mathbf{y}_{1:k-1}). \end{aligned} \tag{7.58}$$

The maximum a posteriori (MAP) estimate of the state $\mathbf{x}_k$ can now be obtained as

$$\begin{aligned} \mathbf{x}_k^{\mathrm{MAP}} &= \arg\max_{\mathbf{x}_k} p(\mathbf{x}_k \mid \mathbf{y}_{1:k}) \\ &= \arg\max_{\mathbf{x}_k} p(\mathbf{y}_k \mid \mathbf{x}_k)\, p(\mathbf{x}_k \mid \mathbf{y}_{1:k-1}). \end{aligned} \tag{7.59}$$

However, it is more convenient to work with the negative logarithms of distributions instead and define

$$L(\mathbf{x}_k) = -\log\left[p(\mathbf{y}_k \mid \mathbf{x}_k)\, p(\mathbf{x}_k \mid \mathbf{y}_{1:k-1})\right], \tag{7.60}$$

which, with Equation (7.56) as the prior and Equation (7.57) as the likelihood, gives

$$\begin{aligned} L(\mathbf{x}_k) &\simeq -\log\left[\mathrm{N}(\mathbf{y}_k \mid \mathbf{h}(\mathbf{x}_k), \mathbf{R}_k)\, \mathrm{N}(\mathbf{x}_k \mid \mathbf{m}_k^-, \mathbf{P}_k^-)\right] \\ &= C + \frac{1}{2}(\mathbf{y}_k - \mathbf{h}(\mathbf{x}_k))^\mathsf{T}\, \mathbf{R}_k^{-1}\, (\mathbf{y}_k - \mathbf{h}(\mathbf{x}_k)) \\ &\quad + \frac{1}{2}(\mathbf{x}_k - \mathbf{m}_k^-)^\mathsf{T}\, [\mathbf{P}_k^-]^{-1}\, (\mathbf{x}_k - \mathbf{m}_k^-), \end{aligned} \tag{7.61}$$

where C is a constant independent of $\mathbf{x}_k$. We can thus solve for the MAP estimate $\mathbf{x}_k^{\mathrm{MAP}}$ from the minimization problem

$$\begin{aligned} \mathbf{x}_k^{\mathrm{MAP}} &= \arg\min_{\mathbf{x}_k} L(\mathbf{x}_k) \\ &\simeq \arg\min_{\mathbf{x}_k} \left[C + \frac{1}{2}(\mathbf{y}_k - \mathbf{h}(\mathbf{x}_k))^\mathsf{T}\, \mathbf{R}_k^{-1}\, (\mathbf{y}_k - \mathbf{h}(\mathbf{x}_k)) \right. \\ &\quad \left. + \frac{1}{2}(\mathbf{x}_k - \mathbf{m}_k^-)^\mathsf{T}\, [\mathbf{P}_k^-]^{-1}\, (\mathbf{x}_k - \mathbf{m}_k^-) \right]. \end{aligned} \tag{7.62}$$

We now describe how to solve for the MAP estimate in Equation (7.62) using the Gauss–Newton method. The Gauss–Newton method (see, e.g., Nocedal and Wright, 2006) is an optimization method that assumes that the objective function can be expressed as a sum of quadratic forms,

$$L_{\mathrm{GN}}(\mathbf{x}_k) = \frac{1}{2}\sum_{j=1}^{N} \mathbf{r}_j^\mathsf{T}(\mathbf{x}_k)\, \mathbf{r}_j(\mathbf{x}_k), \tag{7.63}$$

and then iterates the following steps until convergence:

1. Perform a first order Taylor expansion of $\mathbf{r}_1, \ldots, \mathbf{r}_N$ about the current estimate of the optimum $\mathbf{x}_k$.
2. Analytically find the optimum $\mathbf{x}_k$ of the approximate model obtained by replacing $\mathbf{r}_1, \ldots, \mathbf{r}_N$ with their Taylor expansions.

One way to express the approximation to $L(\mathbf{x}_k)$ in Equation (7.61) as in Equation (7.63) is to select $N = 3$ and

$$\begin{aligned} \mathbf{r}_1(\mathbf{x}_k) &= \sqrt{2C}, \\ \mathbf{r}_2(\mathbf{x}_k) &= \mathbf{R}_k^{-1/2}\, (\mathbf{y}_k - \mathbf{h}(\mathbf{x}_k)), \\ \mathbf{r}_3(\mathbf{x}_k) &= [\mathbf{P}_k^-]^{-1/2}\, (\mathbf{x}_k - \mathbf{m}_k^-). \end{aligned} \tag{7.64}$$

We note that a first order Taylor expansion of $\mathbf{r}_1(\mathbf{x}_k)$, $\mathbf{r}_2(\mathbf{x}_k)$, and $\mathbf{r}_3(\mathbf{x}_k)$ would represent $\mathbf{r}_1(\mathbf{x}_k)$ and $\mathbf{r}_3(\mathbf{x}_k)$ exactly and only introduce an approximation to $\mathbf{r}_2(\mathbf{x}_k)$ due to the linearization of the non-linear function $\mathbf{h}(\mathbf{x}_k)$.

The iterations of Gauss–Newton are started from some initial guess $\mathbf{x}_k^{(0)}$. Then let us assume that the current estimate at iteration $i-1$ is $\mathbf{x}_k^{(i-1)}$. We can then linearize $\mathbf{h}(\mathbf{x}_k)$ by forming a first order Taylor series expansion at the current estimate as follows:

$$\mathbf{h}(\mathbf{x}_k) \simeq \mathbf{h}(\mathbf{x}_k^{(i-1)}) + \mathbf{H}_{\mathbf{x}}(\mathbf{x}_k^{(i-1)}) \, (\mathbf{x}_k - \mathbf{x}_k^{(i-1)}). \tag{7.65}$$

Plugging this back into Equation (7.63) gives the approximation

$$\begin{aligned} L_{\mathrm{GN}}(\mathbf{x}_k) \simeq C &+ \frac{1}{2}(\mathbf{y}_k - \mathbf{h}(\mathbf{x}_k^{(i-1)}) - \mathbf{H}_{\mathbf{x}}(\mathbf{x}_k^{(i-1)}) \, (\mathbf{x}_k - \mathbf{x}_k^{(i-1)}))^\mathsf{T} \, \mathbf{R}_k^{-1} \\ &\quad \times (\mathbf{y}_k - \mathbf{h}(\mathbf{x}_k^{(i-1)}) - \mathbf{H}_{\mathbf{x}}(\mathbf{x}_k^{(i-1)}) \, (\mathbf{x}_k - \mathbf{x}_k^{(i-1)})) \\ &+ \frac{1}{2}(\mathbf{x}_k - \mathbf{m}_k^-)^\mathsf{T} \, [\mathbf{P}_k^-]^{-1} \, (\mathbf{x}_k - \mathbf{m}_k^-). \end{aligned} \tag{7.66}$$

Because the above expression is quadratic in $\mathbf{x}_k$, we can compute its minimum by setting its gradient to zero. The Gauss–Newton method then proceeds to use this minimum as the next iterate $\mathbf{x}_k^{(i)}$. However, instead of explicitly setting the gradient to zero, let us derive the solution in a slightly different way to clarify the connection to our filters.

The approximation on the right-hand side of Equation (7.66) can be seen as the negative log-posterior for a model with the measurement model

$$\tilde{p}(\mathbf{y}_k \mid \mathbf{x}_k) = \mathrm{N}(\mathbf{y}_k \mid \mathbf{h}(\mathbf{x}_k^{(i-1)}) + \mathbf{H}_{\mathbf{x}}(\mathbf{x}_k^{(i-1)}) \, (\mathbf{x}_k - \mathbf{x}_k^{(i-1)}), \mathbf{R}_k), \tag{7.67}$$

which has the form of an affine model with Gaussian noise. As both the prior and likelihood are linear and Gaussian, the minimum of Equation (7.66) will exactly correspond to the posterior mean of $\mathbf{x}_k$ with this measurement model and the prior given in Equation (7.56). We can obtain it by using, for example, the affine Kalman filter update in Theorem 6.9 with the measurement model matrix $\mathbf{H}_k = \mathbf{H}_{\mathbf{x}}(\mathbf{x}_k^{(i-1)})$ and offset $\mathbf{b}_k = \mathbf{h}(\mathbf{x}_k^{(i-1)}) - \mathbf{H}_{\mathbf{x}}(\mathbf{x}_k^{(i-1)}) \, \mathbf{x}_k^{(i-1)}$, which results in

$$\begin{aligned} \mathbf{K}_k^{(i)} &= \mathbf{P}_k^- \, \mathbf{H}_{\mathbf{x}}^\mathsf{T}(\mathbf{x}_k^{(i-1)}) \left[\mathbf{H}_{\mathbf{x}}(\mathbf{x}_k^{(i-1)}) \, \mathbf{P}_k^- \, \mathbf{H}_{\mathbf{x}}^\mathsf{T}(\mathbf{x}_k^{(i-1)}) + \mathbf{R}_k \right]^{-1}, \\ \mathbf{x}_k^{(i)} &= \mathbf{m}_k^- + \mathbf{K}_k^{(i)} \left[\mathbf{y}_k - \mathbf{h}(\mathbf{x}_k^{(i-1)}) - \mathbf{H}_{\mathbf{x}}(\mathbf{x}_k^{(i-1)}) \, (\mathbf{m}_k^- - \mathbf{x}_k^{(i-1)}) \right]. \end{aligned} \tag{7.68}$$

The above iteration can then be repeated until it reaches a stationary point $\mathbf{x}_k^{(*)} \simeq \mathbf{x}_k^{\mathrm{MAP}}$, which we can use as the updated mean estimate $\mathbf{m}_k \simeq$

$\mathbf{x}_k^{(*)}$. The updated covariance can be approximated by the covariance corresponding to the linearized model in Equation (7.67) when evaluated at the mean $\mathbf{m}_k$. The resulting algorithm is the following.

Algorithm 7.9 (Iterated Extended Kalman Filter). *The prediction and update steps of the iterated extended Kalman filter (IEKF) are:*

- *Prediction:*

$$\begin{aligned}\mathbf{m}_k^- &= \mathbf{f}(\mathbf{m}_{k-1}),\\ \mathbf{P}_k^- &= \mathbf{F}_\mathbf{x}(\mathbf{m}_{k-1})\,\mathbf{P}_{k-1}\,\mathbf{F}_\mathbf{x}^\mathsf{T}(\mathbf{m}_{k-1}) + \mathbf{Q}_{k-1}.\end{aligned} \tag{7.69}$$

- *Update:*
  - *Start from an initial guess, for example,* $\mathbf{x}_k^{(0)} = \mathbf{m}_k^-$.
  - *Then for* $i = 1, 2, \ldots, I_{\max}$, *compute:*

$$\begin{aligned}\mathbf{v}_k^{(i)} &= \mathbf{y}_k - \mathbf{h}(\mathbf{x}_k^{(i-1)}) - \mathbf{H}_\mathbf{x}(\mathbf{x}_k^{(i-1)})\,(\mathbf{m}_k^- - \mathbf{x}_k^{(i-1)}),\\ \mathbf{S}_k^{(i)} &= \mathbf{H}_\mathbf{x}(\mathbf{x}_k^{(i-1)})\,\mathbf{P}_k^-\,\mathbf{H}_\mathbf{x}^\mathsf{T}(\mathbf{x}_k^{(i-1)}) + \mathbf{R}_k,\\ \mathbf{K}_k^{(i)} &= \mathbf{P}_k^-\,\mathbf{H}_\mathbf{x}^\mathsf{T}(\mathbf{x}_k^{(i-1)})\left[\mathbf{S}_k^{(i)}\right]^{-1},\\ \mathbf{x}_k^{(i)} &= \mathbf{m}_k^- + \mathbf{K}_k^{(i)}\,\mathbf{v}_k^{(i)},\end{aligned} \tag{7.70}$$

    and put $\mathbf{m}_k = \mathbf{x}_k^{(I_{\max})}$.
  - *Compute the covariance estimate as follows:*

$$\mathbf{P}_k = \mathbf{P}_k^- - \mathbf{K}_k^{(I_{\max})}\,\mathbf{S}_k^{(I_{\max})}\left[\mathbf{K}_k^{(I_{\max})}\right]^\mathsf{T}. \tag{7.71}$$

Above, $\mathbf{F}_\mathbf{x}(\mathbf{m})$ *and* $\mathbf{H}_\mathbf{x}(\mathbf{m})$ *are the Jacobian matrices of functions* $\mathbf{f}(\cdot)$ *and* $\mathbf{h}(\cdot)$ *defined in Equations* (7.28) *and* (7.29), *respectively.*

It is worth noting that the iteration in the IEKF is only done to perform the update step, and it is a very local improvement in the sense that it ignores estimates at other time steps. The iterations can still help, in particular in situations where the measurement likelihood is informative compared to the predicted density (Morelande and García-Fernández, 2013). Later, in Chapters 13 and 14, we discuss iterated extended smoothing algorithms that perform global iteration by iteratively improving the whole trajectory estimate instead of just improving the filter update step.

Example 7.10 (Pendulum tracking with IEKF). *The result of applying IEKF to the pendulum model introduced in Example 7.6 is shown in Figure 7.3. The performance of IEKF in this model is very similar to EKF, and the RMSE is* 0.17, *which is the same as for EKF.*

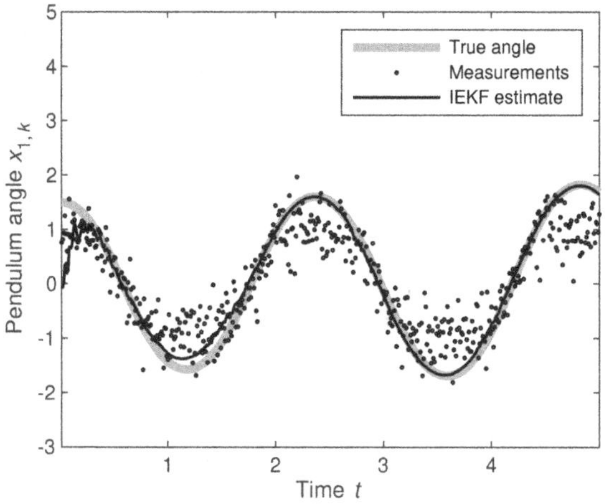

Figure 7.3 Simulated pendulum data and the result of tracking the pendulum angle and angular rate with the IEKF in Example 7.10. The resulting RMSE was 0.17.

Example 7.11 (Coordinated turn model with IEKF). *If we apply the IEKF to the coordinated turn simulation that we considered in Example 7.7, we get basically the same result as for EKF. This is because in that simulation, the dynamic model is quite accurate, and the initial guess is also relatively accurate. However, if we modify the simulation such that the initial guess is further away from the actual position (and increase the prior covariance appropriately), then the EKF and IEKF results do differ. Figure 7.4 illustrates the difference, which is that EKF takes much longer to converge to the true trajectory than IEKF in the beginning of tracking. After the initial transient, both of the methods produce practically the same result.*

7.5 Levenberg–Marquardt, Line-Search, and Related IEKFs

As discussed in Section 7.4 above, the iterated extended Kalman filter (IEKF) can be seen as a Gauss–Newton method for optimization of cost functions of the form

$$\begin{aligned} L(\mathbf{x}_k) \simeq C &+ \frac{1}{2}(\mathbf{y}_k - \mathbf{h}(\mathbf{x}_k))^\mathsf{T} \, \mathbf{R}_k^{-1} \, (\mathbf{y}_k - \mathbf{h}(\mathbf{x}_k)) \\ &+ \frac{1}{2}(\mathbf{x}_k - \mathbf{m}_k^-)^\mathsf{T} \, [\mathbf{P}_k^-]^{-1} \, (\mathbf{x}_k - \mathbf{m}_k^-) \end{aligned} \tag{7.72}$$

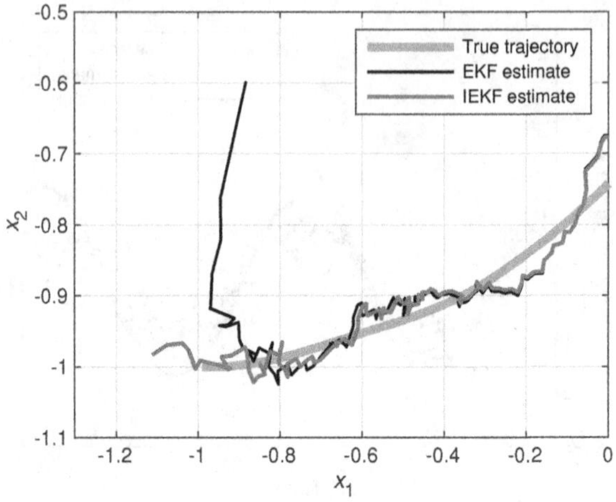

Figure 7.4 The result of applying EKF and IEKF to simulated data from the (polar) coordinated turn model in Example 7.11 when using an inaccurate initial guess. The figure only shows the initial part of the trajectory as the result of the remaining trajectory is identical for EKF and IEKF. The resulting RMSE for EKF was 0.038, while for IEKF it was 0.031, and it can be seen that EKF takes much longer to converge to the trajectory than IEKF.

with respect to $\mathbf{x}_k$. However, there are many other optimization methods that often have better performance than the plain Gauss-Newton method (Nocedal and Wright, 2006). Different kinds of extensions have been discussed in Fatemi et al. (2012) and Skoglund et al. (2015), including:

- The *Levenberg–Marquardt method* corresponds to adding a regularization term to the minimization of the quadratic approximation in Equation (7.66):

$$
\begin{aligned}
L_{\mathrm{GN}}(\mathbf{x}_k) \simeq C &+ \frac{1}{2}(\mathbf{y}_k - \mathbf{h}(\mathbf{x}_k^{(i-1)}) - \mathbf{H}_{\mathbf{x}}(\mathbf{x}_k^{(i-1)})\,(\mathbf{x}_k - \mathbf{x}_k^{(i-1)}))^{\mathsf{T}}\,\mathbf{R}_k^{-1} \\
&\qquad \times (\mathbf{y}_k - \mathbf{h}(\mathbf{x}_k^{(i-1)}) - \mathbf{H}_{\mathbf{x}}(\mathbf{x}_k^{(i-1)})\,(\mathbf{x}_k - \mathbf{x}_k^{(i-1)})) \\
&+ \frac{1}{2}(\mathbf{x}_k - \mathbf{m}_k^-)^{\mathsf{T}}\,[\mathbf{P}_k^-]^{-1}\,(\mathbf{x}_k - \mathbf{m}_k^-) \\
&+ \frac{1}{2}\lambda\,(\mathbf{x}_k - \mathbf{x}_k^{(i-1)})^{\mathsf{T}}\,(\mathbf{x}_k - \mathbf{x}_k^{(i-1)}),
\end{aligned}
\tag{7.73}
$$

which can be implemented within the IEKF update iteration, for example, as an extra measurement at $\mathbf{x}_k^{(i-1)}$ with noise covariance $\mathbf{I}/\lambda$ (cf. Särkkä and Svensson, 2020).

- A *line search* can be implemented by modifying the computation of $\mathbf{x}_k^{(i)}$ in Equation (7.70) to

$$\mathbf{x}_k^{(i)} = \mathbf{x}_k^{(i-1)} + \gamma \, (\mathbf{x}_k^{(i*)} - \mathbf{x}_k^{(i-1)}), \tag{7.74}$$

where $\mathbf{x}_k^{(i*)} = \mathbf{m}_k^- + \mathbf{K}_k^{(i)} \, \mathbf{v}_k^{(i)}$, and the step size parameter γ is selected to approximately minimize $L(\mathbf{x}_k^{(i-1)} + \gamma \, (\mathbf{x}_k^{(i*)} - \mathbf{x}_k^{(i-1)}))$.

- *Quasi-Newton* methods (Skoglund et al., 2015) can be implemented by including an additional correction matrix to the minimization problem, which leads to a similar extension as Levenberg–Marquardt.

7.6 Automatic Differentiation and EKFs

The implementation of EKFs presumes the availability of the Jacobian matrices $\mathbf{F}_{\mathbf{x}}(\cdot)$ and $\mathbf{H}_{\mathbf{x}}(\cdot)$ of the model functions $\mathbf{f}(\cdot)$ and $\mathbf{h}(\cdot)$ as defined in Equations (7.28) and (7.29). To implement the second order EKFs, we further need the Hessians of the model functions defined in Equations (7.54) and (7.55). Although for simple model functions the manual derivation of the Jacobians and Hessian is easy, it can be tedious when the models are, for example, derived as a numerical solutions to partial differential equations or when they are otherwise complicated to handle.

To help with the aforementioned issue, many programming languages and environments, such as MATLAB, Python, and Julia, have support for automatic differentiation (AD, Griewank and Walther, 2008). The idea of AD is to automatically transform the function to be evaluated into a sequence of operations that automatically computes its (exact) derivative along with its value. This is different from numerical differentiation where the derivatives are approximated, for example, by using finite differences, because the derivatives computed with AD are exact.

An obvious use for automatic differentiation (AD) in EKFs is the computation of the Jacobians and Hessians of $\mathbf{f}(\cdot)$ and $\mathbf{h}(\cdot)$. This avoids errors in manual derivation of these quantities, and AD-computed derivatives can also be used to numerically validate the manually computed derivatives. However, a disadvantage of AD is that evaluation of derivatives using it can be computationally slower than using manually implemented derivatives. In the context of parameter estimation (see Chapter 16), AD can be used to compute derivatives of the marginal log-likelihood in order to avoid manual implementation of the parameter gradient recursions.

7.7 Exercises

7.1 Consider the following non-linear state space model:

$$\begin{aligned} x_k &= x_{k-1} - 0.01 \sin(x_{k-1}) + q_{k-1}, \\ y_k &= 0.5 \sin(2\, x_k) + r_k, \end{aligned} \tag{7.75}$$

where q_{k-1} has a variance of 0.01^2 and r_k has a variance of 0.02. Derive the required derivatives for an EKF and implement the EKF for the model. Simulate trajectories from the model, compute the RMSE values, and plot the result.

7.2 In this exercise we consider a classical bearings-only target tracking problem that frequently arises in the context of passive sensor tracking. In this problem there is single target in the scene, and two angular sensors are used for tracking it. The scenario is illustrated in Figure 7.5.

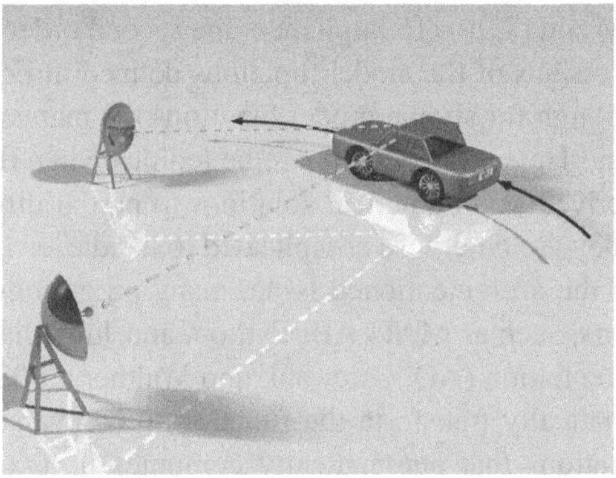

Figure 7.5 In the bearings-only target tracking problem, the sensors generate angle measurements of the target, and the purpose is to determine the target trajectory.

The state of the target at time step k consists of the position (x_k, y_k) and the velocity $(\dot{x}_k, \dot{y}_k)$. The dynamics of the state vector $\mathbf{x}_k = (x_k\ y_k\ \dot{x}_k\ \dot{y}_k)^\mathsf{T}$ are modeled with the discretized Wiener velocity model:

$$\begin{pmatrix} x_k \\ y_k \\ \dot{x}_k \\ \dot{y}_k \end{pmatrix} = \begin{pmatrix} 1 & 0 & \Delta t & 0 \\ 0 & 1 & 0 & \Delta t \\ 0 & 0 & 1 & 0 \\ 0 & 0 & 0 & 1 \end{pmatrix} \begin{pmatrix} x_{k-1} \\ y_{k-1} \\ \dot{x}_{k-1} \\ \dot{y}_{k-1} \end{pmatrix} + \mathbf{q}_{k-1},$$

where $\mathbf{q}_k$ is a zero mean Gaussian process noise with covariance

$$\mathbf{Q} = \begin{pmatrix} q_1^c\,\Delta t^3/3 & 0 & q_1^c\,\Delta t^2/2 & 0 \\ 0 & q_2^c\,\Delta t^3/3 & 0 & q_2^c\,\Delta t^2/2 \\ q_1^c\,\Delta t^2/2 & 0 & q_1^c\,\Delta t & 0 \\ 0 & q_2^c\,\Delta t^2/2 & 0 & q_2^c\,\Delta t \end{pmatrix}.$$

In this scenario the diffusion coefficients are $q_1^c = q_2^c = 0.1$, and the sampling period is $\Delta t = 0.1$. The measurement model for sensor $i \in \{1, 2\}$ is:

$$\theta_k^i = \tan^{-1}\left(\frac{y_k - s_y^i}{x_k - s_x^i}\right) + r_k^i, \tag{7.76}$$

where (s_x^i, s_y^i) is the position of the sensor i, and $r_k^i \sim \mathrm{N}(0, \sigma^2)$ is a Gaussian measurement noise with a standard deviation of $\sigma = 0.05$ radians. At each sampling time, which occurs 10 times per second (i.e., $\Delta t = 0.1$), both of the two sensors produce a measurement.

In the file `exercise7_2.m` (MATLAB) or `Exercise7_2.ipynb` (Python) of the book's companion code repository there is a baseline solution, which computes estimates of the position from the crossing of the measurements and estimates the velocity to be always zero. Your task is to implement an EKF for the problem and compare the results graphically and in the RMSE sense.

(a) Implement an EKF for the bearings-only target tracking problem, which uses the non-linear measurement model (7.76) as its measurement model function (not the crossings). *Hints:*

- Use the function `atan2` in the measurement model instead of `atan` to directly get an answer in the range $[-\pi, \pi]$.
- The two measurements at each measurement time can be processed one at a time, that is, you can simply perform two scalar updates instead of a single two-dimensional measurement update.
- Start by computing the Jacobian matrix of the measurement model function with respect to the state components. Before implementing the filter, check by finite differences or automatic differentiation that the Jacobian matrix is correct.

(b) Compute the RMSE values, and plot figures of the estimates.

7.3 Implement the second order EKF for the model in Exercise 7.1.

7.4 Show that the IEKF with a single iteration is equivalent to the first order EKF.

7.5 Fill in the details of deriving the IEKF update by using the affine Kalman filter update.

7.6 Derive the IEKF update by setting gradient to zero in Equation (7.66) and by using matrix inversion formulas in Corollary A.5.

7.7 Implement IEKF for the model in Exercise 7.1.

7.8 Implement EKF for the Cartesian coordinated turn model in Equations (4.61) and (4.90) with a linear position measurement model. Why would IEKF not make sense for this model?

7.9 Implement IEKF for the bearings-only model in Exercise 7.2.

8

General Gaussian Filtering

In Chapter 7 we used Taylor series expansions to approximate a non-linear state space model as a linear (or affine) state space model. The filtering problem on the linearized model was then essentially solved by using a Kalman filter, which led to extended Kalman filters (EKFs). Furthermore, an iterated use of linearization then led to the iterated extended Kalman filter (IEKF), which can be seen as a special case of the Gauss–Newton optimization method.

In this chapter we present a philosophically different way of forming Gaussian approximations to filter solutions that is based on using moment matching. The idea is to postulate (or assume) the distributional form of the filtering density and then match the moments (in this case mean and covariance) of the approximating density to the true filtering density – at least approximately. This leads to so-called Gaussian filters (Ito and Xiong, 2000), where the non-linear filtering problem is solved using Gaussian assumed density approximations. The generalized framework also enables the use of various powerful Gaussian quadrature and cubature integration methods (Wu et al., 2006; Arasaratnam and Haykin, 2009), including the unscented transform which leads to the unscented Kalman filter (UKF, Julier and Uhlmann, 1995; Julier et al., 2000). The EKF-based filters presented in the previous chapter can be seen as approximations to the general Gaussian filter. In this section we present the Gaussian filtering framework and show how the Gauss–Hermite Kalman filter (GHKF) and the cubature Kalman filter (CKF) can be derived as its approximations. We also show how the UKF can be seen as a generalization of the CKF.

8.1 Gaussian Moment Matching

Recall that in Chapter 7 we derived the EKF in terms of a linear approximation to the non-linear transformation of random variables (e.g., Algorithm 7.1), which is concerned with computing a Gaussian approximation to a pair of random variables $(\mathbf{x}, \mathbf{y})$ defined via

$$\begin{aligned} \mathbf{x} &\sim \mathrm{N}(\mathbf{m}, \mathbf{P}), \\ \mathbf{q} &\sim \mathrm{N}(\mathbf{0}, \mathbf{Q}), \\ \mathbf{y} &= \mathbf{g}(\mathbf{x}) + \mathbf{q}. \end{aligned} \tag{8.1}$$

Instead of resorting to linearization, another approach is now to first compute the mean and covariance of $(\mathbf{x}, \mathbf{y})$ and then form the Gaussian approximation as having these mean and covariance, that is, by using moment matching. We cannot, however, compute the mean and covariance in closed form, but we can indeed write them in terms of integrals over Gaussian distributions. For example, the mean and covariance of $\mathbf{y}$ are given as

$$\begin{aligned} \mathrm{E}[\mathbf{y}] &= \int \mathbf{g}(\mathbf{x})\, \mathrm{N}(\mathbf{x} \mid \mathbf{m}, \mathbf{P})\, \mathrm{d}\mathbf{x}, \\ \mathrm{Cov}[\mathbf{y}] &= \int (\mathbf{g}(\mathbf{x}) - \mathrm{E}[\mathbf{y}])\, (\mathbf{g}(\mathbf{x}) - \mathrm{E}[\mathbf{y}])^\mathsf{T}\, \mathrm{N}(\mathbf{x} \mid \mathbf{m}, \mathbf{P})\, \mathrm{d}\mathbf{x} + \mathbf{Q}. \end{aligned} \tag{8.2}$$

One way to unify various Gaussian approximations to non-linear transforms is to reduce them to approximate computations of Gaussian integrals of the general form

$$\int \mathbf{g}(\mathbf{x})\, \mathrm{N}(\mathbf{x} \mid \mathbf{m}, \mathbf{P})\, \mathrm{d}\mathbf{x}$$

for suitably selected $\mathbf{g}(\cdot)$. If we can compute these, a straightforward way to form the Gaussian approximation for $(\mathbf{x}, \mathbf{y})$ is to simply match the moments of the distributions, which gives the following algorithm.

Algorithm 8.1 (Gaussian moment matching of an additive transform)**.** *The moment matching-based Gaussian approximation to the joint distribution of* $\mathbf{x}$ *and the transformed random variable* $\mathbf{y} = \mathbf{g}(\mathbf{x}) + \mathbf{q}$*, where* $\mathbf{x} \sim \mathrm{N}(\mathbf{m}, \mathbf{P})$ *and* $\mathbf{q} \sim \mathrm{N}(\mathbf{0}, \mathbf{Q})$*, is given by*

$$\begin{pmatrix} \mathbf{x} \\ \mathbf{y} \end{pmatrix} \sim \mathrm{N}\left(\begin{pmatrix} \mathbf{m} \\ \boldsymbol{\mu}_\mathrm{M} \end{pmatrix}, \begin{pmatrix} \mathbf{P} & \mathbf{C}_\mathrm{M} \\ \mathbf{C}_\mathrm{M}^\mathsf{T} & \mathbf{S}_\mathrm{M} \end{pmatrix} \right), \tag{8.3}$$

where

$$\begin{aligned} \boldsymbol{\mu}_\mathrm{M} &= \int \mathbf{g}(\mathbf{x})\, \mathrm{N}(\mathbf{x} \mid \mathbf{m}, \mathbf{P})\, \mathrm{d}\mathbf{x}, \\ \mathbf{S}_\mathrm{M} &= \int (\mathbf{g}(\mathbf{x}) - \boldsymbol{\mu}_\mathrm{M})\, (\mathbf{g}(\mathbf{x}) - \boldsymbol{\mu}_\mathrm{M})^\mathsf{T}\, \mathrm{N}(\mathbf{x} \mid \mathbf{m}, \mathbf{P})\, \mathrm{d}\mathbf{x} + \mathbf{Q}, \\ \mathbf{C}_\mathrm{M} &= \int (\mathbf{x} - \mathbf{m})\, (\mathbf{g}(\mathbf{x}) - \boldsymbol{\mu}_\mathrm{M})^\mathsf{T}\, \mathrm{N}(\mathbf{x} \mid \mathbf{m}, \mathbf{P})\, \mathrm{d}\mathbf{x}. \end{aligned} \tag{8.4}$$

It is now easy to check by substituting the linear approximation $\mathbf{g}(\mathbf{x}) \simeq \mathbf{g}(\mathbf{m}) + \mathbf{G}_{\mathbf{x}}(\mathbf{m})\,(\mathbf{x} - \mathbf{m})$ into the above expression that the integrals reduce to the linear approximations in Algorithm 7.1.

The non-additive version of the transform is the following.

Algorithm 8.2 (Gaussian moment matching of a non-additive transform). *The moment matching-based Gaussian approximation to the joint distribution of* $\mathbf{x}$ *and the transformed random variable* $\mathbf{y} = \mathbf{g}(\mathbf{x}, \mathbf{q})$, *where* $\mathbf{x} \sim \mathrm{N}(\mathbf{m}, \mathbf{P})$ *and* $\mathbf{q} \sim \mathrm{N}(\mathbf{0}, \mathbf{Q})$ *is given by*

$$\begin{pmatrix} \mathbf{x} \\ \mathbf{y} \end{pmatrix} \sim \mathrm{N}\left(\begin{pmatrix} \mathbf{m} \\ \boldsymbol{\mu}_{\mathrm{M}} \end{pmatrix}, \begin{pmatrix} \mathbf{P} & \mathbf{C}_{\mathrm{M}} \\ \mathbf{C}_{\mathrm{M}}^{\mathsf{T}} & \mathbf{S}_{\mathrm{M}} \end{pmatrix} \right), \tag{8.5}$$

where

$$\begin{aligned} \boldsymbol{\mu}_{\mathrm{M}} &= \int \mathbf{g}(\mathbf{x}, \mathbf{q})\, \mathrm{N}(\mathbf{x} \mid \mathbf{m}, \mathbf{P})\, \mathrm{N}(\mathbf{q} \mid \mathbf{0}, \mathbf{Q})\, \mathrm{d}\mathbf{x}\, \mathrm{d}\mathbf{q}, \\ \mathbf{S}_{\mathrm{M}} &= \int (\mathbf{g}(\mathbf{x}, \mathbf{q}) - \boldsymbol{\mu}_{\mathrm{M}})\, (\mathbf{g}(\mathbf{x}, \mathbf{q}) - \boldsymbol{\mu}_{\mathrm{M}})^{\mathsf{T}} \\ &\qquad \times \mathrm{N}(\mathbf{x} \mid \mathbf{m}, \mathbf{P})\, \mathrm{N}(\mathbf{q} \mid \mathbf{0}, \mathbf{Q})\, \mathrm{d}\mathbf{x}\, \mathrm{d}\mathbf{q}, \\ \mathbf{C}_{\mathrm{M}} &= \int (\mathbf{x} - \mathbf{m})\, (\mathbf{g}(\mathbf{x}, \mathbf{q}) - \boldsymbol{\mu}_{\mathrm{M}})^{\mathsf{T}}\, \mathrm{N}(\mathbf{x} \mid \mathbf{m}, \mathbf{P})\, \mathrm{N}(\mathbf{q} \mid \mathbf{0}, \mathbf{Q})\, \mathrm{d}\mathbf{x}\, \mathrm{d}\mathbf{q}. \end{aligned} \tag{8.6}$$

8.2 Gaussian Filter

If we use the moment matching approximations for constructing a filtering algorithm, we get the following *Gaussian assumed density filter* (ADF), which is also called the *Gaussian filter* (Maybeck, 1982b; Ito and Xiong, 2000; Wu et al., 2006). The key idea is to *assume* that the filtering distribution is indeed Gaussian,

$$p(\mathbf{x}_k \mid \mathbf{y}_{1:k}) \simeq \mathrm{N}(\mathbf{x}_k \mid \mathbf{m}_k, \mathbf{P}_k), \tag{8.7}$$

and approximate its mean $\mathbf{m}_k$ and covariance $\mathbf{P}_k$ via moment matching. The Gaussian filter can be used for approximating the filtering distributions of models having the same form as with the EKF, that is, models of the form (7.24) or (7.39).

Algorithm 8.3 (Gaussian filter I). *The prediction and update steps of the additive noise Gaussian (Kalman) filter for models of the form* (7.24) *are:*

- *Prediction:*

$$
\begin{aligned}
\mathbf{m}_k^- &= \int \mathbf{f}(\mathbf{x}_{k-1})\, \mathrm{N}(\mathbf{x}_{k-1} \mid \mathbf{m}_{k-1}, \mathbf{P}_{k-1})\, \mathrm{d}\mathbf{x}_{k-1}, \\
\mathbf{P}_k^- &= \int (\mathbf{f}(\mathbf{x}_{k-1}) - \mathbf{m}_k^-)\, (\mathbf{f}(\mathbf{x}_{k-1}) - \mathbf{m}_k^-)^\mathsf{T} \\
&\qquad \times \mathrm{N}(\mathbf{x}_{k-1} \mid \mathbf{m}_{k-1}, \mathbf{P}_{k-1})\, \mathrm{d}\mathbf{x}_{k-1} + \mathbf{Q}_{k-1}.
\end{aligned} \tag{8.8}
$$

- *Update:*

$$
\begin{aligned}
\boldsymbol{\mu}_k &= \int \mathbf{h}(\mathbf{x}_k)\, \mathrm{N}(\mathbf{x}_k \mid \mathbf{m}_k^-, \mathbf{P}_k^-)\, \mathrm{d}\mathbf{x}_k, \\
\mathbf{S}_k &= \int (\mathbf{h}(\mathbf{x}_k) - \boldsymbol{\mu}_k)\, (\mathbf{h}(\mathbf{x}_k) - \boldsymbol{\mu}_k)^\mathsf{T}\, \mathrm{N}(\mathbf{x}_k \mid \mathbf{m}_k^-, \mathbf{P}_k^-)\, \mathrm{d}\mathbf{x}_k + \mathbf{R}_k, \\
\mathbf{C}_k &= \int (\mathbf{x}_k - \mathbf{m}_k^-)\, (\mathbf{h}(\mathbf{x}_k) - \boldsymbol{\mu}_k)^\mathsf{T}\, \mathrm{N}(\mathbf{x}_k \mid \mathbf{m}_k^-, \mathbf{P}_k^-)\, \mathrm{d}\mathbf{x}_k, \\
\mathbf{K}_k &= \mathbf{C}_k\, \mathbf{S}_k^{-1}, \\
\mathbf{m}_k &= \mathbf{m}_k^- + \mathbf{K}_k\, (\mathbf{y}_k - \boldsymbol{\mu}_k), \\
\mathbf{P}_k &= \mathbf{P}_k^- - \mathbf{K}_k\, \mathbf{S}_k\, \mathbf{K}_k^\mathsf{T}.
\end{aligned} \tag{8.9}
$$

Derivation Let us now follow a similar derivation as we did for the EKF in Section 7.2.

1. If we apply the moment matching in Algorithm 8.1 to $\mathbf{x}_k = \mathbf{f}(\mathbf{x}_{k-1}) + \mathbf{q}_{k-1}$ with $\mathbf{x}_{k-1} \sim \mathrm{N}(\mathbf{m}_{k-1}, \mathbf{P}_{k-1})$, we get

$$
p(\mathbf{x}_{k-1}, \mathbf{x}_k \mid \mathbf{y}_{1:k-1}) \simeq \mathrm{N}\left(\begin{pmatrix} \mathbf{x}_{k-1} \\ \mathbf{x}_k \end{pmatrix} \,\middle|\, \begin{pmatrix} \mathbf{m}_{k-1} \\ \mathbf{m}_k^- \end{pmatrix}, \begin{pmatrix} \mathbf{P}_{k-1} & \mathbf{D}_k \\ \mathbf{D}_k^\mathsf{T} & \mathbf{P}_k^- \end{pmatrix} \right), \tag{8.10}
$$

where $\mathbf{m}_k^-$ and $\mathbf{P}_k^-$ are as defined in Equation (8.8), and

$$
\begin{aligned}
\mathbf{D}_k &= \int (\mathbf{x}_{k-1} - \mathbf{m}_{k-1})\, \left(\mathbf{f}(\mathbf{x}_{k-1}) - \mathbf{m}_k^-\right)^\mathsf{T} \\
&\qquad \times \mathrm{N}(\mathbf{x}_{k-1} \mid \mathbf{m}_{k-1}, \mathbf{P}_{k-1})\, \mathrm{d}\mathbf{x}_{k-1}.
\end{aligned} \tag{8.11}
$$

By Lemma A.3, the marginal distribution for $\mathbf{x}_k$ then has mean $\mathbf{m}_k^-$ and covariance $\mathbf{P}_k^-$, which gives the prediction equations.
2. Similarly we can apply the moment matching to $\mathbf{y}_k = \mathbf{h}(\mathbf{x}_k) + \mathbf{r}_k$, giving

$$
p(\mathbf{x}_k, \mathbf{y}_k \mid \mathbf{y}_{1:k-1}) \simeq \mathrm{N}\left(\begin{pmatrix} \mathbf{x}_k \\ \mathbf{y}_k \end{pmatrix} \,\middle|\, \begin{pmatrix} \mathbf{m}_k^- \\ \boldsymbol{\mu}_k \end{pmatrix}, \begin{pmatrix} \mathbf{P}_k^- & \mathbf{C}_k \\ \mathbf{C}_k^\mathsf{T} & \mathbf{S}_k \end{pmatrix} \right), \tag{8.12}
$$

where $\boldsymbol{\mu}_k$, $\mathbf{C}_k$, and $\mathbf{S}_k$ are as defined in Equations (8.9). Using the Gaussian conditioning in Lemma A.3 then gives a Gaussian distribution with the following mean and covariance:

$$\begin{aligned} \mathbf{m}_k &= \mathbf{m}_k^- + \mathbf{C}_k\, \mathbf{S}_k^{-1} (\mathbf{y}_k - \boldsymbol{\mu}_k), \\ \mathbf{P}_k &= \mathbf{P}_k^- - \mathbf{C}_k\, \mathbf{S}_k^{-1}\, \mathbf{C}_k^\mathsf{T}, \end{aligned} \tag{8.13}$$

which can be further rewritten to give Equations (8.9).

□

The advantage of the moment matching formulation is that it enables the use of many well-known numerical integration methods, such as Gauss–Hermite quadratures and cubature rules (McNamee and Stenger, 1967; Julier and Uhlmann, 1995; Ito and Xiong, 2000; Julier et al., 2000; Wu et al., 2006; Arasaratnam and Haykin, 2009). It is also possible to use other methods, such as the Bayes–Hermite/Gaussian-process quadrature (O'Hagan, 1991; Deisenroth et al., 2009; Särkkä et al., 2016; Karvonen et al., 2019; Prüher et al., 2021) or Monte Carlo integration for approximating the integrals.

The Gaussian filter can be extended to non-additive noise models as follows.

Algorithm 8.4 (Gaussian filter II). *The prediction and update steps of the non-additive noise Gaussian (Kalman) filter for models of the form* (7.39) *are:*

- *Prediction:*

$$\begin{aligned} \mathbf{m}_k^- &= \int \mathbf{f}(\mathbf{x}_{k-1}, \mathbf{q}_{k-1}) \\ &\qquad \times \mathrm{N}(\mathbf{x}_{k-1} \mid \mathbf{m}_{k-1}, \mathbf{P}_{k-1})\, \mathrm{N}(\mathbf{q}_{k-1} \mid \mathbf{0}, \mathbf{Q}_{k-1})\, \mathrm{d}\mathbf{x}_{k-1}\, \mathrm{d}\mathbf{q}_{k-1}, \\ \mathbf{P}_k^- &= \int (\mathbf{f}(\mathbf{x}_{k-1}, \mathbf{q}_{k-1}) - \mathbf{m}_k^-)\, (\mathbf{f}(\mathbf{x}_{k-1}, \mathbf{q}_{k-1}) - \mathbf{m}_k^-)^\mathsf{T} \\ &\qquad \times \mathrm{N}(\mathbf{x}_{k-1} \mid \mathbf{m}_{k-1}, \mathbf{P}_{k-1})\, \mathrm{N}(\mathbf{q}_{k-1} \mid \mathbf{0}, \mathbf{Q}_{k-1})\, \mathrm{d}\mathbf{x}_{k-1}\, \mathrm{d}\mathbf{q}_{k-1}. \end{aligned} \tag{8.14}$$

- *Update:*

$$
\begin{aligned}
\boldsymbol{\mu}_k &= \int \mathbf{h}(\mathbf{x}_k, \mathbf{r}_k) \\
&\quad \times \mathrm{N}(\mathbf{x}_k \mid \mathbf{m}_k^-, \mathbf{P}_k^-)\, \mathrm{N}(\mathbf{r}_k \mid \mathbf{0}, \mathbf{R}_k)\, \mathrm{d}\mathbf{x}_k\, \mathrm{d}\mathbf{r}_k, \\
\mathbf{S}_k &= \int (\mathbf{h}(\mathbf{x}_k, \mathbf{r}_k) - \boldsymbol{\mu}_k)\, (\mathbf{h}(\mathbf{x}_k, \mathbf{r}_k) - \boldsymbol{\mu}_k)^\mathsf{T} \\
&\quad \times \mathrm{N}(\mathbf{x}_k \mid \mathbf{m}_k^-, \mathbf{P}_k^-)\, \mathrm{N}(\mathbf{r}_k \mid \mathbf{0}, \mathbf{R}_k)\, \mathrm{d}\mathbf{x}_k\, \mathrm{d}\mathbf{r}_k, \\
\mathbf{C}_k &= \int (\mathbf{x}_k - \mathbf{m}_k^-)\, (\mathbf{h}(\mathbf{x}_k, \mathbf{r}_k) - \boldsymbol{\mu}_k)^\mathsf{T} \\
&\quad \times \mathrm{N}(\mathbf{x}_k \mid \mathbf{m}_k^-, \mathbf{P}_k^-)\, \mathrm{N}(\mathbf{r}_k \mid \mathbf{0}, \mathbf{R}_k)\, \mathrm{d}\mathbf{x}_k\, \mathrm{d}\mathbf{r}_k, \\
\mathbf{K}_k &= \mathbf{C}_k\, \mathbf{S}_k^{-1}, \\
\mathbf{m}_k &= \mathbf{m}_k^- + \mathbf{K}_k\, (\mathbf{y}_k - \boldsymbol{\mu}_k), \\
\mathbf{P}_k &= \mathbf{P}_k^- - \mathbf{K}_k\, \mathbf{S}_k\, \mathbf{K}_k^\mathsf{T}.
\end{aligned}
\tag{8.15}
$$

The above general Gaussian filters are theoretical constructions rather than practical filtering algorithms. However, there exist many models for which the required integrals can indeed be computed in closed form. But a more practical approach is to compute them numerically. These kinds of methods will be discussed in the next sections.

8.3 Gauss–Hermite Integration

In the Gaussian filter (and later in the smoother), we are interested in approximating Gaussian integrals of the form

$$
\begin{aligned}
&\int \mathbf{g}(\mathbf{x})\, \mathrm{N}(\mathbf{x} \mid \mathbf{m}, \mathbf{P})\, \mathrm{d}\mathbf{x} \\
&\quad = \frac{1}{(2\,\pi)^{n/2}\, |\mathbf{P}|^{1/2}} \int \mathbf{g}(\mathbf{x}) \exp\left(-\frac{1}{2} (\mathbf{x} - \mathbf{m})^\mathsf{T}\, \mathbf{P}^{-1}\, (\mathbf{x} - \mathbf{m}) \right) \mathrm{d}\mathbf{x},
\end{aligned}
\tag{8.16}
$$

where $\mathbf{g}(\mathbf{x})$ is an arbitrary function. In this section, we derive a Gauss–Hermite-based numerical cubature[1] algorithm for computing such integrals. The algorithm is based on direct generalization of the one-dimensional Gauss–Hermite rule into multiple dimensions by taking the Cartesian product of one-dimensional quadratures. The disadvantage of

[1] As one-dimensional integrals are *quadratures*, multi-dimensional integrals have been traditionally called *cubatures*.

the method is that the required number of evaluation points is exponential with respect to the number of dimensions.

In its basic form, one-dimensional Gauss–Hermite quadrature integration refers to the special case of Gaussian quadratures with unit Gaussian weight function $w(x) = \mathrm{N}(x \mid 0, 1)$, that is, to approximations of the form

$$\int_{-\infty}^{\infty} g(x)\, \mathrm{N}(x \mid 0, 1)\, \mathrm{d}x \approx \sum_i W_i\, g(x^{(i)}), \tag{8.17}$$

where $W_i, i = 1, \ldots, p$ are the weights and $x^{(i)}$ are the evaluation points or abscissas – which in the present context are often called sigma points. Note that the quadrature is sometimes defined in terms of the weight function $\exp(-x^2)$, but here we use the "probabilists' definition" above. The two versions of the quadrature are related by simple scaling of variables.

Obviously, there are an infinite number of possible ways to select the weights and evaluation points. In Gauss–Hermite integration, as in all Gaussian quadratures, the weights and sigma points are chosen such that with a polynomial integrand the approximation becomes exact. It turns out that the polynomial order with a given number of points is maximized if we choose the sigma points to be roots of Hermite polynomials. When using the pth order Hermite polynomial $H_p(x)$, the rule will be exact for polynomials up to order $2p - 1$. The required weights can be computed in closed form (see below).

The Hermite polynomial of order p is defined as (these are the so-called "probabilists' Hermite polynomials"):

$$H_p(x) = (-1)^p \exp(x^2/2)\, \frac{\mathrm{d}^p}{\mathrm{d}x^p} \exp(-x^2/2). \tag{8.18}$$

The first few Hermite polynomials are:

$$\begin{aligned} H_0(x) &= 1, \\ H_1(x) &= x, \\ H_2(x) &= x^2 - 1, \\ H_3(x) &= x^3 - 3x, \\ H_4(x) &= x^4 - 6\,x^2 + 3, \end{aligned} \tag{8.19}$$

and further polynomials can be found from the recursion

$$H_{p+1}(x) = x\, H_p(x) - p\, H_{p-1}(x). \tag{8.20}$$

Using the same weights and sigma points, integrals over non-unit Gaussian weights functions $\mathrm{N}(x \mid m, P)$ can be evaluated using a simple change of integration variable:

$$\int_{-\infty}^{\infty} g(x)\, \mathrm{N}(x \mid m, P)\, \mathrm{d}x = \int_{-\infty}^{\infty} g(P^{1/2}\, \xi + m)\, \mathrm{N}(\xi \mid 0, 1)\, \mathrm{d}\xi. \quad (8.21)$$

Gauss–Hermite integration can be written as the following algorithm.

Algorithm 8.5 (Gauss–Hermite quadrature). *The pth order Gauss–Hermite approximation to the one-dimensional integral*

$$\int_{-\infty}^{\infty} g(x)\, \mathrm{N}(x \mid m, P)\, \mathrm{d}x \quad (8.22)$$

can be computed as follows:

1. *Compute the unit sigma points as the roots $\xi^{(i)}, i = 1, \ldots, p$ of the Hermite polynomial $H_p(x)$. Note that we do not need to form the polynomial and then compute its roots, but instead it is numerically more stable to compute the roots as eigenvalues of a suitable tridiagonal matrix (Golub and Welsch, 1969).*
2. *Compute the weights as*

$$W_i = \frac{p!}{p^2\, [H_{p-1}(\xi^{(i)})]^2}. \quad (8.23)$$

3. *Approximate the integral as*

$$\int_{-\infty}^{\infty} g(x)\, \mathrm{N}(x \mid m, P)\, \mathrm{d}x \approx \sum_{i=1}^{p} W_i\, g(P^{1/2}\, \xi^{(i)} + m). \quad (8.24)$$

By generalizing the change of variables idea, we can form approximations to multi-dimensional integrals of the form (8.16). First let $\mathbf{P} = \sqrt{\mathbf{P}}\, \sqrt{\mathbf{P}}^{\mathsf{T}}$, where $\sqrt{\mathbf{P}}$ is the Cholesky factor of the covariance matrix $\mathbf{P}$ or some other similar square root of the covariance matrix. If we define new integration variables $\boldsymbol{\xi}$ by

$$\mathbf{x} = \mathbf{m} + \sqrt{\mathbf{P}}\, \boldsymbol{\xi}, \quad (8.25)$$

we get

$$\int \mathbf{g}(\mathbf{x})\, \mathrm{N}(\mathbf{x} \mid \mathbf{m}, \mathbf{P})\, \mathrm{d}\mathbf{x} = \int \mathbf{g}(\mathbf{m} + \sqrt{\mathbf{P}}\, \boldsymbol{\xi})\, \mathrm{N}(\boldsymbol{\xi} \mid \mathbf{0}, \mathbf{I})\, \mathrm{d}\boldsymbol{\xi}. \quad (8.26)$$

The integration over the multi-dimensional unit Gaussian distribution can be written as an iterated integral over one-dimensional Gaussian distributions, and each of the one-dimensional integrals can be approximated with

Gauss–Hermite quadrature:

$$\begin{aligned}
\int \mathbf{g}(\mathbf{m} + \sqrt{\mathbf{P}}\,\boldsymbol{\xi})\ \mathrm{N}(\boldsymbol{\xi} \mid \mathbf{0}, \mathbf{I})\ \mathrm{d}\boldsymbol{\xi}
&= \int \cdots \int \mathbf{g}(\mathbf{m} + \sqrt{\mathbf{P}}\,\boldsymbol{\xi})\ \mathrm{N}(\xi_1 \mid 0, 1)\ \mathrm{d}\xi_1 \times \cdots \times \mathrm{N}(\xi_n \mid 0, 1)\ \mathrm{d}\xi_n \\
&\approx \sum_{i_1,\ldots,i_n} W_{i_1} \times \cdots \times W_{i_n} \mathbf{g}(\mathbf{m} + \sqrt{\mathbf{P}}\,\boldsymbol{\xi}^{(i_1,\ldots,i_n)}).
\end{aligned} \tag{8.27}$$

The weights $W_{i_k}, k = 1, \ldots, n$ are simply the corresponding one-dimensional Gauss–Hermite weights, and $\boldsymbol{\xi}^{(i_1,\ldots,i_n)}$ is an n-dimensional vector with one-dimensional unit sigma point $\xi^{(i_k)}$ at element k. The algorithm can now be written as follows.

Algorithm 8.6 (Gauss–Hermite cubature). *The pth order Gauss–Hermite approximation to the multi-dimensional integral*

$$\int \mathbf{g}(\mathbf{x})\ \mathrm{N}(\mathbf{x} \mid \mathbf{m}, \mathbf{P})\ \mathrm{d}\mathbf{x} \tag{8.28}$$

can be computed as follows.

1. *Compute the one-dimensional weights $W_i, i = 1, \ldots, p$ and unit sigma points $\xi^{(i)}$ as in the one-dimensional Gauss–Hermite quadrature Algorithm 8.5.*
2. *Form multi-dimensional weights as the products of one-dimensional weights:*

$$\begin{aligned}
W_{i_1,\ldots,i_n} &= W_{i_1} \times \cdots \times W_{i_n} \\
&= \frac{p!}{p^2\,[H_{p-1}(\xi^{(i_1)})]^2} \times \cdots \times \frac{p!}{p^2\,[H_{p-1}(\xi^{(i_n)})]^2},
\end{aligned} \tag{8.29}$$

where each i_k takes values $1, \ldots, p$.

3. *Form multi-dimensional unit sigma points as Cartesian products of the one-dimensional unit sigma points:*

$$\boldsymbol{\xi}^{(i_1,\ldots,i_n)} = \begin{pmatrix} \xi^{(i_1)} \\ \vdots \\ \xi^{(i_n)} \end{pmatrix}. \tag{8.30}$$

4. *Approximate the integral as*

$$\int \mathbf{g}(\mathbf{x})\, \mathrm{N}(\mathbf{x} \mid \mathbf{m}, \mathbf{P})\, \mathrm{d}\mathbf{x} \approx \sum_{i_1,\ldots,i_n} W_{i_1,\ldots,i_n} \mathbf{g}(\mathbf{m} + \sqrt{\mathbf{P}}\, \boldsymbol{\xi}^{(i_1,\ldots,i_n)}), \tag{8.31}$$

where $\sqrt{\mathbf{P}}$ *is a matrix square root defined by* $\mathbf{P} = \sqrt{\mathbf{P}}\, \sqrt{\mathbf{P}}^{\mathsf{T}}$.

The pth order multi-dimensional Gauss–Hermite integration is exact for monomials of the form $x_1^{d_1}\, x_2^{d_2} \cdots x_n^{d_n}$, and their arbitrary linear combinations, where each of the orders $d_i \leq 2p - 1$. The pth order integration rule for an n-dimensional integral requires p^n sigma points, which quickly becomes infeasible when the number of dimensions grows.

8.4 Gauss–Hermite Kalman Filter

The additive form multi-dimensional Gauss–Hermite cubature-based *Gauss–Hermite (Kalman) filter (GHKF)* (Ito and Xiong, 2000), which is also called the *quadrature Kalman filter (QKF)* (Arasaratnam et al., 2007), can be derived by replacing the Gaussian integrals in the Gaussian filter Algorithm 8.3 with the Gauss–Hermite approximations in Algorithm 8.6.

Algorithm 8.7 (Gauss–Hermite Kalman filter)**.** *The additive form Gauss–Hermite Kalman filter (GHKF) algorithm is the following.*

- *Prediction:*
  1. *Form the sigma points as:*

$$\mathcal{X}_{k-1}^{(i_1,\ldots,i_n)} = \mathbf{m}_{k-1} + \sqrt{\mathbf{P}_{k-1}}\, \boldsymbol{\xi}^{(i_1,\ldots,i_n)}, \quad i_1, \ldots, i_n = 1, \ldots, p, \tag{8.32}$$

     where the unit sigma points $\boldsymbol{\xi}^{(i_1,\ldots,i_n)}$ *were defined in Equation* (8.30).
  2. *Propagate the sigma points through the dynamic model:*

$$\hat{\mathcal{X}}_{k}^{(i_1,\ldots,i_n)} = \mathbf{f}(\mathcal{X}_{k-1}^{(i_1,\ldots,i_n)}), \quad i_1, \ldots, i_n = 1, \ldots, p. \tag{8.33}$$

  3. *Compute the predicted mean* $\mathbf{m}_k^-$ *and the predicted covariance* $\mathbf{P}_k^-$*:*

$$\begin{aligned} \mathbf{m}_k^- &= \sum_{i_1,\ldots,i_n} W_{i_1,\ldots,i_n} \hat{\mathcal{X}}_{k}^{(i_1,\ldots,i_n)}, \\ \mathbf{P}_k^- &= \sum_{i_1,\ldots,i_n} W_{i_1,\ldots,i_n} (\hat{\mathcal{X}}_{k}^{(i_1,\ldots,i_n)} - \mathbf{m}_k^-)\, (\hat{\mathcal{X}}_{k}^{(i_1,\ldots,i_n)} - \mathbf{m}_k^-)^{\mathsf{T}} + \mathbf{Q}_{k-1}, \end{aligned} \tag{8.34}$$

where the weights $W_{i_1,\ldots,i_n}$ were defined in Equation (8.29).

- *Update:*

1. *Form the sigma points:*

$$\mathcal{X}_k^{-(i_1,\ldots,i_n)} = \mathbf{m}_k^- + \sqrt{\mathbf{P}_k^-}\,\boldsymbol{\xi}^{(i_1,\ldots,i_n)}, \quad i_1,\ldots,i_n = 1,\ldots,p, \tag{8.35}$$

where the unit sigma points $\boldsymbol{\xi}^{(i_1,\ldots,i_n)}$ were defined in Equation (8.30).

2. *Propagate the sigma points through the measurement model:*

$$\hat{\mathcal{Y}}_k^{(i_1,\ldots,i_n)} = \mathbf{h}(\mathcal{X}_k^{-(i_1,\ldots,i_n)}), \quad i_1,\ldots,i_n = 1,\ldots,p. \tag{8.36}$$

3. *Compute the predicted mean $\boldsymbol{\mu}_k$, the predicted covariance of the measurement $\mathbf{S}_k$, and the cross-covariance of the state and the measurement $\mathbf{C}_k$:*

$$\begin{aligned}
\boldsymbol{\mu}_k &= \sum_{i_1,\ldots,i_n} W_{i_1,\ldots,i_n} \hat{\mathcal{Y}}_k^{(i_1,\ldots,i_n)}, \\
\mathbf{S}_k &= \sum_{i_1,\ldots,i_n} W_{i_1,\ldots,i_n} (\hat{\mathcal{Y}}_k^{(i_1,\ldots,i_n)} - \boldsymbol{\mu}_k)\,(\hat{\mathcal{Y}}_k^{(i_1,\ldots,i_n)} - \boldsymbol{\mu}_k)^\mathsf{T} + \mathbf{R}_k, \\
\mathbf{C}_k &= \sum_{i_1,\ldots,i_n} W_{i_1,\ldots,i_n} (\mathcal{X}_k^{-(i_1,\ldots,i_n)} - \mathbf{m}_k^-)\,(\hat{\mathcal{Y}}_k^{(i_1,\ldots,i_n)} - \boldsymbol{\mu}_k)^\mathsf{T},
\end{aligned} \tag{8.37}$$

where the weights $W_{i_1,\ldots,i_n}$ were defined in Equation (8.29).

4. *Compute the filter gain $\mathbf{K}_k$, the filtered state mean $\mathbf{m}_k$, and the covariance $\mathbf{P}_k$, conditional on the measurement $\mathbf{y}_k$:*

$$\begin{aligned}
\mathbf{K}_k &= \mathbf{C}_k\,\mathbf{S}_k^{-1}, \\
\mathbf{m}_k &= \mathbf{m}_k^- + \mathbf{K}_k\,[\mathbf{y}_k - \boldsymbol{\mu}_k], \\
\mathbf{P}_k &= \mathbf{P}_k^- - \mathbf{K}_k\,\mathbf{S}_k\,\mathbf{K}_k^\mathsf{T}.
\end{aligned} \tag{8.38}$$

The non-additive version can be obtained by applying the Gauss–Hermite quadrature to the non-additive Gaussian filter Algorithm 8.4 in a similar manner. However, due to the rapid growth of computational requirements in state dimension, the augmented form is computationally quite heavy, because it requires roughly double the dimensionality of the integration variable.

Example 8.8 (Pendulum tracking with GHKF). *The result of applying the GHKF to the pendulum model in Example 7.6 is shown in Figure 8.1. Unlike in the EKF, we do not need to derive analytical derivatives or use*

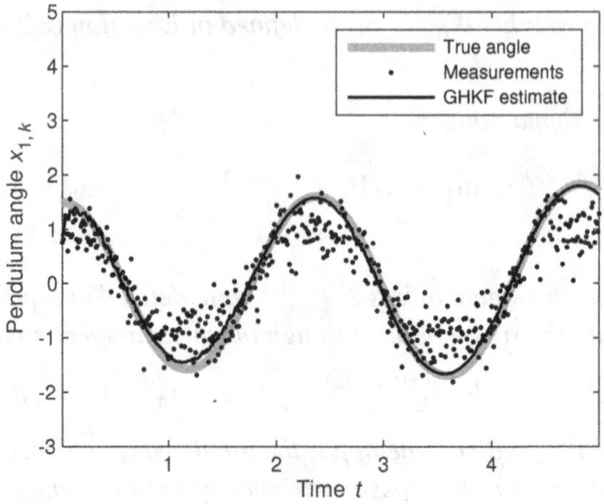

Figure 8.1 Simulated pendulum data and the result of tracking the pendulum with the GHKF (Example 8.8). The resulting RMSE was 0.10, which is lower than the RMSEs of EKF and IEKF (= 0.17).

automatic differentiation for implementing the GHKF. The resulting RMSE is 0.10 *which is lower than the RMSEs of the EKF and IEKF which both were* 0.17.

Example 8.9 (Coordinated turn model with GHKF). *We also applied the GHKF to the coordinated turn model in Example 7.7 and the resulting position RMSE was* 0.030, *which is slightly lower than that of the EKF but still practically the same. The result is shown in Figure 8.2.*

8.5 Spherical Cubature Integration

In this section, we derive the third order spherical cubature rule, which was popularized by Arasaratnam and Haykin (2009). However, this method can in hindsight be seen as a special case of the unscented transform, which we discuss in Section 8.7. Instead of using the derivation of Arasaratnam and Haykin (2009), we use the derivation presented by Wu et al. (2006), due to its simplicity. Although the derivation that we present here is far simpler than the alternative, it is completely equivalent. Furthermore, the derivation presented here can be more easily extended to more complicated spherical

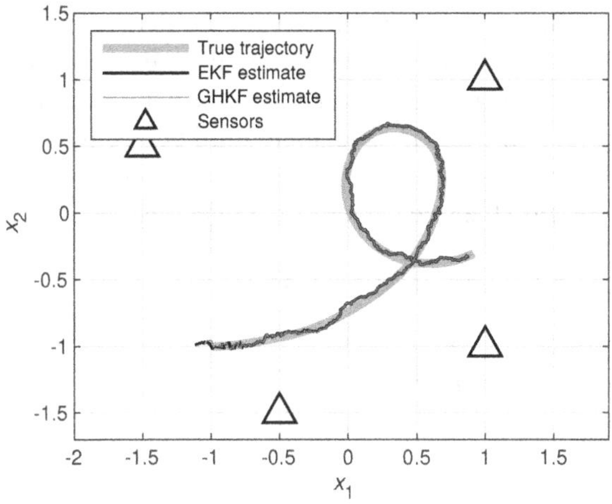

Figure 8.2 The result of applying GHKF to the coordinated turn model in Example 7.7. The resulting RMSE was 0.030, which is almost the same as that of EKF.

cubatures, and it indeed turns out that this was already done a while ago by McNamee and Stenger (1967).

Recall from Section 8.3 that the expectation of a non-linear function over an arbitrary Gaussian distribution $\mathrm{N}(\mathbf{x} \mid \mathbf{m}, \mathbf{P})$ can always be transformed into an expectation over the unit Gaussian distribution $\mathrm{N}(\boldsymbol{\xi} \mid \mathbf{0}, \mathbf{I})$. Thus, we can start by considering the multi-dimensional unit Gaussian integral

$$\int \mathbf{g}(\boldsymbol{\xi}) \, \mathrm{N}(\boldsymbol{\xi} \mid \mathbf{0}, \mathbf{I}) \, \mathrm{d}\boldsymbol{\xi}. \tag{8.39}$$

We now wish to form a $2n$-point approximation of the form

$$\int \mathbf{g}(\boldsymbol{\xi}) \, \mathrm{N}(\boldsymbol{\xi} \mid \mathbf{0}, \mathbf{I}) \, \mathrm{d}\boldsymbol{\xi} \approx W \sum_i \mathbf{g}(c \, \mathbf{u}^{(i)}), \tag{8.40}$$

where the points $\mathbf{u}^{(i)}$ belong to the symmetric set [1] with generator $(1, 0, \ldots, 0)$ (see, e.g., Wu et al., 2006; Arasaratnam and Haykin, 2009):

$$[1] = \left\{ \begin{pmatrix} 1 \\ 0 \\ 0 \\ \vdots \\ 0 \end{pmatrix}, \begin{pmatrix} 0 \\ 1 \\ 0 \\ \vdots \\ 0 \end{pmatrix}, \cdots \begin{pmatrix} -1 \\ 0 \\ 0 \\ \vdots \\ 0 \end{pmatrix}, \begin{pmatrix} 0 \\ -1 \\ 0 \\ \vdots \\ 0 \end{pmatrix}, \cdots \right\}, \tag{8.41}$$

and W is a weight and c is a parameter yet to be determined.

Because the point set is symmetric, the rule is exact for all monomials of the form $x_1^{d_1} x_2^{d_2} \cdots x_n^{d_n}$, if at least one of the exponents d_i is odd. Thus we can construct a rule that is exact up to third degree by determining the coefficients W and c such that it is exact for selections $g_j(\boldsymbol{\xi}) = 1$ and $g_j(\boldsymbol{\xi}) = \xi_j^2$. Because the true values of the integrals are

$$\begin{aligned} \int \mathrm{N}(\boldsymbol{\xi} \mid \mathbf{0}, \mathbf{I}) \, \mathrm{d}\boldsymbol{\xi} &= 1, \\ \int \xi_j^2 \, \mathrm{N}(\boldsymbol{\xi} \mid \mathbf{0}, \mathbf{I}) \, \mathrm{d}\boldsymbol{\xi} &= 1, \end{aligned} \tag{8.42}$$

we get the equations

$$\begin{aligned} W \sum_i 1 &= W \, 2n = 1, \\ W \sum_i [c \, u_j^{(i)}]^2 &= W \, 2c^2 = 1, \end{aligned} \tag{8.43}$$

which have the solutions

$$\begin{aligned} W &= \frac{1}{2n}, \\ c &= \sqrt{n}. \end{aligned} \tag{8.44}$$

That is, we get the following simple rule, which is exact for monomials up to third degree:

$$\int \mathbf{g}(\boldsymbol{\xi}) \, \mathrm{N}(\boldsymbol{\xi} \mid \mathbf{0}, \mathbf{I}) \, \mathrm{d}\boldsymbol{\xi} \approx \frac{1}{2n} \sum_i \mathbf{g}(\sqrt{n} \, \mathbf{u}^{(i)}). \tag{8.45}$$

We can now easily extend the method to arbitrary mean and covariance by using the change of variables in Equations (8.25) and (8.26), and the result is the following algorithm.

Algorithm 8.10 (Spherical cubature integration)**.** *The third order spherical cubature approximation to the multi-dimensional integral*

$$\int \mathbf{g}(\mathbf{x}) \, \mathrm{N}(\mathbf{x} \mid \mathbf{m}, \mathbf{P}) \, \mathrm{d}\mathbf{x} \tag{8.46}$$

can be computed as follows.

1. Compute the unit sigma points as

$$\boldsymbol{\xi}^{(i)} = \begin{cases} \sqrt{n} \, \mathbf{e}_i, & i = 1, \ldots, n, \\ -\sqrt{n} \, \mathbf{e}_{i-n}, & i = n+1, \ldots, 2n, \end{cases} \tag{8.47}$$

where $\mathbf{e}_i$ *denotes a unit vector in the direction of the coordinate axis* i.

2. *Approximate the integral as*

$$\int \mathbf{g}(\mathbf{x})\,\mathrm{N}(\mathbf{x}\mid\mathbf{m},\mathbf{P})\,\mathrm{d}\mathbf{x} \approx \frac{1}{2n}\sum_{i=1}^{2n}\mathbf{g}(\mathbf{m}+\sqrt{\mathbf{P}}\,\boldsymbol{\xi}^{(i)}), \tag{8.48}$$

where $\sqrt{\mathbf{P}}$ *is a matrix square root defined by* $\mathbf{P}=\sqrt{\mathbf{P}}\,\sqrt{\mathbf{P}}^{\mathsf{T}}$.

The derivation presented by Arasaratnam and Haykin (2009) is somewhat more complicated than the derivation of Wu et al. (2006) presented above, as it is based on converting the Gaussian integral into spherical coordinates and then considering the even order monomials. However, Wu et al. (2006) actually did not present the most useful special case given in Algorithm 8.10 but instead presented the method for more general generators [**u**]. The method in the above Algorithm 8.10 has the useful property that its weights are always positive, which is not always true for more general methods (Wu et al., 2006).

Note that "third order" here means a different thing than in the Gauss–Hermite Kalman filter – the pth order Gauss–Hermite filter is exact for monomials up to order $2p-1$, which means that the third order GHKF is exact for monomials up to fifth order. The third order spherical cubature rule is exact only for monomials up to third order. It is also possible to derive symmetric rules that are exact for monomials higher than third order. However, this is no longer possible with a number of sigma points that is linear $O(n)$ in the state dimension (Wu et al., 2006; Arasaratnam and Haykin, 2009). We discuss one such rule, the fifth order rule, in Section 8.9. In that rule, the required number of sigma points is proportional to n^2, the state dimension squared.

As in the case of the unscented transform, being exact up to order three only ensures that the estimate of the mean of $\mathbf{g}(\cdot)$ is exact for polynomials of order three. The covariance will be exact only for polynomials up to order one (linear functions). In this sense the third order spherical cubature rule is actually a first order spherical cubature rule for the covariance.

8.6 Cubature Kalman Filter

When we apply the third order spherical cubature integration rule in Algorithm 8.10 to the Gaussian filter equations in Algorithm 8.3, we get the cubature Kalman filter (CKF) of Arasaratnam and Haykin (2009).

Algorithm 8.11 (Cubature Kalman filter I). *The additive form of the cubature Kalman filter (CKF) algorithm is the following.*

- *Prediction:*
  1. *Form the sigma points as*

$$\mathcal{X}_{k-1}^{(i)} = \mathbf{m}_{k-1} + \sqrt{\mathbf{P}_{k-1}}\, \boldsymbol{\xi}^{(i)}, \qquad i = 1, \ldots, 2n, \tag{8.49}$$

  where the unit sigma points are defined as

$$\boldsymbol{\xi}^{(i)} = \begin{cases} \sqrt{n}\, \mathbf{e}_i, & i = 1, \ldots, n, \\ -\sqrt{n}\, \mathbf{e}_{i-n}, & i = n+1, \ldots, 2n. \end{cases} \tag{8.50}$$

  2. *Propagate the sigma points through the dynamic model:*

$$\hat{\mathcal{X}}_k^{(i)} = \mathbf{f}(\mathcal{X}_{k-1}^{(i)}), \quad i = 1, \ldots, 2n. \tag{8.51}$$

  3. *Compute the predicted mean* $\mathbf{m}_k^-$ *and the predicted covariance* $\mathbf{P}_k^-$*:*

$$\begin{aligned} \mathbf{m}_k^- &= \frac{1}{2n} \sum_{i=1}^{2n} \hat{\mathcal{X}}_k^{(i)}, \\ \mathbf{P}_k^- &= \frac{1}{2n} \sum_{i=1}^{2n} (\hat{\mathcal{X}}_k^{(i)} - \mathbf{m}_k^-)\, (\hat{\mathcal{X}}_k^{(i)} - \mathbf{m}_k^-)^\mathsf{T} + \mathbf{Q}_{k-1}. \end{aligned} \tag{8.52}$$

- *Update:*
  1. *Form the sigma points:*

$$\mathcal{X}_k^{-(i)} = \mathbf{m}_k^- + \sqrt{\mathbf{P}_k^-}\, \boldsymbol{\xi}^{(i)}, \qquad i = 1, \ldots, 2n, \tag{8.53}$$

  where the unit sigma points are defined as in Equation (8.50).

  2. *Propagate sigma points through the measurement model:*

$$\hat{\mathcal{Y}}_k^{(i)} = \mathbf{h}(\mathcal{X}_k^{-(i)}), \quad i = 1 \ldots 2n. \tag{8.54}$$

3. *Compute the predicted mean $\boldsymbol{\mu}_k$, the predicted covariance of the measurement $\mathbf{S}_k$, and the cross-covariance of the state and the measurement $\mathbf{C}_k$:*

$$\begin{aligned}\boldsymbol{\mu}_k &= \frac{1}{2n}\sum_{i=1}^{2n} \hat{\mathcal{Y}}_k^{(i)}, \\ \mathbf{S}_k &= \frac{1}{2n}\sum_{i=1}^{2n}(\hat{\mathcal{Y}}_k^{(i)} - \boldsymbol{\mu}_k)\,(\hat{\mathcal{Y}}_k^{(i)} - \boldsymbol{\mu}_k)^\mathsf{T} + \mathbf{R}_k, \\ \mathbf{C}_k &= \frac{1}{2n}\sum_{i=1}^{2n}(\mathcal{X}_k^{-(i)} - \mathbf{m}_k^-)\,(\hat{\mathcal{Y}}_k^{(i)} - \boldsymbol{\mu}_k)^\mathsf{T}.\end{aligned} \tag{8.55}$$

4. *Compute the filter gain $\mathbf{K}_k$ and the filtered state mean $\mathbf{m}_k$ and covariance $\mathbf{P}_k$, conditional on the measurement $\mathbf{y}_k$:*

$$\begin{aligned}\mathbf{K}_k &= \mathbf{C}_k\,\mathbf{S}_k^{-1}, \\ \mathbf{m}_k &= \mathbf{m}_k^- + \mathbf{K}_k\,[\mathbf{y}_k - \boldsymbol{\mu}_k], \\ \mathbf{P}_k &= \mathbf{P}_k^- - \mathbf{K}_k\,\mathbf{S}_k\,\mathbf{K}_k^\mathsf{T}.\end{aligned} \tag{8.56}$$

By applying the cubature rule to the non-additive Gaussian filter in Algorithm 8.4, we get the following augmented form of the cubature Kalman filter (CKF).

Algorithm 8.12 (Cubature Kalman filter II)**.** *The augmented non-additive form of the cubature Kalman filter (CKF) algorithm is the following.*

- *Prediction:*

1. *Form the matrix of sigma points for the augmented random variable $(\mathbf{x}_{k-1}, \mathbf{q}_{k-1})$:*

$$\tilde{\mathcal{X}}_{k-1}^{(i)} = \tilde{\mathbf{m}}_{k-1} + \sqrt{\tilde{\mathbf{P}}_{k-1}}\,\boldsymbol{\xi}^{(i)'}, \qquad i = 1, \ldots, 2n', \tag{8.57}$$

where

$$\tilde{\mathbf{m}}_{k-1} = \begin{pmatrix}\mathbf{m}_{k-1} \\ \mathbf{0}\end{pmatrix}, \qquad \tilde{\mathbf{P}}_{k-1} = \begin{pmatrix}\mathbf{P}_{k-1} & \mathbf{0} \\ \mathbf{0} & \mathbf{Q}_{k-1}\end{pmatrix}.$$

Here $n' = n + n_q$, where n is the dimensionality of the state $\mathbf{x}_{k-1}$, and n_q is the dimensionality of the noise $\mathbf{q}_{k-1}$. The unit sigma points are defined as

$$\boldsymbol{\xi}^{(i)'} = \begin{cases}\sqrt{n'}\,\mathbf{e}_i, & i = 1, \ldots, n', \\ -\sqrt{n'}\,\mathbf{e}_{i-n'}, & i = n'+1, \ldots, 2n'.\end{cases} \tag{8.58}$$

2. *Propagate the sigma points through the dynamic model:*

$$\hat{\mathcal{X}}_k^{(i)} = \mathbf{f}(\tilde{\mathcal{X}}_{k-1}^{(i),x}, \tilde{\mathcal{X}}_{k-1}^{(i),q}), \quad i = 1, \ldots, 2n', \tag{8.59}$$

where $\tilde{\mathcal{X}}_{k-1}^{(i),x}$ *denotes the first* n *components in* $\tilde{\mathcal{X}}_{k-1}^{(i)}$, *and* $\tilde{\mathcal{X}}_{k-1}^{(i),q}$ *denotes the last* n_q *components.*

3. *Compute the predicted mean* $\mathbf{m}_k^-$ *and the predicted covariance* $\mathbf{P}_k^-$:

$$\begin{aligned}
\mathbf{m}_k^- &= \frac{1}{2n'} \sum_{i=1}^{2n'} \hat{\mathcal{X}}_k^{(i)}, \\
\mathbf{P}_k^- &= \frac{1}{2n'} \sum_{i=1}^{2n'} (\hat{\mathcal{X}}_k^{(i)} - \mathbf{m}_k^-)\,(\hat{\mathcal{X}}_k^{(i)} - \mathbf{m}_k^-)^\mathsf{T}.
\end{aligned} \tag{8.60}$$

- *Update:*

1. *Let* $n'' = n + n_r$, *where* n *is the dimensionality of the state, and* n_r *is the dimensionality of the measurement noise. Form the sigma points for the augmented vector* $(\mathbf{x}_k, \mathbf{r}_k)$ *as follows:*

$$\tilde{\mathcal{X}}_k^{-(i)} = \tilde{\mathbf{m}}_k^- + \sqrt{\tilde{\mathbf{P}}_k^-}\, \boldsymbol{\xi}^{(i)''}, \qquad i = 1, \ldots, 2n'', \tag{8.61}$$

where

$$\tilde{\mathbf{m}}_k^- = \begin{pmatrix} \mathbf{m}_k^- \\ \mathbf{0} \end{pmatrix}, \qquad \tilde{\mathbf{P}}_k^- = \begin{pmatrix} \mathbf{P}_k^- & \mathbf{0} \\ \mathbf{0} & \mathbf{R}_k \end{pmatrix}.$$

The unit sigma points $\boldsymbol{\xi}^{(i)''}$ *are defined as in Equation* (8.58), *but with* n' *replaced by* n''.

2. *Propagate the sigma points through the measurement model:*

$$\hat{\mathcal{Y}}_k^{(i)} = \mathbf{h}(\tilde{\mathcal{X}}_k^{-(i),x}, \tilde{\mathcal{X}}_k^{-(i),r}), \quad i = 1, \ldots, 2n'', \tag{8.62}$$

where $\tilde{\mathcal{X}}_k^{-(i),x}$ *denotes the first* n *components in* $\tilde{\mathcal{X}}_k^{-(i)}$, *and* $\tilde{\mathcal{X}}_k^{-(i),r}$ *denotes the last* n_r *components.*

3. *Compute the predicted mean $\boldsymbol{\mu}_k$, the predicted covariance of the measurement $\mathbf{S}_k$, and the cross-covariance of the state and the measurement $\mathbf{C}_k$:*

$$\begin{aligned}
\boldsymbol{\mu}_k &= \frac{1}{2n''} \sum_{i=1}^{2n''} \hat{\mathcal{Y}}_k^{(i)}, \\
\mathbf{S}_k &= \frac{1}{2n''} \sum_{i=1}^{2n''} (\hat{\mathcal{Y}}_k^{(i)} - \boldsymbol{\mu}_k)\, (\hat{\mathcal{Y}}_k^{(i)} - \boldsymbol{\mu}_k)^\mathsf{T}, \\
\mathbf{C}_k &= \frac{1}{2n''} \sum_{i=1}^{2n''} (\mathcal{X}_k^{-(i),x} - \mathbf{m}_k^-)\, (\hat{\mathcal{Y}}_k^{(i)} - \boldsymbol{\mu}_k)^\mathsf{T}.
\end{aligned} \tag{8.63}$$

4. *Compute the filter gain $\mathbf{K}_k$, the filtered state mean $\mathbf{m}_k$, and the covariance $\mathbf{P}_k$, conditional on the measurement $\mathbf{y}_k$:*

$$\begin{aligned}
\mathbf{K}_k &= \mathbf{C}_k\, \mathbf{S}_k^{-1}, \\
\mathbf{m}_k &= \mathbf{m}_k^- + \mathbf{K}_k\, [\mathbf{y}_k - \boldsymbol{\mu}_k], \\
\mathbf{P}_k &= \mathbf{P}_k^- - \mathbf{K}_k\, \mathbf{S}_k\, \mathbf{K}_k^\mathsf{T}.
\end{aligned} \tag{8.64}$$

Although in the cubature Kalman filter (CKF) literature, the "third order" characteristic of the cubature integration rule is often emphasized (see Arasaratnam and Haykin, 2009), it is important to remember that in the covariance computation, the rule is only exact for first order polynomials. Thus in that sense CKF is a first order method.

Example 8.13 (Pendulum tracking with CKF). *The result of the CKF in the pendulum model (Example 7.6) is shown in Figure 8.3. The RMSE was* 0.11, *which is between the error of the GHKF, which was* 0.10, *and the EKF, which was* 0.17. *However, the result is practically the same as that of the GHKF.*

Example 8.14 (Coordinated turn model with CKF). *The result of applying the CKF to the coordinated turn model considered in Example 7.7 is shown in Figure 8.4. The RMSE was* 0.30, *which is the same as the RMSE of the GHKF.*

8.7 Unscented Transform

Although in this chapter we have started by presenting the GHKF and CKF as instances of numerical integration approximations to the general Gaussian filter, historically, the unscented Kalman filter (UKF, Julier et al.,

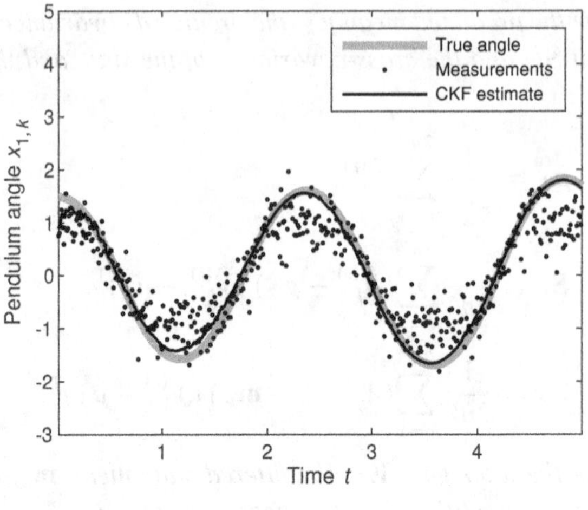

Figure 8.3 Simulated pendulum data and the result of tracking the pendulum described in Example 7.6 with the CKF (Example 8.13). The resulting RMSE was 0.11, which is slightly higher than the error of GHKF (= 0.10) but still lower than EKF (= 0.17).

1995; Wan and Van der Merwe, 2001; Julier and Uhlmann, 2004) was the first general filtering framework that used this approach in non-linear filtering. The UKF can be seen to be based on the unscented transform (UT Julier and Uhlmann, 1995; Julier et al., 2000; Julier, 2002), which in fact is a numerical integration-based approximation to the moment matching transform in Algorithm 8.1. The aims of this and the next section are to explain how the UT and UKF fit into the numerical integration-based Gaussian filtering framework.

In order to arrive at the UT, we notice that we can generalize the numerical integration approach in Section 8.5 by using a $2n + 1$ point approximation, where the origin is also included:

$$\int \mathbf{g}(\boldsymbol{\xi})\, \mathrm{N}(\boldsymbol{\xi} \mid \mathbf{0}, \mathbf{I})\, \mathrm{d}\boldsymbol{\xi} \approx W_0\, \mathbf{g}(\mathbf{0}) + W \sum_i \mathbf{g}(c\, \mathbf{u}^{(i)}). \tag{8.65}$$

We can now solve for the parameters W_0, W, and c such that we get the exact result with selections $g_j(\boldsymbol{\xi}) = 1$ and $g_j(\boldsymbol{\xi}) = \xi_j^2$. The solution can

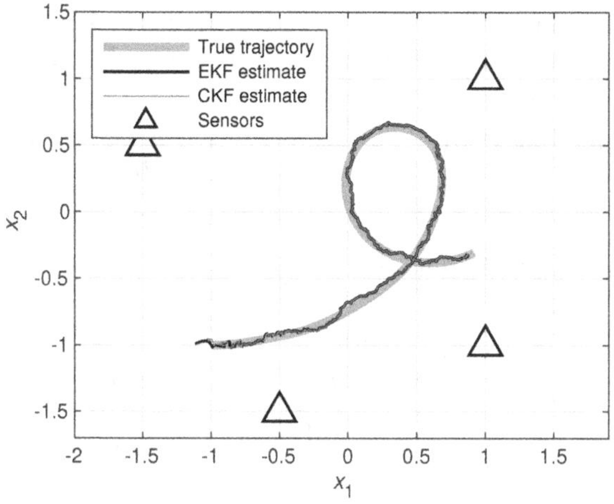

Figure 8.4 The result of applying CKF to the coordinated turn model (see Example 8.14). The resulting RMSE was 0.30, which matches that of GHKF.

be written in the form

$$\begin{aligned} W_0 &= \frac{\kappa}{n+\kappa}, \\ W &= \frac{1}{2(n+\kappa)}, \\ c &= \sqrt{n+\kappa}, \end{aligned} \tag{8.66}$$

where κ is a free parameter. This gives an integration rule that can be written as

$$\begin{aligned} &\int \mathbf{g}(\mathbf{x})\,\mathrm{N}(\mathbf{x} \mid \mathbf{m}, \mathbf{P})\,\mathrm{d}\mathbf{x} \\ &\approx \frac{\kappa}{n+\kappa}\mathbf{g}(\mathbf{m}) + \frac{1}{2(n+\kappa)}\sum_{i=1}^{2n} \mathbf{g}(\mathbf{m} + \sqrt{\mathbf{P}}\,\boldsymbol{\xi}^{(i)}), \end{aligned} \tag{8.67}$$

where

$$\boldsymbol{\xi}^{(i)} = \begin{cases} \sqrt{n+\kappa}\,\mathbf{e}_i, & i = 1,\dots,n, \\ -\sqrt{n+\kappa}\,\mathbf{e}_{i-n}, & i = n+1,\dots,2n. \end{cases} \tag{8.68}$$

However, we in fact have $\sqrt{\mathbf{P}}\,\mathbf{e}_i = [\sqrt{\mathbf{P}}]_i$, where $[\sqrt{\mathbf{P}}]_i$ denotes the ith column of the matrix $\sqrt{\mathbf{P}}$. Therefore, we can rewrite the integration rule as

$$\int \mathbf{g}(\mathbf{x})\, \mathrm{N}(\mathbf{x} \mid \mathbf{m}, \mathbf{P})\, \mathrm{d}\mathbf{x} \approx \frac{\kappa}{n+\kappa} \mathbf{g}(\mathcal{X}^{(0)}) + \frac{1}{2(n+\kappa)} \sum_{i=1}^{2n} \mathbf{g}(\mathcal{X}^{(i)}), \tag{8.69}$$

where we have defined sigma (i.e., evaluation) points

$$\begin{aligned} \mathcal{X}^{(0)} &= \mathbf{m}, \\ \mathcal{X}^{(i)} &= \mathbf{m} + \sqrt{n+\kappa} \left[\sqrt{\mathbf{P}}\right]_i, \quad i = 1, \ldots, n, \\ \mathcal{X}^{(i+n)} &= \mathbf{m} - \sqrt{n+\kappa} \left[\sqrt{\mathbf{P}}\right]_i, \quad i = 1, \ldots, n. \end{aligned} \tag{8.70}$$

This rule already appears in the original article on the unscented transform (Julier and Uhlmann, 1995), and it is equivalent to the cubature rule in Algorithm 8.10 when $\kappa = 0$. We can equivalently interpret the integration rule such that we first compute transformed sigma points by propagating them though the function $\mathcal{Y}^{(i)} = \mathbf{g}(\mathcal{X}^{(i)})$ for $i = 0, \ldots, 2n$, and then use

$$\int \mathbf{g}(\mathbf{x})\, \mathrm{N}(\mathbf{x} \mid \mathbf{m}, \mathbf{P})\, \mathrm{d}\mathbf{x} \approx \frac{\kappa}{n+\kappa} \mathcal{Y}^{(0)} + \frac{1}{2(n+\kappa)} \sum_{i=1}^{2n} \mathcal{Y}^{(i)}. \tag{8.71}$$

In the so-called scaled unscented transform (Julier, 2002), this rule is modified by evaluating the integrand in the following sigma-points instead of the original ones:

$$\mathcal{X}^{(i)\prime} = \mathcal{X}^{(0)} + \alpha \left(\mathcal{X}^{(i)} - \mathcal{X}^{(0)}\right), \tag{8.72}$$

where $\alpha > 0$ is an additional parameter. This corresponds to replacing $\sqrt{n+\kappa}$ in the sigma-point computation with $\alpha\sqrt{n+\kappa}$. To find the corresponding weights that still lead to a method that is exact to third order, we need to find a modified κ, which we denote as λ, such that

$$\alpha\sqrt{n+\kappa} = \sqrt{n+\lambda}, \tag{8.73}$$

which gives

$$\lambda = \alpha^2\,(n+\kappa) - n. \tag{8.74}$$

Furthermore, we can observe that there is no inherent need to use the same weights for the mean and covariance computation. In this regard, Julier (2002) suggest the introduction of an additional parameter $\beta > 0$, which only appears in the covariance computation, to account for high order moments of the distribution. In particular, the choice of $\beta = 2$ is in a certain sense optimal for a Gaussian distribution. With this addition we get different weights for the mean (m) and covariance (c) computation:

$$
\begin{aligned}
W_0^{(\mathrm{m})} &= \frac{\lambda}{n+\lambda}, \\
W_0^{(\mathrm{c})} &= \frac{\lambda}{n+\lambda} + (1 - \alpha^2 + \beta), \\
W_i^{(\mathrm{m})} &= \frac{1}{2(n+\lambda)}, \quad i = 1, \ldots, 2n, \\
W_i^{(\mathrm{c})} &= \frac{1}{2(n+\lambda)}, \quad i = 1, \ldots, 2n.
\end{aligned} \tag{8.75}
$$

If we apply the resulting integration rule to the moment matching transform in Algorithm 8.1, we get the following unscented transform.

Algorithm 8.15 (Unscented approximation of an additive transform)**.** *The unscented transform-based Gaussian approximation to the joint distribution of* $\mathbf{x}$ *and the transformed random variable* $\mathbf{y} = \mathbf{g}(\mathbf{x}) + \mathbf{q}$, *where* $\mathbf{x} \sim \mathrm{N}(\mathbf{m}, \mathbf{P})$ *and* $\mathbf{q} \sim \mathrm{N}(\mathbf{0}, \mathbf{Q})$ *is given by*

$$
\begin{pmatrix} \mathbf{x} \\ \mathbf{y} \end{pmatrix} \sim \mathrm{N}\left(\begin{pmatrix} \mathbf{m} \\ \boldsymbol{\mu}_{\mathrm{U}} \end{pmatrix}, \begin{pmatrix} \mathbf{P} & \mathbf{C}_{\mathrm{U}} \\ \mathbf{C}_{\mathrm{U}}^{\mathsf{T}} & \mathbf{S}_{\mathrm{U}} \end{pmatrix} \right), \tag{8.76}
$$

where the submatrices can be computed as follows.

1. *Form the set of* $2n + 1$ *sigma points as follows:*

$$
\begin{aligned}
\mathcal{X}^{(0)} &= \mathbf{m}, \\
\mathcal{X}^{(i)} &= \mathbf{m} + \sqrt{n+\lambda} \left[\sqrt{\mathbf{P}} \right]_i, \\
\mathcal{X}^{(i+n)} &= \mathbf{m} - \sqrt{n+\lambda} \left[\sqrt{\mathbf{P}} \right]_i, \quad i = 1, \ldots, n,
\end{aligned} \tag{8.77}
$$

   where the parameter λ *is defined in Equation* (8.74).

2. *Propagate the sigma points through the non-linear function* $\mathbf{g}(\cdot)$*:*

$$
\mathcal{Y}^{(i)} = \mathbf{g}(\mathcal{X}^{(i)}), \quad i = 0, \ldots, 2n.
$$

3. The submatrices are then given as:

$$\begin{aligned}
\boldsymbol{\mu}_{\mathrm{U}} &= \sum_{i=0}^{2n} W_i^{(\mathrm{m})}\, \mathcal{Y}^{(i)}, \\
\mathbf{S}_{\mathrm{U}} &= \sum_{i=0}^{2n} W_i^{(\mathrm{c})}\, (\mathcal{Y}^{(i)} - \boldsymbol{\mu}_{\mathrm{U}})\, (\mathcal{Y}^{(i)} - \boldsymbol{\mu}_{\mathrm{U}})^{\mathsf{T}} + \mathbf{Q}, \\
\mathbf{C}_{\mathrm{U}} &= \sum_{i=0}^{2n} W_i^{(\mathrm{c})}\, (\mathcal{X}^{(i)} - \mathbf{m})\, (\mathcal{Y}^{(i)} - \boldsymbol{\mu}_{\mathrm{U}})^{\mathsf{T}},
\end{aligned} \tag{8.78}$$

where the constant weights $W_i^{(\mathrm{m})}$ and $W_i^{(\mathrm{c})}$ were defined in the Equation (8.75).

The unscented transform approximation to a transformation of the form $\mathbf{y} = \mathbf{g}(\mathbf{x}, \mathbf{q})$ can be derived by considering the augmented random variable $\tilde{\mathbf{x}} = (\mathbf{x}, \mathbf{q})$ as the random variable in the transform, which corresponds to using the numerical integration scheme above in the non-additive moment matching transform in Algorithm 8.2. The resulting algorithm is the following.

Algorithm 8.16 (Unscented approximation of a non-additive transform). *The (augmented) unscented transform-based Gaussian approximation to the joint distribution of* $\mathbf{x}$ *and the transformed random variable* $\mathbf{y} = \mathbf{g}(\mathbf{x}, \mathbf{q})$ *when* $\mathbf{x} \sim \mathrm{N}(\mathbf{m}, \mathbf{P})$ *and* $\mathbf{q} \sim \mathrm{N}(\mathbf{0}, \mathbf{Q})$ *is given by*

$$\begin{pmatrix} \mathbf{x} \\ \mathbf{y} \end{pmatrix} \sim \mathrm{N}\left(\begin{pmatrix} \mathbf{m} \\ \boldsymbol{\mu}_{\mathrm{U}} \end{pmatrix}, \begin{pmatrix} \mathbf{P} & \mathbf{C}_{\mathrm{U}} \\ \mathbf{C}_{\mathrm{U}}^{\mathsf{T}} & \mathbf{S}_{\mathrm{U}} \end{pmatrix} \right), \tag{8.79}$$

where the sub-matrices can be computed as follows. Let the dimensionalities of $\mathbf{x}$ *and* $\mathbf{q}$ *be* n *and* n_q*, respectively, and let* $n' = n + n_q$.

1. Form the sigma points for the augmented random variable $\tilde{\mathbf{x}} = (\mathbf{x}, \mathbf{q})$:

$$\begin{aligned}
\tilde{\mathcal{X}}^{(0)} &= \tilde{\mathbf{m}}, \\
\tilde{\mathcal{X}}^{(i)} &= \tilde{\mathbf{m}} + \sqrt{n' + \lambda'} \left[\sqrt{\tilde{\mathbf{P}}}\right]_i, \\
\tilde{\mathcal{X}}^{(i+n')} &= \tilde{\mathbf{m}} - \sqrt{n' + \lambda'} \left[\sqrt{\tilde{\mathbf{P}}}\right]_i, \quad i = 1, \ldots, n',
\end{aligned} \tag{8.80}$$

where the parameter λ' *is defined as in Equation* (8.74)*, but with* n *replaced by* n'*, and the augmented mean and covariance are defined by*

$$\tilde{\mathbf{m}} = \begin{pmatrix} \mathbf{m} \\ \mathbf{0} \end{pmatrix}, \qquad \tilde{\mathbf{P}} = \begin{pmatrix} \mathbf{P} & \mathbf{0} \\ \mathbf{0} & \mathbf{Q} \end{pmatrix}.$$

2. *Propagate the sigma points through the function:*

$$\tilde{\mathcal{Y}}^{(i)} = \mathbf{g}(\tilde{\mathcal{X}}^{(i),x}, \tilde{\mathcal{X}}^{(i),q}), \quad i = 0, \ldots, 2n',$$

where $\tilde{\mathcal{X}}^{(i),x}$ *and* $\tilde{\mathcal{X}}^{(i),q}$ *denote the parts of the augmented sigma point* i *that correspond to* $\mathbf{x}$ *and* $\mathbf{q}$*, respectively.*

3. *Compute the predicted mean* $\boldsymbol{\mu}_{\mathrm{U}}$*, the predicted covariance* $\mathbf{S}_{\mathrm{U}}$*, and the cross-covariance* $\mathbf{C}_{\mathrm{U}}$*:*

$$\begin{aligned}
\boldsymbol{\mu}_{\mathrm{U}} &= \sum_{i=0}^{2n'} W_i^{(\mathrm{m})'} \, \tilde{\mathcal{Y}}^{(i)}, \\
\mathbf{S}_{\mathrm{U}} &= \sum_{i=0}^{2n'} W_i^{(\mathrm{c})'} \, (\tilde{\mathcal{Y}}^{(i)} - \boldsymbol{\mu}_{\mathrm{U}}) \, (\tilde{\mathcal{Y}}^{(i)} - \boldsymbol{\mu}_{\mathrm{U}})^{\mathsf{T}}, \\
\mathbf{C}_{\mathrm{U}} &= \sum_{i=0}^{2n'} W_i^{(\mathrm{c})'} \, (\tilde{\mathcal{X}}^{(i),x} - \mathbf{m}) \, (\tilde{\mathcal{Y}}^{(i)} - \boldsymbol{\mu}_{\mathrm{U}})^{\mathsf{T}},
\end{aligned}$$

where the definitions of the weights $W_i^{(\mathrm{m})'}$ *and* $W_i^{(\mathrm{c})'}$ *are the same as in Equation* (8.75)*, but with* n *replaced by* n' *and* λ *replaced by* λ'.

The unscented transform is also a third order method in the sense that the estimate of the mean of $\mathbf{g}(\cdot)$ is exact for polynomials up to order three. That is, if $\mathbf{g}(\cdot)$ is indeed a multi-variate polynomial of order three, the mean is exact. However, the covariance approximation is exact only for first order polynomials, because the square of a second order polynomial is already a polynomial of order four, and the unscented transform (UT) does not compute the exact result for fourth order polynomials. In this sense the UT is only a first order method. With suitable selection of parameters ($\kappa = 3 - n$), it is possible to get some, but not all, of the fourth order terms appearing in the covariance computation correct also for quadratic functions.

8.8 Unscented Kalman Filter

The *unscented Kalman filter* (UKF) (Julier et al., 1995; Wan and Van der Merwe, 2001; Julier and Uhlmann, 2004) is an approximate filtering framework that utilizes the unscented transform to approximate the general Gaussian filters in Algorithms 8.3 and 8.4, and it can thus be applied to models of the form (7.24) or (7.39). The UKF forms a Gaussian approximation to the filtering distribution:

$$p(\mathbf{x}_k \mid \mathbf{y}_{1:k}) \simeq \mathrm{N}(\mathbf{x}_k \mid \mathbf{m}_k, \mathbf{P}_k), \tag{8.81}$$

where $\mathbf{m}_k$ and $\mathbf{P}_k$ are the mean and covariance computed by the algorithm.

Algorithm 8.17 (Unscented Kalman filter I). *In the additive form of the unscented Kalman filter (UKF) algorithm, which can be applied to additive models of the form* (7.24), *the following operations are performed at each measurement step* $k = 1, 2, 3, \ldots$

- *Prediction:*

1. *Form the sigma points:*

$$\begin{aligned} \mathcal{X}_{k-1}^{(0)} &= \mathbf{m}_{k-1}, \\ \mathcal{X}_{k-1}^{(i)} &= \mathbf{m}_{k-1} + \sqrt{n+\lambda} \left[\sqrt{\mathbf{P}_{k-1}}\right]_i, \\ \mathcal{X}_{k-1}^{(i+n)} &= \mathbf{m}_{k-1} - \sqrt{n+\lambda} \left[\sqrt{\mathbf{P}_{k-1}}\right]_i, \quad i = 1, \ldots, n, \end{aligned} \tag{8.82}$$

where the parameter λ *is defined in Equation* (8.74).

2. *Propagate the sigma points through the dynamic model:*

$$\hat{\mathcal{X}}_k^{(i)} = \mathbf{f}(\mathcal{X}_{k-1}^{(i)}), \quad i = 0, \ldots, 2n. \tag{8.83}$$

3. *Compute the predicted mean* $\mathbf{m}_k^-$ *and the predicted covariance* $\mathbf{P}_k^-$*:*

$$\begin{aligned} \mathbf{m}_k^- &= \sum_{i=0}^{2n} W_i^{(\mathrm{m})} \hat{\mathcal{X}}_k^{(i)}, \\ \mathbf{P}_k^- &= \sum_{i=0}^{2n} W_i^{(\mathrm{c})} (\hat{\mathcal{X}}_k^{(i)} - \mathbf{m}_k^-) (\hat{\mathcal{X}}_k^{(i)} - \mathbf{m}_k^-)^\mathsf{T} + \mathbf{Q}_{k-1}, \end{aligned} \tag{8.84}$$

where the weights $W_i^{(\mathrm{m})}$ *and* $W_i^{(\mathrm{c})}$ *were defined in Equation* (8.75).

- *Update:*

1. *Form the sigma points:*

$$\begin{aligned} \mathcal{X}_k^{-(0)} &= \mathbf{m}_k^-, \\ \mathcal{X}_k^{-(i)} &= \mathbf{m}_k^- + \sqrt{n+\lambda} \left[\sqrt{\mathbf{P}_k^-}\right]_i, \\ \mathcal{X}_k^{-(i+n)} &= \mathbf{m}_k^- - \sqrt{n+\lambda} \left[\sqrt{\mathbf{P}_k^-}\right]_i, \quad i = 1, \ldots, n. \end{aligned} \tag{8.85}$$

2. *Propagate sigma points through the measurement model:*

$$\hat{\mathcal{Y}}_k^{(i)} = \mathbf{h}(\mathcal{X}_k^{-(i)}), \quad i = 0, \ldots, 2n. \tag{8.86}$$

3. *Compute the predicted mean $\boldsymbol{\mu}_k$, the predicted covariance of the measurement $\mathbf{S}_k$, and the cross-covariance of the state and the measurement $\mathbf{C}_k$:*

$$
\begin{aligned}
\boldsymbol{\mu}_k &= \sum_{i=0}^{2n} W_i^{(\mathrm{m})} \, \hat{\mathcal{Y}}_k^{(i)}, \\
\mathbf{S}_k &= \sum_{i=0}^{2n} W_i^{(\mathrm{c})} \, (\hat{\mathcal{Y}}_k^{(i)} - \boldsymbol{\mu}_k) \, (\hat{\mathcal{Y}}_k^{(i)} - \boldsymbol{\mu}_k)^\mathsf{T} + \mathbf{R}_k, \\
\mathbf{C}_k &= \sum_{i=0}^{2n} W_i^{(\mathrm{c})} \, (\mathcal{X}_k^{-(i)} - \mathbf{m}_k^-) \, (\hat{\mathcal{Y}}_k^{(i)} - \boldsymbol{\mu}_k)^\mathsf{T}.
\end{aligned}
\tag{8.87}
$$

4. *Compute the filter gain $\mathbf{K}_k$, the filtered state mean $\mathbf{m}_k$, and the covariance $\mathbf{P}_k$, conditional on the measurement $\mathbf{y}_k$:*

$$
\begin{aligned}
\mathbf{K}_k &= \mathbf{C}_k \, \mathbf{S}_k^{-1}, \\
\mathbf{m}_k &= \mathbf{m}_k^- + \mathbf{K}_k \, [\mathbf{y}_k - \boldsymbol{\mu}_k], \\
\mathbf{P}_k &= \mathbf{P}_k^- - \mathbf{K}_k \, \mathbf{S}_k \, \mathbf{K}_k^\mathsf{T}.
\end{aligned}
\tag{8.88}
$$

The filtering equations above can be derived in an analogous manner to the EKF or Gaussian filter equations, but the unscented transform-based approximations are used instead of the linear approximations or moment matching.

The non-additive form of the UKF (Julier and Uhlmann, 2004) can be derived by augmenting the process and measurement noises to the state vector and then using the UT approximation for performing prediction and update steps simultaneously. Alternatively, we can first augment the state vector with process noise, then approximate the prediction step, and after that do the same with measurement noise on the update step. This latter form corresponds to the moment matching-based Gaussian filter, whereas the former is different (though it would be possible to derive a general Gaussian filter using that kind of moment matching as well). The different algorithms and ways of doing this in practice are analyzed in the article by Wu et al. (2005). However, if we directly apply the non-additive UT in Algorithm 8.16 separately to the prediction and update steps, we get the following algorithm.

Algorithm 8.18 (Unscented Kalman filter II). *In the augmented form of the unscented Kalman filter (UKF) algorithm, which can be applied to non-additive models of the form* (7.39), *the following operations are performed at each measurement step $k = 1, 2, 3, \ldots$*

- *Prediction:*
  1. *Form the sigma points for the augmented random variable* $(\mathbf{x}_{k-1}, \mathbf{q}_{k-1})$*:*

$$
\begin{aligned}
\tilde{\mathcal{X}}_{k-1}^{(0)} &= \tilde{\mathbf{m}}_{k-1}, \\
\tilde{\mathcal{X}}_{k-1}^{(i)} &= \tilde{\mathbf{m}}_{k-1} + \sqrt{n' + \lambda'} \left[\sqrt{\tilde{\mathbf{P}}_{k-1}} \right]_i, \\
\tilde{\mathcal{X}}_{k-1}^{(i+n')} &= \tilde{\mathbf{m}}_{k-1} - \sqrt{n' + \lambda'} \left[\sqrt{\tilde{\mathbf{P}}_{k-1}} \right]_i, \quad i = 1, \ldots, n',
\end{aligned}
\tag{8.89}
$$

  where

$$
\tilde{\mathbf{m}}_{k-1} = \begin{pmatrix} \mathbf{m}_{k-1} \\ \mathbf{0} \end{pmatrix}, \qquad \tilde{\mathbf{P}}_{k-1} = \begin{pmatrix} \mathbf{P}_{k-1} & \mathbf{0} \\ \mathbf{0} & \mathbf{Q}_{k-1} \end{pmatrix}.
$$

  Here $n' = n + n_q$*, where* n *is the dimensionality of the state* $\mathbf{x}_{k-1}$*, and* n_q *is the dimensionality of the noise* $\mathbf{q}_{k-1}$*. The parameter* λ' *is defined as in Equation* (8.74)*, but with* n *replaced by* n'*.*
  2. *Propagate the sigma points through the dynamic model:*

$$
\hat{\mathcal{X}}_k^{(i)} = \mathbf{f}(\tilde{\mathcal{X}}_{k-1}^{(i),x}, \tilde{\mathcal{X}}_{k-1}^{(i),q}), \quad i = 0, \ldots, 2n', \tag{8.90}
$$

  where $\tilde{\mathcal{X}}_{k-1}^{(i),x}$ *denotes the first* n *components in* $\tilde{\mathcal{X}}_{k-1}^{(i)}$*, and* $\tilde{\mathcal{X}}_{k-1}^{(i),q}$ *denotes the last* n_q *components.*
  3. *Compute the predicted mean* $\mathbf{m}_k^-$ *and the predicted covariance* $\mathbf{P}_k^-$*:*

$$
\begin{aligned}
\mathbf{m}_k^- &= \sum_{i=0}^{2n'} W_i^{(\mathrm{m})'} \, \hat{\mathcal{X}}_k^{(i)}, \\
\mathbf{P}_k^- &= \sum_{i=0}^{2n'} W_i^{(\mathrm{c})'} \, (\hat{\mathcal{X}}_k^{(i)} - \mathbf{m}_k^-) \, (\hat{\mathcal{X}}_k^{(i)} - \mathbf{m}_k^-)^\mathsf{T},
\end{aligned}
\tag{8.91}
$$

  where the weights $W_i^{(\mathrm{m})'}$ *and* $W_i^{(\mathrm{c})'}$ *are the same as in Equation* (8.75)*, but with* n *replaced by* n' *and* λ *by* λ'*.*

- *Update:*
  1. *Form the sigma points for the augmented random variable* $(\mathbf{x}_k, \mathbf{r}_k)$*:*

$$
\begin{aligned}
\tilde{\mathcal{X}}_k^{-(0)} &= \tilde{\mathbf{m}}_k^-, \\
\tilde{\mathcal{X}}_k^{-(i)} &= \tilde{\mathbf{m}}_k^- + \sqrt{n'' + \lambda''} \left[\sqrt{\tilde{\mathbf{P}}_k^-} \right]_i, \\
\tilde{\mathcal{X}}_k^{-(i+n'')} &= \tilde{\mathbf{m}}_k^- - \sqrt{n'' + \lambda''} \left[\sqrt{\tilde{\mathbf{P}}_k^-} \right]_i, \quad i = 1, \ldots, n'',
\end{aligned}
\tag{8.92}
$$

where

$$\tilde{\mathbf{m}}_k^- = \begin{pmatrix} \mathbf{m}_k^- \\ \mathbf{0} \end{pmatrix}, \qquad \tilde{\mathbf{P}}_k^- = \begin{pmatrix} \mathbf{P}_k^- & \mathbf{0} \\ \mathbf{0} & \mathbf{R}_k \end{pmatrix}.$$

Here we have defined $n'' = n + n_r$, *where* n *is the dimensionality of the state* $\mathbf{x}_k$, *and* n_r *is the dimensionality of the noise* $\mathbf{r}_k$. *The parameter* λ'' *is defined as in Equation* (8.74), *but with* n *replaced by* n''.

2. *Propagate the sigma points through the measurement model:*

$$\hat{\mathcal{Y}}_k^{(i)} = \mathbf{h}(\tilde{\mathcal{X}}_k^{-(i),x}, \tilde{\mathcal{X}}_k^{-(i),r}), \quad i = 0, \ldots, 2n'', \tag{8.93}$$

where $\tilde{\mathcal{X}}_k^{-(i),x}$ *denotes the first* n *components in* $\tilde{\mathcal{X}}_k^{-(i)}$, *and* $\tilde{\mathcal{X}}_k^{-(i),r}$ *denotes the last* n_r *components.*

3. *Compute the predicted mean* $\boldsymbol{\mu}_k$, *the predicted covariance of the measurement* $\mathbf{S}_k$, *and the cross-covariance of the state and the measurement* $\mathbf{C}_k$*:*

$$\begin{aligned}
\boldsymbol{\mu}_k &= \sum_{i=0}^{2n''} W_i^{(\mathrm{m})''} \, \hat{\mathcal{Y}}_k^{(i)}, \\
\mathbf{S}_k &= \sum_{i=0}^{2n''} W_{i-1}^{(\mathrm{c})''} \, (\hat{\mathcal{Y}}_k^{(i)} - \boldsymbol{\mu}_k) \, (\hat{\mathcal{Y}}_k^{(i)} - \boldsymbol{\mu}_k)^\mathsf{T}, \\
\mathbf{C}_k &= \sum_{i=0}^{2n''} W_i^{(\mathrm{c})''} \, (\mathcal{X}_k^{-(i),x} - \mathbf{m}_k^-) \, (\hat{\mathcal{Y}}_k^{(i)} - \boldsymbol{\mu}_k)^\mathsf{T},
\end{aligned} \tag{8.94}$$

where the weights $W_i^{(\mathrm{m})''}$ *and* $W_i^{(\mathrm{c})''}$ *are the same as in Equation* (8.75), *but with* n *replaced by* n'' *and* λ *by* λ''.

4. *Compute the filter gain* $\mathbf{K}_k$ *and the filtered state mean* $\mathbf{m}_k$ *and covariance* $\mathbf{P}_k$, *conditional to the measurement* $\mathbf{y}_k$*:*

$$\begin{aligned}
\mathbf{K}_k &= \mathbf{C}_k \, \mathbf{S}_k^{-1}, \\
\mathbf{m}_k &= \mathbf{m}_k^- + \mathbf{K}_k \, [\mathbf{y}_k - \boldsymbol{\mu}_k], \\
\mathbf{P}_k &= \mathbf{P}_k^- - \mathbf{K}_k \, \mathbf{S}_k \, \mathbf{K}_k^\mathsf{T}.
\end{aligned} \tag{8.95}$$

The UKF was originally proposed as an improvement to the EKF (Julier et al., 1995). In that regard, the advantage of the UKF over the EKF is that the UKF is not based on a linear approximation at a single point but uses further points in approximating the non-linearity. As discussed in Julier and

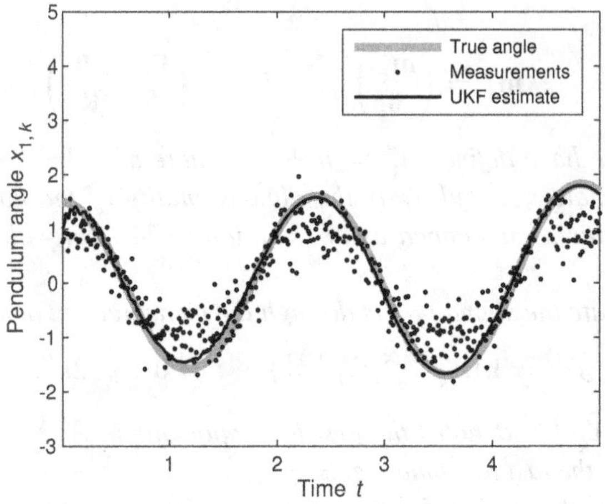

Figure 8.5 Simulated pendulum data and the result of tracking the pendulum described in Example 7.6 with the UKF (Example 8.19). The resulting RMSE was 0.10, which is practically the same as with the GHKF (Figure 8.1).

Uhlmann (2004), the unscented transform is able to capture the higher order moments caused by the non-linear transform better than Taylor series-based approximations. However, as already pointed out in the previous section, although the mean estimate of the UKF is exact for polynomials up to order three, the covariance computation is only exact for polynomials up to first order. In the UKF, the dynamic and model functions are also not required to be formally differentiable nor do their Jacobian matrices need to be computed. The disadvantage over the EKF is that the UKF often requires slightly more computational operations than the EKF.

Example 8.19 (Pendulum tracking with UKF). *The result of applying the UKF to the pendulum model and simulated data in Example 7.6 is shown in Figure 8.5. The resulting RMSE was* 0.10*, which is the same as with GHKF while better than the EKF (*= 0.17*) and CKF (*= 0.11*).*

Example 8.20 (Coordinated turn model with UKF). *Figure 8.6 shows the result of applying the UKF to the coordinated turn model in Example 7.7. The resulting RMSE was* 0.30*, which is the same as with the CKF and GHKF.*

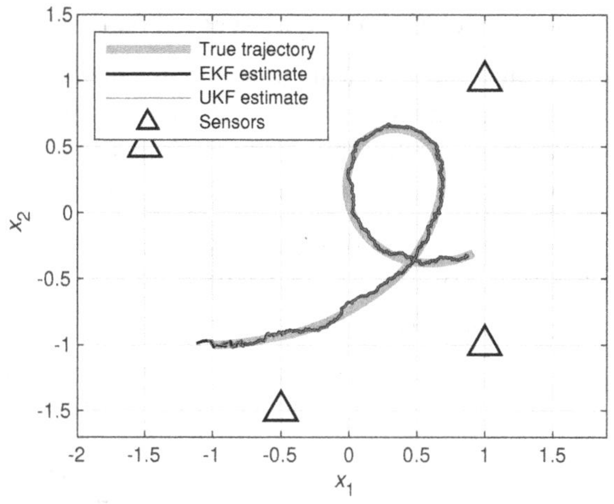

Figure 8.6 The result of applying UKF to the coordinated turn model (see Example 8.20). The resulting RMSE is the same as that of CKF and GHKF.

It is worth noting that although the unscented Kalman filters (UKFs) in the forms that we have described them in this section appear to differ from the other numerical integration filters, such as the Gauss–Hermite Kalman filter in Algorithm 8.7 and the cubature Kalman filter in Algorithm 8.11, in the sigma-point forming step, the difference is only superficial. The use of columns of the Chokesly factors of the covariances in forming the sigma points in Algorithm 8.17, for example, is in fact equivalent to using the following unit sigma points:

$$\boldsymbol{\xi}^{(i)} = \begin{cases} \sqrt{n+\lambda}\,\mathbf{e}_i, & i = 1,\ldots,n, \\ -\sqrt{n+\lambda}\,\mathbf{e}_{i-n}, & i = n+1,\ldots,2n, \end{cases} \tag{8.96}$$

because $\sqrt{\mathbf{P}}\,(c\,\mathbf{e}_i) = c\,[\sqrt{\mathbf{P}}]_i$. Furthermore, the UKF uses different sets of numerical integration weights $W_i^{(\mathrm{m})}$ and $W_i^{(\mathrm{c})}$ for the mean and covariance. This is in reality is not a unique feature of the UKF either, because in any numerical integration-based filter we could use different numerical integration rules for the mean and covariance – or in fact for all the integrals appearing in the equations. However, to simplify the story, we have not discussed this explicitly in the case of other filters.

8.9 Higher Order Cubature/Unscented Kalman Filters

We can also generalize the spherical cubature/unscented integration rules discussed in Section 8.5 and 8.7 to higher than third order (McNamee and Stenger, 1967; Wu et al., 2006). This can be done by using symmetric sets of higher order. We have already seen two kinds of symmetric sets:

- Set $[0] = \{(0,\ldots,0)\}$, which only contains the origin.
- Set $[u] = \{(u,0,\ldots,0),(-u,0,\ldots,0),(0,u,\ldots,0),\ldots\}$, whose special case was given in (8.41) with $u = 1$.

With these definitions, the unscented rule in (8.67) can be written in the form

$$\int \mathbf{g}(\boldsymbol{\xi})\,\mathrm{N}(\boldsymbol{\xi} \mid \mathbf{0}, \mathbf{I})\,\mathrm{d}\boldsymbol{\xi} \approx W_0 \sum_{\boldsymbol{\xi}\in[0]} g(\boldsymbol{\xi}) + W \sum_{\boldsymbol{\xi}\in[c]} \mathbf{g}(\boldsymbol{\xi}), \tag{8.97}$$

where W_0, W, and c are given by Equation (8.66). The cubature rule in Equation (8.45) is a special case of this of the form

$$\int \mathbf{g}(\boldsymbol{\xi})\,\mathrm{N}(\boldsymbol{\xi} \mid \mathbf{0}, \mathbf{I})\,\mathrm{d}\boldsymbol{\xi} \approx W \sum_{\boldsymbol{\xi}\in[c]} \mathbf{g}(\boldsymbol{\xi}), \tag{8.98}$$

which is obtained by setting $\kappa = 0$.

It turns out that a fifth order symmetric cubature rule can now be constructed in the form

$$\begin{aligned}\int \mathbf{g}(\boldsymbol{\xi})\,\mathrm{N}(\boldsymbol{\xi} \mid \mathbf{0}, \mathbf{I})\,\mathrm{d}\boldsymbol{\xi} \\ \approx W_0 \sum_{\boldsymbol{\xi}\in[0]} \mathbf{g}(\boldsymbol{\xi}) + W^{(1)} \sum_{\boldsymbol{\xi}\in[u]} \mathbf{g}(\boldsymbol{\xi}) + W^{(2)} \sum_{\boldsymbol{\xi}\in[u,u]} \mathbf{g}(\boldsymbol{\xi}),\end{aligned} \tag{8.99}$$

where we have introduced the additional symmetric set $[u, u]$, which contains all vectors that have only two non-zero elements having values u or $-u$. As an example, in three dimensions we have

$$\begin{aligned}[1,1] = \Bigg\{ &\begin{pmatrix}1\\1\\0\end{pmatrix}, \begin{pmatrix}-1\\-1\\0\end{pmatrix}, \begin{pmatrix}1\\-1\\0\end{pmatrix}, \begin{pmatrix}-1\\1\\0\end{pmatrix}, \begin{pmatrix}1\\0\\1\end{pmatrix}, \begin{pmatrix}-1\\0\\-1\end{pmatrix}, \\ &\begin{pmatrix}1\\0\\-1\end{pmatrix}, \begin{pmatrix}-1\\0\\1\end{pmatrix}, \begin{pmatrix}0\\1\\1\end{pmatrix}, \begin{pmatrix}0\\-1\\-1\end{pmatrix}, \begin{pmatrix}0\\1\\-1\end{pmatrix}, \begin{pmatrix}0\\-1\\1\end{pmatrix} \Bigg\}.\end{aligned} \tag{8.100}$$

As shown in McNamee and Stenger (1967) and Wu et al. (2006), this rule can integrate monomials up to fifth order provided that we select $u = \sqrt{3}$,

$W_0 = 1 + (n^2 - 7n)/18$, $W_1 = (4 - n)/18$, and $W_2 = 1/36$. This leads to the following integration method.

Algorithm 8.21 (Fifth order cubature/unscented integration). *The fifth order spherical cubature/unscented approximation to the multi-dimensional integral*

$$\int \mathbf{g}(\mathbf{x}) \, \mathrm{N}(\mathbf{x} \mid \mathbf{m}, \mathbf{P}) \, \mathrm{d}\mathbf{x} \tag{8.101}$$

can be computed as follows.

1. *Compute the unit sigma points as*

$$\begin{aligned} \boldsymbol{\xi}^{(0)} &= \mathbf{0}, \\ \boldsymbol{\xi}^{(i)} &= i\text{'th vector in symmetric set } [\sqrt{3}], \\ \boldsymbol{\xi}^{(i+2n)} &= i\text{'th vector in symmetric set } [\sqrt{3}, \sqrt{3}]. \end{aligned} \tag{8.102}$$

2. *Compute the weights as*

$$\begin{aligned} W_0 &= 1 + (n^2 - 7n)/18, \\ W_i &= (4 - n)/18, \quad i = 1, \ldots, 2n, \\ W_{i+2n} &= 1/36, \quad i = 1, 2, \ldots . \end{aligned} \tag{8.103}$$

3. *Approximate the integral as*

$$\int \mathbf{g}(\mathbf{x}) \, \mathrm{N}(\mathbf{x} \mid \mathbf{m}, \mathbf{P}) \, \mathrm{d}\mathbf{x} \approx \sum_i W_i \, \mathbf{g}(\mathbf{m} + \sqrt{\mathbf{P}} \, \boldsymbol{\xi}^{(i)}), \tag{8.104}$$

where $\sqrt{\mathbf{P}}$ is a matrix square root defined by $\mathbf{P} = \sqrt{\mathbf{P}} \sqrt{\mathbf{P}}^{\mathsf{T}}$.

A disadvantage of this rule is that it leads to negative weights when $n \geq 5$. This is a typical challenge with higher order symmetric rules and can lead to non-positive definite covariance matrices and hence unstable filters.

If we use this rule in the Gaussian filter (Algorithm 8.3), we obtain the following fifth order cubature/unscented Kalman filter for additive noise models. The non-additive version could be derived similarly, as we did for the CKF.

Algorithm 8.22 (Fifth order cubature/unscented Kalman filter). *The additive form of the fifth order cubature/unscented Kalman filter (UKF5) algorithm is the following.*

- *Prediction:*

1. *Form the sigma points as:*

$$\mathcal{X}_{k-1}^{(i)} = \mathbf{m}_{k-1} + \sqrt{\mathbf{P}_{k-1}}\, \boldsymbol{\xi}^{(i)}, \qquad i = 0, 1, 2, \ldots, \tag{8.105}$$

where the unit sigma points are defined by Equation (8.102).

2. *Propagate the sigma points through the dynamic model:*

$$\hat{\mathcal{X}}_k^{(i)} = \mathbf{f}(\mathcal{X}_{k-1}^{(i)}), \quad i = 0, 1, 2, \ldots . \tag{8.106}$$

3. *Compute the predicted mean* $\mathbf{m}_k^-$ *and the predicted covariance* $\mathbf{P}_k^-$*:*

$$\begin{aligned} \mathbf{m}_k^- &= \sum_i W_i\, \hat{\mathcal{X}}_k^{(i)}, \\ \mathbf{P}_k^- &= \sum_i W_i\, (\hat{\mathcal{X}}_k^{(i)} - \mathbf{m}_k^-)\, (\hat{\mathcal{X}}_k^{(i)} - \mathbf{m}_k^-)^\mathsf{T} + \mathbf{Q}_{k-1}, \end{aligned} \tag{8.107}$$

where the weights W_i *are given by Equation* (8.103).

- *Update:*

1. *Form the sigma points:*

$$\mathcal{X}_k^{-(i)} = \mathbf{m}_k^- + \sqrt{\mathbf{P}_k^-}\, \boldsymbol{\xi}^{(i)}, \qquad i = 0, 1, 2, \ldots, \tag{8.108}$$

where the unit sigma points are defined as in Equation (8.102).

2. *Propagate sigma points through the measurement model:*

$$\hat{\mathcal{Y}}_k^{(i)} = \mathbf{h}(\mathcal{X}_k^{-(i)}), \quad i = 0, 1, 2, \ldots . \tag{8.109}$$

3. *Compute the predicted mean* $\boldsymbol{\mu}_k$*, the predicted covariance of the measurement* $\mathbf{S}_k$*, and the cross-covariance of the state and the measurement* $\mathbf{C}_k$*:*

$$\begin{aligned} \boldsymbol{\mu}_k &= \sum_i W_i\, \hat{\mathcal{Y}}_k^{(i)}, \\ \mathbf{S}_k &= \sum_i W_i\, (\hat{\mathcal{Y}}_k^{(i)} - \boldsymbol{\mu}_k)\, (\hat{\mathcal{Y}}_k^{(i)} - \boldsymbol{\mu}_k)^\mathsf{T} + \mathbf{R}_k, \\ \mathbf{C}_k &= \sum_i W_i\, (\mathcal{X}_k^{-(i)} - \mathbf{m}_k^-)\, (\hat{\mathcal{Y}}_k^{(i)} - \boldsymbol{\mu}_k)^\mathsf{T}, \end{aligned} \tag{8.110}$$

where the weights W_i *are defined as in* (8.103).

4. *Compute the filter gain* $\mathbf{K}_k$ *and the filtered state mean* $\mathbf{m}_k$ *and covariance* $\mathbf{P}_k$*, conditional on the measurement* $\mathbf{y}_k$*:*

$$\begin{aligned} \mathbf{K}_k &= \mathbf{C}_k\, \mathbf{S}_k^{-1}, \\ \mathbf{m}_k &= \mathbf{m}_k^- + \mathbf{K}_k\, [\mathbf{y}_k - \boldsymbol{\mu}_k], \\ \mathbf{P}_k &= \mathbf{P}_k^- - \mathbf{K}_k\, \mathbf{S}_k\, \mathbf{K}_k^\mathsf{T}. \end{aligned} \tag{8.111}$$

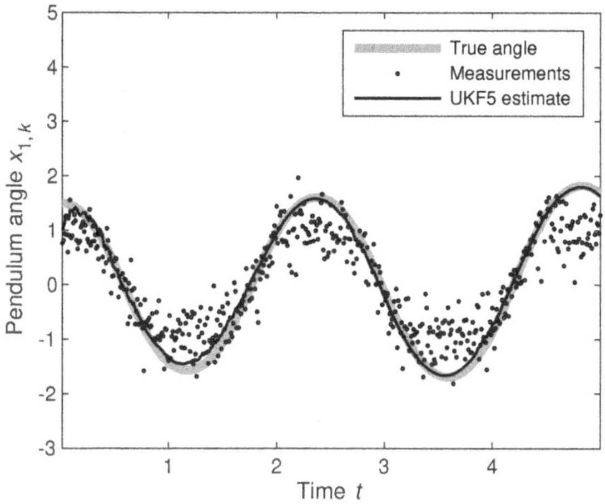

Figure 8.7 Simulated pendulum data and the result of tracking the pendulum described in Example 7.6 with the UKF5 (Example 8.23). The resulting RMSE was 0.10 which is the same as that of GHKF and UKF.

It would also be possible to derive higher order filters in this class by applying the higher order rules given in McNamee and Stenger (1967). In fact, by simply changing the weights and unit sigma points in the above algorithm, it can serve as a prototype of implementing filters using an arbitrary integration rule having this form (e.g., McNamee and Stenger, 1967; Ito and Xiong, 2000; Julier et al., 2000; Wu et al., 2006; Arasaratnam and Haykin, 2009; Särkkä et al., 2016; Karvonen et al., 2019).

Example 8.23 (Pendulum tracking with UKF5). *The result of applying the UKF5 to the pendulum model and simulated data in Example 7.6 is shown in Figure 8.7. The resulting RMSE of* 0.10 *is practically the same as for the GHKF and UKF.*

Example 8.24 (Coordinated turn model with UKF5). *Figure 8.8 shows the result of applying fifth order cubature/unscented Kalman filter (UKF5) to the coordinated turn model introduced in Example 7.7. The result is similar to the results of the GHKF, CKF, and UKF with RMSE of* 0.30.

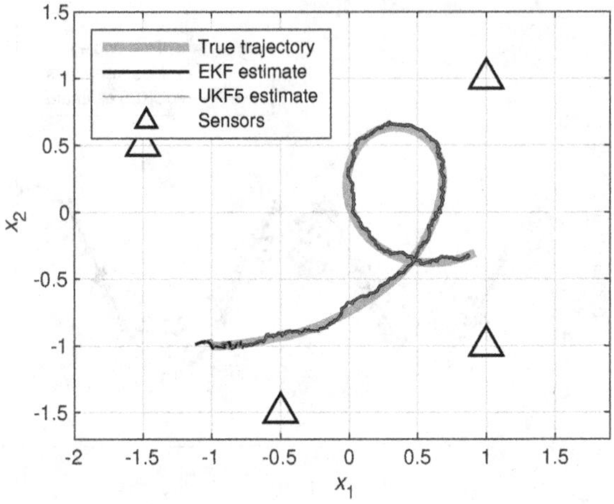

Figure 8.8 The result of applying the fifth order cubature/unscented Kalman filter to the coordinated turn model. The result is practically the same as with the other numerical integration filters.

8.10 Exercises

8.1 Show that the selection $\kappa = 3 - n$ in Equation (8.67) causes fourth order terms ξ_i^4 to be integrated exactly when $\mathbf{m} = \mathbf{0}$ and $\mathbf{P} = \mathbf{I}$. What happens to the weights if $n > 3$?

8.2 Show that when the function is linear, both the unscented transform and the spherical cubature integration rule give the exact result.

8.3 Show that one-dimensional Hermite polynomials are orthogonal with respect to the inner product

$$\langle f, g \rangle = \int f(x)\, g(x)\, \mathrm{N}(x \mid 0, 1)\, \mathrm{d}x. \tag{8.112}$$

8.4 Show that multivariate Hermite polynomials defined as

$$H_{i_1,\ldots,i_n}(\mathbf{x}) = H_{i_1}(x_1) \times \cdots \times H_{i_n}(x_n) \tag{8.113}$$

are orthogonal with respect to the inner product

$$\langle f, g \rangle = \int f(\mathbf{x})\, g(\mathbf{x})\, \mathrm{N}(\mathbf{x} \mid \mathbf{0}, \mathbf{I})\, \mathrm{d}\mathbf{x}. \tag{8.114}$$

8.5 Implement a GHKF for the model in Exercise 7.1. Plot the results and compare the RMSE values with the EKF.

8.6 Write down the cubature transform, that is, the moment transformation resulting from approximating the integrals in Algorithm 8.1 with spherical cubature integration in Algorithm 8.10.

8.7 Implement a CKF and UKF for the model in Exercise 7.1. Plot the results and compare the RMSE values with the EKF and GHKF. Can you find parameter values for the methods that cause the UKF, GHKF, and CKF methods to become identical?

8.8 Implement a CKF for the bearings-only target tracking problem in Exercise 7.2. Compare the performance with the EKF.

8.9 Implement a UKF for the Cartesian coordinated turn model in Equations (4.61) and (4.90) with a linear position measurement model. Also compare its performance with the EKF in Exercise 7.8.

8.10 Check numerically, by comparing to the closed form expressions for integrals of polynomials, that the integration rules introduced in Sections 8.5 and 8.7 are exact (only) up to third order and the rule in Section 8.9 up to fifth order.

8.11 Implement the fifth order filter in Algorithm 8.22 to the model in Exercise 7.1. Compare its performance with the other filters.

8.12 Write down the fifth order unscented transform, that is, the moment transformation resulting from approximating the integrals in Algorithm 8.1 with the fifth order integration rule in Algorithm 8.21.

9

Gaussian Filtering by Enabling Approximations

We have already seen several Gaussian filters for filtering in non-linear systems, with different premises and motivations. In Chapter 7 we encountered extended Kalman filters (EKFs), which first linearize the dynamic and measurement models and then apply a Kalman filter to the linearized model. In Chapter 8 we looked into moment matching-based Gaussian filters, which compute the integrals needed in the moment matching using numerical integration. It turns out that the moment matching-based Gaussian filters can also be viewed as different strategies to linearize measurement and dynamic models (Lefebvre et al., 2002; Arasaratnam et al., 2007; Tronarp et al., 2018). We refer to these linearizations as enabling approximations since they enable us to use the closed-form Kalman filter prediction and update equations. In fact, most filters that recursively approximate the posterior as Gaussian share this property, which means that they combine two steps:

1. Approximate measurement and dynamic models as linear (or affine) with additive Gaussian noise.
2. Use the Kalman filter equations to perform prediction and update.

Perhaps the simplest example is the EKF, which performs a first order Taylor expansion of both models. In this chapter, we elaborate on this perspective and explain why all the Gaussian filters presented earlier arguably contain the above two steps (sometimes implicitly). Considering that the second step is identical for all filters, the main difference between different filters is how the models are linearized. Understanding the Gaussian filters in terms of these enabling approximations provides a unifying perspective, sheds new light on the family of Gaussian filters, and enables us to develop new filters, such as posterior linearization filters which we discuss in Chapter 10.

9.1 Enabling Linearization

Let us again take a look at filtering in non-linear Gaussian models of the form

$$\begin{aligned} \mathbf{x}_k &= \mathbf{f}(\mathbf{x}_{k-1}) + \mathbf{q}_{k-1}, & \mathbf{q}_{k-1} &\sim \mathrm{N}(\mathbf{0}, \mathbf{Q}_{k-1}), \\ \mathbf{y}_k &= \mathbf{h}(\mathbf{x}_k) + \mathbf{r}_k, & \mathbf{r}_k &\sim \mathrm{N}(\mathbf{0}, \mathbf{R}_k), \end{aligned} \tag{9.1}$$

which we already considered in Chapters 7 and 8. Let us now assume for a while that we have some way to approximate this state space model as affine[1] and Gaussian:

$$\begin{aligned} \mathbf{x}_k &\simeq \mathbf{A}_{k-1}\,\mathbf{x}_{k-1} + \mathbf{a}_{k-1} + \tilde{\mathbf{e}}_{k-1}, \\ \mathbf{y}_k &\simeq \mathbf{H}_k\,\mathbf{x}_k + \mathbf{b}_k + \tilde{\mathbf{v}}_k, \end{aligned} \tag{9.2}$$

where $\tilde{\mathbf{e}}_{k-1} \sim \mathrm{N}(\mathbf{0}, \mathbf{\Lambda}_{k-1})$ and $\tilde{\mathbf{v}}_k \sim \mathrm{N}(\mathbf{0}, \mathbf{\Omega}_k)$, whereas $\mathbf{A}_{k-1}$ and $\mathbf{H}_k$ are constant matrices, and $\mathbf{a}_{k-1}$ and $\mathbf{b}_k$ are constant vectors. If the noise covariances $\mathbf{\Lambda}_{k-1}$ and $\mathbf{\Omega}_k$ are full rank, we can also express Equation (9.2) in terms of conditional probability density functions

$$\begin{aligned} p(\mathbf{x}_k \mid \mathbf{x}_{k-1}) &\simeq \mathrm{N}(\mathbf{x}_k \mid \mathbf{A}_{k-1}\,\mathbf{x}_{k-1} + \mathbf{a}_{k-1}, \mathbf{\Lambda}_{k-1}), \\ p(\mathbf{y}_k \mid \mathbf{x}_k) &\simeq \mathrm{N}(\mathbf{y}_k \mid \mathbf{H}_k\,\mathbf{x}_k + \mathbf{b}_k, \mathbf{\Omega}_k). \end{aligned} \tag{9.3}$$

Once we have introduced the approximations in Equation (9.2), the prediction and update steps have simple, closed-form expressions given by the affine Kalman filter in Theorem 6.9.

Algorithm 9.1 (Gaussian filtering with enabling linearizations). *The prediction and update steps of a Gaussian filter for computing the filtering means $\mathbf{m}_k$ and covariances $\mathbf{P}_k$ that makes use of enabling linearizations are:*

- *Prediction: Find $\mathbf{A}_{k-1}$, $\mathbf{a}_{k-1}$, and $\mathbf{\Lambda}_{k-1}$ such that*

$$\mathbf{x}_k \simeq \mathbf{A}_{k-1}\,\mathbf{x}_{k-1} + \mathbf{a}_{k-1} + \tilde{\mathbf{e}}_{k-1}, \tag{9.4}$$

  where $\tilde{\mathbf{e}}_{k-1} \sim \mathrm{N}(\mathbf{0}, \mathbf{\Lambda}_{k-1})$. Compute the predicted moments

$$\begin{aligned} \mathbf{m}_k^- &= \mathbf{A}_{k-1}\,\mathbf{m}_{k-1} + \mathbf{a}_{k-1}, \\ \mathbf{P}_k^- &= \mathbf{A}_{k-1}\,\mathbf{P}_{k-1}\,\mathbf{A}_{k-1}^\mathsf{T} + \mathbf{\Lambda}_{k-1}. \end{aligned} \tag{9.5}$$

- *Update: Find $\mathbf{H}_k$, $\mathbf{b}_k$, and $\mathbf{\Omega}_k$ such that*

$$\mathbf{y}_k \simeq \mathbf{H}_k\,\mathbf{x}_k + \mathbf{b}_k + \tilde{\mathbf{v}}_k, \tag{9.6}$$

[1] Recall that $\mathbf{A}\,\mathbf{x} + \mathbf{a}$ is an affine function in $\mathbf{x}$ for all values of $\mathbf{a}$ but that it is only linear if $\mathbf{a} = 0$.

where $\tilde{\mathbf{v}}_k \sim \mathrm{N}(\mathbf{0}, \boldsymbol{\Omega}_k)$. *Compute the updated moments*

$$\begin{aligned} \boldsymbol{\mu}_k &= \mathbf{H}_k \, \mathbf{m}_k^- + \mathbf{b}_k, \\ \mathbf{S}_k &= \mathbf{H}_k \, \mathbf{P}_k^- \, \mathbf{H}_k^\mathsf{T} + \boldsymbol{\Omega}_k, \\ \mathbf{K}_k &= \mathbf{P}_k^- \, \mathbf{H}_k^\mathsf{T} \, \mathbf{S}_k^{-1}, \\ \mathbf{m}_k &= \mathbf{m}_k^- + \mathbf{K}_k \, (\mathbf{y}_k - \boldsymbol{\mu}_k), \\ \mathbf{P}_k &= \mathbf{P}_k^- - \mathbf{K}_k \, \mathbf{S}_k \, \mathbf{K}_k^\mathsf{T}. \end{aligned} \tag{9.7}$$

It is easy to verify that both the EKF and the IEKF are examples of Algorithm 9.1 for models with additive and non-additive noise. Let us verify this in the additive noise case, that is, when the state space model takes the form of Equation (9.1). The prediction step is identical in the EKF and the IEKF, and they both use $\mathbf{A}_{k-1} = \mathbf{F}_\mathbf{x}(\mathbf{m}_{k-1})$, $\mathbf{a}_{k-1} = \mathbf{f}(\mathbf{m}_{k-1}) - \mathbf{F}_\mathbf{x}(\mathbf{m}_{k-1})\, \mathbf{m}_{k-1}$ and $\boldsymbol{\Lambda}_{k-1} = \mathbf{Q}_{k-1}$. This corresponds to a first order Taylor expansion of $\mathbf{f}_{k-1}(\mathbf{x}_{k-1})$ at $\mathbf{x}_{k-1} = \mathbf{m}_{k-1}$:

$$\mathbf{f}(\mathbf{x}_{k-1}) \simeq \mathbf{f}(\mathbf{m}_{k-1}) + \mathbf{F}_\mathbf{x}(\mathbf{m}_{k-1})\,(\mathbf{x}_{k-1} - \mathbf{m}_{k-1}). \tag{9.8}$$

In the update step, both filters rely on a first order Taylor expansion of $\mathbf{h}(\mathbf{x}_k)$. Introducing $\bar{\mathbf{m}}_k$ as the linearization point, both filters use $\mathbf{H}_k = \mathbf{H}_\mathbf{x}(\bar{\mathbf{m}}_k)$, $\mathbf{b}_k = \mathbf{h}(\bar{\mathbf{m}}_k) - \mathbf{H}_\mathbf{x}(\bar{\mathbf{m}}_k)\,\bar{\mathbf{m}}_k$ and $\boldsymbol{\Omega}_k = \mathbf{R}_k$. The difference is that the EKF linearizes $\mathbf{h}(\mathbf{x}_k)$ at the predicted mean $\bar{\mathbf{m}}_k = \mathbf{m}_k^-$, whereas the IEKF seeks to set $\bar{\mathbf{m}}_k$ to the maximum a posteriori estimate of $\mathbf{x}_k$.

Taylor expansions such as Equation (9.8) are commonly used in the literature, but the enabling approximations of the form of Equation (9.2) are more flexible than that. To start with, we may want to approximate $\mathbf{f}(\mathbf{x}_{k-1})$ and $\mathbf{h}(\mathbf{x}_k)$ accurately for a range of values, which means that the Taylor expansion may not be optimal. Also, even though these functions are deterministic, we are free to select the noise covariances $\boldsymbol{\Lambda}_{k-1}$ and $\boldsymbol{\Omega}_k$ larger than the original noise covariances $\mathbf{Q}_{k-1}$ and $\mathbf{R}_k$ to acknowledge errors introduced when linearizing $\mathbf{f}(\mathbf{x}_{k-1})$ and $\mathbf{h}(\mathbf{x}_k)$.

In this chapter, we consider three families of probabilistic state space models. Apart from the additive noise models as in Equation (9.1), we also study models with non-additive Gaussian noise:

$$\begin{aligned} \mathbf{x}_k &= \mathbf{f}(\mathbf{x}_{k-1}, \mathbf{q}_{k-1}), & \mathbf{q}_{k-1} &\sim \mathrm{N}(\mathbf{0}, \mathbf{Q}_{k-1}), \\ \mathbf{y}_k &= \mathbf{h}(\mathbf{x}_k, \mathbf{r}_k), & \mathbf{r}_k &\sim \mathrm{N}(\mathbf{0}, \mathbf{R}_k). \end{aligned} \tag{9.9}$$

We also consider a third formulation where the model is expressed in terms of conditional probability distributions:

$$\begin{aligned} \mathbf{x}_k &\sim p(\mathbf{x}_k \mid \mathbf{x}_{k-1}), \\ \mathbf{y}_k &\sim p(\mathbf{y}_k \mid \mathbf{x}_k). \end{aligned} \tag{9.10}$$

This general formulation enables us to handle models that are inconvenient to express using Equations (9.1) or (9.9).

In this chapter (and the next), we describe principles for selecting the enabling approximations in Equation (9.2) and explain how they can be applied to each of the above families of models. As we will see, some of these principles give rise to filters that are identical to the general Gaussian filters in Chapter 8, whereas other principles give rise to new filters.

9.2 Statistical Linearization

Statistical linearization (SL, see, e.g., Gelb, 1974) is a type of linearization that can be used in Algorithm 9.1 for models with additive or non-additive noise, where the relation between two random variables $\mathbf{x}$ and $\mathbf{y}$ is approximated as affine and noise free, $\mathbf{y} \simeq \mathbf{A}\,\mathbf{x} + \mathbf{b}$.

Definition 9.2 (Statistical linearization). *If* $\mathbf{x}$ *and* $\mathbf{y}$ *are two random variables,* $\mathbf{y} \simeq \mathbf{A}\,\mathbf{x} + \mathbf{b}$ *is called a* statistical linearization *if* $\mathbf{A}$ *and* $\mathbf{b}$ *are selected to minimize the mean squared error (MSE)*

$$\mathrm{MSE}(\mathbf{A}, \mathbf{b}) = \mathrm{E}\left[(\mathbf{y} - \mathbf{A}\,\mathbf{x} - \mathbf{b})^\mathsf{T} (\mathbf{y} - \mathbf{A}\,\mathbf{x} - \mathbf{b})\right], \tag{9.11}$$

where the matrix $\mathbf{A}$ *and the vector* $\mathbf{b}$ *are deterministic variables.*

Theorem 9.3 (Statistical linearization I). *Suppose* $\mathbf{x}$ *and* $\mathbf{y}$ *are two random variables and that* $\mathbf{m} = \mathrm{E}[\mathbf{x}]$, $\mathbf{P} = \mathrm{Cov}[\mathbf{x}]$, $\boldsymbol{\mu}_S = \mathrm{E}[\mathbf{y}]$, *and* $\mathbf{C}_S = \mathrm{Cov}[\mathbf{x}, \mathbf{y}]$. *The approximation* $\mathbf{y} \simeq \mathbf{A}\,\mathbf{x} + \mathbf{b}$ *minimizes the mean squared error (MSE)*

$$\mathrm{MSE}(\mathbf{A}, \mathbf{b}) = \mathrm{E}\left[(\mathbf{y} - \mathbf{A}\,\mathbf{x} - \mathbf{b})^\mathsf{T} (\mathbf{y} - \mathbf{A}\,\mathbf{x} - \mathbf{b})\right] \tag{9.12}$$

when

$$\begin{aligned} \mathbf{A} &= \mathbf{C}_S^\mathsf{T}\, \mathbf{P}^{-1}, \\ \mathbf{b} &= \boldsymbol{\mu}_S - \mathbf{A}\,\mathbf{m}. \end{aligned} \tag{9.13}$$

Proof The result follows from setting the derivatives of $\mathrm{MSE}(\mathbf{A}, \mathbf{b})$ with respect to $\mathbf{A}$ and $\mathbf{b}$ to zero. □

Corollary 9.4 (Gaussian approximation from statistical linearization). *When* $\mathbf{x} \sim \mathrm{N}(\mathbf{m}, \mathbf{P})$, $\boldsymbol{\mu}_\mathrm{S} = \mathrm{E}[\mathbf{y}]$, *and* $\mathbf{C}_\mathrm{S} = \mathrm{Cov}[\mathbf{x}, \mathbf{y}]$, *the statistical linearization* $\mathbf{y} \simeq \mathbf{A}\,\mathbf{x} + \mathbf{b}$ *gives rise to a Gaussian approximation*

$$\begin{pmatrix} \mathbf{x} \\ \mathbf{y} \end{pmatrix} \sim \mathrm{N}\left(\begin{pmatrix} \mathbf{m} \\ \boldsymbol{\mu}_\mathrm{S} \end{pmatrix}, \begin{pmatrix} \mathbf{P} & \mathbf{C}_\mathrm{S} \\ \mathbf{C}_\mathrm{S}^\mathsf{T} & \mathbf{S}_\mathrm{S} \end{pmatrix} \right), \tag{9.14}$$

where $\mathbf{S}_\mathrm{S} = \mathbf{A}\,\mathbf{P}\,\mathbf{A}^\mathsf{T}$.

Proof We obtain this result from Theorem 9.3 by combining the expressions for $\mathbf{A}$ and $\mathbf{b}$ with $\mathbf{y} \simeq \mathbf{A}\,\mathbf{x} + \mathbf{b}$. We get, for instance, $\mathrm{Cov}[\mathbf{x}, \mathbf{A}\,\mathbf{x} + \mathbf{b}] = \mathrm{Cov}[\mathbf{x}, \mathbf{x}]\,\mathbf{A}^\mathsf{T} = \mathbf{C}_\mathrm{S}$. □

To make statistical linearization somewhat more concrete, let us define the random variable $\mathbf{y}$ as follows:

$$\begin{aligned} \mathbf{x} &\sim \mathrm{N}(\mathbf{m}, \mathbf{P}), \\ \mathbf{y} &= \mathbf{g}(\mathbf{x}), \end{aligned} \tag{9.15}$$

which corresponds the transformation problem first considered in Section 7.1 and later in Chapter 8. The statistical linearization for Equation (9.15) is given in Corollary 9.5.

Corollary 9.5 (Statistical linearization II). *The approximation* $\mathbf{g}(\mathbf{x}) \simeq \mathbf{A}\,\mathbf{x} + \mathbf{b}$, *with* $\mathbf{x}$ *and* $\mathbf{g}(\mathbf{x})$ *defined in Equation* (9.15), *minimizes the mean squared error*

$$\mathrm{MSE}(\mathbf{A}, \mathbf{b}) = \mathrm{E}\left[(\mathbf{g}(\mathbf{x}) - \mathbf{A}\,\mathbf{x} - \mathbf{b})^\mathsf{T}\,(\mathbf{g}(\mathbf{x}) - \mathbf{A}\,\mathbf{x} - \mathbf{b})\right] \tag{9.16}$$

when

$$\begin{aligned} \mathbf{A} &= \mathrm{E}[(\mathbf{x} - \mathbf{m})\,\mathbf{g}(\mathbf{x})^\mathsf{T}]^\mathsf{T}\,\mathbf{P}^{-1}, \\ \mathbf{b} &= \mathrm{E}[\mathbf{g}(\mathbf{x})] - \mathbf{A}\,\mathbf{m}, \end{aligned} \tag{9.17}$$

where all expected values are taken with respect to $\mathbf{x} \sim \mathrm{N}(\mathbf{m}, \mathbf{P})$.

Proof The result is a special case of Theorem 9.3. Specifically, for $\mathbf{y} = \mathbf{g}(\mathbf{x})$ the expressions in (9.17) follow from (9.13) since $\boldsymbol{\mu}_\mathrm{S} = \mathrm{E}[\mathbf{y}] = \mathrm{E}[\mathbf{g}(\mathbf{x})]$ and

$$\begin{aligned} \mathbf{C}_\mathrm{S} &= \mathrm{E}[(\mathbf{x} - \mathbf{m})\,(\mathbf{g}(\mathbf{x}) - \mathrm{E}[\mathbf{g}(\mathbf{x})])^\mathsf{T}] \\ &= \mathrm{E}[(\mathbf{x} - \mathbf{m})\,\mathbf{g}(\mathbf{x})^\mathsf{T}] + \underbrace{\mathrm{E}[(\mathbf{x} - \mathbf{m})]}_{=\mathbf{0}}\,\mathrm{E}[\mathbf{g}(\mathbf{x})]^\mathsf{T} = \mathrm{E}[(\mathbf{x} - \mathbf{m})\,\mathbf{g}(\mathbf{x})^\mathsf{T}]. \end{aligned} \tag{9.18}$$

□

Note that $\mathrm{E}[(\mathbf{x} - \mathbf{m})\,\mathbf{g}(\mathbf{x})^\mathsf{T}]^\mathsf{T} = \mathrm{E}[\mathbf{g}(\mathbf{x})\,(\mathbf{x} - \mathbf{m})^\mathsf{T}]$, which is used to simplify and rewrite several of the expressions below.

We can now form an approximation to an additive transform analogously to the previous chapters but now by using the statistical linearization approximation.

Corollary 9.6 (Statistical linearization of an additive transform). *When* $\mathbf{x} \sim \mathrm{N}(\mathbf{m}, \mathbf{P})$ *and* $\mathbf{y} = \mathbf{g}(\mathbf{x}) + \mathbf{q}$ *with* $\mathbf{q} \sim \mathrm{N}(\mathbf{0}, \mathbf{Q})$, *statistical linearization gives rise to a Gaussian approximation*

$$\begin{pmatrix} \mathbf{x} \\ \mathbf{y} \end{pmatrix} \sim \mathrm{N}\left(\begin{pmatrix} \mathbf{m} \\ \boldsymbol{\mu}_\mathrm{S} \end{pmatrix}, \begin{pmatrix} \mathbf{P} & \mathbf{C}_\mathrm{S} \\ \mathbf{C}_\mathrm{S}^\mathsf{T} & \mathbf{S}_\mathrm{S} \end{pmatrix} \right), \tag{9.19}$$

where

$$\begin{aligned} \boldsymbol{\mu}_\mathrm{S} &= \mathrm{E}[\mathbf{g}(\mathbf{x})], \\ \mathbf{S}_\mathrm{S} &= \mathrm{E}[\mathbf{g}(\mathbf{x})\,(\mathbf{x} - \mathbf{m})^\mathsf{T}]\,\mathbf{P}^{-1}\,\mathrm{E}[(\mathbf{x} - \mathbf{m})\,\mathbf{g}(\mathbf{x})^\mathsf{T}] + \mathbf{Q}, \\ \mathbf{C}_\mathrm{S} &= \mathrm{E}[(\mathbf{x} - \mathbf{m})\,\mathbf{g}(\mathbf{x})^\mathsf{T}], \end{aligned} \tag{9.20}$$

with the expectations taken with respect to $\mathbf{x} \sim \mathrm{N}(\mathbf{m}, \mathbf{P})$.

Proof This results directly from Corollary 9.4. □

It is now useful to compare the approximation above to the moment matching approximation in Algorithm 8.1. We can rewrite the above explicitly in terms of integrals as follows:

$$\begin{aligned} \boldsymbol{\mu}_\mathrm{S} &= \int \mathbf{g}(\mathbf{x})\,\mathrm{N}(\mathbf{x} \mid \mathbf{m}, \mathbf{P})\,\mathrm{d}\mathbf{x}, \\ \mathbf{S}_\mathrm{S} &= \left[\int (\mathbf{g}(\mathbf{x}) - \boldsymbol{\mu}_\mathrm{S})\,(\mathbf{x} - \mathbf{m})^\mathsf{T}\,\mathrm{N}(\mathbf{x} \mid \mathbf{m}, \mathbf{P})\,\mathrm{d}\mathbf{x} \right] \mathbf{P}^{-1} \\ &\quad \times \left[\int (\mathbf{x} - \mathbf{m})\,(\mathbf{g}(\mathbf{x}) - \boldsymbol{\mu}_\mathrm{S})^\mathsf{T}\,\mathrm{N}(\mathbf{x} \mid \mathbf{m}, \mathbf{P})\,\mathrm{d}\mathbf{x} \right] + \mathbf{Q}, \\ \mathbf{C}_\mathrm{S} &= \int (\mathbf{x} - \mathbf{m})\,(\mathbf{g}(\mathbf{x}) - \boldsymbol{\mu}_\mathrm{S})^\mathsf{T}\,\mathrm{N}(\mathbf{x} \mid \mathbf{m}, \mathbf{P})\,\mathrm{d}\mathbf{x}. \end{aligned} \tag{9.21}$$

By comparing to Equation (8.4) we see that we indeed have $\boldsymbol{\mu}_\mathrm{S} = \boldsymbol{\mu}_\mathrm{M}$ and $\mathbf{C}_\mathrm{S} = \mathbf{C}_\mathrm{M}$, but $\mathbf{S}_\mathrm{S} \neq \mathbf{S}_\mathrm{M}$. In fact, we always have $\mathbf{S}_\mathrm{S} \preceq \mathbf{S}_\mathrm{M}$, that is, the approximation to the covariance of $\mathbf{y}$ produced by SL is generally smaller than the approximation produced by moment matching, and thus it is likely to underestimate the uncertainty in $\mathbf{y}$. This is a problem caused the deterministic approximation used by SL, and we come back to a solution to this problem in Section 9.4 by introducing statistical linear regression (SLR), which can be seen as stochastic version of SL.

Statistical linearization for additive noise models is closely related to a first order Taylor expansion. If the function $\mathbf{g}(\mathbf{x})$ is differentiable, we can use the following well-known property of Gaussian random variables to highlight the similarities:

$$\mathrm{E}\left[\mathbf{g}(\mathbf{x})\,(\mathbf{x}-\mathbf{m})^{\mathsf{T}}\right] = \mathrm{E}\left[\mathbf{G}_{\mathbf{x}}(\mathbf{x})\right]\,\mathbf{P}, \tag{9.22}$$

where $\mathrm{E}[\cdot]$ denotes the expected value with respect to $\mathbf{x} \sim \mathrm{N}(\mathbf{m}, \mathbf{P})$, and $\mathbf{G}_{\mathbf{x}}(\mathbf{x})$ is the Jacobian matrix of $\mathbf{g}(\mathbf{x})$. Given this result, the matrix in the SL simplifies to

$$\mathbf{A} = \mathrm{E}[\mathbf{g}(\mathbf{x})\,(\mathbf{x}-\mathbf{m})^{\mathsf{T}}]\,\mathbf{P}^{-1} = \mathrm{E}\left[\mathbf{G}_{\mathbf{x}}(\mathbf{x})\right]. \tag{9.23}$$

Using this expression, the similarities between a first order Taylor expansion

$$\mathbf{g}(\mathbf{x}) \simeq \mathbf{G}_{\mathbf{x}}(\mathbf{m})\,(\mathbf{x}-\mathbf{m}) + \mathbf{g}(\mathbf{m}) \tag{9.24}$$

and an SL

$$\mathbf{g}(\mathbf{x}) \simeq \mathrm{E}[\mathbf{G}_{\mathbf{x}}(\mathbf{x})]\,(\mathbf{x}-\mathbf{m}) + \mathrm{E}[\mathbf{g}(\mathbf{x})] \tag{9.25}$$

become even more evident. That is, the Taylor expansion uses the Jacobian $\mathbf{G}_{\mathbf{x}}(\mathrm{E}[\mathbf{x}])$ whereas SL uses $\mathrm{E}[\mathbf{G}_{\mathbf{x}}(\mathbf{x})]$, and where the Taylor expansion uses $\mathbf{g}(\mathrm{E}[\mathbf{x}])$, SL uses $\mathrm{E}[\mathbf{g}(\mathbf{x})]$.

Example 9.7 (Statistical linearization). *Let us compare a first order Taylor expansion about $x = 1$ with two different statistical linearizations (SLs) of a scalar function $g(x) = x^3$. In this example, the SLs assume that $x \sim \mathrm{N}(m, P)$ where $m = 1$, but the first SL uses $P = 0.1$ whereas the second uses $P = 1$. We can obtain the SLs by computing the integrals*

$$\begin{aligned} A &= \mathrm{E}[G_x(x)] \\ &= \int 3x^2\,\mathrm{N}(x \mid m, P)\,\mathrm{d}x, \\ \mu &= \int x^3\,\mathrm{N}(x \mid m, P)\,\mathrm{d}x \end{aligned} \tag{9.26}$$

and then setting $g(x) \simeq A\,(x-m) + \mu$.

The involved functions are illustrated in Figure 9.1. As can be seen, the SL with $P = 0.1$ is similar to the Taylor expansion, and they are both accurate near $x = m = 1$. The SLs are optimal on average (in the MSE sense) when x is distributed as $\mathrm{N}(m, P)$, and the SL with $P = 1$ seeks to approximate $g(x) = x^3$ for a larger range of x values. We note that

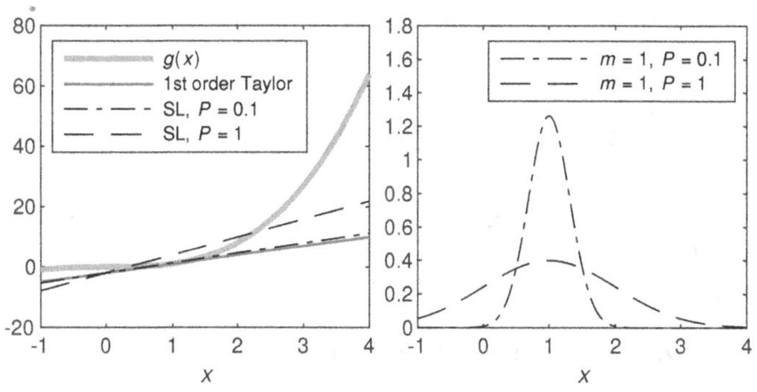

(a) (b)

Figure 9.1 (a) The function $g(x) = x^3$ and the three statistical linearizations in Example 9.7. (b) The two prior densities.

all three linearizations are different even though they are all performed around $x = 1$.

We can also use SL as an enabling approximation for the non-additive models in (9.9), that is, to approximate $\mathbf{f}(\mathbf{x}_{k-1}, \mathbf{q}_{k-1})$ and $\mathbf{h}(\mathbf{x}_k, \mathbf{r}_k)$. To describe the method, suppose $\mathbf{x} \sim \mathrm{N}(\mathbf{m}, \mathbf{P})$ and $\mathbf{q} \sim \mathrm{N}(\mathbf{0}, \mathbf{Q})$ are independent random variables and that we would like to approximate $\mathbf{g}(\mathbf{x}, \mathbf{q})$. It is possible to directly use SL and Theorem 9.3 to find $\mathbf{g}(\mathbf{x}, \mathbf{q}) \simeq \mathbf{A}\,\mathbf{x} + \mathbf{b}$. In this case, the noise $\mathbf{q}$ may influence how we select $\mathbf{A}$ and $\mathbf{b}$, but $\mathbf{A}\,\mathbf{x} + \mathbf{b}$ is still a deterministic function, and it is generally undesirable to approximate the dynamic or measurement models as noise free.

A more suitable approach is to introduce

$$\mathbf{z} = \begin{pmatrix} \mathbf{x} \\ \mathbf{q} \end{pmatrix} \tag{9.27}$$

and to find an SL

$$\begin{aligned} \mathbf{y} &\simeq \mathbf{A}^z\,\mathbf{z} + \mathbf{b} \\ &= \begin{pmatrix} \mathbf{A} & \mathbf{A}^q \end{pmatrix} \begin{pmatrix} \mathbf{x} \\ \mathbf{q} \end{pmatrix} + \mathbf{b} \\ &= \mathbf{A}\,\mathbf{x} + \mathbf{A}^q\,\mathbf{q} + \mathbf{b}, \end{aligned} \tag{9.28}$$

where $\mathbf{A}^z = \begin{pmatrix} \mathbf{A} & \mathbf{A}^q \end{pmatrix}$. Here $\mathbf{A}^q\,\mathbf{q}$ represents the additive noise which means that the approximation in (9.28) is not noise free. For instance, with this approach, SL of the dynamic model yields an approximation

$$\mathbf{f}(\mathbf{x}_{k-1}, \mathbf{q}_{k-1}) \simeq \mathbf{A}\,\mathbf{x}_{k-1} + \mathbf{A}^{q}\,\mathbf{q}_{k-1} + \mathbf{a}_{k-1}. \tag{9.29}$$

Setting $\tilde{\mathbf{e}}_{k-1} = \mathbf{A}^{q}\,\mathbf{q}_{k-1} \sim \mathrm{N}(\mathbf{0}, \mathbf{A}^{q}\,\mathbf{Q}_{k-1}\,(\mathbf{A}^{q})^{\mathsf{T}})$ implies that $\mathbf{\Lambda}_{k-1} = \mathbf{A}^{q}\,\mathbf{Q}_{k-1}\,(\mathbf{A}^{q})^{\mathsf{T}}$.

The approximation $\mathbf{g}(\mathbf{x}, \mathbf{q}) \simeq \mathbf{A}\,\mathbf{x} + \mathbf{A}^{q}\,\mathbf{q} + \mathbf{b}$ is called a statistical linearization if $\mathbf{A}$, $\mathbf{A}^{q}$, and $\mathbf{b}$ are selected to minimize the MSE, as described in Corollary 9.8.

Corollary 9.8 (Statistical linearization III). *The approximation* $\mathbf{y} = \mathbf{g}(\mathbf{x}, \mathbf{q}) \simeq \mathbf{A}\,\mathbf{x} + \mathbf{A}^{q}\,\mathbf{q} + \mathbf{b}$ *minimizes the MSE*

$$\begin{aligned}\mathrm{MSE}(\mathbf{A}, \mathbf{A}^{q}, \mathbf{b}) = \mathrm{E}\big[&(\mathbf{g}(\mathbf{x}, \mathbf{q}) - \mathbf{A}\,\mathbf{x} - \mathbf{A}^{q}\,\mathbf{q} - \mathbf{b})^{\mathsf{T}} \\ &\times (\mathbf{g}(\mathbf{x}, \mathbf{q}) - \mathbf{A}\,\mathbf{x} - \mathbf{A}^{q}\,\mathbf{q} - \mathbf{b})\big]\end{aligned} \tag{9.30}$$

when

$$\begin{aligned}\mathbf{A} &= \mathrm{Cov}[\mathbf{y}, \mathbf{x}]\,\mathbf{P}^{-1} = \mathrm{E}[\mathbf{g}(\mathbf{x}, \mathbf{q})\,(\mathbf{x} - \mathbf{m})^{\mathsf{T}}]\,\mathbf{P}^{-1}, \\ \mathbf{A}^{q} &= \mathrm{Cov}[\mathbf{y}, \mathbf{q}]\,\mathbf{Q}^{-1} = \mathrm{E}[\mathbf{g}(\mathbf{x}, \mathbf{q})\,\mathbf{q}^{\mathsf{T}}]\,\mathbf{Q}^{-1}, \\ \mathbf{b} &= \mathrm{E}[\mathbf{g}(\mathbf{x}, \mathbf{q})] - \mathbf{A}\,\mathbf{m},\end{aligned} \tag{9.31}$$

where all expected values are taken with respect to $\mathbf{x} \sim \mathrm{N}(\mathbf{m}, \mathbf{P})$*, and* $\mathbf{q} \sim \mathrm{N}(\mathbf{0}, \mathbf{Q})$.

Proof The corollary follows from Theorem 9.3, see Exercise 9.10 for details. □

Statistical linearization of $\mathbf{g}(\mathbf{x})$ can also be seen as the first order Fourier–Hermite series expansion of the function (Sarmavuori and Särkkä, 2012) with respect to the Gaussian distribution $\mathbf{x} \sim \mathrm{N}(\mathbf{m}, \mathbf{P})$. This point of view enables the extension of statistical linearization to quadratic approximations and beyond, but these approximations are not discussed here in detail.

9.3 Statistically Linearized Filter

A filter that uses statistical linearization as an enabling approximation is called a statistically linearized filter (SLF). An SLF (Gelb, 1974), or quasi-linear filter (Stengel, 1994), is similar to the EKF, except that statistical linearizations are used instead of the Taylor series approximations.

If we use the SL in Corollary 9.5 for the additive noise model in Equation (9.1) and combine it with Algorithm 9.1, we obtain the following algorithm.

Algorithm 9.9 (Statistically linearized filter I). *The prediction and update steps of the additive noise statistically linearized (Kalman) filter for the model in Equation* (9.1) *are:*

- *Prediction:*

$$\begin{aligned} \mathbf{m}_k^- &= \mathrm{E}[\mathbf{f}(\mathbf{x}_{k-1})], \\ \mathbf{P}_k^- &= \mathrm{E}[\mathbf{f}(\mathbf{x}_{k-1})\,\delta\mathbf{x}_{k-1}^\mathsf{T}]\,\mathbf{P}_{k-1}^{-1}\,\mathrm{E}[\mathbf{f}(\mathbf{x}_{k-1})\,\delta\mathbf{x}_{k-1}^\mathsf{T}]^\mathsf{T} + \mathbf{Q}_{k-1}, \end{aligned} \tag{9.32}$$

  where $\delta\mathbf{x}_{k-1} = \mathbf{x}_{k-1} - \mathbf{m}_{k-1}$ *and the expectations are taken with respect to the variable* $\mathbf{x}_{k-1} \sim \mathrm{N}(\mathbf{m}_{k-1}, \mathbf{P}_{k-1})$.
- *Update:*

$$\begin{aligned} \boldsymbol{\mu}_k &= \mathrm{E}[\mathbf{h}(\mathbf{x}_k)], \\ \mathbf{S}_k &= \mathrm{E}[\mathbf{h}(\mathbf{x}_k)\,\delta\tilde{\mathbf{x}}_k^\mathsf{T}]\,(\mathbf{P}_k^-)^{-1}\,\mathrm{E}[\mathbf{h}(\mathbf{x}_k)\,\delta\tilde{\mathbf{x}}_k^\mathsf{T}]^\mathsf{T} + \mathbf{R}_k, \\ \mathbf{K}_k &= \mathrm{E}[\mathbf{h}(\mathbf{x}_k)\,\delta\tilde{\mathbf{x}}_k^\mathsf{T}]^\mathsf{T}\,\mathbf{S}_k^{-1}, \\ \mathbf{m}_k &= \mathbf{m}_k^- + \mathbf{K}_k\,(\mathbf{y}_k - \boldsymbol{\mu}_k), \\ \mathbf{P}_k &= \mathbf{P}_k^- - \mathbf{K}_k\,\mathbf{S}_k\,\mathbf{K}_k^\mathsf{T}, \end{aligned} \tag{9.33}$$

  where $\delta\tilde{\mathbf{x}}_k = \mathbf{x}_k - \mathbf{m}_k^-$ *and the expectations are taken with respect to the variable* $\mathbf{x}_k \sim \mathrm{N}(\mathbf{m}_k^-, \mathbf{P}_k^-)$.

We can also combine Corollary 9.8 with Algorithm 9.1 to obtain the statistically linearized filter for the non-additive noise models in (9.9).

Algorithm 9.10 (Statistically linearized filter II). *The prediction and update steps of the non-additive-noise statistically linearized (Kalman) filter for the model in Equation* (9.9) *are:*

- *Prediction:*

$$\begin{aligned} \mathbf{m}_k^- &= \mathrm{E}[\mathbf{f}(\mathbf{x}_{k-1}, \mathbf{q}_{k-1})], \\ \mathbf{P}_k^- &= \mathrm{E}[\mathbf{f}(\mathbf{x}_{k-1}, \mathbf{q}_{k-1})\,\delta\mathbf{x}_{k-1}^\mathsf{T}]\,\mathbf{P}_{k-1}^{-1}\,\mathrm{E}[\mathbf{f}(\mathbf{x}_{k-1}, \mathbf{q}_{k-1})\,\delta\mathbf{x}_{k-1}^\mathsf{T}]^\mathsf{T} \\ &\quad + \mathrm{E}[\mathbf{f}(\mathbf{x}_{k-1}, \mathbf{q}_{k-1})\,\mathbf{q}_{k-1}^\mathsf{T}]\,\mathbf{Q}_{k-1}^{-1}\,\mathrm{E}[\mathbf{f}(\mathbf{x}_{k-1}, \mathbf{q}_{k-1})\,\mathbf{q}_{k-1}^\mathsf{T}]^\mathsf{T}, \end{aligned} \tag{9.34}$$

  where $\delta\mathbf{x}_{k-1} = \mathbf{x}_{k-1} - \mathbf{m}_{k-1}$ *and the expectations are taken with respect to the variables* $\mathbf{x}_{k-1} \sim \mathrm{N}(\mathbf{m}_{k-1}, \mathbf{P}_{k-1})$ *and* $\mathbf{q}_{k-1} \sim \mathrm{N}(\mathbf{0}, \mathbf{Q}_{k-1})$.

- *Update:*

$$
\begin{aligned}
\boldsymbol{\mu}_k &= \mathrm{E}[\mathbf{h}(\mathbf{x}_k, \mathbf{r}_k)], \\
\mathbf{S}_k &= \mathrm{E}[\mathbf{h}(\mathbf{x}_k, \mathbf{r}_k)\, \delta\tilde{\mathbf{x}}_k^\mathsf{T}]\, (\mathbf{P}_k^-)^{-1}\, \mathrm{E}[\mathbf{h}(\mathbf{x}_k, \mathbf{r}_k)\, \delta\tilde{\mathbf{x}}_k^\mathsf{T}]^\mathsf{T} \\
&\quad + \mathrm{E}[\mathbf{h}(\mathbf{x}_k, \mathbf{r}_k)\, \mathbf{r}_k^\mathsf{T}]\, \mathbf{R}_k^{-1}\, \mathrm{E}[\mathbf{h}(\mathbf{x}_k, \mathbf{r}_k)\, \mathbf{r}_k^\mathsf{T}]^\mathsf{T}, \\
\mathbf{K}_k &= \mathrm{E}[\mathbf{h}(\mathbf{x}_k, \mathbf{r}_k)\, \delta\tilde{\mathbf{x}}_k^\mathsf{T}]^\mathsf{T}\, \mathbf{S}_k^{-1}, \\
\mathbf{m}_k &= \mathbf{m}_k^- + \mathbf{K}_k\, (\mathbf{y}_k - \boldsymbol{\mu}_k), \\
\mathbf{P}_k &= \mathbf{P}_k^- - \mathbf{K}_k\, \mathbf{S}_k\, \mathbf{K}_k^\mathsf{T},
\end{aligned}
\tag{9.35}
$$

where $\delta\tilde{\mathbf{x}}_k = \mathbf{x}_k - \mathbf{m}_k^-$ *and the expectations are taken with respect to the variables* $\mathbf{x}_k \sim \mathrm{N}(\mathbf{m}_k^-, \mathbf{P}_k^-)$ *and* $\mathbf{r}_k \sim \mathrm{N}(\mathbf{0}, \mathbf{R}_k)$.

The advantage of the SLF over the EKF is that it is a more global approximation than the EKF, because the linearization is not only based on the local region around the mean but on a whole range of function values (see (9.24) and (9.25) for a comparison between a Taylor expansion and SL in the additive noise case). In addition, the non-linearities do not have to be differentiable since the expressions in our filters do not involve Jacobians. However, if the non-linearities *are* differentiable, then we can use the Gaussian random variable property in (9.22) to rewrite the equations in an EKF-like form, as indicated by (9.23).

An important disadvantage with the SLF described above is that the expected values of the non-linear functions have to be computed in closed form. Naturally, this is not possible for all functions. Fortunately, the expected values involved are of such a type that one is likely to find many of them tabulated in older physics and control engineering books (see, e.g., Gelb and Vander Velde, 1968). Also, when the expected values do not have closed-form expressions, we can instead use a Gaussian quadrature technique, such as those presented in Chapter 8, see Exercise 9.3 for an example. However, if we are willing to compute the expectations numerically anyway, it is better to use statistical linear regression instead, which we discuss next.

9.4 Statistical Linear Regression

Statistical linearization (SL) provides a deterministic approximation to the relation between two random variables $\mathbf{x}$ and $\mathbf{y}$. However, even if $\mathbf{y} = \mathbf{g}(\mathbf{x})$, such that the true relation between $\mathbf{x}$ and $\mathbf{y}$ is deterministic, it can still be valuable to approximate the relation as an affine function with Gaussian noise, to reflect the fact that we have introduced errors when approximating the function as affine. Additionally, in Section 9.2 from Equations (9.21) we also saw that the deterministic approximation produced by SL failed to match the moment matching approximation for stochastic transformation of the form $\mathbf{y} = \mathbf{g}(\mathbf{x}) + \mathbf{q}$ for the covariance of $\mathbf{y}$, which is sometimes a problem.

In this section, we describe statistical linear regression (SLR), which is an enabling approximation that includes a strategy for selecting the noise covariance (Arasaratnam et al., 2007) accounting for the approximation error. The fact that SLR does not yield a deterministic approximation makes it suitable for combination with all the considered model families (see Equations (9.1), (9.9), and (9.10)). It also resolves the aforementioned problem of approximating the covariance of $\mathbf{y}$ properly.

In SLR, the affine approximation is selected as in a statistical linearization, and the noise covariance is the covariance of the approximation error.

Definition 9.11 (Statistical linear regression). *If* $\mathbf{x}$ *and* $\mathbf{y}$ *are two random variables,* $\mathbf{y} \simeq \mathbf{A}\,\mathbf{x} + \mathbf{b} + \tilde{\mathbf{e}}$, $\tilde{\mathbf{e}} \sim \mathrm{N}(\mathbf{0}, \boldsymbol{\Lambda})$ *is called a* statistical linear regression *if* $\mathbf{A}$ *and* $\mathbf{b}$ *are selected to minimize the mean squared error*

$$\mathrm{MSE}(\mathbf{A}, \mathbf{b}) = \mathrm{E}\left[(\mathbf{y} - \mathbf{A}\,\mathbf{x} - \mathbf{b})^\mathsf{T} (\mathbf{y} - \mathbf{A}\,\mathbf{x} - \mathbf{b})\right], \tag{9.36}$$

where $\mathbf{A}$ *and* $\mathbf{b}$ *are deterministic variables, whereas the noise covariance is selected as*

$$\boldsymbol{\Lambda} = \mathrm{E}\left[(\mathbf{y} - \mathbf{A}\,\mathbf{x} - \mathbf{b})\,(\mathbf{y} - \mathbf{A}\,\mathbf{x} - \mathbf{b})^\mathsf{T}\right]. \tag{9.37}$$

Theorem 9.12 (Statistical linear regression I). *Suppose* $\mathbf{x}$ *and* $\mathbf{y}$ *are two random variables, and let* $\mathbf{m} = \mathrm{E}[\mathbf{x}]$, $\mathbf{P} = \mathrm{Cov}[\mathbf{x}]$, $\boldsymbol{\mu}_\mathrm{R} = \mathrm{E}[\mathbf{y}]$, $\mathbf{S}_\mathrm{R} = \mathrm{Cov}[\mathbf{y}]$, *and* $\mathbf{C}_\mathrm{R} = \mathrm{Cov}[\mathbf{x}, \mathbf{y}]$. *In the statistical linear regression (SLR),* $\mathbf{y} \simeq \mathbf{A}\,\mathbf{x} + \mathbf{b} + \tilde{\mathbf{e}}$, $\tilde{\mathbf{e}} \sim \mathrm{N}(\mathbf{0}, \boldsymbol{\Lambda})$, *we then have*

$$\begin{aligned}
\mathbf{A} &= \mathbf{C}_\mathrm{R}^\mathsf{T}\,\mathbf{P}^{-1}, \\
\mathbf{b} &= \boldsymbol{\mu}_\mathrm{R} - \mathbf{A}\,\mathbf{m}, \\
\boldsymbol{\Lambda} &= \mathbf{S}_\mathrm{R} - \mathbf{A}\,\mathbf{P}\,\mathbf{A}^\mathsf{T}.
\end{aligned} \tag{9.38}$$

Proof The expressions for $\mathbf{A}$ and $\mathbf{b}$ follow from Theorem 9.3. The expression for $\boldsymbol{\Lambda}$ follows from its definition

$$\begin{aligned}
\boldsymbol{\Lambda} &= \mathrm{Cov}[\mathbf{y} - \mathbf{A}\,\mathbf{x} - \mathbf{b}] \\
&= \mathrm{E}\left[(\mathbf{y} - \boldsymbol{\mu}_\mathrm{R} - \mathbf{A}\,(\mathbf{x} - \mathbf{m}))\,(\mathbf{y} - \boldsymbol{\mu}_\mathrm{R} - \mathbf{A}\,(\mathbf{x} - \mathbf{m}))^\mathsf{T}\right] \\
&= \mathrm{Cov}[\mathbf{y}] + \mathbf{A}\,\mathbf{P}\,\mathbf{A}^\mathsf{T} - \mathbf{A}\,\mathbf{C}_\mathrm{R} - \mathbf{C}_\mathrm{R}^\mathsf{T}\,\mathbf{A}^\mathsf{T} \\
&= \mathbf{S}_\mathrm{R} - \mathbf{A}\,\mathbf{P}\,\mathbf{A}^\mathsf{T},
\end{aligned} \tag{9.39}$$

where $\boldsymbol{\mu}_{\mathrm{R}} = \mathrm{E}[\mathbf{y}]$, and the last equality follows from $\mathbf{C}_{\mathrm{R}} = \mathbf{P}\,\mathbf{A}^{\mathsf{T}}$. □

To understand why the SLR selects the noise covariance using (9.37), we note that this preserves the first two moments of the joint distribution of $\mathbf{x}$ and $\mathbf{y}$:

Corollary 9.13 (Gaussian approximation from SLR). *When* $\mathbf{x} \sim \mathrm{N}(\mathbf{m}, \mathbf{P})$, $\boldsymbol{\mu}_{\mathrm{R}} = \mathrm{E}[\mathbf{y}]$, $\mathbf{C}_{\mathrm{R}} = \mathrm{Cov}[\mathbf{x}, \mathbf{y}]$, *and* $\mathbf{S}_{\mathrm{R}} = \mathrm{Cov}[\mathbf{y}]$, *the statistical linearization* $\mathbf{y} \simeq \mathbf{A}\,\mathbf{x} + \mathbf{b} + \tilde{\mathbf{e}}$, $\tilde{\mathbf{e}} \sim \mathrm{N}(\mathbf{0}, \boldsymbol{\Lambda})$ *gives rise to a Gaussian approximation*

$$\begin{pmatrix} \mathbf{x} \\ \mathbf{y} \end{pmatrix} \sim \mathrm{N}\left(\begin{pmatrix} \mathbf{m} \\ \boldsymbol{\mu}_{\mathrm{R}} \end{pmatrix}, \begin{pmatrix} \mathbf{P} & \mathbf{C}_{\mathrm{R}} \\ \mathbf{C}_{\mathrm{R}}^{\mathsf{T}} & \mathbf{S}_{\mathrm{R}} \end{pmatrix} \right). \tag{9.40}$$

Proof We know that $\mathbf{y} \simeq \mathbf{A}\,\mathbf{x} + \mathbf{b} + \tilde{\mathbf{e}}$ is Gaussian when $\mathbf{x}$ and $\tilde{\mathbf{e}}$ are Gaussian, and it follows from Theorem 9.12 that

$$\begin{aligned}
\mathrm{E}\left[\begin{pmatrix} \mathbf{x} \\ \mathbf{A}\,\mathbf{x} + \mathbf{b} + \tilde{\mathbf{e}} \end{pmatrix} \right] &= \begin{pmatrix} \mathbf{m} \\ \mathbf{A}\,\mathbf{m} + \mathbf{b} \end{pmatrix} \\
&= \begin{pmatrix} \mathbf{m} \\ \boldsymbol{\mu}_{\mathrm{R}} \end{pmatrix}, \\
\mathrm{Cov}\left[\begin{pmatrix} \mathbf{x} \\ \mathbf{A}\mathbf{x} + \mathbf{b} + \tilde{\mathbf{e}} \end{pmatrix} \right] &= \begin{pmatrix} \mathbf{P} & \mathbf{P}\,\mathbf{A}^{\mathsf{T}} \\ \mathbf{A}\,\mathbf{P} & \mathbf{A}\,\mathbf{P}\,\mathbf{A}^{\mathsf{T}} + \boldsymbol{\Lambda} \end{pmatrix} \\
&= \begin{pmatrix} \mathbf{P} & \mathbf{C}_{\mathrm{R}} \\ \mathbf{C}_{\mathrm{R}}^{\mathsf{T}} & \mathbf{S}_{\mathrm{R}} \end{pmatrix}.
\end{aligned} \tag{9.41}$$

□

Importantly, when $\mathbf{x} \sim \mathrm{N}(\mathbf{m}, \mathbf{P})$, the SLR approximates the joint distribution of $\mathbf{x}$ and $\mathbf{y}$ as Gaussian where the first two moments match the original joint distribution. When applied to a transformation of the form $\mathbf{y} = \mathbf{g}(\mathbf{x}) + \mathbf{q}$, $\mathbf{q} \sim \mathrm{N}(\mathbf{0}, \mathbf{Q})$, the SLR approximation exactly reproduces the moment matching approximation in Algorithm 8.1 as can be seen from the following.

Corollary 9.14 (SLR of an additive transform). *Suppose* $\mathbf{y} = \mathbf{g}(\mathbf{x}) + \mathbf{q}$, *where* $\mathbf{x} \sim \mathrm{N}(\mathbf{m}, \mathbf{P})$ *and* $\mathbf{q} \sim \mathrm{N}(\mathbf{0}, \mathbf{Q})$. *The SLR forms the approximation* $\mathbf{y} \simeq \mathbf{A}\,\mathbf{x} + \mathbf{b} + \tilde{\mathbf{e}}$, $\tilde{\mathbf{e}} \sim \mathrm{N}(\mathbf{0}, \boldsymbol{\Lambda})$, *where, as in Equation* (9.38), *we have*

$$\begin{aligned}
\mathbf{A} &= \mathbf{C}_{\mathrm{R}}^{\mathsf{T}}\,\mathbf{P}^{-1}, \\
\mathbf{b} &= \boldsymbol{\mu}_{\mathrm{R}} - \mathbf{A}\,\mathbf{m}, \\
\boldsymbol{\Lambda} &= \mathbf{S}_{\mathrm{R}} - \mathbf{A}\,\mathbf{P}\,\mathbf{A}^{\mathsf{T}},
\end{aligned} \tag{9.42}$$

where

$$\mu_R = \int g(x)\, N(x \mid m, P)\, dx,$$
$$C_R = \int (x - m)\,(g(x) - \mu_R)^\top\, N(x \mid m, P)\, dx, \tag{9.43}$$
$$S_R = \int (g(x) - \mu_R)\,(g(x) - \mu_R)^\top\, N(x \mid m, P)\, dx + Q.$$

Furthermore, the corresponding joint Gaussian approximation for (x, y) *is given by*

$$\begin{pmatrix} x \\ y \end{pmatrix} \sim N\left(\begin{pmatrix} m \\ \mu_R \end{pmatrix}, \begin{pmatrix} P & C_R \\ C_R^\top & S_R \end{pmatrix} \right). \tag{9.44}$$

As we can see, the expressions for μ_R, C_R, and S_R in Equation (9.43) are identical to the expressions for μ_M, C_M, and S_M in Equation (8.4). SLR therefore provides the same approximation to (x, y) as the Gaussian moment matching approach used in Chapter 8 with $\mu_R = \mu_M$, $C_R = C_M$, and $S_R = S_M$. This also implies that all the numerical integration methods such as Gauss–Hermite and spherically symmetric cubature methods can be used to numerically compute the integrals in Equations (9.43) in the same way as we did in Chapter 8.

To compare the SLR approximation $y \simeq A\,x + b + \tilde{e}$, $\tilde{e} \sim N(0, \Lambda)$ with the SL approximation $y \simeq A\,x + b$, we note that A and b are identical in both approximations. The difference is that the SLR includes the additional noise $\tilde{e}$, which represents the errors introduced when approximating y as an affine function of x. In terms of the Gaussian approximations of (x, y) that SL and SLR provide (see Corollaries 9.4 and 9.13), it holds that $\mu_R = \mu_S$ and $C_R = C_S$. Considering that $S_S = A\,P\,A^\top$ and $S_R = A\,P\,A^\top + \Lambda$, it follows that $S_R \succeq S_S$, because the error covariance Λ is always positive semi-definite. That is, compared to SL, SLR introduces an additional noise to ensure that the approximation of (x, y) matches the original moments.

When we construct an enabling approximation of $y = g(x) + q$ using SL, we first obtain A and b by performing SL on $g(x)$, and we then select $\Lambda_{SL} = Q$. Note that SL is only used to approximate $g(x)$ and that the noise is simply added to the SL approximation. Using SLR we directly linearize $y = g(x) + q \simeq A\,x + b + \tilde{e}$, $\tilde{e} \sim N(0, \Lambda)$. We obtain the same A and b as the SL, but the noise covariance is different, $\Lambda \neq \Lambda_{SL}$. Since q has zero mean and x and q are independent, it follows that

$$\begin{aligned} \Lambda &= E[(g(x) + q - A\,x - b)\,(g(x) + q - A\,x - b)^\top] \\ &= E[(g(x) - A\,x - b)\,(g(x) - A\,x - b)^\top] + Q. \end{aligned} \tag{9.45}$$

Since $\mathbf{\Lambda}_{\text{SL}} = \mathbf{Q}$, we conclude that SLR adds the matrix $\mathrm{E}[(\mathbf{g}(\mathbf{x}) - \mathbf{A}\,\mathbf{x} - \mathbf{b})\,(\mathbf{g}(\mathbf{x}) - \mathbf{A}\,\mathbf{x} - \mathbf{b})^{\mathsf{T}}]$ to the noise covariance used by SL. Note that this additional term represents the errors introduced when approximating $\mathbf{g}(\mathbf{x})$ as affine.

Although SLR yields the same Gaussian approximation to $(\mathbf{x}, \mathbf{y})$ as the moment matching approach in Chapter 8, it is still useful in its own right. First, we obtain a unified description of all our Gaussian filters as different types of linearizations of state space models. Second, the SLR perspective provides insights that can be used to develop further linearizations and filters as we see in Chapter 10.

To illustrate the differences between SLR and SL, we can consider an example without additional noise.

Example 9.15 (Statistical linear regression). *Let us return to the functions and random variables in Example 9.7. If we wish to perform SLR on the same random variables, and find* $y = g(x) = x^3 \simeq A\,x + b + e$, $e \sim \mathrm{N}(0, \Lambda)$, *we would obtain the functions* $A\,x + b$ *illustrated in Figure 9.1, since* A *and* b *are identical in SL and SLR.*

In contrast to SL, SLR also contains a noise term $e \sim \mathrm{N}(0, \Lambda)$. *In this particular example, the noise term has the standard deviation* $\sqrt{\Lambda} \approx 4.86$ *when* $P = 1$ *and* $\sqrt{\Lambda} \approx 0.43$ *when* $P = 0.1$. *We note that the error is smaller when* P *is smaller. The reason for this is that* Λ *is the variance of* $g(x) - A\,x - b$ *when* $x \sim \mathrm{N}(m, P)$, *and that error is generally smaller closer to* m. *In fact, as* $P \to 0$, *both the SL and the SLR converge to the noise-free first order Taylor expansion* $y \approx g(m) + G_x(m)\,(x - m)$.

We can also use SLR for models with non-additive noise (9.9) and with conditional distributions (9.10). For these models, we can formulate expressions for $\boldsymbol{\mu}_{\text{R}}$, $\mathbf{C}_{\text{R}}$, and $\mathbf{S}_{\text{R}}$ directly from their definitions, but we also use conditional moments (Tronarp et al., 2018) to find alternative expressions for them; for additive noise models both versions are identical.

Let us now assume that instead of a functional relationship such as $\mathbf{y} = \mathbf{g}(\mathbf{x}) + \mathbf{q}$, we have only been given a conditional distribution $p(\mathbf{y} \mid \mathbf{x})$. Although this distribution can be arbitrarily complicated and non-Gaussian (cf. Chapter 5), we can still often compute its conditional moments

$$\begin{aligned} \boldsymbol{\mu}_{\text{R}}(\mathbf{x}) &= \mathrm{E}[\mathbf{y} \mid \mathbf{x}] = \int \mathbf{y}\, p(\mathbf{y} \mid \mathbf{x})\,\mathrm{d}\mathbf{y}, \\ \mathbf{S}_{\text{R}}(\mathbf{x}) &= \mathrm{Cov}[\mathbf{y} \mid \mathbf{x}] = \int (\mathbf{y} - \boldsymbol{\mu}_{\text{R}}(\mathbf{x}))\,(\mathbf{y} - \boldsymbol{\mu}_{\text{R}}(\mathbf{x}))^{\mathsf{T}}\, p(\mathbf{y} \mid \mathbf{x})\,\mathrm{d}\mathbf{y}. \end{aligned} \tag{9.46}$$

For example, when $\mathbf{y} = \mathbf{g}(\mathbf{x}) + \mathbf{q}$ with $\mathbf{q} \sim \mathrm{N}(\mathbf{0}, \mathbf{Q})$, we have $\boldsymbol{\mu}_{\mathrm{R}}(\mathbf{x}) = \mathbf{g}(\mathbf{x})$ and $\mathbf{S}_{\mathrm{R}}(\mathbf{x}) = \mathbf{Q}$. SLR can be reformulated in terms of the conditional moments as follows.

Theorem 9.16 (Conditional moments form of SLR joint moments). *In the statistical linear regression (SLR), see Theorem 9.12, the moments $\boldsymbol{\mu}_{\mathrm{R}}$, $\mathbf{C}_{\mathrm{R}}$, and $\mathbf{S}_{\mathrm{R}}$ can be expressed as functions of the conditional moments $\boldsymbol{\mu}_{\mathrm{R}}(\mathbf{x}) = \mathrm{E}[\mathbf{y} \mid \mathbf{x}]$ and $\mathbf{S}_{\mathrm{R}}(\mathbf{x}) = \mathrm{Cov}[\mathbf{y} \mid \mathbf{x}]$ as follows:*

$$\begin{aligned} \boldsymbol{\mu}_{\mathrm{R}} &= \mathrm{E}[\boldsymbol{\mu}_{\mathrm{R}}(\mathbf{x})], \\ \mathbf{C}_{\mathrm{R}} &= \mathrm{E}[(\mathbf{x} - \mathbf{m})\,(\boldsymbol{\mu}_{\mathrm{R}}(\mathbf{x}) - \boldsymbol{\mu}_{\mathrm{R}})^{\mathsf{T}}], \\ \mathbf{S}_{\mathrm{R}} &= \mathrm{Cov}[\boldsymbol{\mu}_{\mathrm{R}}(\mathbf{x})] + \mathrm{E}[\mathbf{S}_{\mathrm{R}}(\mathbf{x})]. \end{aligned} \tag{9.47}$$

Proof The relation $\boldsymbol{\mu}_{\mathrm{R}} = \mathrm{E}[\boldsymbol{\mu}_{\mathrm{R}}(\mathbf{x})]$ is a direct consequence of the law of iterated expectations (LIE): $\mathrm{E}[\mathbf{y}] = \mathrm{E}[\mathrm{E}[\mathbf{y} \mid \mathbf{x}]]$. Combining LIE with the definition of $\mathbf{C}_{\mathrm{R}}$ gives

$$\begin{aligned} \mathbf{C}_{\mathrm{R}} &= \mathrm{E}[\mathrm{E}[(\mathbf{x} - \mathbf{m})\,(\mathbf{y} - \boldsymbol{\mu}_{\mathrm{R}})^{\mathsf{T}} \mid \mathbf{x}]] \\ &= \mathrm{E}[(\mathbf{x} - \mathbf{m})\,(\boldsymbol{\mu}_{\mathrm{R}}(\mathbf{x}) - \boldsymbol{\mu}_{\mathrm{R}})^{\mathsf{T}}]. \end{aligned} \tag{9.48}$$

Finally, we can use LIE to derive the expression for $\mathbf{S}_{\mathrm{R}}$ as follows:

$$\begin{aligned} \mathbf{S}_{\mathrm{R}} &= \mathrm{E}[\mathrm{E}[(\mathbf{y} - \boldsymbol{\mu}_{\mathrm{R}})\,(\mathbf{y} - \boldsymbol{\mu}_{\mathrm{R}})^{\mathsf{T}} \mid \mathbf{x}]] \\ &= \mathrm{E}[\mathrm{E}[(\mathbf{y} - \boldsymbol{\mu}_{\mathrm{R}}(\mathbf{x}) + \boldsymbol{\mu}_{\mathrm{R}}(\mathbf{x}) - \boldsymbol{\mu}_{\mathrm{R}})\,(\mathbf{y} - \boldsymbol{\mu}_{\mathrm{R}}(\mathbf{x}) + \boldsymbol{\mu}_{\mathrm{R}}(\mathbf{x}) - \boldsymbol{\mu}_{\mathrm{R}})^{\mathsf{T}} \mid \mathbf{x}]] \\ &= \mathrm{E}[\mathrm{E}[(\mathbf{y} - \boldsymbol{\mu}_{\mathrm{R}}(\mathbf{x}))\,(\mathbf{y} - \boldsymbol{\mu}_{\mathrm{R}}(\mathbf{x}))^{\mathsf{T}} \mid \mathbf{x}]] \\ &\quad + \mathrm{E}[\mathrm{E}[(\boldsymbol{\mu}_{\mathrm{R}}(\mathbf{x}) - \boldsymbol{\mu}_{\mathrm{R}})\,(\boldsymbol{\mu}_{\mathrm{R}}(\mathbf{x}) - \boldsymbol{\mu}_{\mathrm{R}})^{\mathsf{T}} \mid \mathbf{x}]] \\ &= \mathrm{E}[\mathbf{S}_{\mathrm{R}}(\mathbf{x})] + \mathrm{Cov}[\boldsymbol{\mu}_{\mathrm{R}}(\mathbf{x})]. \end{aligned} \tag{9.49}$$

□

To obtain enabling approximations for the non-additive noise models in (9.9), we show how to perform SLR on $\mathbf{y} \sim \mathbf{g}(\mathbf{x}, \mathbf{q})$, $\mathbf{q} \sim \mathrm{N}(\mathbf{0}, \mathbf{Q})$.

Corollary 9.17 (Non-additive transform in conditional moment form). *Let $\mathbf{y} = \mathbf{g}(\mathbf{x}, \mathbf{q})$, where $\mathbf{x} \sim \mathrm{N}(\mathbf{m}, \mathbf{P})$ and $\mathbf{q} \sim \mathrm{N}(\mathbf{0}, \mathbf{Q})$ are independent random variables. The SLR, $\mathbf{y} \simeq \mathbf{A}\,\mathbf{x} + \mathbf{b} + \tilde{\mathbf{e}}$, $\tilde{\mathbf{e}} \sim \mathrm{N}(\mathbf{0}, \boldsymbol{\Lambda})$, is given by (9.38) where $\boldsymbol{\mu}_{\mathrm{R}}$, $\mathbf{C}_{\mathrm{R}}$, and $\mathbf{S}_{\mathrm{R}}$ can be expressed in two different forms. First,* *their definitions give*

$$\begin{aligned}
\boldsymbol{\mu}_{\mathrm{R}} &= \int \mathbf{g}(\mathbf{x},\mathbf{q})\, \mathrm{N}(\mathbf{x} \mid \mathbf{m},\mathbf{P})\, \mathrm{N}(\mathbf{q} \mid \mathbf{0},\mathbf{Q})\, \mathrm{d}\mathbf{x}\, \mathrm{d}\mathbf{q}, \\
\mathbf{C}_{\mathrm{R}} &= \int (\mathbf{x}-\mathbf{m})\, (\mathbf{g}(\mathbf{x},\mathbf{q}) - \boldsymbol{\mu}_{\mathrm{R}})^{\mathsf{T}}\, \mathrm{N}(\mathbf{x} \mid \mathbf{m},\mathbf{P})\, \mathrm{N}(\mathbf{q} \mid \mathbf{0},\mathbf{Q})\, \mathrm{d}\mathbf{x}\, \mathrm{d}\mathbf{q}, \\
\mathbf{S}_{\mathrm{R}} &= \int (\mathbf{g}(\mathbf{x},\mathbf{q}) - \boldsymbol{\mu}_{\mathrm{R}})\, (\mathbf{g}(\mathbf{x},\mathbf{q}) - \boldsymbol{\mu}_{\mathrm{R}})^{\mathsf{T}}\, \mathrm{N}(\mathbf{x} \mid \mathbf{m},\mathbf{P})\, \mathrm{N}(\mathbf{q} \mid \mathbf{0},\mathbf{Q})\, \mathrm{d}\mathbf{x}\, \mathrm{d}\mathbf{q}.
\end{aligned} \tag{9.50}$$

Second, from (9.47) *it follows that using the conditional moments*

$$\begin{aligned}
\boldsymbol{\mu}_{\mathrm{R}}(\mathbf{x}) &= \int \mathbf{g}(\mathbf{x},\mathbf{q})\, \mathrm{N}(\mathbf{q} \mid \mathbf{0},\mathbf{Q})\, \mathrm{d}\mathbf{q}, \\
\mathbf{S}_{\mathrm{R}}(\mathbf{x}) &= \int (\mathbf{g}(\mathbf{x},\mathbf{q}) - \boldsymbol{\mu}_{\mathrm{R}}(\mathbf{x}))\, (\mathbf{g}(\mathbf{x},\mathbf{q}) - \boldsymbol{\mu}_{\mathrm{R}}(\mathbf{x}))^{\mathsf{T}}\, \mathrm{N}(\mathbf{q} \mid \mathbf{0},\mathbf{Q})\, \mathrm{d}\mathbf{q},
\end{aligned} \tag{9.51}$$

we get

$$\begin{aligned}
\boldsymbol{\mu}_{\mathrm{R}} &= \int \boldsymbol{\mu}_{\mathrm{R}}(\mathbf{x})\, \mathrm{N}(\mathbf{x} \mid \mathbf{m},\mathbf{P})\, \mathrm{d}\mathbf{x}, \\
\mathbf{C}_{\mathrm{R}} &= \int (\mathbf{x}-\mathbf{m})\, (\boldsymbol{\mu}_{\mathrm{R}}(\mathbf{x}) - \boldsymbol{\mu}_{\mathrm{R}})^{\mathsf{T}}\, \mathrm{N}(\mathbf{x} \mid \mathbf{m},\mathbf{P})\, \mathrm{d}\mathbf{x}, \\
\mathbf{S}_{\mathrm{R}} &= \int \big[(\boldsymbol{\mu}_{\mathrm{R}}(\mathbf{x}) - \boldsymbol{\mu}_{\mathrm{R}})\, (\boldsymbol{\mu}_{\mathrm{R}}(\mathbf{x}) - \boldsymbol{\mu}_{\mathrm{R}})^{\mathsf{T}}\, \mathrm{N}(\mathbf{x} \mid \mathbf{m},\mathbf{P}) \\
&\qquad + \mathbf{S}_{\mathrm{R}}(\mathbf{x})\, \mathrm{N}(\mathbf{x} \mid \mathbf{m},\mathbf{P}) \big]\, \mathrm{d}\mathbf{x}.
\end{aligned} \tag{9.52}$$

The expressions in Corollary 9.17 can be used to develop Gaussian filters by approximating the integrals in (9.50) or (9.51) and (9.52) using, for instance, a sigma point method. A potential disadvantage with using (9.51) and (9.52) is that the integrals are nested and that (9.51) should be evaluated for every $\mathbf{x}$ in (9.52). In comparison, (9.50) jointly integrates over both $\mathbf{x}$ and $\mathbf{q}$.

Finally, the SLR can also used for the conditional distribution models in (9.10). To this end, we describe the expressions for the variables in Theorem 9.12 for models with conditional distributions.

Corollary 9.18 (Conditional moments of conditional distributions)**.** *Let* $\mathbf{x} \sim \mathrm{N}(\mathbf{m},\mathbf{P})$ *and* $\mathbf{y} \sim p(\mathbf{y} \mid \mathbf{x})$. *In the SLR,* $\mathbf{y} \simeq \mathbf{A}\,\mathbf{x} + \mathbf{b} + \tilde{\mathbf{e}}$, $\tilde{\mathbf{e}} \sim \mathrm{N}(\mathbf{0}, \boldsymbol{\Lambda})$, *we have* $\mathbf{A}$, $\mathbf{b}$, *and* $\mathbf{\Lambda}$ *as in* (9.38), *where*

$$\begin{aligned} \boldsymbol{\mu}_\mathrm{R} &= \int \mathbf{y}\, p(\mathbf{y} \mid \mathbf{x})\, \mathrm{N}(\mathbf{x} \mid \mathbf{m}, \mathbf{P})\, \mathrm{d}\mathbf{y}\, \mathrm{d}\mathbf{x}, \\ \mathbf{C}_\mathrm{R} &= \int (\mathbf{x} - \mathbf{m})\, (\mathbf{y} - \boldsymbol{\mu}_\mathrm{R})^\mathsf{T}\, p(\mathbf{y} \mid \mathbf{x})\, \mathrm{N}(\mathbf{x} \mid \mathbf{m}, \mathbf{P})\, \mathrm{d}\mathbf{y}\, \mathrm{d}\mathbf{x}, \\ \mathbf{S}_\mathrm{R} &= \int (\mathbf{y} - \boldsymbol{\mu}_\mathrm{R})\, (\mathbf{y} - \boldsymbol{\mu}_\mathrm{R})^\mathsf{T}\, p(\mathbf{y} \mid \mathbf{x})\, \mathrm{N}(\mathbf{x} \mid \mathbf{m}, \mathbf{P})\, \mathrm{d}\mathbf{y}\, \mathrm{d}\mathbf{x}. \end{aligned} \tag{9.53}$$

We can also express these moments using (9.52), *where the conditional moments are given by*

$$\begin{aligned} \boldsymbol{\mu}_\mathrm{R}(\mathbf{x}) &= \int \mathbf{y}\, p(\mathbf{y} \mid \mathbf{x})\, \mathrm{d}\mathbf{y}, \\ \mathbf{S}_\mathrm{R}(\mathbf{x}) &= \int (\mathbf{y} - \boldsymbol{\mu}_\mathrm{R}(\mathbf{x}))\, (\mathbf{y} - \boldsymbol{\mu}_\mathrm{R}(\mathbf{x}))^\mathsf{T}\, p(\mathbf{y} \mid \mathbf{x})\, \mathrm{d}\mathbf{y}. \end{aligned} \tag{9.54}$$

It should be noted that the random variable $\mathbf{y}$ may also be discrete valued. In that case, the integrals with respect to $\mathbf{y}$ in Corollary 9.18 should be replaced by summations. Also, the integrals with respect to $\mathbf{x}$ can be approximated using Gaussian quadrature methods (such as a sigma point method), whereas the integrals with respect to $\mathbf{y}$ may require other approximations unless $p(\mathbf{y} \mid \mathbf{x})$ is also Gaussian.

Interestingly, the expressions in (9.52) are valid for all three model families. The difference is that the expressions for the conditional moments vary. For the additive noise case, we have $\boldsymbol{\mu}_\mathrm{R}(\mathbf{x}) = \mathbf{g}(\mathbf{x})$ and $\mathbf{S}_\mathrm{R}(\mathbf{x}) = \mathbf{Q}$ (and (9.52) then simplifies to (9.43)); for the non-additive models, these functions are described in (9.51) and for conditional distributions in (9.54). The expressions in (9.52) are particularly useful when the conditional moments have closed form solutions, such as in the following example.

Example 9.19 (SLR with Poisson observations). *Let the state x be distributed as* $\mathrm{N}(m, P)$, *and suppose* $\exp(x)$ *represents the size of a population. For population models, it is common to assume Poisson-distributed observations*

$$y \sim \mathrm{Po}(\alpha \exp(x)). \tag{9.55}$$

Recall from Chapter 5 that the probability density (or mass) function of a Poisson-distributed random variable with parameter $\lambda(x)$ is

$$\mathrm{Po}(y \mid \lambda(x)) = \exp(-\lambda(x))\, \frac{\lambda(x)^y}{y!}, \tag{9.56}$$

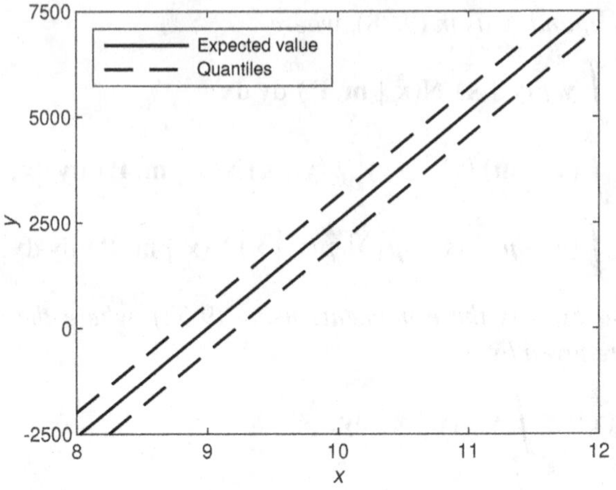

Figure 9.2 The figure illustrates $y \simeq A\,x + b + e$ in Example 9.19. The solid line shows $A\,x + b$, whereas the dashed lines show $A\,x + b \pm \sqrt{\Lambda}$, where $A\,x + b$ is the expected value of $y \mid x$, whereas the quantiles $A\,x + b \pm \sqrt{\Lambda}$ represent values one standard deviation above and below the mean.

and both its mean and variance are $\lambda(x)$. That is, for this model the conditional moments in (9.54) *have closed form solutions, and it holds that $\mu(x) = \mathrm{E}[y \mid x]$ and $P_y(x)$ are both $\lambda(x) = \alpha \exp(x)$.*

Consequently, to perform SLR we can use (9.52) *to identify the involved moments*

$$\begin{aligned} \mu &= \int \alpha \exp(x)\, \mathrm{N}(x \mid m, P)\, \mathrm{d}x, \\ P_{xy} &= \int (x - m)\, (\alpha \exp(x) - \mu)\, \mathrm{N}(x \mid m, P)\, \mathrm{d}x, \\ P_y &= \int \left(\alpha \exp(x) + (\alpha \exp(x) - \mu)^2\right) \mathrm{N}(x \mid m, P)\, \mathrm{d}x. \end{aligned} \tag{9.57}$$

The above expected values can be approximated numerically or solved analytically, see Exercise 9.5. From the obtained values, we compute the linearization parameters $A = P_{xy}/P$, $b = \mu - A\,m$ and $\Lambda = P_y - A^2\,P$. The resulting linearization when $m = 10$ and $P = 0.3$ is shown in Figure 9.2.

9.5 Statistical Linear Regression Filters

Shortly after the introduction of the unscented Kalman filter (Julier et al., 2000), the filter was re-derived using statistical linear regression (SLR) (Lefebvre et al., 2002). SLR was later used to re-derive the Gauss–Hermite Kalman filter (Arasaratnam et al., 2007). More recently, SLR has also been used to develop filters and smoothers for state space models expressed using conditional moments (Tronarp et al., 2018).

Let us first present a naive version of the statistical linear regression filter (SLRF) for additive noise models, where we combine Corollary 9.14 with Algorithm 9.1 without any simplifications. This algorithm will help us understand and explain several of the upcoming algorithms.

Algorithm 9.20 (Statistical linear regression filter I). *The prediction and update steps of an additive noise statistical linear regression (Kalman) filter for the model in Equation* (9.1) *are:*

- *Prediction:*

$$\begin{aligned} \mathbf{A}_{k-1} &= (\mathbf{P}_k^{\mathrm{xx}})^\mathsf{T}\, \mathbf{P}_{k-1}^{-1}, \\ \mathbf{a}_{k-1} &= \boldsymbol{\mu}_k^- - \mathbf{A}_{k-1}\, \mathbf{m}_{k-1}, \\ \mathbf{\Lambda}_{k-1} &= \mathbf{P}_k^{\mathrm{x}} - \mathbf{A}_{k-1}\, \mathbf{P}_{k-1}\, \mathbf{A}_{k-1}^\mathsf{T}, \end{aligned} \tag{9.58}$$

where

$$\begin{aligned} \boldsymbol{\mu}_k^- &= \int \mathbf{f}(\mathbf{x})\, \mathrm{N}(\mathbf{x} \mid \mathbf{m}_{k-1}, \mathbf{P}_{k-1})\, \mathrm{d}\mathbf{x}, \\ \mathbf{P}_k^{\mathrm{xx}} &= \int (\mathbf{x} - \mathbf{m}_{k-1})\, (\mathbf{f}(\mathbf{x}) - \boldsymbol{\mu}_k^-)^\mathsf{T}\, \mathrm{N}(\mathbf{x} \mid \mathbf{m}_{k-1}, \mathbf{P}_{k-1})\, \mathrm{d}\mathbf{x}, \\ \mathbf{P}_k^{\mathrm{x}} &= \int (\mathbf{f}(\mathbf{x}) - \boldsymbol{\mu}_k^-)\, (\mathbf{f}(\mathbf{x}) - \boldsymbol{\mu}_k^-)^\mathsf{T}\, \mathrm{N}(\mathbf{x} \mid \mathbf{m}_{k-1}, \mathbf{P}_{k-1})\, \mathrm{d}\mathbf{x} + \mathbf{Q}_{k-1}. \end{aligned} \tag{9.59}$$

Compute the predicted moments

$$\begin{aligned} \mathbf{m}_k^- &= \mathbf{A}_{k-1}\, \mathbf{m}_{k-1} + \mathbf{a}_{k-1}, \\ \mathbf{P}_k^- &= \mathbf{A}_{k-1}\, \mathbf{P}_{k-1}\, \mathbf{A}_{k-1}^\mathsf{T} + \mathbf{\Lambda}_{k-1}. \end{aligned} \tag{9.60}$$

- *Update:*

$$\begin{aligned} \mathbf{H}_k &= (\mathbf{P}_k^{\mathrm{xy}})^\mathsf{T}\, (\mathbf{P}_k^-)^{-1}, \\ \mathbf{b}_k &= \boldsymbol{\mu}_k^+ - \mathbf{H}_k\, \mathbf{m}_k^-, \\ \mathbf{\Omega}_k &= \mathbf{P}_k^{\mathrm{y}} - \mathbf{H}_k\, \mathbf{P}_k^-\, \mathbf{H}_k^\mathsf{T}, \end{aligned} \tag{9.61}$$

where

$$\boldsymbol{\mu}_k^+ = \int \mathbf{h}_k(\mathbf{x})\, \mathrm{N}(\mathbf{x} \mid \mathbf{m}_k^-, \mathbf{P}_k^-)\, \mathrm{d}\mathbf{x},$$
$$\mathbf{P}_k^{\mathrm{xy}} = \int (\mathbf{x} - \mathbf{m}_k^-)\, (\mathbf{h}_k(\mathbf{x}) - \boldsymbol{\mu}_k^+)^\mathsf{T}\, \mathrm{N}(\mathbf{x} \mid \mathbf{m}_k^-, \mathbf{P}_k^-)\, \mathrm{d}\mathbf{x},$$
$$\mathbf{P}_k^{\mathrm{y}} = \int (\mathbf{h}(\mathbf{x}) - \boldsymbol{\mu}_k^+)\, (\mathbf{h}(\mathbf{x}) - \boldsymbol{\mu}_k^+)^\mathsf{T}\, \mathrm{N}(\mathbf{x} \mid \mathbf{m}_k^-, \mathbf{P}_k^-)\, \mathrm{d}\mathbf{x} + \mathbf{R}_k. \tag{9.62}$$

Compute the updated moments

$$\begin{aligned}
\boldsymbol{\mu}_k &= \mathbf{H}_k\, \mathbf{m}_k^- + \mathbf{b}_k, \\
\mathbf{S}_k &= \mathbf{H}_k\, \mathbf{P}_k^-\, \mathbf{H}_k^\mathsf{T} + \boldsymbol{\Omega}_k, \\
\mathbf{K}_k &= \mathbf{P}_k^-\, \mathbf{H}_k^\mathsf{T}\, \mathbf{S}_k^{-1}, \\
\mathbf{m}_k &= \mathbf{m}_k^- + \mathbf{K}_k\, (\mathbf{y}_k - \boldsymbol{\mu}_k), \\
\mathbf{P}_k &= \mathbf{P}_k^- - \mathbf{K}_k\, \mathbf{S}_k\, \mathbf{K}_k^\mathsf{T}.
\end{aligned} \tag{9.63}$$

If we analyse the equations in Algorithm 9.20, we notice that some of the calculations are redundant. Specifically, we can show that $\mathbf{m}_k^- = \boldsymbol{\mu}_k^-$ and $\mathbf{P}_k^- = \mathbf{P}_k^{\mathrm{x}}$ in the prediction step, and that $\boldsymbol{\mu}_k = \boldsymbol{\mu}_k^+$ and $\mathbf{S}_k = \mathbf{P}_k^{\mathrm{y}}$ in the update step. If we also introduce $\mathbf{C}_k = \mathbf{P}_k^{\mathrm{xy}}$, the algorithm can be simplified to the following filter.

Algorithm 9.21 (Statistical linear regression filter II). *The prediction and update steps of an additive noise statistical linear regression (Kalman) filter for the model in Equation* (9.1) *are:*

- *Prediction:*

$$\mathbf{m}_k^- = \int \mathbf{f}(\mathbf{x})\, \mathrm{N}(\mathbf{x} \mid \mathbf{m}_{k-1}, \mathbf{P}_{k-1})\, \mathrm{d}\mathbf{x},$$
$$\mathbf{P}_k^- = \int (\mathbf{f}(\mathbf{x}) - \mathbf{m}_k^-)\, (\mathbf{f}(\mathbf{x}) - \mathbf{m}_k^-)^\mathsf{T}\, \mathrm{N}(\mathbf{x} \mid \mathbf{m}_{k-1}, \mathbf{P}_{k-1})\, \mathrm{d}\mathbf{x} + \mathbf{Q}_{k-1}. \tag{9.64}$$

- *Update:*

$$\begin{aligned}
\boldsymbol{\mu}_k &= \int \mathbf{h}_k(\mathbf{x})\, \mathrm{N}(\mathbf{x} \mid \mathbf{m}_k^-, \mathbf{P}_k^-)\, \mathrm{d}\mathbf{x}, \\
\mathbf{C}_k &= \int (\mathbf{x} - \mathbf{m}_k^-)\, (\mathbf{h}_k(\mathbf{x}) - \boldsymbol{\mu}_k)^\mathsf{T}\, \mathrm{N}(\mathbf{x} \mid \mathbf{m}_k^-, \mathbf{P}_k^-)\, \mathrm{d}\mathbf{x}, \\
\mathbf{S}_k &= \int (\mathbf{h}(\mathbf{x}) - \boldsymbol{\mu}_k)\, (\mathbf{h}(\mathbf{x}) - \boldsymbol{\mu}_k)^\mathsf{T}\, \mathrm{N}(\mathbf{x} \mid \mathbf{m}_k^-, \mathbf{P}_k^-)\, \mathrm{d}\mathbf{x} + \mathbf{R}_k.
\end{aligned} \tag{9.65}$$

Compute the updated moments

$$\begin{aligned}\mathbf{K}_k &= \mathbf{C}_k\,\mathbf{S}_k^{-1},\\ \mathbf{m}_k &= \mathbf{m}_k^- + \mathbf{K}_k\,(\mathbf{y}_k - \boldsymbol{\mu}_k),\\ \mathbf{P}_k &= \mathbf{P}_k^- - \mathbf{K}_k\,\mathbf{S}_k\,\mathbf{K}_k^\mathsf{T}.\end{aligned} \tag{9.66}$$

Importantly, Algorithm 9.21 is identical to Algorithm 8.3, and we have therefore demonstrated that the general Gaussian filters derived from Algorithm 8.3 can be viewed as a combination of SLR and Algorithm 9.1. That is, the general Gaussian filters for additive noise models presented in Chapter 8 (e.g., UKF, CKF, GHKF) implicitly use SLR as an enabling approximation.

We can also use SLR to develop filters for the non-additive noise models in Equation (9.9) and for the general conditional distribution models in Equation (9.10). To modify Algorithm 9.20 to work for non-additive or conditional models, only the expressions for the moments in Equations (9.59) and (9.62) change, and the rest of the algorithm remains identical. Interestingly, if we modify Algorithm 9.21 to handle non-additive noise, by plugging in the expressions from (9.50), we obtain Algorithm 8.4 after some simplifications, see Exercise 9.8. That is, both Algorithm 8.3 and 8.4 correspond to SLR filters, which confirms that all the general Gaussian filters presented in Chapter 8 implicitly use SLR as an enabling approximation.

Let us now consider a general model in the form of Equation (9.10), which is expressed in terms of the conditional distributions describing the dynamic and measurement models:

$$\begin{aligned}\mathbf{x}_k &\sim p(\mathbf{x}_k \mid \mathbf{x}_{k-1}),\\ \mathbf{y}_k &\sim p(\mathbf{y}_k \mid \mathbf{x}_k).\end{aligned} \tag{9.67}$$

Further assume that we can compute the following conditional moments (analytically or numerically):

$$\begin{aligned}
\boldsymbol{\mu}_k^-(\mathbf{x}_{k-1}) &= \mathrm{E}\,[\mathbf{x}_k \mid \mathbf{x}_{k-1}] \\
&= \int \mathbf{x}_k \, p(\mathbf{x}_k \mid \mathbf{x}_{k-1}) \, \mathrm{d}\mathbf{x}_k, \\
\mathbf{P}_k^{\mathrm{x}}(\mathbf{x}_{k-1}) &= \mathrm{Cov}\,[\mathbf{x}_k \mid \mathbf{x}_{k-1}] \\
&= \int (\mathbf{x}_k - \boldsymbol{\mu}_k^-(\mathbf{x}_{k-1}))\,(\mathbf{x}_k - \boldsymbol{\mu}_k^-(\mathbf{x}_{k-1}))^\mathsf{T} \, p(\mathbf{x}_k \mid \mathbf{x}_{k-1}) \, \mathrm{d}\mathbf{x}_k, \\
\boldsymbol{\mu}_k(\mathbf{x}_k) &= \mathrm{E}\,[\mathbf{y}_k \mid \mathbf{x}_k] \\
&= \int \mathbf{y}_k \, p(\mathbf{y}_k \mid \mathbf{x}_k) \, \mathrm{d}\mathbf{y}_k, \\
\mathbf{P}_k^{\mathrm{y}}(\mathbf{x}_k) &= \mathrm{Cov}\,[\mathbf{y}_k \mid \mathbf{x}_k] \\
&= \int (\mathbf{y}_k - \boldsymbol{\mu}_k(\mathbf{x}_k))\,(\mathbf{y}_k - \boldsymbol{\mu}_k(\mathbf{x}_k))^\mathsf{T} \, p(\mathbf{y}_k \mid \mathbf{x}_k) \, \mathrm{d}\mathbf{y}_k.
\end{aligned} \tag{9.68}$$

We now get the following SLR-based filter, which uses the conditional moments above.

Algorithm 9.22 (Statistical linear regression filter III). *The prediction and update steps of the conditional moments formulation of the SLR filter are:*

- *Prediction:*

$$\begin{aligned}
\mathbf{m}_k^- &= \int \boldsymbol{\mu}_k^-(\mathbf{x}_{k-1}) \, \mathrm{N}(\mathbf{x}_{k-1} \mid \mathbf{m}_{k-1}, \mathbf{P}_{k-1}) \, \mathrm{d}\mathbf{x}_{k-1}, \\
\mathbf{P}_k^- &= \int (\boldsymbol{\mu}_k^-(\mathbf{x}_{k-1}) - \mathbf{m}_k^-)\,(\boldsymbol{\mu}_k^-(\mathbf{x}_{k-1}) - \mathbf{m}_k^-)^\mathsf{T} \\
&\qquad \times \mathrm{N}(\mathbf{x}_{k-1} \mid \mathbf{m}_{k-1}, \mathbf{P}_{k-1}) \, \mathrm{d}\mathbf{x}_{k-1} \\
&\quad + \int \mathbf{P}_k^{\mathrm{x}}(\mathbf{x}_{k-1}) \, \mathrm{N}(\mathbf{x}_{k-1} \mid \mathbf{m}_{k-1}, \mathbf{P}_{k-1}) \, \mathrm{d}\mathbf{x}_{k-1}.
\end{aligned} \tag{9.69}$$

- *Update:*

$$
\begin{aligned}
\mu_k &= \int \mu_k(\mathbf{x}_k)\, \mathrm{N}(\mathbf{x}_k \mid \mathbf{m}_k^-, \mathbf{P}_k^-)\, \mathrm{d}\mathbf{x}_k, \\
\mathbf{C}_k &= \int (\mathbf{x}_k - \mathbf{m}_k^-)\, (\mu_k(\mathbf{x}_k) - \mu_k)^\mathsf{T}\, \mathrm{N}(\mathbf{x}_k \mid \mathbf{m}_k^-, \mathbf{P}_k^-)\, \mathrm{d}\mathbf{x}_k, \\
\mathbf{S}_k &= \int (\mu_k(\mathbf{x}_k) - \mu_k)\, (\mu_k(\mathbf{x}_k) - \mu_k)^\mathsf{T}\, \mathrm{N}(\mathbf{x}_k \mid \mathbf{m}_k^-, \mathbf{P}_k^-)\, \mathrm{d}\mathbf{x}_k \\
&\quad + \int \mathbf{P}_k^\mathrm{y}(\mathbf{x}_k)\, \mathrm{N}(\mathbf{x}_k \mid \mathbf{m}_k^-, \mathbf{P}_k^-)\, \mathrm{d}\mathbf{x}_k.
\end{aligned}
\tag{9.70}
$$

Compute the updated mean and covariance

$$
\begin{aligned}
\mathbf{K}_k &= \mathbf{C}_k\, \mathbf{S}_k^{-1}, \\
\mathbf{m}_k &= \mathbf{m}_k^- + \mathbf{K}_k\, (\mathbf{y}_k - \mu_k), \\
\mathbf{P}_k &= \mathbf{P}_k^- - \mathbf{K}_k\, \mathbf{S}_k\, \mathbf{K}_k^\mathsf{T}.
\end{aligned}
\tag{9.71}
$$

For example, in the special case of the non-additive Gaussian noise model in Equation (9.9), we have

$$
\begin{aligned}
\mu_k^-(\mathbf{x}_{k-1}) &= \int \mathbf{f}(\mathbf{x}_{k-1}, \mathbf{q}_{k-1})\, \mathrm{N}(\mathbf{q}_{k-1} \mid \mathbf{0}, \mathbf{Q}_{k-1})\, \mathrm{d}\mathbf{q}_{k-1}, \\
\mathbf{P}_k^\mathrm{x}(\mathbf{x}_{k-1}) &= \int (\mathbf{f}(\mathbf{x}_{k-1}, \mathbf{q}_{k-1}) - \mu_k^-(\mathbf{x}_{k-1})) \\
&\quad \times (\mathbf{f}(\mathbf{x}_{k-1}, \mathbf{q}_{k-1}) - \mu_k^-(\mathbf{x}_{k-1}))^\mathsf{T} \\
&\quad \times \mathrm{N}(\mathbf{q}_{k-1} \mid \mathbf{0}, \mathbf{Q}_{k-1})\, \mathrm{d}\mathbf{q}_{k-1}
\end{aligned}
\tag{9.72}
$$

and

$$
\begin{aligned}
\mu_k(\mathbf{x}_k) &= \int \mathbf{h}(\mathbf{x}_k, \mathbf{r}_k)\, \mathrm{N}(\mathbf{r}_k \mid \mathbf{0}, \mathbf{R}_k)\, \mathrm{d}\mathbf{r}_k, \\
\mathbf{P}_k^\mathrm{y}(\mathbf{x}_k) &= \int (\mathbf{h}(\mathbf{x}_k, \mathbf{r}_k) - \mu_k(\mathbf{x}_k))\, (\mathbf{h}(\mathbf{x}_k, \mathbf{r}_k) - \mu_k(\mathbf{x}_k))^\mathsf{T} \\
&\quad \times \mathrm{N}(\mathbf{r}_k \mid \mathbf{0}, \mathbf{R}_k)\, \mathrm{d}\mathbf{r}_k.
\end{aligned}
\tag{9.73}
$$

The conditional moment formulations arguably give rise to algorithms with more involved expressions, but they also have some advantages. For example, to modify Algorithm 9.22 to handle additive or conditional distribution models, we only need to modify (9.72) and (9.73), which is useful from a software engineering point of view.

In situations when we have conditional distribution models as described in Equation (9.10) (and (9.67)), but the conditional moments in (9.68) are inconvenient to compute, we can instead use Corollary 9.18 to derive the following Gaussian filter.

Algorithm 9.23 (Statistical linear regression filter IV). *The prediction and update steps of a conditional distribution statistical linear regression (Kalman) filter for the model in Equation* (9.10) *are:*

- *Prediction:*

$$\begin{aligned}
\mathbf{m}_k^- &= \int \mathbf{x}_k \, p(\mathbf{x}_k \mid \mathbf{x}_{k-1}) \, \mathrm{N}(\mathbf{x}_{k-1} \mid \mathbf{m}_{k-1}, \mathbf{P}_{k-1}) \, \mathrm{d}\mathbf{x}_k \, \mathrm{d}\mathbf{x}_{k-1}, \\
\mathbf{P}_k^- &= \int (\mathbf{x}_k - \mathbf{m}_k^-) \, (\mathbf{x}_k - \mathbf{m}_k^-)^\mathsf{T} \\
&\quad \times p(\mathbf{x}_k \mid \mathbf{x}_{k-1}) \, \mathrm{N}(\mathbf{x}_{k-1} \mid \mathbf{m}_{k-1}, \mathbf{P}_{k-1}) \, \mathrm{d}\mathbf{x}_k \, \mathrm{d}\mathbf{x}_{k-1}.
\end{aligned} \tag{9.74}$$

- *Update:*

$$\begin{aligned}
\boldsymbol{\mu}_k &= \int \mathbf{y}_k \, p(\mathbf{y}_k \mid \mathbf{x}_k) \, \mathrm{N}(\mathbf{x}_k \mid \mathbf{m}_k^-, \mathbf{P}_k^-) \, \mathrm{d}\mathbf{y}_k \, \mathrm{d}\mathbf{x}_k, \\
\mathbf{C}_k &= \int (\mathbf{x}_k - \mathbf{m}_k^-) \, (\mathbf{y}_k - \boldsymbol{\mu}_k)^\mathsf{T} \\
&\quad \times p(\mathbf{y}_k \mid \mathbf{x}_k) \, \mathrm{N}(\mathbf{x}_k \mid \mathbf{m}_k^-, \mathbf{P}_k^-) \, \mathrm{d}\mathbf{y}_k \, \mathrm{d}\mathbf{x}_k, \\
\mathbf{S}_k &= \int (\mathbf{y}_k - \boldsymbol{\mu}_k^-) \, (\mathbf{y}_k - \boldsymbol{\mu}_k)^\mathsf{T} \\
&\quad \times p(\mathbf{y}_k \mid \mathbf{x}_k) \, \mathrm{N}(\mathbf{x}_k \mid \mathbf{m}_k^-, \mathbf{P}_k^-) \, \mathrm{d}\mathbf{y}_k \, \mathrm{d}\mathbf{x}_k.
\end{aligned} \tag{9.75}$$

Compute the updated moments

$$\begin{aligned}
\mathbf{K}_k &= \mathbf{C}_k \, \mathbf{S}_k^{-1}, \\
\mathbf{m}_k &= \mathbf{m}_k^- + \mathbf{K}_k \, (\mathbf{y}_k - \boldsymbol{\mu}_k), \\
\mathbf{P}_k &= \mathbf{P}_k^- - \mathbf{K}_k \, \mathbf{S}_k \, \mathbf{K}_k^\mathsf{T}.
\end{aligned} \tag{9.76}$$

Let us next take a look at how SLR filters can be implemented using sigma-point and Monte Carlo approximations.

9.6 Practical SLR Filters

From practical algorithm point of view, the final conclusions of this chapter are a bit dull: to implement SLR filters, the key issue is to compute the moments $\boldsymbol{\mu}_k$, $\mathbf{C}_k$, and $\mathbf{S}_k$. Given these moments, we can, in principle, form a statistical linear regression-based linearization and apply an affine Kalman filter. However, it turns out that this procedure is equivalent to just using these same moments in a Gaussian filter, which we already introduced in

Chapter 8, and thus by using methods such as unscented transform, Gauss–Hermite integration, and spherical cubature integration, we recover the corresponding filters already introduced earlier. Still, during the development of SLR methods we managed to extend the classes of models that SLR filters and hence Gaussian filters can be applied to. In this section, we expand on the set of Gaussian filters by introducing tractable versions of some remaining SLR filters.

One of the model classes that we did not consider in the Gaussian filtering chapter is general conditional distribution models of the form given in Equation (9.10). In order to implement a general form of SLR-based filtering for these models using numerical integration methods (i.e., sigma-point methods) as in Chapter 8, we need to have a method to compute the following conditional moments introduced in Equation (9.68):

$$\begin{aligned} \boldsymbol{\mu}_k^-(\mathbf{x}_{k-1}) &= \mathrm{E}\left[\mathbf{x}_k \mid \mathbf{x}_{k-1}\right], \\ \mathbf{P}_k^x(\mathbf{x}_{k-1}) &= \mathrm{Cov}\left[\mathbf{x}_k \mid \mathbf{x}_{k-1}\right], \\ \boldsymbol{\mu}_k(\mathbf{x}_k) &= \mathrm{E}\left[\mathbf{y}_k \mid \mathbf{x}_k\right], \\ \mathbf{P}_k^y(\mathbf{x}_k) &= \mathrm{Cov}\left[\mathbf{y}_k \mid \mathbf{x}_k\right]. \end{aligned} \tag{9.77}$$

For example, if the model is of the additive form in Equation (9.1), we have simple closed form expressions $\boldsymbol{\mu}_k^-(\mathbf{x}_{k-1}) = \mathbf{f}(\mathbf{x}_{k-1})$, $\mathbf{P}_k^x(\mathbf{x}_{k-1}) = \mathbf{Q}_{k-1}$, $\boldsymbol{\mu}_k(\mathbf{x}_k) = \mathbf{h}(\mathbf{x}_k)$, and $\mathbf{P}_k^y(\mathbf{x}_k) = \mathbf{R}_k$. For the non-additive models in the form of Equation (9.9), we can use numerical integration approximations to the integrals in Equations (9.72) and (9.73) to evaluate them. For many models with non-Gaussian noise (see Section 5.2), we have either closed form expressions or accurate approximations of these moments.

Given the moment functions or their approximations, we can construct a conditional moment-based filter that uses sigma points and can be used for state space models of the types in (9.1), (9.9), and (9.10). In the following, we present that algorithm and express the sigma point method in a generic form. The unit sigma points $\boldsymbol{\xi}^{(i)}$ and weights W_i for $i = 1, \ldots, m$ are determined by the numerical integration method at hand (cf. Chapter 8).

Algorithm 9.24 (Sigma-point conditional moment Kalman filter). *The sigma point conditional moment Kalman filter (SPCMKF) is the following.*

- *Prediction:*

  1. *Form the sigma points as:*

$$\mathcal{X}_{k-1}^{(i)} = \mathbf{m}_{k-1} + \sqrt{\mathbf{P}_{k-1}}\, \boldsymbol{\xi}^{(i)}, \qquad i = 1, \ldots, m. \tag{9.78}$$

2. *Propagate the sigma points through the conditional mean and covariance functions:*

$$\begin{aligned}\boldsymbol{\mu}_k^{-(i)} &= \boldsymbol{\mu}_k^{-}(\mathcal{X}_{k-1}^{(i)}), \quad i = 1, \ldots, m, \\ \mathbf{P}_k^{x,(i)} &= \mathbf{P}_k^{x}(\mathcal{X}_{k-1}^{(i)}), \quad i = 1, \ldots, m.\end{aligned} \tag{9.79}$$

3. *Compute the predicted mean* $\mathbf{m}_k^-$ *and the predicted covariance* $\mathbf{P}_k^-$*:*

$$\begin{aligned}\mathbf{m}_k^- &= \sum_{i=1}^{m} W_i\, \boldsymbol{\mu}_k^{-(i)}, \\ \mathbf{P}_k^- &= \sum_{i=1}^{m} W_i \left(\mathbf{P}_k^{x,(i)} + (\boldsymbol{\mu}_k^{-(i)} - \mathbf{m}_k^-)\,(\boldsymbol{\mu}_k^{-(i)} - \mathbf{m}_k^-)^\top\right).\end{aligned} \tag{9.80}$$

- *Update:*

1. *Form the sigma points:*

$$\mathcal{X}_k^{-(i)} = \mathbf{m}_k^- + \sqrt{\mathbf{P}_k^-}\, \boldsymbol{\xi}^{(i)}, \qquad i = 1, \ldots, m. \tag{9.81}$$

2. *Propagate the sigma points through the conditional mean and covariance functions:*

$$\begin{aligned}\boldsymbol{\mu}_k^{(i)} &= \boldsymbol{\mu}_k(\mathcal{X}_k^{-(i)}), \quad i = 1, \ldots, m, \\ \mathbf{P}_k^{y,(i)} &= \mathbf{P}_k^{y}(\mathcal{X}_k^{-(i)}), \quad i = 1, \ldots, m.\end{aligned} \tag{9.82}$$

3. *Compute the predicted mean* $\boldsymbol{\mu}_k$*, the predicted covariance of the measurement* $\mathbf{S}_k$*, and the cross-covariance of the state and the measurement* $\mathbf{C}_k$*:*

$$\begin{aligned}\boldsymbol{\mu}_k &= \sum_{i=1}^{m} W_i\, \boldsymbol{\mu}_k^{(i)}, \\ \mathbf{S}_k &= \sum_{i=1}^{m} W_i \left(\mathbf{P}_k^{y,(i)} + (\boldsymbol{\mu}_k^{(i)} - \boldsymbol{\mu}_k)\,(\boldsymbol{\mu}_k^{(i)} - \boldsymbol{\mu}_k)^\top\right), \\ \mathbf{C}_k &= \sum_{i=1}^{m} W_i\, (\mathcal{X}_k^{-(i)} - \mathbf{m}_k^-)\,(\boldsymbol{\mu}_k^{(i)} - \boldsymbol{\mu}_k)^\top.\end{aligned} \tag{9.83}$$

4. *Compute the filter gain* $\mathbf{K}_k$ *and the filtered state mean* $\mathbf{m}_k$ *and covariance* $\mathbf{P}_k$*, conditional on the measurement* $\mathbf{y}_k$*:*

$$\begin{aligned}\mathbf{K}_k &= \mathbf{C}_k\, \mathbf{S}_k^{-1}, \\ \mathbf{m}_k &= \mathbf{m}_k^- + \mathbf{K}_k\, [\mathbf{y}_k - \boldsymbol{\mu}_k], \\ \mathbf{P}_k &= \mathbf{P}_k^- - \mathbf{K}_k\, \mathbf{S}_k\, \mathbf{K}_k^\top.\end{aligned} \tag{9.84}$$

The extension of the Gaussian filtering framework to general conditional distribution models in Equation (9.10) also allows us to develop the following simple Monte Carlo Kalman filter that uses Monte Carlo sampling (see Section 11.1) to approximate the integrals in the SLR/Gaussian filter. It is worth noting that this filter still approximates the filtering distribution as Gaussian, although in Chapter 11 we will get familiar with particle filtering where this approximation is dropped.

Algorithm 9.25 (Monte Carlo Kalman filter). *The conditional distribution form of the Monte Carlo Kalman filter (MCKF) is the following.*

- *Prediction:*
  1. *Generate samples of* $\mathbf{x}_{k-1}$*:*
  $$\mathbf{x}_{k-1}^{(i)} \sim \mathrm{N}(\mathbf{m}_{k-1}, \mathbf{P}_{k-1}), \qquad i = 1, \ldots, N. \tag{9.85}$$
  2. *Propagate the samples through the dynamic model:*
  $$\hat{\mathbf{x}}_k^{(i)} \sim p(\mathbf{x}_k \mid \mathbf{x}_{k-1}^{(i)}), \quad i = 1, \ldots, N. \tag{9.86}$$
  3. *Compute the predicted mean* $\mathbf{m}_k^-$ *and the predicted covariance* $\mathbf{P}_k^-$*:*
  $$\begin{aligned} \mathbf{m}_k^- &= \frac{1}{N} \sum_{i=1}^{N} \hat{\mathbf{x}}_k^{(i)}, \\ \mathbf{P}_k^- &= \frac{1}{N} \sum_{i=1}^{N} (\hat{\mathbf{x}}_k^{(i)} - \mathbf{m}_k^-) \, (\hat{\mathbf{x}}_k^{(i)} - \mathbf{m}_k^-)^\mathsf{T}. \end{aligned} \tag{9.87}$$
- *Update:*
  1. *Generate samples of* $\mathbf{x}_k$*:*
  $$\mathbf{x}_k^{-(i)} \sim \mathrm{N}(\mathbf{m}_k^-, \mathbf{P}_k^-), \qquad i = 1, \ldots, N. \tag{9.88}$$
  2. *Propagate samples through the measurement model:*
  $$\mathbf{y}_k^{(i)} \sim p(\mathbf{y}_k \mid \mathbf{x}_k^{-(i)}), \quad i = 1, \ldots, N. \tag{9.89}$$

3. *Compute the predicted mean* $\boldsymbol{\mu}_k$, *the predicted covariance of the measurement* $\mathbf{S}_k$, *and the cross-covariance of the state and the measurement* $\mathbf{C}_k$:

$$\begin{aligned}
\boldsymbol{\mu}_k &= \frac{1}{N}\sum_{i=1}^{N} \mathbf{y}_k^{(i)}, \\
\mathbf{S}_k &= \frac{1}{N}\sum_{i=1}^{N} (\mathbf{y}_k^{(i)} - \boldsymbol{\mu}_k)\,(\mathbf{y}_k^{(i)} - \boldsymbol{\mu}_k)^\mathsf{T}, \\
\mathbf{C}_k &= \frac{1}{N}\sum_{i=1}^{N} (\mathbf{x}_k^{-(i)} - \mathbf{m}_k^-)\,(\mathbf{y}_k^{(i)} - \boldsymbol{\mu}_k)^\mathsf{T}.
\end{aligned} \tag{9.90}$$

4. *Compute the filter gain* $\mathbf{K}_k$ *and the filtered state mean* $\mathbf{m}_k$ *and covariance* $\mathbf{P}_k$, *conditional on the measurement* $\mathbf{y}_k$:

$$\begin{aligned}
\mathbf{K}_k &= \mathbf{C}_k\,\mathbf{S}_k^{-1}, \\
\mathbf{m}_k &= \mathbf{m}_k^- + \mathbf{K}_k\,[\mathbf{y}_k - \boldsymbol{\mu}_k], \\
\mathbf{P}_k &= \mathbf{P}_k^- - \mathbf{K}_k\,\mathbf{S}_k\,\mathbf{K}_k^\mathsf{T}.
\end{aligned} \tag{9.91}$$

The above MCKF is theoretically appealing since it can approximate the moments required to perform SLR arbitrarily well and since it can handle state space models expressed using conditional distributions.

Let us compare the SPCMKF and the MCKF presented above in a small example.

Example 9.26 (Gaussian filtering with Poisson observations)**.** *Let the state sequence be a scalar random walk,*

$$x_k = x_{k-1} + q_{k-1}, \quad q_{k-1} \sim \mathrm{N}(0, Q), \tag{9.92}$$

and let $\exp(x_k)$ *represent the size of a population. As in Example 9.19, we observe a Poisson-distributed variable,*

$$y_k \sim \mathrm{Po}(\lambda\,\exp(x_k)), \tag{9.93}$$

where λ *is the probability that a specific event occurs to a single individual at time step* k. *Figure 9.3 shows the result of running the MCKF (with* $n = 10{,}000$ *samples) and SPCMKF on one sequence with the parameters* $\lambda = 0.1$ *and* $Q = 0.1$ *and an initial distribution with mean* $m_0 = 10$ *and variance* $P_0 = 0.1$. *In this simulation, we use the cubature rule to select the points and weights in the SPCMKF, which means that the SPCMKF is a CKF. The result from running a Monte Carlo experiment with* 500 *sequences of length* 25 *gave root mean squared errors of* 0.084 *for the*

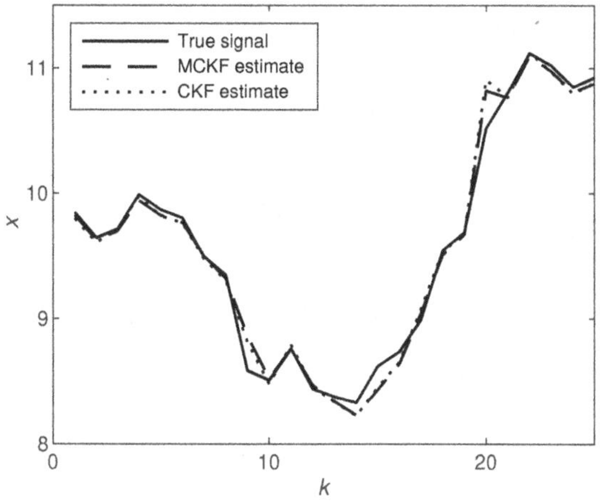

Figure 9.3 Simulated population data and estimated posterior means using MCKF and CKF in Example 9.26.

MCKF and 0.091 *for the CKF. That is, MCKF is more computationally demanding to run but also gave slightly better estimates.*

Finally, for concreteness, we present a sigma point version of Algorithm 9.24 for non-additive models of the form of Equation (9.9), that is, a sigma-point algorithm to perform SLR filtering for non-additive noise models. This filtering method uses the following approximations for the conditional moments in Equation (9.72):

$$\begin{aligned}
\boldsymbol{\mu}_k^-(\mathbf{x}_{k-1}) &\approx \sum_{j=1}^{m'} W_j' \, \mathbf{f}(\mathbf{x}_{k-1}, \sqrt{\mathbf{Q}_{k-1}}\, \boldsymbol{\xi}^{(j)'}), \\
\mathbf{P}_k^{\mathrm{x}}(\mathbf{x}_{k-1}) &\approx \sum_{j=1}^{m'} W_j' \, (\mathbf{f}(\mathbf{x}_{k-1}, \sqrt{\mathbf{Q}_{k-1}}\, \boldsymbol{\xi}^{(j)'}) - \boldsymbol{\mu}_k^-(\mathbf{x}_{k-1})) \\
&\quad \times (\mathbf{f}(\mathbf{x}_{k-1}, \sqrt{\mathbf{Q}_{k-1}}\, \boldsymbol{\xi}^{(j)'}) - \boldsymbol{\mu}_k^-(\mathbf{x}_{k-1}))^{\mathsf{T}},
\end{aligned} \tag{9.94}$$

and the following in Equation (9.73):

$$
\begin{aligned}
\boldsymbol{\mu}_k(\mathbf{x}_k) &\approx \sum_{j=1}^{m''} W_j'' \, \mathbf{h}(\mathbf{x}_k, \sqrt{\mathbf{R}_k} \, \boldsymbol{\xi}^{(j)''}), \\
\mathbf{P}_k^y(\mathbf{x}_k) &\approx \sum_{j=1}^{m''} W_j'' \, (\mathbf{h}(\mathbf{x}_k, \sqrt{\mathbf{R}_k} \, \boldsymbol{\xi}^{(j)''}) - \boldsymbol{\mu}_k(\mathbf{x}_k)) \\
&\qquad \times (\mathbf{h}(\mathbf{x}_k, \sqrt{\mathbf{R}_k} \, \boldsymbol{\xi}^{(j)''}) - \boldsymbol{\mu}_k(\mathbf{x}_k))^\mathsf{T},
\end{aligned}
\tag{9.95}
$$

where the unit sigma-points $\boldsymbol{\xi}^{(j)'}, \boldsymbol{\xi}^{(j)''}$ and weights W_j', W_j'' are determined by the numerical integration methods for integrating over the noises $\mathbf{q}_{k-1}$ and $\mathbf{r}_k$. The resulting algorithm is the following.

Algorithm 9.27 (Sigma-point non-additive conditional moment filter). *The sigma-point non-additive noise conditional moment filter is the following.*

- *Prediction:*
  1. *Form the sigma points as:*

$$
\begin{aligned}
\mathcal{X}_{k-1}^{(i)} &= \mathbf{m}_{k-1} + \sqrt{\mathbf{P}_{k-1}} \, \boldsymbol{\xi}^{(i)}, & i &= 1, \ldots, m, \\
\mathcal{Q}_{k-1}^{(j)} &= \sqrt{\mathbf{Q}_{k-1}} \, \boldsymbol{\xi}^{(j)'}, & j &= 1, \ldots, m'.
\end{aligned}
\tag{9.96}
$$

  2. *Propagate the sigma points through the dynamic model:*

$$
\hat{\mathcal{X}}_k^{(i,j)} = \mathbf{f}(\mathcal{X}_{k-1}^{(i)}, \mathcal{Q}_{k-1}^{(j)}),
\tag{9.97}
$$

  for $i = 1, \ldots, m$ and $j = 1, \ldots, m'$.
  3. *Compute the conditional moments $\boldsymbol{\mu}_k^{-(i)} = \boldsymbol{\mu}_k(\mathcal{X}_{k-1}^{(i)})$ and $\mathbf{P}_k^{x,(i)} = \mathbf{P}_k^x(\mathcal{X}_{k-1}^{(i)})$:*

$$
\begin{aligned}
\boldsymbol{\mu}_k^{-(i)} &= \sum_{j=1}^{m'} W_j' \, \hat{\mathcal{X}}_k^{(i,j)}, \\
\mathbf{P}_k^{x,(i)} &= \sum_{j=1}^{m'} W_j' \, (\hat{\mathcal{X}}_k^{(i,j)} - \boldsymbol{\mu}_k^{-(i)}) \, (\hat{\mathcal{X}}_k^{(i,j)} - \boldsymbol{\mu}_k^{-(i)})^\mathsf{T},
\end{aligned}
\tag{9.98}
$$

  for $i = 1, \ldots, m$.

4. *Compute the predicted mean* $\mathbf{m}_k^-$ *and the predicted covariance* $\mathbf{P}_k^-$*:*

$$
\begin{aligned}
\mathbf{m}_k^- &= \sum_{i=1}^{m} W_i \, \boldsymbol{\mu}_k^{-(i)}, \\
\mathbf{P}_k^- &= \sum_{i=1}^{m} W_i \left(\mathbf{P}_k^{x,(i)} + (\boldsymbol{\mu}_k^{-(i)} - \mathbf{m}_k^-) \, (\boldsymbol{\mu}_k^{-(i)} - \mathbf{m}_k^-)^\mathsf{T} \right).
\end{aligned}
\tag{9.99}
$$

- *Update:*

1. *Form the sigma points:*

$$
\begin{aligned}
\mathcal{X}_k^{-(i)} &= \mathbf{m}_k^- + \sqrt{\mathbf{P}_k^-} \, \boldsymbol{\xi}^{(i)}, \qquad & i &= 1, \ldots, m, \\
\mathcal{R}_k^{(j)} &= \sqrt{\mathbf{R}_k} \, \boldsymbol{\xi}^{(j)''}, \qquad & j &= 1, \ldots, m''.
\end{aligned}
\tag{9.100}
$$

2. *Propagate the sigma points through the measurement model:*

$$
\mathcal{Y}_k^{(i,j)} = \mathbf{h}(\mathcal{X}_k^{-(i)}, \mathcal{R}_k^{(j)}),
\tag{9.101}
$$

for $i = 1, \ldots, m$ *and* $j = 1, \ldots, m''$.

3. *Compute the conditional mean and covariance functions* $\boldsymbol{\mu}_k^{(i)} = \boldsymbol{\mu}_k(\mathcal{X}_k^{-(i)})$ *and* $\mathbf{P}_k^{y,(i)} = \mathbf{P}_k^y(\mathcal{X}_k^{-(i)})$*:*

$$
\begin{aligned}
\boldsymbol{\mu}_k^{(i)} &= \sum_{j=1}^{m''} W_j'' \, \mathcal{Y}_k^{(i,j)}, \\
\mathbf{P}_k^{y,(i)} &= \sum_{j=1}^{m''} W_j'' \, (\mathcal{Y}_k^{(i,j)} - \boldsymbol{\mu}_k^{(i)})(\mathcal{Y}_k^{(i,j)} - \boldsymbol{\mu}_k^{(i)})^\mathsf{T},
\end{aligned}
\tag{9.102}
$$

for $i = 1, \ldots, m$.

4. *Compute the predicted mean* $\boldsymbol{\mu}_k$*, the predicted covariance of the measurement* $\mathbf{S}_k$*, and the cross-covariance of the state and the measurement* $\mathbf{C}_k$*:*

$$
\begin{aligned}
\boldsymbol{\mu}_k &= \sum_{i=1}^{m} W_i \, \boldsymbol{\mu}_k^{(i)}, \\
\mathbf{S}_k &= \sum_{i=1}^{m} W_i \left(\mathbf{P}_k^{y,(i)} + (\boldsymbol{\mu}_k^{(i)} - \boldsymbol{\mu}_k) \, (\boldsymbol{\mu}_k^{(i)} - \boldsymbol{\mu}_k)^\mathsf{T} \right), \\
\mathbf{C}_k &= \sum_{i=1}^{m} W_i \, (\mathcal{X}_k^{-(i)} - \mathbf{m}_k^-) \, (\boldsymbol{\mu}_k^{(i)} - \boldsymbol{\mu}_k)^\mathsf{T}.
\end{aligned}
\tag{9.103}
$$

5. *Compute the filter gain* $\mathbf{K}_k$ *and the filtered state mean* $\mathbf{m}_k$ *and covariance* $\mathbf{P}_k$*, conditional on the measurement* $\mathbf{y}_k$*:*

$$\begin{aligned}
\mathbf{K}_k &= \mathbf{C}_k \, \mathbf{S}_k^{-1}, \\
\mathbf{m}_k &= \mathbf{m}_k^- + \mathbf{K}_k \, [\mathbf{y}_k - \boldsymbol{\mu}_k], \\
\mathbf{P}_k &= \mathbf{P}_k^- - \mathbf{K}_k \, \mathbf{S}_k \, \mathbf{K}_k^\top.
\end{aligned} \tag{9.104}$$

Algorithm 9.27 is an alternative to the sigma point algorithms for non-additive noise models presented in Chapter 8. One difference between these alternatives is how we form the sigma points. In Chapter 8, we used sigma points to jointly integrate over $\mathbf{x}_{k-1}$ and $\mathbf{q}_{k-1}$ at the prediction step and to jointly integrate over $\mathbf{x}_k$ and $\mathbf{r}_k$ at the update step, and we did that using augmented random variables. For instance, if n is the dimensionality of the state and n_r is the dimensionality of the measurement noise, the sigma points used in the update step of the non-additive filters had dimensionality $n + n_r$ and represented the augmented state $(\mathbf{x}_k, \mathbf{y}_k)$. In Algorithm 9.27, we instead use separate sigma points for the state and the noise, and we perform the integrations in a nested fashion.

9.7 Relation to Other Gaussian Filters

We have now seen that all the Gaussian filters presented earlier in this book (e.g., EKF, IEKF, UKF, CKF, GHKF) can be understood from the perspective of enabling approximations. The EKF and IEKF both perform analytic linearization of the models but use different linearization points, whereas the general Gaussian filters in Chapter 8 implicitly perform SLR to approximate the models.

The fact that we managed to rederive the general Gaussian filters using SLR is fundamentally a consequence of the properties of the underlying approximations. In general Gaussian filters, we approximate the joint distribution of two random variables $\mathbf{x}$ and $\mathbf{y}$ as Gaussian by matching the moments in the original distribution. Using SLR, we write $\mathbf{y} \simeq \mathbf{A}\,\mathbf{x} + \mathbf{b} + \tilde{\mathbf{e}}$, $\tilde{\mathbf{e}} \sim \mathrm{N}(\mathbf{0}, \boldsymbol{\Lambda})$, which yields an approximation to the conditional distribution

$$p(\mathbf{y} \mid \mathbf{x}) \simeq \mathrm{N}(\mathbf{y} \mid \mathbf{A}\,\mathbf{x} + \mathbf{b}, \boldsymbol{\Lambda}). \tag{9.105}$$

That is, instead of directly approximating $p(\mathbf{x}, \mathbf{y})$ as Gaussian, SLR approximates $p(\mathbf{y} \mid \mathbf{x})$. However, the parameters $\mathbf{A}$, $\mathbf{b}$, and $\boldsymbol{\Lambda}$ are still selected to ensure that the resulting approximation of the joint distribution $p(\mathbf{x}, \mathbf{y})$ matches the original distribution, see (9.41). Consequently, when

$\mathbf{x}$ is Gaussian, SLR yields a Gaussian approximation of the joint distribution that exactly matches the approximation used in the general Gaussian filters. For additive noise models, we can also compare (9.43) in Corollary 9.14 with (8.4) in Algorithm 8.1 and verify that the involved integrals are identical.

The general Gaussian filters presented in Chapter 8 approximate the joint distribution of $(\mathbf{x}_k, \mathbf{x}_{k-1})$ as Gaussian during the prediction step and the joint distribution of $(\mathbf{x}_k, \mathbf{y}_k)$ as Gaussian during the update step. Based on that approximation, the prediction and update steps are performed in closed form. Similarly, the SLR filters presented in Section 9.5 linearize the models during both the prediction and update steps and then perform filtering in closed form. Considering that the SLR used by the SLR filters corresponds to the same Gaussian approximations used to derive the filters in Chapter 8, it is not surprising that the two approaches give rise to identical filters.

The literature contains several examples of filters that make use of higher order expansions of the involved models. For instance, the second order extended Kalman filter (EKF2) (Bar-Shalom et al., 2001; Roth and Gustafsson, 2011), the Fourier–Hermite Kalman filter (FHKF) (Sarmavuori and Särkkä, 2012), and the Gaussian process quadrature filter (GPQF) (Särkkä et al., 2016) all approximate $\mathbf{f}(\cdot)$ and $\mathbf{h}(\cdot)$ using different types of series expansions. These approximations are then used to compute certain integrals.

Interestingly, it is possible to argue that these higher order filters are instantiations of SLRF filters described above. The reason for this is that the series expansions are introduced to enable us to approximate the integrals that appear in the general Gaussian filters that we studied in Chapter 8, and the higher order filters (implicitly or explicitly) approximate the joint distributions of $(\mathbf{x}_k, \mathbf{x}_{k-1})$ and $(\mathbf{y}_k, \mathbf{x}_k)$ as Gaussian.

From the perspective of enabling approximations, this means that these filters (EKF2, FHKF, and GPQF) implicitly make use of SLR to linearize the models. That is, even though the higher order series expansions may approximate the models accurately, we cannot use them as enabling approximations since they do not enable us to compute the posterior distributions in closed form. Instead we still use the SLR as enabling approximation and only use the higher order expansions to approximate the involved integrals.

9.8 Exercises

9.1 Consider the following non-linear state space model:

$$\begin{aligned} x_k &= x_{k-1} - 0.01\,\sin(x_{k-1}) + q_{k-1}, \\ y_k &= 0.5\,\sin(2\,x_k) + r_k, \end{aligned} \tag{9.106}$$

where q_{k-1} has a variance of 0.01^2 and r_k has a variance of 0.02, which we saw already in Exercise 7.1. Derive the required expected values for an SLF, and implement the SLF for the model. *Hint:* Use the imaginary part of the inverse Fourier transform of the Gaussian distribution. Simulate data from the model, compute the RMSE values, plot the results, and compare the performance to the EKF.

9.2 In this exercise your task is to derive the derivative form of the statistically linearized filter (SLF).

(a) Prove using integration by parts the following identity for a Gaussian random variable $\mathbf{x}$, differentiable non-linear function $\mathbf{g}(\mathbf{x})$, and its Jacobian matrix $\mathbf{G}_x(\mathbf{x}) = \partial \mathbf{g}(\mathbf{x})/\partial \mathbf{x}$:

$$\mathrm{E}[\mathbf{g}(\mathbf{x})\,(\mathbf{x}-\mathbf{m})^\mathsf{T}] = \mathrm{E}[\mathbf{G}_x(\mathbf{x})]\,\mathbf{P}, \tag{9.107}$$

where $\mathrm{E}[\cdot]$ denotes the expected value with respect to $\mathrm{N}(\mathbf{x} \mid \mathbf{m}, \mathbf{P})$. *Hint:* $\frac{\partial}{\partial \mathbf{x}} \mathrm{N}(\mathbf{x} \mid \mathbf{m}, \mathbf{P}) = -\mathbf{P}^{-1}(\mathbf{x}-\mathbf{m})\,\mathrm{N}(\mathbf{x} \mid \mathbf{m}, \mathbf{P})$.

(b) Prove the following. Let

$$\boldsymbol{\mu}(\mathbf{m}) = \mathrm{E}[\mathbf{g}(\mathbf{x})], \tag{9.108}$$

where $\mathrm{E}[\cdot]$ denotes the expected value with respect to $\mathrm{N}(\mathbf{x} \mid \mathbf{m}, \mathbf{P})$. Then

$$\frac{\partial \boldsymbol{\mu}(\mathbf{m})}{\partial \mathbf{m}} = \mathrm{E}[\mathbf{G}_x(\mathbf{x})]. \tag{9.109}$$

(c) Write down the additive form of the SLF equations in an alternative form, where you have eliminated all the cross terms of the form $\mathrm{E}[\mathbf{f}(\mathbf{x}_{k-1})\,\delta\mathbf{x}_{k-1}^\mathsf{T}]$ and $\mathrm{E}[\mathbf{h}(\mathbf{x}_k)\,\delta\mathbf{x}_k^\mathsf{T}]^\mathsf{T}$, using the result in (a).

(d) How can you utilize the result (b) when using the alternative form of the SLF? Check that you get the same equations for the SLF in the previous exercise using this alternative form of the SLF.

9.3 Algorithm 9.9 can be combined with a Gaussian quadrature method to approximate the expected values $\mathrm{E}[\mathbf{f}(\mathbf{x}_{k-1})]$, $\mathrm{E}[\mathbf{f}(\mathbf{x}_{k-1})\,\delta\mathbf{x}_{k-1}^\mathsf{T}]$, $\mathrm{E}[\mathbf{h}(\mathbf{x}_k)]$ and $\mathrm{E}[\mathbf{h}(\mathbf{x}_k)\,\delta\mathbf{x}_k^\mathsf{T}]$. Formulate an algorithm that combines Algorithm 9.9 with spherical cubature integration described in Algorithm 8.10.
Hint: You can form sigma points as in Algorithm 8.11 and then use them to approximate the four expected values mentioned above.

9.4 Implement the cubature-based version of the SL formulated in Exercise 9.3 for the model in Exercise 7.1. Plot the results and compare the RMSE values to the SLF from Exercise 7.1.

9.5 Find analytical solutions to the integrals in (9.57).
Hint: One can show that

$$\exp(\beta x)\,\mathrm{N}(x \mid m, P) = \mathrm{N}(x \mid m + P\beta, P)\exp(\beta m + P\beta^2/2).$$

9.6 Implement the Gauss–Hermite integration-based SLR filter in Algorithm 9.20 for the model in Exercise 7.1, and check that its result matches the GHKF for the same model.

9.7 Continue Exercise 9.1 by computing all the expectations needed for the SLR filter in Algorithm 9.20 in closed form. Compare its performance numerically with the EKF and GHKF.

9.8 Formulate an SLR filter for the non-additive noise models in (9.9), where we first linearize the models and then perform the Kalman filter prediction and update. That is, formulate a version of Algorithm 9.20 for non-additive noise models. Also, verify that it can be simplified to Algorithm 8.4.
Hint: Considering that we want an algorithm that can be simplified to Algorithm 8.4 you should use (9.50), and avoid the conditional moments formulation. See also Appendix A.9.

9.9 Formulate an algorithm that combines Algorithm 9.20 with spherical cubature integration described in Algorithm 8.10.

9.10 Verify that Corollary 9.8 holds.
Hint: Apply Theorem 9.3 to the variables $\mathbf{y} = \mathbf{g}(\mathbf{x}, \mathbf{q})$ and $\mathbf{z}$ in (9.28).

9.11 Consider the bicycle model with the mean dynamic model given by Equations (4.66) and the covariance approximation constructed in Exercise 4.10. Implement the conditional moment filter in Algorithm 9.24 for the model when linear position measurements are used. You can use, for example, the cubature integration method. Simulate data from the model, and compute the RMSE of the filter.

10

Posterior Linearization Filtering

In Chapter 9, we demonstrated that all the Gaussian filters presented earlier can be understood as methods that first linearize the models and then use the Kalman filter equations to solve the resulting state estimation problem. How we linearize the models may significantly influence the filter performance, and we can approximate the models using any affine function with additive Gaussian noise. In this chapter, we present the concepts of posterior linearization and iterated posterior linearization filtering, first introduced in García-Fernández et al. (2015) and further developed, for example, in Tronarp et al. (2018), and explain how they can be used to develop practical Gaussian filters.

10.1 Generalized Statistical Linear Regression

In the SLR filters that we encountered in the previous chapter, we implicitly used the prediction and update distributions of $\mathbf{x}_k$ as the linearization distributions in statistical linear regression (SLR). This then led to filters that turned out to be equivalent to moment matching-based Gaussian filters. However, there is no inherent reason why we should linearize with respect to this "true" distribution of $\mathbf{x}_k$. Instead, the linearization distribution can even be completely decoupled from the distribution that $\mathbf{x}_k$ has with respect to the estimation problem at hand. In fact, linearization was decoupled from the prediction and update distributions already in the iterated extended Kalman filter (IEKF), which we saw in Section 7.4, as the linearization point was optimized by iteration in the IEKF update. A similar philosophy can also be applied in SLR.

Let us start by defining generalized SLR, which is defined for any linearization distribution $\pi(\mathbf{x})$.

Definition 10.1 (Generalized statistical linear regression). *Let* $\pi(\mathbf{x})$ *be an arbitrary linearization distribution for* $\mathbf{x}$. *If* $\mathbf{x}$ *and* $\mathbf{y}$ *are two random variables,* $\mathbf{y} \simeq \mathbf{A}\,\mathbf{x} + \mathbf{b} + \tilde{\mathbf{e}}$, $\tilde{\mathbf{e}} \sim \mathrm{N}(\mathbf{0}, \mathbf{\Lambda})$ *is called a* statistical linear regression *with respect to* $\pi(\mathbf{x})$ *if* $\mathbf{A}$ *and* $\mathbf{b}$ *are selected to minimize the mean squared error*

$$\mathrm{MSE}(\mathbf{A}, \mathbf{b}) = \mathrm{E}_\pi\left[(\mathbf{y} - \mathbf{A}\,\mathbf{x} - \mathbf{b})^\mathsf{T}\,(\mathbf{y} - \mathbf{A}\,\mathbf{x} - \mathbf{b})\right], \tag{10.1}$$

where E_π *denotes expectation with respect* π, $\mathbf{A}$ *and* $\mathbf{b}$ *are deterministic variables, and the noise covariance is selected as*

$$\mathbf{\Lambda} = \mathrm{E}_\pi\left[(\mathbf{y} - \mathbf{A}\,\mathbf{x} - \mathbf{b})\,(\mathbf{y} - \mathbf{A}\,\mathbf{x} - \mathbf{b})^\mathsf{T}\right]. \tag{10.2}$$

With this definition we get the following linearization. Below, we have used a slight abuse of notation by using subscript π to denote means and covariances when the distribution of $\mathbf{x}$ is replaced with π.

Theorem 10.2 (Generalized statistical linear regression). *Let* $\pi(\mathbf{x})$ *be an arbitrary linearization distribution for* $\mathbf{x}$. *Suppose* $\mathbf{x}$ *and* $\mathbf{y}$ *are two random variables, and let* $\mathbf{m}_\pi = \mathrm{E}_\pi[\mathbf{x}]$, $\mathbf{P}_\pi = \mathrm{Cov}_\pi[\mathbf{x}]$, $\boldsymbol{\mu}_\mathrm{G} = \mathrm{E}_\pi[\mathbf{y}]$, $\mathbf{S}_\mathrm{G} = \mathrm{Cov}_\pi[\mathbf{y}]$, *and* $\mathbf{C}_\mathrm{G} = \mathrm{Cov}_\pi[\mathbf{x}, \mathbf{y}]$. *In the generalized statistical linear regression (GSLR) with respect to* $\pi(\mathbf{x})$ *for* $\mathbf{y} \simeq \mathbf{A}\,\mathbf{x} + \mathbf{b} + \tilde{\mathbf{e}}$, $\tilde{\mathbf{e}} \sim \mathrm{N}(\mathbf{0}, \mathbf{\Lambda})$, *we then have*

$$\begin{aligned}
\mathbf{A} &= \mathbf{C}_\mathrm{G}^\mathsf{T}\,\mathbf{P}_\pi^{-1}, \\
\mathbf{b} &= \boldsymbol{\mu}_\mathrm{G} - \mathbf{A}\,\mathbf{m}_\pi, \\
\mathbf{\Lambda} &= \mathbf{S}_\mathrm{G} - \mathbf{A}\,\mathbf{P}_\pi\,\mathbf{A}^\mathsf{T}.
\end{aligned} \tag{10.3}$$

Proof The result can be derived in the same way as in Theorem 9.12. □

We can now apply GSLR to a transformation of the form $\mathbf{y} = \mathbf{g}(\mathbf{x}) + \mathbf{q}$, which gives the following.

Corollary 10.3 (Generalized SLR of an additive transform). *Suppose* $\mathbf{y} = \mathbf{g}(\mathbf{x}) + \mathbf{q}$, *where* $\mathbf{x} \sim \mathrm{N}(\mathbf{m}, \mathbf{P})$ *and* $\mathbf{q} \sim \mathrm{N}(\mathbf{0}, \mathbf{Q})$. *If we use generalized statistical linear regression (GSLR) with respect to a linearization distribution* $\pi(\mathbf{x})$ *to form the approximation* $\mathbf{y} \simeq \mathbf{A}\,\mathbf{x} + \mathbf{b} + \tilde{\mathbf{e}}$, $\tilde{\mathbf{e}} \sim \mathrm{N}(\mathbf{0}, \mathbf{\Lambda})$, *we get*

$$\begin{aligned}
\mathbf{A} &= \mathbf{C}_\mathrm{G}^\mathsf{T}\,\mathbf{P}_\pi^{-1}, \\
\mathbf{b} &= \boldsymbol{\mu}_\mathrm{G} - \mathbf{A}\,\mathbf{m}_\pi, \\
\mathbf{\Lambda} &= \mathbf{S}_\mathrm{G} - \mathbf{A}\,\mathbf{P}_\pi\,\mathbf{A}^\mathsf{T},
\end{aligned} \tag{10.4}$$

where

$$\begin{aligned}
\mathbf{m}_\pi &= \int \mathbf{x}\, \pi(\mathbf{x})\, \mathrm{d}\mathbf{x}, \\
\mathbf{P}_\pi &= \int (\mathbf{x} - \mathbf{m}_\pi)\,(\mathbf{x} - \mathbf{m}_\pi)^\mathsf{T}\, \pi(\mathbf{x})\, \mathrm{d}\mathbf{x}, \\
\boldsymbol{\mu}_\mathrm{G} &= \int \mathbf{g}(\mathbf{x})\, \pi(\mathbf{x})\, \mathrm{d}\mathbf{x}, \\
\mathbf{C}_\mathrm{G} &= \int (\mathbf{x} - \mathbf{m}_\pi)\,(\mathbf{g}(\mathbf{x}) - \boldsymbol{\mu}_\mathrm{G})^\mathsf{T}\, \pi(\mathbf{x})\, \mathrm{d}\mathbf{x}, \\
\mathbf{S}_\mathrm{G} &= \int (\mathbf{g}(\mathbf{x}) - \boldsymbol{\mu}_\mathrm{G})\,(\mathbf{g}(\mathbf{x}) - \boldsymbol{\mu}_\mathrm{G})^\mathsf{T}\, \pi(\mathbf{x})\, \mathrm{d}\mathbf{x} + \mathbf{Q}.
\end{aligned} \tag{10.5}$$

The corresponding joint Gaussian approximation for $(\mathbf{x}, \mathbf{y})$ *is given by*

$$\begin{pmatrix} \mathbf{x} \\ \mathbf{y} \end{pmatrix} \sim \mathrm{N}\left(\begin{pmatrix} \mathbf{m} \\ \mathbf{A}\,\mathbf{m} + \mathbf{b} \end{pmatrix}, \begin{pmatrix} \mathbf{P} & \mathbf{P}\,\mathbf{A}^\mathsf{T} \\ \mathbf{A}\,\mathbf{P} & \mathbf{A}\,\mathbf{P}\,\mathbf{A}^\mathsf{T} + \mathbf{Q} \end{pmatrix} \right). \tag{10.6}$$

The key thing above is that it no longer reduces to moment matching, because the distribution $\mathbf{x} \sim \mathrm{N}(\mathbf{m}, \mathbf{P})$ is different from the linearization distribution $\pi(\mathbf{x})$. This also has the side effect that the moments of the joint Gaussian approximation no longer exactly match the true ones when we have $\mathbf{x} \sim \mathrm{N}(\mathbf{m}, \mathbf{P})$. However, we can use this to our advantage, because we do not actually want the linearization to be optimal for the prior distribution for $\mathbf{x}$, but instead, we are more interested in the posterior distribution of $\mathbf{x}$. With this in mind, it makes sense to attempt to use the posterior distribution as the linearization distribution instead of the prior of $\mathbf{x}$. We proceed to discuss this next.

10.2 Posterior Linearization

Statistical linearization (SL) and statistical linear regression (SLR) are appealing because they seek linearizations that are accurate *on average* across the different state values that may appear. However, once we have observed measurements, our belief (distribution) on the different state values will change. Thus, although initially it makes sense to linearize with respect to the prior distribution of $\mathbf{x}$, after obtaining measurements, a more appropriate choice for the linearization distribution in SL and SLR would be the posterior distribution. In this section we introduce the posterior linearization, which aims to do that. We restrict our discussion to SLR as we already saw in the previous chapter that SL can be seen as a deterministic version of SLR where we simply ignore the approximation error.

Definition 10.4 (Posterior linearization). *Let* $\mathbf{x}$ *and* $\mathbf{y}$ *be two random variables, and assume that we observe* $\mathbf{y}$*, which then leads to the posterior distribution* $p(\mathbf{x} \mid \mathbf{y})$*. The procedure of putting* $\pi(\mathbf{x}) = p(\mathbf{x} \mid \mathbf{y})$ *and forming the approximation* $\mathbf{y} \simeq \mathbf{A}\,\mathbf{x} + \mathbf{b} + \tilde{\mathbf{e}}$, $\tilde{\mathbf{e}} \sim \mathrm{N}(\mathbf{0}, \boldsymbol{\Lambda})$ *with generalized statistical linear regression in Definition 10.1 with respect to* $\pi(\mathbf{x})$ *is called* posterior linearization *(PL).*

The application of the above definition to a transformation of the form $\mathbf{y} = \mathbf{g}(\mathbf{x}) + \mathbf{q}$, where $\mathbf{x} \sim \mathrm{N}(\mathbf{m}, \mathbf{P})$ and $\mathbf{q} \sim \mathrm{N}(\mathbf{0}, \mathbf{Q})$, then leads to Corollary 10.3 with $\pi(\mathbf{x}) = p(\mathbf{x} \mid \mathbf{y})$. Clearly there is a catch – to do the linearization we need to have access to the posterior distribution, and the whole aim of linearization is to approximately compute this distribution, which we do not know. However, we come back to this dilemma in the next section where we introduce iterated posterior linearization.

The intuition behind posterior linearization is the same as the original SLR: we want an approximation that is accurate on average across the state values that we expect to see; but posterior linearization tries to leverage what we have learned from the measurements and seeks a linearization that is accurate in the support of the posterior density. Specifically, $\mathbf{A}$ and $\mathbf{b}$ are selected to approximate $\mathbf{y}$ accurately on average across the values of $\mathbf{x}$ that we expect to see according to our posterior distribution of $\mathbf{x}$, whereas the original SLR seeks an approximation that is accurate when we average according to the prior. In what follows we, therefore, refer to the two alternatives as prior SLR and posterior SLR, respectively, where the posterior SLR is also referred to as a posterior linearization.

The motivation for linearizing our models is to compute the posterior distribution of the state $\mathbf{x}$. If we have a Gaussian prior, $p(\mathbf{x}) = \mathrm{N}(\mathbf{x} \mid \mathbf{m}, \mathbf{P})$, and we observe $\mathbf{y} = \mathbf{A}\,\mathbf{x} + \mathbf{b} + \tilde{\mathbf{e}}$ where $\tilde{\mathbf{e}} \sim \mathrm{N}(\mathbf{0}, \boldsymbol{\Lambda})$, the posterior moments $(\mathbf{m}^+, \mathbf{P}^+)$ can be computed using the calculations:

$$
\begin{aligned}
\boldsymbol{\mu} &= \mathbf{A}\,\mathbf{m} + \mathbf{b}, \\
\mathbf{S} &= \mathbf{A}\,\mathbf{P}\,\mathbf{A}^\mathsf{T} + \boldsymbol{\Lambda}, \\
\mathbf{K} &= \mathbf{P}\,\mathbf{A}^\mathsf{T}\,\mathbf{S}^{-1}, \\
\mathbf{m}^+ &= \mathbf{m} + \mathbf{K}\,(\mathbf{y} - \boldsymbol{\mu}), \\
\mathbf{P}^+ &= \mathbf{P} - \mathbf{K}\,\mathbf{S}\,\mathbf{K}^\mathsf{T}.
\end{aligned}
\tag{10.7}
$$

For both the posterior SLR (i.e., PL) and the prior SLR, we use Equations (10.7) to compute the posterior, but the linearization parameters are different. What happens in the prior SLR is that the linearization is not accurate (moment matched) when the state $\mathbf{x}$ is distributed according to the posterior distribution – it is only accurate for the prior distribution of $\mathbf{x}$. However, PL is maximally accurate when $\mathbf{x}$ is distributed according to the posterior distribution. These properties are related to the Kullback–Leibler divergence optimality properties of prior and posterior SLR, which are discussed in García-Fernández et al. (2015) as well as in Section 10.6.

Example 10.5 (Posterior statistical linear regression). *Consider the scalar variables in Example 9.7, where we have* $x \sim \mathrm{N}(1, 0.1)$ *and* $g(x) = x^3$. *To ensure that the posterior density is not degenerate, we assume that*

$$y = g(x) + q, \tag{10.8}$$

where the additive noise is Gaussian: $q \sim \mathrm{N}(0, 0.1)$.

Figure 10.1(a) illustrates the affine components, $A\,x + b$, *of the prior and posterior SLRs when we have observed the value* $y = 5$. *The posterior density* $p(x \mid y)$ *for this observation is centered around* $x \simeq 1.7$, *and the posterior linearization selects* $A\,x + b$ *to approximate* $g(x)$ *accurately close to that value, whereas the prior SLR selects* $A\,x + b$ *to approximate* $g(x)$ *accurately around the prior mean* $x = 1$. *The standard deviations of the linearization noise are* $\sqrt{\Lambda} \simeq 0.45$ *for the prior SLR and* $\sqrt{\Lambda} \simeq 0.01$ *for the posterior SLR. The reason* Λ *is so much smaller for the posterior SLR is that the posterior density* $p(x \mid y)$ *has much smaller variance than the prior density* $p(x)$ *in this example.*

The resulting densities are illustrated in Figure 10.1(b). The posterior computed using the posterior SLR overlaps with the true posterior $p(x \mid y)$, *whereas the posterior computed using the prior SLR has a larger variance and a different mean. At least in this example, the posterior SLR yields a more accurate approximation to the posterior than the prior SLR.*

10.3 Iterated Posterior Linearization

The concept of posterior linearization (PL) introduced in the previous section has the dilemma that in order to form it, we would already need to have access to the posterior distribution. However, we can construct a practical iterative algorithm by always making use of the best available approximation to the posterior when linearizing the models. This is called iterated posterior linearization (IPL).

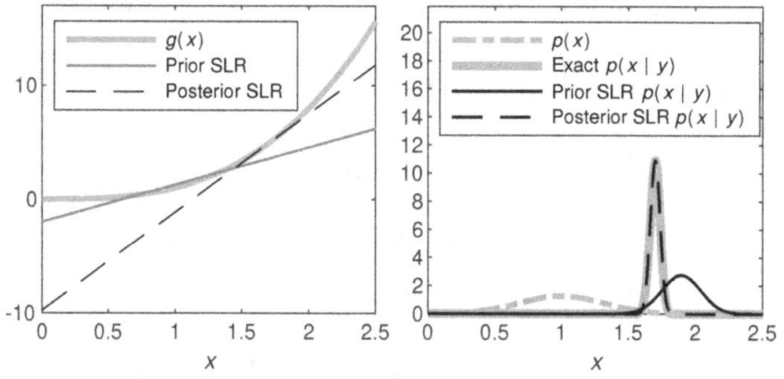

(a) (b)

Figure 10.1 Illustrations related to Example 10.5. (a) The function $g(x) = x^3$ and the affine approximations $y = Ax + b$ in the prior SLR and posterior SLR, respectively. (b) The prior density $p(x)$, the true posterior $p(x \mid y)$, and the approximate posteriors $p(x \mid y)$ computed using prior SLR and posterior SLR, respectively.

Algorithm 10.6 (Iterated posterior linearization). *Given a model* $\mathbf{y} = \mathbf{g}(\mathbf{x}) + \mathbf{q}$, $\mathbf{q} \sim \mathrm{N}(\mathbf{0}, \mathbf{Q})$, *a prior density* $\mathbf{x} \sim \mathrm{N}(\mathbf{m}, \mathbf{P})$, *and observation* $\mathbf{y}$, *the iterated posterior linearization (IPL) algorithm is:*

- *Start from the prior density* $\mathbf{m}^{(0)} = \mathbf{m}$, $\mathbf{P}^{(0)} = \mathbf{P}$.
- *For* $i = 1, 2, 3, \ldots$ *do the following steps:*
  1. Linearize: *Find the SLR parameters* $\mathbf{A}^{(i)}$, $\mathbf{b}^{(i)}$, *and* $\mathbf{\Lambda}^{(i)}$ *such that* $\mathbf{y} \simeq \mathbf{A}^{(i)}\,\mathbf{x} + \mathbf{b}^{(i)} + \tilde{\mathbf{e}}$, $\tilde{\mathbf{e}} \sim \mathrm{N}(\mathbf{0}, \mathbf{\Lambda}^{(i)})$:

$$\begin{aligned} \mathbf{A}^{(i)} &= [\mathbf{P}^{\mathrm{xy},(i-1)}]^{\mathsf{T}}\,[\mathbf{P}^{(i-1)}]^{-1}, \\ \mathbf{b}^{(i)} &= \boldsymbol{\mu}^{+,(i-1)} - \mathbf{A}^{(i)}\,\mathbf{m}^{(i-1)}, \\ \mathbf{\Lambda}^{(i)} &= \mathbf{P}^{\mathrm{y},(i-1)} - \mathbf{A}^{(i)}\,\mathbf{P}^{(i-1)}\,[\mathbf{A}^{(i)}]^{\mathsf{T}}, \end{aligned} \tag{10.9}$$

  where the expectations in $\mathbf{P}^{\mathrm{xy},(i-1)} = \mathrm{Cov}[\mathbf{x}, \mathbf{y}]$, $\boldsymbol{\mu}^{+,(i-1)} = \mathrm{E}[\mathbf{y}]$, *and* $\mathbf{P}^{\mathrm{y},(i-1)} = \mathrm{Cov}[\mathbf{y}]$ *are taken with respect to our current best approximation to the posterior distribution,* $\mathbf{x} \sim \mathrm{N}(\mathbf{m}^{(i-1)}, \mathbf{P}^{(i-1)})$*:*

$$
\begin{aligned}
\boldsymbol{\mu}^{+,(i-1)} &= \int \mathbf{g}(\mathbf{x})\, \mathrm{N}(\mathbf{x} \mid \mathbf{m}^{(i-1)}, \mathbf{P}^{(i-1)})\, \mathrm{d}\mathbf{x}, \\
\mathbf{P}^{\mathrm{xy},(i-1)} &= \int (\mathbf{x} - \mathbf{m}^{(i-1)})\, (\mathbf{g}(\mathbf{x}) - \boldsymbol{\mu}^{+,(i-1)})^\mathsf{T} \\
&\qquad \times \mathrm{N}(\mathbf{x} \mid \mathbf{m}^{(i-1)}, \mathbf{P}^{(i-1)})\, \mathrm{d}\mathbf{x}, \\
\mathbf{P}^{\mathrm{y},(i-1)} &= \int (\mathbf{g}(\mathbf{x}) - \boldsymbol{\mu}^{+,(i-1)})\, (\mathbf{g}(\mathbf{x}) - \boldsymbol{\mu}^{+,(i-1)})^\mathsf{T} \\
&\qquad \times \mathrm{N}(\mathbf{x} \mid \mathbf{m}^{(i-1)}, \mathbf{P}^{(i-1)})\, \mathrm{d}\mathbf{x} + \mathbf{Q}.
\end{aligned}
\tag{10.10}
$$

2. Update: *Compute the updated moments* $\mathbf{m}^{(i)}$, $\mathbf{P}^{(i)}$ *such that* $p(\mathbf{x} \mid \mathbf{y}) \simeq \mathrm{N}(\mathbf{x} \mid \mathbf{m}^{(i)}, \mathbf{P}^{(i)})$*:*

$$
\begin{aligned}
\boldsymbol{\mu}^{(i)} &= \mathbf{A}^{(i)}\, \mathbf{m} + \mathbf{b}^{(i)}, \\
\mathbf{S}^{(i)} &= \mathbf{A}^{(i)}\, \mathbf{P}\, [\mathbf{A}^{(i)}]^\mathsf{T} + \boldsymbol{\Lambda}^{(i)}, \\
\mathbf{K}^{(i)} &= \mathbf{P}\, [\mathbf{A}^{(i)}]^\mathsf{T}\, \mathbf{S}^{-1}, \\
\mathbf{m}^{(i)} &= \mathbf{m} + \mathbf{K}^{(i)}\, (\mathbf{y} - \boldsymbol{\mu}^{(i)}), \\
\mathbf{P}^{(i)} &= \mathbf{P} - \mathbf{K}^{(i)}\, \mathbf{S}^{(i)}\, [\mathbf{K}^{(i)}]^\mathsf{T}.
\end{aligned}
\tag{10.11}
$$

We note that IPL always performs SLR with respect to the best available approximation to the posterior, $p(\mathbf{x} \mid \mathbf{y}) \simeq \mathrm{N}(\mathbf{x} \mid \mathbf{m}^{(i-1)}, \mathbf{P}^{(i-1)})$. Before performing the first iteration, the best available approximation is the prior density. The algorithm is iterative since we end each iteration with a new approximation to the posterior.

Importantly, the update step of IPL in Equation (10.11) updates the *prior density* $p(\mathbf{x}) = \mathrm{N}(\mathbf{x} \mid \mathbf{m}, \mathbf{P})$ using the most recent linearization of the model, $\mathbf{y} \simeq \mathbf{A}^{(i)}\mathbf{x} + \mathbf{b}^{(i)} + \tilde{\mathbf{e}}$, $\tilde{\mathbf{e}} \sim \mathrm{N}(\mathbf{0}, \boldsymbol{\Lambda}^{(i)})$. That is, the most recent approximation to the posterior $p(\mathbf{x} \mid \mathbf{y}) \simeq \mathrm{N}(\mathbf{x} \mid \mathbf{m}^{(i-1)}, \mathbf{P}^{(i-1)})$ is only used to linearize the model, and it is not explicitly used in the update step.

Example 10.7 (Iterated posterior linearization). *Let us revisit Example 10.5, where we had a prior* $x \sim \mathrm{N}(1, 0.1)$ *and where we observed* $y = x^3 + q$, $q \sim \mathrm{N}(0, 0.1)$.

The result of applying IPL to this problem is illustrated in Figure 10.2. After the first iteration, IPL yields the same approximation to the posterior as prior SLR, since we initialize IPL using the prior (see also Figure 10.1(b) for a comparison). Note that prior SLR yields the same result as the moment matching approach used in general Gaussian filters. In this example, IPL already provides an accurate approximation to the posterior

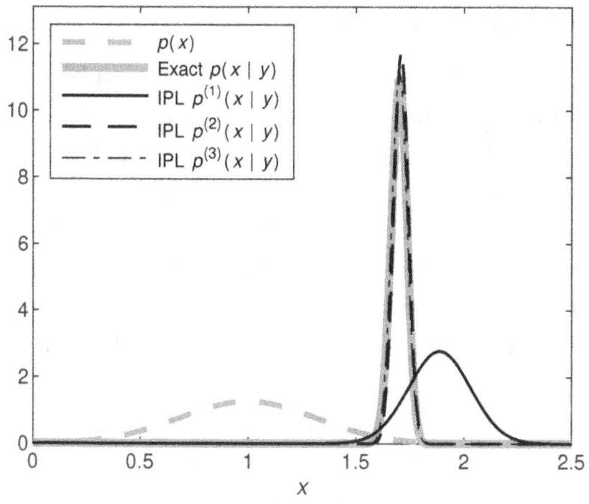

Figure 10.2 An illustration of the densities in Example 10.7. The densities $p^{(i)}(x \mid y)$ are the posterior density approximations obtained using IPL at iterations $i = 1, 2, 3$.

after two iterations, and after the third iteration, it is difficult to distinguish the true posterior from the approximation obtained using IPL.

IPL provides a strategy for selecting the enabling linearization, which can be used to construct Gaussian filters and smoothers. In this chapter, we focus on the filtering problem and present practical algorithms for different model families. In the previous chapter we already discussed different strategies of computing the expectation and covariance defined by integrals in Equations (10.10), not only for additive noise models covered by Algorithm 10.6 but also for non-additive noise models and general conditional distribution models. These strategies can also be directly used instead of Equations (10.10) to apply IPL to these more general model classes.

We can also see that if the expectation and covariances in Equations (10.10) are approximately computed with Taylor series-based linearization at the current guess instead of the Gaussian integrals, we obtain a Gauss–Newton method, which is also the basis of IEKF introduced in Section 7.4. Furthermore, if the posterior variance is small, then the moment matching approximation and Taylor series expansion coincide, and we again get the Gauss–Newton method. When the measurement noise is large, then the posterior distribution resembles the prior, and the prior and posterior SLR (i.e., PL) are likely to produce similar results.

10.4 Iterated Posterior Linearization Filter

The iterated posterior linearization (IPL) strategy can be used in different contexts, and in this chapter, we explore how it can be used in the measurement update of a Gaussian filter. In the filters that we describe here, we perform prediction as in the general Gaussian filters from Chapter 8 (which are closely related to the SLR filters from Chapter 9), and we then use IPL in the update step. Our objective is to present iterated posterior linearization filters (IPLFs) for additive noise models, for non-additive noise models, and for the conditional distribution models in Equations (9.1), (9.9), and (9.10), respectively.

In the iterated posterior linearization filter, the prediction step is approximated in the same way as in conventional Gaussian filters, but the update is done using iterated posterior linearization. In the IPLF update, we iterate to improve our approximations to the posterior moments $\mathbf{m}_k^{(i)}, \mathbf{P}_k^{(i)}$ for $i = 0, 1, 2, \ldots$. The resulting algorithm for additive models is the following.

Algorithm 10.8 (Iterated posterior linearization filter I). *The prediction and update steps of an additive noise iterated posterior linearization (Kalman) filter for model in Equation* (9.1) *are:*

- *Prediction:*
  Compute the predicted moments

$$\begin{aligned}\mathbf{m}_k^- &= \int \mathbf{f}(\mathbf{x})\, \mathrm{N}(\mathbf{x} \mid \mathbf{m}_{k-1}, \mathbf{P}_{k-1})\, \mathrm{d}\mathbf{x}, \\ \mathbf{P}_k^- &= \int (\mathbf{f}(\mathbf{x}) - \mathbf{m}_k^-)\, (\mathbf{f}(\mathbf{x}) - \mathbf{m}_k^-)^\mathsf{T}\, \mathrm{N}(\mathbf{x} \mid \mathbf{m}_{k-1}, \mathbf{P}_{k-1})\, \mathrm{d}\mathbf{x} + \mathbf{Q}_{k-1}.\end{aligned} \tag{10.12}$$

- *Update:*
  Start from the predicted density $\mathbf{m}_k^{(0)} = \mathbf{m}_k^-$, $\mathbf{P}_k^{(0)} = \mathbf{P}_k^-$.
  For $i = 1, 2, 3, \ldots$ *do the following steps:*

1. Compute the moments with respect to $\mathbf{x}_k \sim \mathrm{N}(\mathbf{m}_k^{(i-1)}, \mathbf{P}_k^{(i-1)})$:

$$\begin{aligned}
\boldsymbol{\mu}_k^{+,(i-1)} &= \int \mathbf{h}_k(\mathbf{x})\, \mathrm{N}(\mathbf{x} \mid \mathbf{m}_k^{(i-1)}, \mathbf{P}_k^{(i-1)})\, \mathrm{d}\mathbf{x}, \\
\mathbf{P}_k^{\mathrm{xy},(i-1)} &= \int (\mathbf{x} - \mathbf{m}_k^{(i-1)})\, (\mathbf{h}_k(\mathbf{x}) - \boldsymbol{\mu}_k^{+,(i-1)})^\mathsf{T} \\
&\quad \times \mathrm{N}(\mathbf{x} \mid \mathbf{m}_k^{(i-1)}, \mathbf{P}_k^{(i-1)})\, \mathrm{d}\mathbf{x}, \\
\mathbf{P}_k^{\mathrm{y},(i-1)} &= \int (\mathbf{h}(\mathbf{x}) - \boldsymbol{\mu}_k^{+,(i-1)})\, (\mathbf{h}(\mathbf{x}) - \boldsymbol{\mu}_k^{+,(i-1)})^\mathsf{T} \\
&\quad \times \mathrm{N}(\mathbf{x} \mid \mathbf{m}_k^{(i-1)}, \mathbf{P}_k^{(i-1)})\, \mathrm{d}\mathbf{x} + \mathbf{R}_k.
\end{aligned} \tag{10.13}$$

2. Linearize measurement model:

$$\begin{aligned}
\mathbf{H}_k^{(i)} &= [\mathbf{P}_k^{\mathrm{xy},(i-1)}]^\mathsf{T}\, [\mathbf{P}_k^{(i-1)}]^{-1}, \\
\mathbf{b}_k^{(i)} &= \boldsymbol{\mu}_k^{+,(i-1)} - \mathbf{H}_k^{(i)}\, \mathbf{m}_k^{(i-1)}, \\
\boldsymbol{\Omega}_k^{(i)} &= \mathbf{P}_k^{\mathrm{y},(i-1)} - \mathbf{H}_k^{(i)}\, \mathbf{P}_k^{(i-1)}\, (\mathbf{H}_k^{(i)})^\mathsf{T}.
\end{aligned} \tag{10.14}$$

3. Perform the Kalman update using the linearized model:

$$\begin{aligned}
\boldsymbol{\mu}_k^{(i)} &= \mathbf{H}_k^{(i)}\, \mathbf{m}_k^- + \mathbf{b}_k^{(i)}, \\
\mathbf{S}_k^{(i)} &= \mathbf{H}_k^{(i)}\, \mathbf{P}_k^-\, (\mathbf{H}_k^{(i)})^\mathsf{T} + \boldsymbol{\Omega}_k^{(i)}, \\
\mathbf{K}_k^{(i)} &= \mathbf{P}_k^-\, (\mathbf{H}_k^{(i)})^\mathsf{T}\, (\mathbf{S}_k^{(i)})^{-1}, \\
\mathbf{m}_k^{(i)} &= \mathbf{m}_k^- + \mathbf{K}_k^{(i)}\, (\mathbf{y}_k - \boldsymbol{\mu}_k^{(i)}), \\
\mathbf{P}_k^{(i)} &= \mathbf{P}_k^- - \mathbf{K}_k^{(i)}\, \mathbf{S}_k^{(i)}\, (\mathbf{K}_k^{(i)})^\mathsf{T}.
\end{aligned} \tag{10.15}$$

At convergence, set $\mathbf{m}_k = \mathbf{m}_k^{(i)}$ *and* $\mathbf{P}_k = \mathbf{P}_k^{(i)}$.

The prediction step in Algorithm 10.8 is identical to the prediction step in Algorithms 8.3 and 9.21. Note that the prediction in step Algorithm 9.20 also yields the same result but that the other algorithms make use of simplified expressions. The update step in Algorithm 10.8 is based on the IPL. The update equations used in each iteration are closely related to the update step in Algorithm 9.20, where we perform linearization with respect to $\mathbf{x}_k \sim \mathrm{N}(\mathbf{m}_k^-, \mathbf{P}_k^-)$, and which can then be simplified to Algorithm 9.21. However, we now linearize with respect to $\mathbf{x}_k \sim \mathrm{N}(\mathbf{m}_k^{(i-1)}, \mathbf{P}_k^{(i-1)})$, which prevents us from doing such simplification.

The above IPLF algorithm is closely related to the IEKF described in Algorithm 7.9. Both algorithms rely on the intuition that we should use the posterior to select the linearization. The difference is that the IEKF performs a first-order Taylor expansion about $\mathbf{m}_k^{(i-1)}$, whereas the IPLF performs SLR with respect to $\mathbf{x}_k \sim \mathrm{N}(\mathbf{m}_k^{(i-1)}, \mathbf{P}_k^{(i-1)})$. We recall that an SLR seeks a linearization that is accurate on average across the selected distribution, which has benefits compared to a Taylor expansion in some contexts.

We can now present a non-additive noise generalization of Algorithm 10.8. It turns out that in the IPL-based update step, only the moment computations in Equations (10.13) change, and the linearization and posterior mean and covariance update in Equations (10.14) and (10.15) remain intact.

Algorithm 10.9 (Iterated posterior linearization filter II)**.** *The prediction and update steps of a non-additive noise iterated posterior linearization (Kalman) filter are:*

- *Prediction:*
  Compute the predicted moments

$$
\begin{aligned}
\mathbf{m}_k^- &= \int \mathbf{f}(\mathbf{x}_{k-1}, \mathbf{q}_{k-1}) \\
&\qquad \times \mathrm{N}(\mathbf{x}_{k-1} \mid \mathbf{m}_{k-1}, \mathbf{P}_{k-1})\, \mathrm{N}(\mathbf{q}_{k-1} \mid \mathbf{0}, \mathbf{Q}_{k-1})\, \mathrm{d}\mathbf{x}_{k-1}\, \mathrm{d}\mathbf{q}_{k-1}, \\
\mathbf{P}_k^- &= \int (\mathbf{f}(\mathbf{x}_{k-1}, \mathbf{q}_{k-1}) - \mathbf{m}_k^-)\, (\mathbf{f}(\mathbf{x}_{k-1}, \mathbf{q}_{k-1}) - \mathbf{m}_k^-)^\mathsf{T} \\
&\qquad \times \mathrm{N}(\mathbf{x}_{k-1} \mid \mathbf{m}_{k-1}, \mathbf{P}_{k-1})\, \mathrm{N}(\mathbf{q}_{k-1} \mid \mathbf{0}, \mathbf{Q}_{k-1})\, \mathrm{d}\mathbf{x}_{k-1}\, \mathrm{d}\mathbf{q}_{k-1}.
\end{aligned}
\tag{10.16}
$$

- *Update:*
  Start from the predicted density $\mathbf{m}_k^{(0)} = \mathbf{m}_k^-$, $\mathbf{P}_k^{(0)} = \mathbf{P}_k^-$.
  For $i = 1, 2, 3, \ldots$ *do the following steps:*

1. Compute the moments:

$$\begin{aligned}
\mu_k^{+,(i-1)} &= \int \mathbf{h}(\mathbf{x}_k, \mathbf{r}_k) \\
&\qquad \times \mathrm{N}(\mathbf{x}_k \mid \mathbf{m}_k^{(i-1)}, \mathbf{P}_k^{(i-1)})\, \mathrm{N}(\mathbf{r}_k \mid \mathbf{0}, \mathbf{R}_k)\, \mathrm{d}\mathbf{x}_k\, \mathrm{d}\mathbf{r}_k, \\
\mathbf{P}_k^{\mathrm{xy},(i-1)} &= \int (\mathbf{x}_k - \mathbf{m}_k^{(i-1)})\, (\mathbf{h}(\mathbf{x}_k, \mathbf{q}_k) - \mu_k^{+,(i-1)})^\mathsf{T} \\
&\qquad \times \mathrm{N}(\mathbf{x}_k \mid \mathbf{m}_k^{(i-1)}, \mathbf{P}_k^{(i-1)})\, \mathrm{N}(\mathbf{r}_k \mid \mathbf{0}, \mathbf{R}_k)\, \mathrm{d}\mathbf{x}_k\, \mathrm{d}\mathbf{r}_k, \\
\mathbf{P}_k^{\mathrm{y},(i-1)} &= \int (\mathbf{h}(\mathbf{x}_k, \mathbf{r}_k) - \mu_k^{+,(i-1)})\, (\mathbf{h}(\mathbf{x}_k, \mathbf{r}_k) - \mu_k^{+})^\mathsf{T} \\
&\qquad \times \mathrm{N}(\mathbf{x}_k \mid \mathbf{m}_k^{(i-1)}, \mathbf{P}_k^{(i-1)})\, \mathrm{N}(\mathbf{r}_k \mid \mathbf{0}, \mathbf{R}_k)\, \mathrm{d}\mathbf{x}_k\, \mathrm{d}\mathbf{r}_k.
\end{aligned} \tag{10.17}$$

2. Linearize measurement model using Equations (10.14).

3. Perform the Kalman update using Equations (10.15).

At convergence, set $\mathbf{m}_k = \mathbf{m}_k^{(i)}$ *and* $\mathbf{P}_k = \mathbf{P}_k^{(i)}$.

That is, the IPLF for non-additive noise is similar to the IPLF for additive noise, but the expressions for the involved moments are different. These moments can also be expressed using conditional moments. If we assume that the conditional moments in Equation (9.68) can be computed (analytically or numerically), we can use the following version of the IPLF.

Algorithm 10.10 (Iterated posterior linearization filter III)**.** *The conditional moments form of the iterated posterior linearization (Kalman) filter are:*

- *Prediction:*
  Compute the predicted moments

$$\begin{aligned}
\mathbf{m}_k^- &= \int \mu_k^-(\mathbf{x}_{k-1})\, \mathrm{N}(\mathbf{x}_{k-1} \mid \mathbf{m}_{k-1}, \mathbf{P}_{k-1})\, \mathrm{d}\mathbf{x}_{k-1}, \\
\mathbf{P}_k^- &= \int (\mu_k^-(\mathbf{x}_{k-1}) - \mathbf{m}_k^-)\, (\mu_k^-(\mathbf{x}_{k-1}) - \mathbf{m}_k^-)^\mathsf{T} \\
&\quad \times \mathrm{N}(\mathbf{x}_{k-1} \mid \mathbf{m}_{k-1}, \mathbf{P}_{k-1})\, \mathrm{d}\mathbf{x}_{k-1} \\
&\quad + \int \mathbf{P}_k^{\mathrm{x}}(\mathbf{x}_{k-1})\, \mathrm{N}(\mathbf{x}_{k-1} \mid \mathbf{m}_{k-1}, \mathbf{P}_{k-1})\, \mathrm{d}\mathbf{x}_{k-1}.
\end{aligned} \tag{10.18}$$

- *Update:*
  Start from the predicted density $\mathbf{m}_k^{(0)} = \mathbf{m}_k^-$, $\mathbf{P}_k^{(0)} = \mathbf{P}_k^-$.
  For $i = 1, 2, 3, \ldots$ *do the following steps:*
  1. Compute the moments:

$$\begin{aligned}
\boldsymbol{\mu}_k^{+,(i-1)} &= \int \boldsymbol{\mu}_k(\mathbf{x}_k) \, \mathrm{N}(\mathbf{x}_k \mid \mathbf{m}_k^{(i-1)}, \mathbf{P}_k^{(i-1)}) \, \mathrm{d}\mathbf{x}_k, \\
\mathbf{P}_k^{\mathrm{xy},(i-1)} &= \int (\mathbf{x}_k - \mathbf{m}_k^{(i-1)}) \, (\boldsymbol{\mu}_k(\mathbf{x}_k) - \boldsymbol{\mu}_k^{+,(i-1)})^\mathsf{T} \\
&\qquad \times \mathrm{N}(\mathbf{x}_k \mid \mathbf{m}_k^{(i-1)}, \mathbf{P}_k^{(i-1)}) \, \mathrm{d}\mathbf{x}_k, \\
\mathbf{P}_k^{\mathrm{y},(i-1)} &= \int (\boldsymbol{\mu}_k(\mathbf{x}_k) - \boldsymbol{\mu}_k^{+,(i-1)}) \, (\boldsymbol{\mu}_k(\mathbf{x}_k) - \boldsymbol{\mu}_k^{+,(i-1)})^\mathsf{T} \\
&\qquad \times \mathrm{N}(\mathbf{x}_k \mid \mathbf{m}_k^{(i-1)}, \mathbf{P}_k^{(i-1)}) \, \mathrm{d}\mathbf{x}_k \\
&\qquad + \int \mathbf{P}_k^{\mathrm{y}}(\mathbf{x}_k) \, \mathrm{N}(\mathbf{x}_k \mid \mathbf{m}_k^{(i-1)}, \mathbf{P}_k^{(i-1)}) \, \mathrm{d}\mathbf{x}_k.
\end{aligned} \tag{10.19}$$

  2. Linearize measurement model using Equations (10.14).
  3. Perform the Kalman update using Equations (10.15).

  At convergence, set $\mathbf{m}_k = \mathbf{m}_k^{(i)}$ *and* $\mathbf{P}_k = \mathbf{P}_k^{(i)}$.

Algorithm 10.10 can be used for all the considered model types, as long as the conditional moments in Equation (9.68) can be computed. For additive noise models (see Equation (9.1)), it is trivial to compute the conditional moments, and Algorithm 10.10 then simplifies to Algorithm 10.8. For non-additive noise models (see Equation (9.9)) and conditional distribution models (see Equation (9.10)), the expressions for the conditional moments are given by Equations (A.55) and (A.63), and Equations (A.56) and (A.64), respectively. Note that we sometimes also encounter non-additive noise models for which the conditional moments take closed-form expressions. Important examples of such models are the dynamic models from Chapter 4 for which we have a state dependent covariance matrix such that $p(\mathbf{x}_k \mid \mathbf{x}_{k-1}) = \mathrm{N}(\mathbf{x}_k \mid \mathbf{f}_{k-1}(\mathbf{x}_{k-1}), \mathbf{Q}_{k-1}(\mathbf{x}_{k-1}))$.

When the model is given in the form of conditional distributions as in Equation (9.10), we can also use the following form of the IPLF.

Algorithm 10.11 (Iterated posterior linearization filter IV)**.** *The conditional distribution form of the iterated posterior linearization (Kalman) filter is:*

- *Prediction:*
  Compute the predicted moments

$$\mathbf{m}_k^- = \int \mathbf{x}_k \, p(\mathbf{x}_k \mid \mathbf{x}_{k-1}) \, \mathrm{N}(\mathbf{x}_{k-1} \mid \mathbf{m}_{k-1}, \mathbf{P}_{k-1}) \, \mathrm{d}\mathbf{x}_k \, \mathrm{d}\mathbf{x}_{k-1},$$
$$\mathbf{P}_k^- = \int (\mathbf{x}_k - \mathbf{m}_k^-) \, (\mathbf{x}_k - \mathbf{m}_k^-)^\mathsf{T} \times p(\mathbf{x}_k \mid \mathbf{x}_{k-1}) \, \mathrm{N}(\mathbf{x}_{k-1} \mid \mathbf{m}_{k-1}, \mathbf{P}_{k-1}) \, \mathrm{d}\mathbf{x}_k \, \mathrm{d}\mathbf{x}_{k-1}. \tag{10.20}$$

- *Update:*

  Start from the predicted density $\mathbf{m}_k^{(0)} = \mathbf{m}_k^-$, $\mathbf{P}_k^{(0)} = \mathbf{P}_k^-$.
  For $i = 1, 2, 3, \ldots$ *do the following steps:*

  1. Compute the moments:

$$\boldsymbol{\mu}_k^{+,(i-1)} = \int \mathbf{y}_k \, p(\mathbf{y}_k \mid \mathbf{x}_k) \, \mathrm{N}(\mathbf{x}_k \mid \mathbf{m}_k^{(i-1)}, \mathbf{P}_k^{(i-1)}) \, \mathrm{d}\mathbf{y}_k \, \mathrm{d}\mathbf{x}_k,$$
$$\mathbf{P}_k^{xy,(i-1)} = \int (\mathbf{x}_k - \mathbf{m}_k^{(i-1)}) \, (\mathbf{y}_k - \boldsymbol{\mu}_k^{+,(i-1)})^\mathsf{T} \times p(\mathbf{y}_k \mid \mathbf{x}_k) \, \mathrm{N}(\mathbf{x}_k \mid \mathbf{m}_k^{(i-1)}, \mathbf{P}_k^{(i-1)}) \, \mathrm{d}\mathbf{y}_k \, \mathrm{d}\mathbf{x}_k,$$
$$\mathbf{P}_k^{y,(i-1)} = \int (\mathbf{y}_k - \boldsymbol{\mu}_k^{+,(i-1)}) \, (\mathbf{y}_k - \boldsymbol{\mu}_k^{+,(i-1)})^\mathsf{T} \times p(\mathbf{y}_k \mid \mathbf{x}_k) \, \mathrm{N}(\mathbf{x}_k \mid \mathbf{m}_k^{(i-1)}, \mathbf{P}_k^{(i-1)}) \, \mathrm{d}\mathbf{y}_k \, \mathrm{d}\mathbf{x}_k. \tag{10.21}$$

  2. Linearize measurement model using Equations (10.14).
  3. Perform the Kalman update using Equations (10.15).

  At convergence, set $\mathbf{m}_k = \mathbf{m}_k^{(i)}$ *and* $\mathbf{P}_k = \mathbf{P}_k^{(i)}$.

We have now presented one IPLF for each of the three model types, as well as one IPLF for the conditional moments formulation.

10.5 Practical Iterated Posterior Linearization Filters

To convert the above IPLFs into practical algorithms, we need to solve (or approximate) the involved integrals. In this section, we present general sigma point versions of the additive noise and non-additive noise IPLFs and a Monte Carlo IPLF for conditional distribution models. We start by defining an IPLF for the additive noise model in Equations (9.1).

Algorithm 10.12 (Sigma-point iterated posterior linearization filter I)**.** *The prediction and update steps of an additive noise sigma-point iterated posterior linearization (Kalman) filter are:*

- *Prediction:*
  1. *Form the sigma points as:*

$$\mathcal{X}_{k-1}^{(j)} = \mathbf{m}_{k-1} + \sqrt{\mathbf{P}_{k-1}}\, \boldsymbol{\xi}^{(j)}, \qquad j = 1, \ldots, m, \tag{10.22}$$

  and select the weights $W_1, \ldots, W_m$.

  2. *Propagate the sigma points through the dynamic model:*

$$\mathcal{X}_k^{-(j)} = \mathbf{f}(\mathcal{X}_{k-1}^{(j)}), \qquad j = 1, \ldots, m. \tag{10.23}$$

  3. *Compute the predicted mean $\mathbf{m}_k^-$ and the predicted covariance $\mathbf{P}_k^-$:*

$$\begin{aligned} \mathbf{m}_k^- &= \sum_{j=1}^{m} W_j\, \mathcal{X}_k^{-(j)}, \\ \mathbf{P}_k^- &= \sum_{j=1}^{m} W_j\, (\mathcal{X}_k^{-(j)} - \mathbf{m}_k^-)\, (\mathcal{X}_k^{-(j)} - \mathbf{m}_k^-)^\mathsf{T} + \mathbf{Q}_{k-1}. \end{aligned} \tag{10.24}$$

- *Update:*
  Start from the predicted density $\mathbf{m}_k^{(0)} = \mathbf{m}_k^-$, $\mathbf{P}_k^{(0)} = \mathbf{P}_k^-$.
  For $i = 1, 2, 3, \ldots$ do the following steps:
  1. Form the sigma points:

$$\mathcal{X}_k^{(j)} = \mathbf{m}_k^{(i-1)} + \sqrt{\mathbf{P}_k^{(i-1)}}\, \boldsymbol{\xi}^{(j)}, \qquad j = 1, \ldots, m, \tag{10.25}$$

  and select the corresponding weights $W_1, \ldots, W_m$.

  2. Propagate the sigma points through the measurement model:

$$\mathcal{Y}_k^{(j)} = \mathbf{h}(\mathcal{X}_k^{(j)}), \quad j = 1, \ldots, m. \tag{10.26}$$

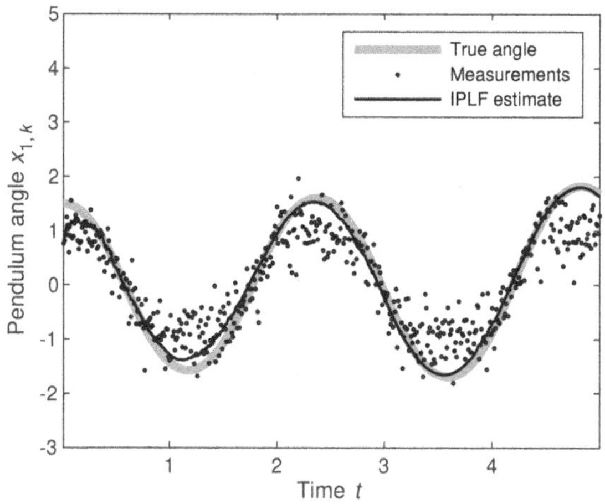

Figure 10.3 Simulated pendulum data and the result of tracking the pendulum with the IPLF (Example 10.13). The resulting RMSE was 0.14 which is lower than the RMSEs of the EKF and IEKF (= 0.17) but higher than many of the non-iterated Gaussian filters.

3. Compute the required moments:

$$\begin{aligned}
\boldsymbol{\mu}_k^{+,(i-1)} &= \sum_{j=1}^{m} W_j \, \mathcal{Y}_k^{(j)}, \\
\mathbf{P}_k^{\mathrm{xy},(i-1)} &= \sum_{j=1}^{m} W_j \, (\mathcal{X}_k^{(j)} - \mathbf{m}_k^{(i-1)}) \, (\mathcal{Y}_k^{(j)} - \boldsymbol{\mu}_k^{+,(i-1)})^{\mathsf{T}}, \\
\mathbf{P}_k^{\mathrm{y},(i-1)} &= \sum_{j=1}^{m} W_j \, (\mathcal{Y}_k^{(j)} - \boldsymbol{\mu}_k^{+,(i-1)}) \, (\mathcal{Y}_k^{(j)} - \boldsymbol{\mu}_k^{+,(i-1)})^{\mathsf{T}} + \mathbf{R}_k.
\end{aligned} \tag{10.27}$$

4. Linearize measurement model using Equations (10.14).
5. Perform the Kalman update using Equations (10.15).

At convergence, set $\mathbf{m}_k = \mathbf{m}_k^{(i)}$ *and* $\mathbf{P}_k = \mathbf{P}_k^{(i)}$.

The prediction step in Algorithm 10.12 is analogous to the prediction step for the sigma points presented in Chapter 8 and, depending on how we select the sigma points, can become identical to the prediction step in the corresponding algorithm in that chapter (e.g., CKF, GHKF, or UKF).

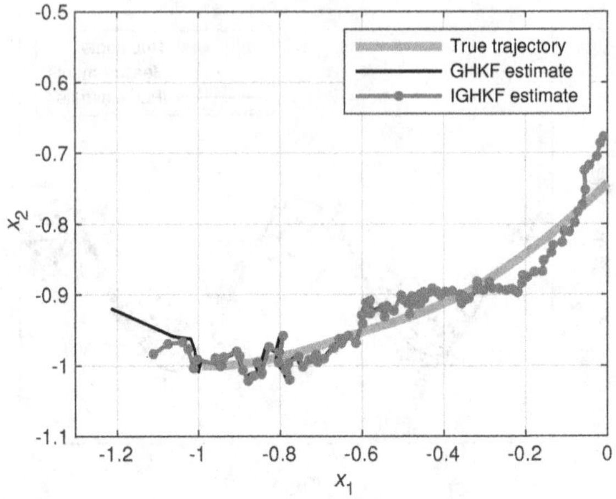

Figure 10.4 The result of applying GHKF and IPLF with Gauss–Hermite integration to simulated data from the (polar) coordinated turn model in Example 7.11 when using an inaccurate initial guess (see Example 10.14). The figure only shows the initial part of trajectory as the result on the rest of the trajectory is identical for both of the filters. The resulting RMSE for GHKF was 0.032 whereas for IPLF it was marginally lower, 0.031, which can be explained by the faster convergence at the start.

Example 10.13 (Pendulum tracking with IPLF). *Figure 10.3 shows the result of applying a Gauss–Hermite integration-based IPLF to the pendulum problem first considered in Example 7.6. Although the result of the filter has a lower error than the EKF and IEKF, it appears that in this model (and data) the iteration does not improve the result over non-iterated Gaussian filters.*

Example 10.14 (Coordinated turn model with posterior linearization filters). *Figure 10.4 shows the result of IPLF with Gauss–Hermite integration to the coordinated turn model problem considered in Example 7.11. The iteration has the same effect as in EKF versus IEKF – the iteration helps to initialize the tracking from the inaccurate initial condition, after which the result is practically identical to the corresponding non-iterated filter.*

To present a sigma point IPLF for non-additive noise models in (9.9), we mimic the strategy used in Algorithm 8.12 and construct sigma points on an augmented state space.

Algorithm 10.15 (Sigma-point iterated posterior linearization filter II). *The prediction and update steps of a non-additive noise sigma point iterated posterior linearization (Kalman) filter are:*

- *Prediction:*
  1. *Form sigma points for the augmented random variable* $(\mathbf{x}_{k-1}, \mathbf{q}_{k-1})$ *as:*
  $$\tilde{\mathcal{X}}_{k-1}^{(j)} = \tilde{\mathbf{m}}_{k-1} + \sqrt{\tilde{\mathbf{P}}_{k-1}}\, \boldsymbol{\xi}^{(j)}, \qquad j = 1, \ldots, m, \tag{10.28}$$
  where
  $$\tilde{\mathbf{m}}_{k-1} = \begin{pmatrix} \mathbf{m}_{k-1} \\ \mathbf{0} \end{pmatrix}, \qquad \tilde{\mathbf{P}}_{k-1} = \begin{pmatrix} \mathbf{P}_{k-1} & \mathbf{0} \\ \mathbf{0} & \mathbf{Q}_{k-1} \end{pmatrix},$$
  and select the weights $W_1, \ldots, W_m$.
  2. *Propagate the sigma points through the dynamic model:*
  $$\mathcal{X}_k^{-(j)} = \mathbf{f}(\tilde{\mathcal{X}}_{k-1}^{(j),x}, \tilde{\mathcal{X}}_{k-1}^{(j),q}), \qquad j = 1, \ldots, m, \tag{10.29}$$
  where $\tilde{\mathcal{X}}_{k-1}^{(j),x}$ *denotes the first* $n = \dim(\mathbf{x}_{k-1})$ *components in* $\tilde{\mathcal{X}}_{k-1}^{(j)}$ *and* $\tilde{\mathcal{X}}_{k-1}^{(j),q}$ *denotes the last* $n_q = \dim(\mathbf{q}_{k-1})$ *components.*
  3. *Compute the predicted mean* $\mathbf{m}_k^-$ *and the predicted covariance* $\mathbf{P}_k^-$*:*
  $$\begin{aligned} \mathbf{m}_k^- &= \sum_{j=1}^{m} W_j\, \mathcal{X}_k^{-(j)}, \\ \mathbf{P}_k^- &= \sum_{j=1}^{m} W_j\, (\mathcal{X}_k^{-(j)} - \mathbf{m}_k^-)\,(\mathcal{X}_k^{-(j)} - \mathbf{m}_k^-)^\mathsf{T}. \end{aligned} \tag{10.30}$$

- *Update:*
  Start from the predicted density $\mathbf{m}_k^{(0)} = \mathbf{m}_k^-$, $\mathbf{P}_k^{(0)} = \mathbf{P}_k^-$.
  For $i = 1, 2, 3, \ldots$ *do the following steps:*
  1. Form the sigma points for the augmented vector $(\mathbf{x}_k, \mathbf{r}_k)$ as:
  $$\tilde{\mathcal{X}}_k^{(j)} = \tilde{\mathbf{m}}_k^{(i-1)} + \sqrt{\tilde{\mathbf{P}}_k^{(i-1)}}\, \boldsymbol{\xi}^{(j)}, \qquad j = 1, \ldots, m, \tag{10.31}$$
  where
  $$\tilde{\mathbf{m}}_k^{(i-1)} = \begin{pmatrix} \mathbf{m}_k^{(i-1)} \\ \mathbf{0} \end{pmatrix}, \qquad \tilde{\mathbf{P}}_k^{-} = \begin{pmatrix} \mathbf{P}_k^{(i-1)} & \mathbf{0} \\ \mathbf{0} & \mathbf{R}_k \end{pmatrix},$$

and select the corresponding weights $W_1, \ldots, W_m$.

2. Propagate the sigma points through the measurement model:

$$\mathcal{Y}_k^{(j)} = \mathbf{h}(\tilde{\mathcal{X}}_k^{(j),x}, \tilde{\mathcal{X}}_k^{(j),r}), \quad j = 1, \ldots, m, \tag{10.32}$$

where $\tilde{\mathcal{X}}_k^{(j),x}$ denotes the first n components in $\tilde{\mathcal{X}}_k^{(j)}$ and $\tilde{\mathcal{X}}_k^{(j),r}$ denotes the last $n_r = \dim(\mathbf{r}_k)$ components.

3. Compute the required moments:

$$\begin{aligned} \boldsymbol{\mu}_k^{+,(i-1)} &= \sum_{j=1}^{m} W_j \, \mathcal{Y}_k^{(j)}, \\ \mathbf{P}_k^{\mathrm{xy},(i-1)} &= \sum_{j=1}^{m} W_j \, (\tilde{\mathcal{X}}_k^{(j),x} - \mathbf{m}_k^{(i-1)}) \, (\mathcal{Y}_k^{(j)} - \boldsymbol{\mu}_k^{+,(i-1)})^\mathsf{T}, \\ \mathbf{P}_k^{\mathrm{y},(i-1)} &= \sum_{j=1}^{m} W_j \, (\mathcal{Y}_k^{(j)} - \boldsymbol{\mu}_k^{+,(i-1)}) \, (\mathcal{Y}_k^{(j)} - \boldsymbol{\mu}_k^{+,(i-1)})^\mathsf{T}. \end{aligned} \tag{10.33}$$

4. Linearize measurement model using Equations (10.14).
5. Perform the Kalman update using Equations (10.15).

At convergence, set $\mathbf{m}_k = \mathbf{m}_k^{(i)}$ *and* $\mathbf{P}_k = \mathbf{P}_k^{(i)}$.

We can also construct a sigma point version of the conditional moments IPLF described in Algorithm 10.10.

Algorithm 10.16 (Sigma-point iterated posterior linearization filter III). *The prediction and update steps of a conditional moments sigma point iterated posterior linearization (Kalman) filter are:*

- *Prediction:*

1. *Form the sigma points as:*

$$\mathcal{X}_{k-1}^{(j)} = \mathbf{m}_{k-1} + \sqrt{\mathbf{P}_{k-1}} \, \boldsymbol{\xi}^{(j)}. \qquad j = 1, \ldots, m. \tag{10.34}$$

2. *Propagate the sigma points through the conditional mean and covariance functions:*

$$\begin{aligned} \boldsymbol{\mu}_k^{-(j)} &= \boldsymbol{\mu}_k^{-}(\mathcal{X}_{k-1}^{(j)}), \quad j = 1, \ldots, m, \\ \mathbf{P}_k^{x,(j)} &= \mathbf{P}_k^{x}(\mathcal{X}_{k-1}^{(j)}), \quad j = 1, \ldots, m. \end{aligned} \tag{10.35}$$

3. *Compute the predicted mean* $\mathbf{m}_k^-$ *and the predicted covariance* $\mathbf{P}_k^-$*:*

$$
\begin{aligned}
\mathbf{m}_k^- &= \sum_{j=1}^{m} W_j \, \boldsymbol{\mu}_k^{-(j)}, \\
\mathbf{P}_k^- &= \sum_{j=1}^{m} W_j \left(\mathbf{P}_k^{x,(j)} + (\boldsymbol{\mu}_k^{-(j)} - \mathbf{m}_k^-) \, (\boldsymbol{\mu}_k^{-(j)} - \mathbf{m}_k^-)^\mathsf{T} \right).
\end{aligned}
\tag{10.36}
$$

- *Update:*
  Start from the predicted density $\mathbf{m}_k^{(0)} = \mathbf{m}_k^-$, $\mathbf{P}_k^{(0)} = \mathbf{P}_k^-$.
  For $i = 1, 2, 3, \ldots$ *do the following steps:*

1. Form the sigma points:

$$
\mathcal{X}_k^{-(j)} = \mathbf{m}_k^{(i-1)} + \sqrt{\mathbf{P}_k^{(i-1)}} \, \boldsymbol{\xi}^{(j)}, \qquad j = 1, \ldots, m, \tag{10.37}
$$

and select the corresponding weights $W_1, \ldots, W_m$.

2. *Propagate the sigma points through the conditional mean and covariance functions:*

$$
\begin{aligned}
\boldsymbol{\mu}_k^{(j)} &= \boldsymbol{\mu}_k(\mathcal{X}_k^{-(j)}), \quad j = 1, \ldots, m, \\
\mathbf{P}_k^{y,(j)} &= \mathbf{P}_k^{y}(\mathcal{X}_k^{-(j)}), \quad j = 1, \ldots, m.
\end{aligned}
\tag{10.38}
$$

3. Compute the required moments:

$$
\begin{aligned}
\boldsymbol{\mu}_k^{+,(i-1)} &= \sum_{j=1}^{m} W_j \, \boldsymbol{\mu}_k^{(j)}, \\
\mathbf{P}_k^{xy,(i-1)} &= \sum_{j=1}^{m} W_j \, (\mathcal{X}_k^{-(j)} - \mathbf{m}_k^{(i-1)}) \, (\boldsymbol{\mu}_k^{(j)} - \boldsymbol{\mu}_k^{+,(i-1)})^\mathsf{T}, \\
\mathbf{P}_k^{y,(i-1)} &= \sum_{i=1}^{m} W_j \left(\mathbf{P}_k^{y,(j)} + (\boldsymbol{\mu}_k^{(j)} - \boldsymbol{\mu}_k^{+,(i-1)}) \, (\boldsymbol{\mu}_k^{(j)} - \boldsymbol{\mu}_k^{+,(i-1)})^\mathsf{T} \right).
\end{aligned}
\tag{10.39}
$$

4. Linearize measurement model using Equations (10.14).
5. Perform the Kalman update using Equations (10.15).

   At convergence, set $\mathbf{m}_k = \mathbf{m}_k^{(i)}$ *and* $\mathbf{P}_k = \mathbf{P}_k^{(i)}$.

Finally, we present an IPLF for the conditional distribution models in (9.10), where we make use of Monte Carlo sampling to approximate the involved integrals. Please note that in this case one has to be careful with convergence of the inner iteration, because of its stochasticity.

Algorithm 10.17 (Monte Carlo iterated posterior linearization filter)**.** *The prediction and update steps of a Monte Carlo sigma point iterated posterior linearization (Kalman) filter for conditional distribution models are:*

- *Prediction:*
  1. *Generate samples of* $\mathbf{x}_{k-1}$ *as:*

$$\mathbf{x}_{k-1}^{(j)} \sim \mathrm{N}(\mathbf{m}_{k-1}, \mathbf{P}_{k-1}), \qquad j = 1, \ldots, N. \tag{10.40}$$

  2. *Propagate the samples through the dynamic model:*

$$\mathbf{x}_k^{-(j)} \sim p(\mathbf{x}_k \mid \mathbf{x}_{k-1}^{(j)}), \quad j = 1, \ldots, N. \tag{10.41}$$

  3. *Compute the predicted mean* $\mathbf{m}_k^-$ *and the predicted covariance* $\mathbf{P}_k^-$*:*

$$\begin{aligned}
\mathbf{m}_k^- &= \frac{1}{N} \sum_{j=1}^{N} \mathbf{x}_k^{-(j)}, \\
\mathbf{P}_k^- &= \frac{1}{N} \sum_{j=1}^{N} (\mathbf{x}_k^{-(j)} - \mathbf{m}_k^-)\,(\mathbf{x}_k^{-(j)} - \mathbf{m}_k^-)^\mathsf{T}.
\end{aligned} \tag{10.42}$$

- *Update:*
  Start from the predicted density $\mathbf{m}_k^{(0)} = \mathbf{m}_k^-$, $\mathbf{P}_k^{(0)} = \mathbf{P}_k^-$.
  For $i = 1, 2, 3, \ldots$ *do the following steps:*
  1. Generate samples of $\mathbf{x}_k$ from our current approximation to the posterior:

$$\mathbf{x}_k^{(j)} \sim \mathrm{N}(\mathbf{m}_k^{(i-1)}, \mathbf{P}_k^{(i-1)}), \qquad j = 1, \ldots, N. \tag{10.43}$$

  2. Propagate the samples through the measurement model:

$$\mathbf{y}_k^{(j)} \sim p(\mathbf{y}_k \mid \mathbf{x}_k^{(j)}), \quad j = 1, \ldots, N. \tag{10.44}$$

  3. Compute the required moments:

$$\begin{aligned}
\boldsymbol{\mu}_k^{+,(i-1)} &= \frac{1}{N} \sum_{j=1}^{N} \mathbf{y}_k^{(j)}, \\
\mathbf{P}_k^{xy,(i-1)} &= \frac{1}{N} \sum_{j=1}^{N} (\mathbf{x}_k^{(j)} - \mathbf{m}_k^{(i-1)})\,(\mathbf{y}_k^{(j)} - \boldsymbol{\mu}_k^{+,(i-1)})^\mathsf{T}, \\
\mathbf{P}_k^{y,(i-1)} &= \frac{1}{N} \sum_{i=1}^{N} (\mathbf{y}_k^{(j)} - \boldsymbol{\mu}_k^{+,(i-1)})\,(\mathbf{y}_k^{(j)} - \boldsymbol{\mu}_k^{+,(i-1)})^\mathsf{T}.
\end{aligned} \tag{10.45}$$

4. Linearize measurement model using Equations (10.14).
5. Perform the Kalman update using Equations (10.15).

   At convergence, set $\mathbf{m}_k = \mathbf{m}_k^{(i)}$ *and* $\mathbf{P}_k = \mathbf{P}_k^{(i)}$.

10.6 Optimality Properties of Different Linearizations

So far in this book we have seen a selection of different deterministic, stochastic, and iterative linearizations. It is now useful to discuss in which sense each of these is optimal. The first encountered linearization of a function $\mathbf{g}(\mathbf{x})$ in Section 7.1 was a Taylor series expansion of it on a given point $\mathbf{m}$:

$$\mathbf{g}(\mathbf{x}) \simeq \mathbf{g}(\mathbf{m}) + \mathbf{G_x}(\mathbf{m})\,(\mathbf{x} - \mathbf{m}), \tag{10.46}$$

where $\mathbf{G_x}$ is the Jacobian matrix of $\mathbf{g}$. This linearization can be seen to be optimal when we wish to approximate the function $\mathbf{g}$ very close (infinitesimally close) to the point $\mathbf{x} = \mathbf{m}$. The linearization is exact at this point, and its accuracy quickly decreases when going away from the point.

A slightly less local linearization, the statistical linearization (SL), was introduced in Section 9.2, and it was formulated as the linearization of the form

$$\mathbf{g}(\mathbf{x}) \simeq \mathbf{A}\,\mathbf{x} + \mathbf{b}, \tag{10.47}$$

which minimizes the mean squared error

$$\mathrm{MSE}(\mathbf{A}, \mathbf{b}) = \mathrm{E}\left[(\mathbf{g}(\mathbf{x}) - \mathbf{A}\,\mathbf{x} - \mathbf{b})^\mathsf{T}\,(\mathbf{g}(\mathbf{x}) - \mathbf{A}\,\mathbf{x} - \mathbf{b})\right], \tag{10.48}$$

where $\mathbf{x} \sim \mathrm{N}(\mathbf{m}, \mathbf{P})$. This linearization is less local than the Taylor series expansion in the sense that it is formed with respect to the distribution $\mathrm{N}(\mathbf{m}, \mathbf{P})$ instead of a single point $\mathbf{x} = \mathbf{m}$. However, as discussed on page 174, the Taylor and SL are closely related as SL can be seen as a Taylor series expansion where $\mathbf{g}(\mathbf{m})$ and $\mathbf{G_x}(\mathbf{m})$ in Equation (10.46) are replaced with their expectations over $\mathrm{N}(\mathbf{m}, \mathbf{P})$. This also implies that SL becomes exactly the Taylor series expansion when $\mathbf{P} \to \mathbf{0}$, because in this limit the expectation becomes simply evaluation at the mean.

In Section 9.4 we encountered statistical linear regression (SLR), which strictly speaking only makes sense as an approximation to a stochastic mapping such as $\mathbf{y} = \mathbf{g}(\mathbf{x}) + \mathbf{q}$, where $\mathbf{q} \sim \mathrm{N}(\mathbf{0}, \mathbf{Q})$. A linearization was then found of the form

$$\mathbf{y} \simeq \mathbf{A}\,\mathbf{x} + \mathbf{b} + \tilde{\mathbf{e}}, \tag{10.49}$$

where $\tilde{\mathbf{e}} \sim \mathrm{N}(\mathbf{0}, \boldsymbol{\Lambda})$. In Section 9.4 the SLR linearization was formed to match the moments of $(\mathbf{x}, \mathbf{y})$. However, it turns out that there is an error criterion that this linearization also minimizes (see, e.g., García-Fernández et al., 2015). Let us first notice that $\mathbf{y} = \mathbf{g}(\mathbf{x}) + \mathbf{q}$ and $\mathbf{y} = \mathbf{A}\,\mathbf{x} + \mathbf{b} + \tilde{\mathbf{e}}$ correspond to the conditional distribution models

$$\begin{aligned} p(\mathbf{y} \mid \mathbf{x}) &= \mathrm{N}(\mathbf{y} \mid \mathbf{g}(\mathbf{x}), \mathbf{Q}), \\ q(\mathbf{y} \mid \mathbf{x}) &= \mathrm{N}(\mathbf{y} \mid \mathbf{A}\,\mathbf{x} + \mathbf{b}, \boldsymbol{\Lambda}), \end{aligned} \tag{10.50}$$

respectively. We can now compute the Kullback–Leibler divergence between the two Gaussian distributions, which gives

$$\begin{aligned} \mathrm{KL}(p||q) &= \int \log\left(\frac{p(\mathbf{y} \mid \mathbf{x})}{q(\mathbf{y} \mid \mathbf{x})}\right) p(\mathbf{y} \mid \mathbf{x})\,\mathrm{d}\mathbf{y} \\ &= \frac{1}{2}\bigg[\mathrm{tr}\left\{\boldsymbol{\Lambda}^{-1}\,\mathbf{Q}\right\} - n + \log\left(\frac{|\boldsymbol{\Lambda}|}{|\mathbf{Q}|}\right) \\ &\qquad + (\mathbf{g}(\mathbf{x}) - \mathbf{A}\,\mathbf{x} - \mathbf{b})^\mathsf{T}\,\boldsymbol{\Lambda}^{-1}\,(\mathbf{g}(\mathbf{x}) - \mathbf{A}\,\mathbf{x} - \mathbf{b}) \bigg], \end{aligned} \tag{10.51}$$

which is still a function of $\mathbf{x}$. Taking expectation with respect to $\mathbf{x} \sim \mathrm{N}(\mathbf{m}, \mathbf{P})$ then leads to the criterion

$$\begin{aligned} \mathrm{MKL}(\mathbf{A}, \mathbf{b}, \boldsymbol{\Lambda}) = \frac{1}{2}\,\mathrm{E}\bigg[&\mathrm{tr}\left\{\boldsymbol{\Lambda}^{-1}\,\mathbf{Q}\right\} - n + \log\left(\frac{|\boldsymbol{\Lambda}|}{|\mathbf{Q}|}\right) \\ &+ (\mathbf{g}(\mathbf{x}) - \mathbf{A}\,\mathbf{x} - \mathbf{b})^\mathsf{T}\,\boldsymbol{\Lambda}^{-1}\,(\mathbf{g}(\mathbf{x}) - \mathbf{A}\,\mathbf{x} - \mathbf{b}) \bigg], \end{aligned} \tag{10.52}$$

which turns out to be the error criterion that SLR minimizes (see Exercise 10.7). The generalized SLR discussed in Section 10.1 minimizes the same criterion, except that the distribution of $\mathbf{x}$ is replaced with an arbitrary distribution $\pi(\mathbf{x})$.

In Chapters 9 and 10 we generalized the SLR to conditional distributions of the form $p(\mathbf{y} \mid \mathbf{x})$ that might not in general have a Gaussian form such as $p(\mathbf{y} \mid \mathbf{x}) = \mathrm{N}(\mathbf{y} \mid \mathbf{g}(\mathbf{x}), \mathbf{Q})$. It might appear that a suitable generalization of SLR would be to replace the Kullback–Leibler divergence in Equation (10.51) by plugging in $p(\mathbf{y} \mid \mathbf{x})$. However, this is not correct, but instead, the correct generalization is obtained by approximating the conditional distribution as

$$p(\mathbf{y} \mid \mathbf{x}) \simeq \mathrm{N}(\mathbf{y} \mid \boldsymbol{\mu}_\mathrm{R}(\mathbf{x}), \mathbf{S}_\mathrm{R}(\mathbf{x})), \tag{10.53}$$

where $\boldsymbol{\mu}_{\mathrm{R}}(\mathbf{x})$ and $\mathbf{S}_{\mathrm{R}}(\mathbf{x})$ are the conditional moments defined as in Equation (9.46). This then leads to the conditional distribution forms of SLR-based filters that we discussed in the mentioned chapters.

In (this) Chapter 10 we finally discussed posterior linearization, which corresponds to the selection $\pi(\mathbf{x}) = p(\mathbf{x} \mid \mathbf{y})$ in SLR. To find this linearization we had to resort to iteration (iterated posterior linearization). The posterior linearization has the optimality property that the posterior mean and covariance using the linearized model match those of the exact posterior up to the Gaussian approximations needed on the way to computing the posterior. Thus in this sense posterior linearization has its maximum accuracy at the posterior distribution. In this regard the linearization used in the IEKF in Section 7.4 can be seen as a type of posterior linearization that maximizes the accuracy of Taylor series expansion under this same philosophy by linearizing at the MAP estimate of the state.

So far in this book we have only applied the SLR and posterior linearization methods to filters, but in Chapter 14 we proceed to apply them to non-linear smoothing as well.

10.7 Exercises

10.1 Show that Equation (10.3) reduces to moment matching only when $\pi(\mathbf{x})$ is the actual distribution of $\mathbf{x}$.

10.2 Consider the posterior linearization of $y = x^3 + r$, $r \sim N(0, R)$, where $R = 0.1$, with respect to $x \sim \mathrm{N}(m, P)$, where $m = 1$ and $P = 0.1$. First perform the linearization for $y = 5$, and then repeat it for a few other values of y. To approximate scalar integrals we can use Riemann sums,

$$\int f(x)\,\mathrm{d}x \approx \sum_{k=-\infty}^{\infty} f(k\Delta x)\,\Delta x, \tag{10.54}$$

where Δx is the step length (say, 0.01). In practice, it is sufficient to sum k from n_1 to n_2 where $f(k\Delta x) \approx 0$ for $k < n_1$ and $k > n_2$.

(a) Use Riemann sums to approximate

$$I = \int \mathrm{N}(y \mid x^3, R)\,\mathrm{N}(x \mid m, P)\,\mathrm{d}x, \tag{10.55}$$

and plot the posterior density

$$p(x \mid y) = \frac{\mathrm{N}(y \mid x^3, R)\,\mathrm{N}(x \mid m, P)}{I}. \tag{10.56}$$

Note that this becomes an accurate approximation to the posterior for small values of Δx. If you prefer, you can instead use Monte Carlo sampling to approximate the integrals in this exercise.

(b) Use Riemann sums to perform posterior linearization. Then plot x^3 and the deterministic part, $Ax + b$, of the posterior linearization. Does the linearization appear to be accurate in the support of the posterior density?
Hint: You need to compute the integrals in Corollary 10.3. To this end, it is helpful to notice that we have

$$\begin{aligned} g(x) &= x^3, \\ \pi(x) &= \frac{1}{I} \mathrm{N}(y \mid x^3, R)\, \mathrm{N}(x \mid m, P). \end{aligned} \tag{10.57}$$

(c) Use the linearized model to compute the posterior density. Plot the result along with the posterior density computed in (a).
Hint: Note that the linearized model is used to approximate the measurement model, which is then used to (Kalman) update the prior $\mathrm{N}(x \mid m, P)$.

(d) Perform a prior linearization, that is, perform an SLR where x is instead distributed according to the prior, $x \sim \mathrm{N}(m, P)$. Plot the deterministic part $A\,x + b$ along with x^3 and the deterministic part of the posterior linearization in (b).

(e) Use the prior linearization to approximate the posterior density. Plot the result along with the densities from (b) and (c).

10.3 Use Riemann sums to perform iterated posterior linearization for the model in Exercise 10.2. It should be sufficient to run the algorithm for a handful of iterations. Plot $A\,x + b$ and the approximation of the posterior density , at different iterations, along with x^3 and the accurate approximation of the posterior density from Exercise 10.2, respectively.

10.4 Formulate an IPL algorithm that makes use of the spherical cubature rule to approximate integrals. Use the algorithm to approximate the posterior density in Exercise 10.2 and compare with the results from Exercise 10.3.

10.5 Implement a GHKF-based IPLF for the model in Exercise 7.1. Plot the results and compare the RMSE values with other filters.

10.6 Implement a spherical cubature rule-based IPLF for the bearings-only target tracking problem in Exercise 7.2. Compare the performance with the EKF and CKF.

10.7 Show that SLR minimizes the error criterion in Equation (10.52).

11

Particle Filtering

Although in many filtering problems Gaussian approximations work well, sometimes the filtering distributions can be, for example, multi-modal, or some of the state components might be discrete, in which case Gaussian approximations are not appropriate. In such cases sequential importance resampling-based particle filters can be a better alternative. This chapter is concerned with particle filters, which are methods for forming Monte Carlo approximations to the solutions of the Bayesian filtering equations.

11.1 Monte Carlo Approximations in Bayesian Inference

In Bayesian inference, including Bayesian filtering, the main inference problem can often be reduced to computation of the following kind of expectations over the posterior distribution:[1]

$$\mathrm{E}[\mathbf{g}(\mathbf{x}) \mid \mathbf{y}_{1:T}] = \int \mathbf{g}(\mathbf{x})\, p(\mathbf{x} \mid \mathbf{y}_{1:T})\, \mathrm{d}\mathbf{x}, \tag{11.1}$$

where $\mathbf{g} : \mathbb{R}^n \to \mathbb{R}^m$ is an arbitrary function, and $p(\mathbf{x} \mid \mathbf{y}_{1:T})$ is the posterior probability density of $\mathbf{x}$ given the measurements $\mathbf{y}_1, \ldots, \mathbf{y}_T$. Now the problem is that such an integral can be evaluated in closed form only in a few special cases, and generally, numerical methods have to be used.

Monte Carlo methods provide a numerical method for calculating integrals of the form (11.1). Monte Carlo refers to a general class of methods where closed form computation of statistical quantities is replaced by drawing samples from the distribution and estimating the quantities by sample averages.

[1] In this section we formally treat $\mathbf{x}$ as a continuous random variable with a density, but the analogous results apply to discrete random variables.

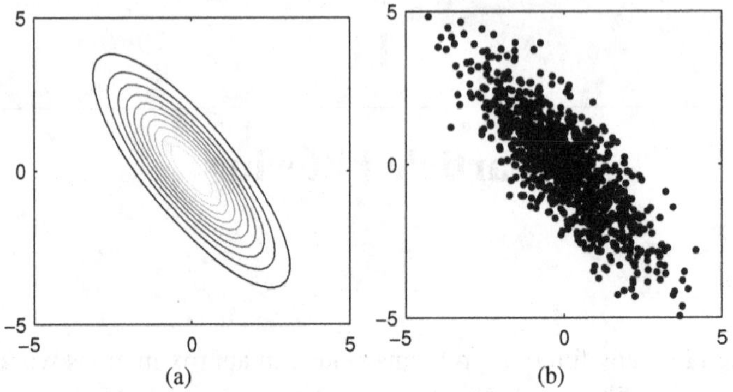

Figure 11.1 (a) Two-dimensional Gaussian distribution. (b) Monte Carlo representation of the same Gaussian distribution.

In a (perfect) Monte Carlo approximation, we draw N independent random samples $\mathbf{x}^{(i)} \sim p(\mathbf{x} \mid \mathbf{y}_{1:T})$, $i = 1, \ldots, N$ and estimate the expectation as

$$\mathrm{E}[\mathbf{g}(\mathbf{x}) \mid \mathbf{y}_{1:T}] \approx \frac{1}{N} \sum_{i=1}^{N} \mathbf{g}(\mathbf{x}^{(i)}). \tag{11.2}$$

Thus Monte Carlo methods approximate the target distribution by a set of samples that are distributed according to the target density. Figure 11.1 represents a two-dimensional Gaussian distribution and its Monte Carlo representation.

The convergence of the Monte Carlo approximation is guaranteed by the law of large numbers (LLN) and the central limit theorem (CLT, see, e.g., Liu, 2001), and the error term is $O(N^{-1/2})$, regardless of the dimensionality of $\mathbf{x}$. This invariance with respect to dimensionality is unique to Monte Carlo methods and makes them superior to practically all other numerical methods when the dimensionality of $\mathbf{x}$ is considerable – at least in theory, not necessarily in practice (see Daum and Huang, 2003; Snyder et al., 2008).

11.2 Importance Sampling

Often, in practical Bayesian models, it is not possible to obtain samples directly from $p(\mathbf{x} \mid \mathbf{y}_{1:T})$ due to its complicated functional form. In *importance sampling* (IS) (see, e.g., Liu, 2001) we use an approximate distribution called the importance distribution $\pi(\mathbf{x} \mid \mathbf{y}_{1:T})$, from which we can easily draw samples. Importance sampling is based on the following decomposition of the expectation over the posterior probability density $p(\mathbf{x} \mid \mathbf{y}_{1:T})$:

$$\int \mathbf{g}(\mathbf{x})\, p(\mathbf{x} \mid \mathbf{y}_{1:T})\, d\mathbf{x} = \int \left[\mathbf{g}(\mathbf{x}) \frac{p(\mathbf{x} \mid \mathbf{y}_{1:T})}{\pi(\mathbf{x} \mid \mathbf{y}_{1:T})} \right] \pi(\mathbf{x} \mid \mathbf{y}_{1:T})\, d\mathbf{x}, \tag{11.3}$$

where the importance density $\pi(\mathbf{x} \mid \mathbf{y}_{1:T})$ is required to be non-zero whenever $p(\mathbf{x} \mid \mathbf{y}_{1:T})$ is non-zero, that is, the *support* of $\pi(\mathbf{x} \mid \mathbf{y}_{1:T})$ needs to be greater than or equal to the support of $p(\mathbf{x} \mid \mathbf{y}_{1:T})$. As the above expression is just the expectation of the term in the brackets over the distribution $\pi(\mathbf{x} \mid \mathbf{y}_{1:T})$, we can form a Monte Carlo approximation to it by drawing N samples from the importance distribution:

$$\mathbf{x}^{(i)} \sim \pi(\mathbf{x} \mid \mathbf{y}_{1:T}), \qquad i = 1, \ldots, N \tag{11.4}$$

and by forming the approximation as

$$\begin{aligned} \mathrm{E}[\mathbf{g}(\mathbf{x}) \mid \mathbf{y}_{1:T}] &\approx \frac{1}{N} \sum_{i=1}^{N} \frac{p(\mathbf{x}^{(i)} \mid \mathbf{y}_{1:T})}{\pi(\mathbf{x}^{(i)} \mid \mathbf{y}_{1:T})}\, \mathbf{g}(\mathbf{x}^{(i)}) \\ &= \sum_{i=1}^{N} \tilde{w}^{(i)}\, \mathbf{g}(\mathbf{x}^{(i)}), \end{aligned} \tag{11.5}$$

where the weights have been defined as

$$\tilde{w}^{(i)} = \frac{1}{N} \frac{p(\mathbf{x}^{(i)} \mid \mathbf{y}_{1:T})}{\pi(\mathbf{x}^{(i)} \mid \mathbf{y}_{1:T})}. \tag{11.6}$$

Figure 11.2 illustrates the idea of importance sampling. We sample from the importance distribution, which is an approximation to the target distribution. Because the distribution of samples is not exact, we need to correct the approximation by associating a weight with each of the samples.

The disadvantage of this direct importance sampling is that we should be able to evaluate $p(\mathbf{x}^{(i)} \mid \mathbf{y}_{1:T})$ in order to use it directly. Recall that by Bayes' rule, the evaluation of the posterior probability density can be

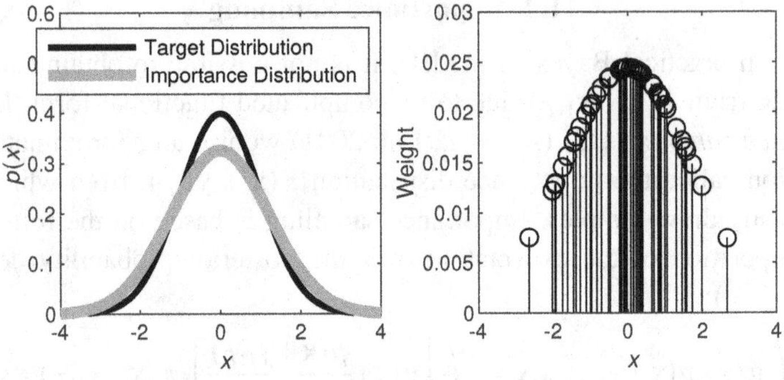

(a) (b)

Figure 11.2 (a) The importance distribution approximates the target distribution. (b) Weights are associated with each of the samples to correct the approximation.

written as

$$p(\mathbf{x}^{(i)} \mid \mathbf{y}_{1:T}) = \frac{p(\mathbf{y}_{1:T} \mid \mathbf{x}^{(i)})\, p(\mathbf{x}^{(i)})}{\int p(\mathbf{y}_{1:T} \mid \mathbf{x})\, p(\mathbf{x})\, \mathrm{d}\mathbf{x}}. \tag{11.7}$$

The likelihood $p(\mathbf{y}_{1:T} \mid \mathbf{x}^{(i)})$ and prior terms $p(\mathbf{x}^{(i)})$ are usually easy to evaluate, but often the integral in the denominator – the normalization constant – cannot be computed. To overcome this problem, we can form an importance sampling approximation to the expectation integral by also approximating the normalization constant by importance sampling. For this purpose we can decompose the expectation integral and form the approximation as follows:

$$
\begin{aligned}
\mathrm{E}[\mathbf{g}(\mathbf{x}) \mid \mathbf{y}_{1:T}] &= \int \mathbf{g}(\mathbf{x})\, p(\mathbf{x} \mid \mathbf{y}_{1:T})\, \mathrm{d}\mathbf{x} \\
&= \frac{\int \mathbf{g}(\mathbf{x})\, p(\mathbf{y}_{1:T} \mid \mathbf{x})\, p(\mathbf{x})\, \mathrm{d}\mathbf{x}}{\int p(\mathbf{y}_{1:T} \mid \mathbf{x})\, p(\mathbf{x})\, \mathrm{d}\mathbf{x}} \\
&= \frac{\int \left[\frac{p(\mathbf{y}_{1:T} \mid \mathbf{x})\, p(\mathbf{x})}{\pi(\mathbf{x} \mid \mathbf{y}_{1:T})}\, \mathbf{g}(\mathbf{x}) \right] \pi(\mathbf{x} \mid \mathbf{y}_{1:T})\, \mathrm{d}\mathbf{x}}{\int \left[\frac{p(\mathbf{y}_{1:T} \mid \mathbf{x})\, p(\mathbf{x})}{\pi(\mathbf{x} \mid \mathbf{y}_{1:T})} \right] \pi(\mathbf{x} \mid \mathbf{y}_{1:T})\, \mathrm{d}\mathbf{x}} \\
&\approx \frac{\frac{1}{N} \sum_{i=1}^{N} \frac{p(\mathbf{y}_{1:T} \mid \mathbf{x}^{(i)})\, p(\mathbf{x}^{(i)})}{\pi(\mathbf{x}^{(i)} \mid \mathbf{y}_{1:T})}\, \mathbf{g}(\mathbf{x}^{(i)})}{\frac{1}{N} \sum_{j=1}^{N} \frac{p(\mathbf{y}_{1:T} \mid \mathbf{x}^{(j)})\, p(\mathbf{x}^{(j)})}{\pi(\mathbf{x}^{(j)} \mid \mathbf{y}_{1:T})}} \\
&= \sum_{i=1}^{N} \underbrace{\left[\frac{\frac{p(\mathbf{y}_{1:T} \mid \mathbf{x}^{(i)})\, p(\mathbf{x}^{(i)})}{\pi(\mathbf{x}^{(i)} \mid \mathbf{y}_{1:T})}}{\sum_{j=1}^{N} \frac{p(\mathbf{y}_{1:T} \mid \mathbf{x}^{(j)})\, p(\mathbf{x}^{(j)})}{\pi(\mathbf{x}^{(j)} \mid \mathbf{y}_{1:T})}} \right]}_{w^{(i)}} \mathbf{g}(\mathbf{x}^{(i)}).
\end{aligned}
\tag{11.8}
$$

Thus we get the following algorithm.

Algorithm 11.1 (Importance sampling). *Given a measurement model $p(\mathbf{y}_{1:T} \mid \mathbf{x})$ and a prior $p(\mathbf{x})$, we can form an importance sampling approximation to the posterior as follows.*

1. *Draw N samples from the importance distribution:*

$$\mathbf{x}^{(i)} \sim \pi(\mathbf{x} \mid \mathbf{y}_{1:T}), \qquad i = 1, \ldots, N. \tag{11.9}$$

2. *Compute the unnormalized weights by*

$$w^{*(i)} = \frac{p(\mathbf{y}_{1:T} \mid \mathbf{x}^{(i)})\, p(\mathbf{x}^{(i)})}{\pi(\mathbf{x}^{(i)} \mid \mathbf{y}_{1:T})} \tag{11.10}$$

and the normalized weights by

$$w^{(i)} = \frac{w^{*(i)}}{\sum_{j=1}^{N} w^{*(j)}}. \tag{11.11}$$

3. *The approximation to the posterior expectation of $\mathbf{g}(\mathbf{x})$ is then given as*

$$\mathrm{E}[\mathbf{g}(\mathbf{x}) \mid \mathbf{y}_{1:T}] \approx \sum_{i=1}^{N} w^{(i)}\, \mathbf{g}(\mathbf{x}^{(i)}). \tag{11.12}$$

The approximation to the posterior probability density formed by the above algorithm can then be formally written as

$$p(\mathbf{x} \mid \mathbf{y}_{1:T}) \approx \sum_{i=1}^{N} w^{(i)}\, \delta(\mathbf{x} - \mathbf{x}^{(i)}), \tag{11.13}$$

where $\delta(\cdot)$ is the Dirac delta function.

11.3 Sequential Importance Sampling

Sequential importance sampling (SIS) (see, e.g., Doucet et al., 2001) is a sequential version of importance sampling. The SIS algorithm can be used for generating importance sampling approximations to filtering distributions of generic state space models of the form

$$\begin{aligned} \mathbf{x}_k &\sim p(\mathbf{x}_k \mid \mathbf{x}_{k-1}), \\ \mathbf{y}_k &\sim p(\mathbf{y}_k \mid \mathbf{x}_k), \end{aligned} \tag{11.14}$$

where $\mathbf{x}_k \in \mathbb{R}^n$ is the state at time step k. and $\mathbf{y}_k \in \mathbb{R}^m$ is the measurement. The state and measurements may contain both discrete and continuous components.

The SIS algorithm uses a weighted set of *particles* $\{(w_k^{(i)}, \mathbf{x}_k^{(i)}) : i = 1, \ldots, N\}$, that is, samples from an importance distribution and their weights, to represent the filtering distribution $p(\mathbf{x}_k \mid \mathbf{y}_{1:k})$ such that at every time step k the approximation to the expectation of an arbitrary function $\mathbf{g}(\cdot)$ can be calculated as the weighted sample average

$$\mathrm{E}[\mathbf{g}(\mathbf{x}_k) \mid \mathbf{y}_{1:k}] \approx \sum_{i=1}^{N} w_k^{(i)} \mathbf{g}(\mathbf{x}_k^{(i)}). \tag{11.15}$$

Equivalently, SIS can be interpreted as forming an approximation to the filtering distribution as

$$p(\mathbf{x}_k \mid \mathbf{y}_{1:k}) \approx \sum_{i=1}^{N} w_k^{(i)} \delta(\mathbf{x}_k - \mathbf{x}_k^{(i)}). \tag{11.16}$$

To derive the algorithm, we consider the full posterior distribution of states $\mathbf{x}_{0:k}$ given the measurements $\mathbf{y}_{1:k}$. By using the Markov properties of the model, we get the following recursion for the posterior distribution:

$$\begin{aligned} p(\mathbf{x}_{0:k} \mid \mathbf{y}_{1:k}) &\propto p(\mathbf{y}_k \mid \mathbf{x}_{0:k}, \mathbf{y}_{1:k-1})\, p(\mathbf{x}_{0:k} \mid \mathbf{y}_{1:k-1}) \\ &= p(\mathbf{y}_k \mid \mathbf{x}_k)\, p(\mathbf{x}_k \mid \mathbf{x}_{0:k-1}, \mathbf{y}_{1:k-1})\, p(\mathbf{x}_{0:k-1} \mid \mathbf{y}_{1:k-1}) \\ &= p(\mathbf{y}_k \mid \mathbf{x}_k)\, p(\mathbf{x}_k \mid \mathbf{x}_{k-1})\, p(\mathbf{x}_{0:k-1} \mid \mathbf{y}_{1:k-1}). \end{aligned} \tag{11.17}$$

Using a similar rationale as in the previous section, we can now construct an importance sampling method that draws samples from a given importance distribution $\mathbf{x}_{0:k}^{(i)} \sim \pi(\mathbf{x}_{0:k} \mid \mathbf{y}_{1:k})$ and computes the importance weights by

$$w_k^{(i)} \propto \frac{p(\mathbf{y}_k \mid \mathbf{x}_k^{(i)})\, p(\mathbf{x}_k^{(i)} \mid \mathbf{x}_{k-1}^{(i)})\, p(\mathbf{x}_{0:k-1}^{(i)} \mid \mathbf{y}_{1:k-1})}{\pi(\mathbf{x}_{0:k}^{(i)} \mid \mathbf{y}_{1:k})}. \tag{11.18}$$

If we form the importance distribution for the states $\mathbf{x}_k$ recursively as follows:

$$\pi(\mathbf{x}_{0:k} \mid \mathbf{y}_{1:k}) = \pi(\mathbf{x}_k \mid \mathbf{x}_{0:k-1}, \mathbf{y}_{1:k})\, \pi(\mathbf{x}_{0:k-1} \mid \mathbf{y}_{1:k-1}), \tag{11.19}$$

then the expression for the weights can be written as

$$w_k^{(i)} \propto \frac{p(\mathbf{y}_k \mid \mathbf{x}_k^{(i)})\, p(\mathbf{x}_k^{(i)} \mid \mathbf{x}_{k-1}^{(i)})}{\pi(\mathbf{x}_k^{(i)} \mid \mathbf{x}_{0:k-1}^{(i)}, \mathbf{y}_{1:k})}\, \frac{p(\mathbf{x}_{0:k-1}^{(i)} \mid \mathbf{y}_{1:k-1})}{\pi(\mathbf{x}_{0:k-1}^{(i)} \mid \mathbf{y}_{1:k-1})}. \tag{11.20}$$

Let us now assume that we have already drawn the samples $\mathbf{x}_{0:k-1}^{(i)}$ from the importance distribution $\pi(\mathbf{x}_{0:k-1} \mid \mathbf{y}_{1:k-1})$ and computed the corresponding importance weights $w_{k-1}^{(i)}$. We can now draw samples $\mathbf{x}_{0:k}^{(i)}$ from the importance distribution $\pi(\mathbf{x}_{0:k} \mid \mathbf{y}_{1:k})$ by drawing the new state samples for the step k as $\mathbf{x}_k^{(i)} \sim \pi(\mathbf{x}_k \mid \mathbf{x}_{0:k-1}^{(i)}, \mathbf{y}_{1:k})$. The importance weights from the previous step are proportional to the last term in Equation (11.20):

$$w_{k-1}^{(i)} \propto \frac{p(\mathbf{x}_{0:k-1}^{(i)} \mid \mathbf{y}_{1:k-1})}{\pi(\mathbf{x}_{0:k-1}^{(i)} \mid \mathbf{y}_{1:k-1})}, \tag{11.21}$$

and thus the weights satisfy the recursion

$$w_k^{(i)} \propto \frac{p(\mathbf{y}_k \mid \mathbf{x}_k^{(i)})\, p(\mathbf{x}_k^{(i)} \mid \mathbf{x}_{k-1}^{(i)})}{\pi(\mathbf{x}_k^{(i)} \mid \mathbf{x}_{0:k-1}^{(i)}, \mathbf{y}_{1:k})}\, w_{k-1}^{(i)}. \tag{11.22}$$

The generic sequential importance sampling algorithm can now be described as follows.

Algorithm 11.2 (Sequential importance sampling). *The steps of SIS are the following:*

- *Draw N samples* $\mathbf{x}_0^{(i)}$ *from the prior*

$$\mathbf{x}_0^{(i)} \sim p(\mathbf{x}_0), \qquad i = 1, \ldots, N, \tag{11.23}$$

  and set $w_0^{(i)} = 1/N$ *for all* $i = 1, \ldots, N$.
- *For each* $k = 1, \ldots, T$, *do the following.*
  1. *Draw samples* $\mathbf{x}_k^{(i)}$ *from the importance distributions*

$$\mathbf{x}_k^{(i)} \sim \pi(\mathbf{x}_k \mid \mathbf{x}_{0:k-1}^{(i)}, \mathbf{y}_{1:k}), \qquad i = 1, \ldots, N. \tag{11.24}$$

  2. *Calculate new weights according to*

$$w_k^{(i)} \propto w_{k-1}^{(i)} \frac{p(\mathbf{y}_k \mid \mathbf{x}_k^{(i)})\, p(\mathbf{x}_k^{(i)} \mid \mathbf{x}_{k-1}^{(i)})}{\pi(\mathbf{x}_k^{(i)} \mid \mathbf{x}_{0:k-1}^{(i)}, \mathbf{y}_{1:k})}, \tag{11.25}$$

  and normalize them to sum to unity.

Note that it is convenient to select the importance distribution to be Markovian in the sense that

$$\pi(\mathbf{x}_k \mid \mathbf{x}_{0:k-1}, \mathbf{y}_{1:k}) = \pi(\mathbf{x}_k \mid \mathbf{x}_{k-1}, \mathbf{y}_{1:k}). \tag{11.26}$$

With this form of importance distribution we do not need to store the whole histories $\mathbf{x}_{0:k}^{(i)}$ in the SIS algorithm, only the current states $\mathbf{x}_k^{(i)}$. This form is also convenient in sequential importance resampling (SIR) discussed in the next section, because we do not need to worry about the state histories during the resampling step as in the SIR particle smoother (see Section 15.1). Thus in the following section we assume that the importance distribution has indeed been selected to have the above Markovian form.

11.4 Resampling

One problem in the SIS algorithm described in the previous section is that we can easily encounter the situation where almost all the particles have zero or nearly zero weights. This is called the *degeneracy problem* in particle filtering literature, and it prevented practical applications of particle filters for many years.

The degeneracy problem can be solved by using a *resampling* procedure. It refers to a procedure where we draw N new samples from the discrete distribution defined by the weights and replace the old set of N samples with this new set. This procedure can be written as the following algorithm.

Algorithm 11.3 (Resampling). *The resampling procedure can be described as follows.*

1. *Interpret each weight $w_k^{(i)}$ as the probability of obtaining the sample index i in the set $\{\mathbf{x}_k^{(i)} : i = 1, \ldots, N\}$.*
2. *Draw N samples from that discrete distribution, and replace the old sample set with this new one.*
3. *Set all weights to the constant value $w_k^{(i)} = 1/N$.*

What resampling actually does is the following. Recall that sequential importance sampling (SIR) described in Section 11.3 forms a weighted set of samples $\{(w_k^{(i)}, \mathbf{x}_k^{(i)}) : i = 1, \ldots, N\}$ such that the posterior expectation of a function $\mathbf{g}(\mathbf{x}_k)$ can be approximated as

$$\mathrm{E}[\mathbf{g}(\mathbf{x}_k) \mid \mathbf{y}_{1:k}] \approx \sum_{i=1}^{N} w_k^{(i)} \, \mathbf{g}(\mathbf{x}_k^{(i)}). \tag{11.27}$$

Let us now sample an index A with probabilities $(w_k^{(1)}, \ldots, w_k^{(N)})$. Then the expected value over the index is simply the original one:

$$\mathrm{E}_A[\mathbf{g}(\mathbf{x}_k^{(A)}) \mid \mathbf{y}_{1:k}] = \sum_{i=1}^{N} p(A = i) \, \mathbf{g}(\mathbf{x}_k^{(i)}) = \sum_{i=1}^{N} w_k^{(i)} \, \mathbf{g}(\mathbf{x}_k^{(i)}). \tag{11.28}$$

If we sample N indices $A_{1:N} = (A_1, \ldots, A_N)$ and compute their average, we get an approximation

$$\mathrm{E}[\mathbf{g}(\mathbf{x}_k) \mid \mathbf{y}_{1:k}] \approx \frac{1}{N} \sum_{i=1}^{N} \mathbf{g}(\mathbf{x}_k^{(A_i)}), \tag{11.29}$$

which has the form of a non-weighted (plain) Monte Carlo. The expected value over $A_{1:N}$ is still the same (i.e., resampling is unbiased):

$$\mathrm{E}_{A_{1:N}} \left[\frac{1}{N} \sum_{i=1}^{N} \mathbf{g}(\mathbf{x}_k^{(A_i)}) \right] = \sum_{i=1}^{N} w_k^{(i)} \, \mathbf{g}(\mathbf{x}_k^{(i)}). \tag{11.30}$$

The idea of resampling is to replace the weighted sample set $(w^{(i)}, \mathbf{x}^{(i)})$ with the newly indexed sample set $\mathbf{x}^{(A_i)}$, which then also effectively resets the weights to $w_k^{(i)} = 1/N$. This procedure has the effect of removing particles with very small weights and duplicating particles with large weights. Although the theoretical distribution represented by the weighted set of samples does not change, resampling introduces additional variance to estimates. This variance introduced by the resampling procedure can be reduced by proper choice of the resampling method.

A naive way to do resampling is to draw N times independently from the multinomial distribution defined by the weights $(w_k^{(1)}, \ldots, w_k^{(N)})$. However, this is inefficient and has high variance and is therefore not recommended. Instead, it is advisable to use more sophisticated resampling algorithms, of which Chopin and Papaspiliopoulos (2020) provide a unified formulation. Resampling algorithms (or at least many of them) can be seen as different ways of a drawing sorted array of uniform random variables $u^{(i)}$ for $i = 1, \ldots, N$ such that $u^{(1)} < \cdots < u^{(N)}$, which are then used in the following inverse transform sampler of indices $A_{1:N}$.

Algorithm 11.4 (Inverse transform sampling for resampling). *The resampling indices $A_1, \ldots, A_N$ can be drawn as follows. Given the weights $w^{(i)}$ and ordered uniforms $u^{(i)}$ for $i = 1, \ldots, N$, do the following:*

- *Set $m = 1$ and $s = w^{(1)}$.*
- *For $n = 1, \ldots, N$ do*
  - *While $s < u^{(n)}$ do*
    - *Set $m = m + 1$.*
    - *Set $s = s + w^{(m)}$.*
  - *Set $A_n = m$.*

Using Algorithm 11.4, we can implement different resampling methods (Chopin and Papaspiliopoulos, 2020). The multinomial resampling method becomes the following.

Algorithm 11.5 (Multinomial resampling). *Given the importance weights $w^{(1)}, \ldots, w^{(N)}$, we can draw the resampling indices $A_1, \ldots, A_N$ by using multinomial resampling as follows.*

- *Draw $v^{(j)} \sim \mathrm{U}(0, 1)$ for $j = 1, \ldots, N + 1$.*
- *Compute the cumulative sum $t^{(j)} = -\sum_{i=1}^{j} \log v^{(i)}$.*
- *Put $u^{(i)} = t^{(i)} / t^{(N+1)}$.*
- *Run the sampling routine in Algorithm 11.4 with the weights $w^{(i)}$ and uniforms $u^{(i)}$ for $i = 1, \ldots, N$.*

However, multinomial resampling has high variance and therefore is not recommended for many uses. A lower variance alternative is stratified resampling, which is the following.

Algorithm 11.6 (Stratified resampling). *Given the importance weights $w^{(1)}, \ldots, w^{(N)}$, we can draw the resampling indices $A_1, \ldots, A_N$ by using stratified resampling as follows.*

- *For* $n = 1, \ldots, N$ *do*
  - *Sample* $u^{(n)} \sim \mathrm{U}((n-1)/N, n/N)$.
- *Run the sampling routine in Algorithm 11.4 with the weights* $w^{(i)}$ *and uniforms* $u^{(i)}$ *for* $i = 1, \ldots, N$.

One more choice for resampling is systematic resampling, which can be implemented as follows.

Algorithm 11.7 (Systematic resampling). *Given the importance weights* $w^{(1)}, \ldots, w^{(N)}$, *we can draw the resampling indices* $A_1, \ldots, A_N$ *by using systematic resampling as follows.*

- *Sample* $u \sim \mathrm{U}(0, 1)$.
- *For* $n = 1, \ldots, N$ *do*
  - *Set* $u^{(n)} = (n-1)/N + u/N$.
- *Run the sampling routine in Algorithm 11.4 with the weights* $w^{(i)}$ *and uniforms* $u^{(i)}$ *for* $i = 1, \ldots, N$.

All the aforementioned resampling methods are unbiased and therefore valid choices for resampling. For practical use, Chopin and Papaspiliopoulos (2020) recommend systematic resampling, although its theoretical properties are not as clear as those of the other methods. In this sense stratified resampling is a safe choice as it is ensured to have a lower variance than multinomial resampling (Chopin and Papaspiliopoulos, 2020).

11.5 Particle Filter

Adding a resampling step to the sequential importance sampling algorithm leads to *sequential importance resampling (SIR)*[2] (Gordon et al., 1993; Kitagawa, 1996; Doucet et al., 2001; Ristic et al., 2004; Chopin and Papaspiliopoulos, 2020), which is the algorithm usually referred to as the *particle filter*. In SIR, resampling is not usually performed at every time step, but only when it is actually needed. One way of implementing this is to do resampling on every nth step, where n is some predefined constant. Another way, which is used here, is *adaptive resampling*. In this method, the "effective" number of particles, which is estimated from the variance of the particle weights (Liu and Chen, 1995), is used for monitoring the need

[2] *Sequential importance resampling* is also often referred to as *sampling importance resampling or* sequential importance sampling resampling.

for resampling. The estimate for the effective number of particles (or effective sample size, Chopin and Papaspiliopoulos, 2020) can be computed as

$$n_{\text{eff}} \approx \frac{1}{\sum_{i=1}^{N} \left(w_k^{(i)}\right)^2}, \tag{11.31}$$

where $w_k^{(i)}$ is the normalized weight of particle i at the time step k (Liu and Chen, 1995). Resampling is performed when the effective number of particles is significantly less than the total number of particles, for example, $n_{\text{eff}} < N/10$, where N is the total number of particles.

Algorithm 11.8 (Sequential importance resampling). *The sequential importance resampling (SIR) algorithm, which is also called the particle filter (PF), is the following.*

- *Draw N samples $\mathbf{x}_0^{(i)}$ from the prior*

$$\mathbf{x}_0^{(i)} \sim p(\mathbf{x}_0), \qquad i = 1, \ldots, N, \tag{11.32}$$

and set $w_0^{(i)} = 1/N$, for all $i = 1, \ldots, N$.
- *For each $k = 1, \ldots, T$ do the following:*
  1. *Draw samples $\mathbf{x}_k^{(i)}$ from the importance distributions*

$$\mathbf{x}_k^{(i)} \sim \pi(\mathbf{x}_k \mid \mathbf{x}_{k-1}^{(i)}, \mathbf{y}_{1:k}), \qquad i = 1, \ldots, N. \tag{11.33}$$

  2. *Calculate new weights according to*

$$w_k^{(i)} \propto w_{k-1}^{(i)} \frac{p(\mathbf{y}_k \mid \mathbf{x}_k^{(i)})\, p(\mathbf{x}_k^{(i)} \mid \mathbf{x}_{k-1}^{(i)})}{\pi(\mathbf{x}_k^{(i)} \mid \mathbf{x}_{k-1}^{(i)}, \mathbf{y}_{1:k})}, \tag{11.34}$$

  and normalize them to sum to unity.
  3. *If the effective number of particles* (11.31) *is too low, perform resampling.*

The SIR algorithm forms the following approximation to the filtering distribution:

$$p(\mathbf{x}_k \mid \mathbf{y}_{1:k}) \approx \sum_{i=1}^{N} w_k^{(i)} \delta(\mathbf{x}_k - \mathbf{x}_k^{(i)}), \tag{11.35}$$

and thus the expectation of an arbitrary function $\mathbf{g}(\cdot)$ can be approximated as

$$\mathrm{E}[\mathbf{g}(\mathbf{x}_k) \mid \mathbf{y}_{1:k}] \approx \sum_{i=1}^{N} w_k^{(i)} \mathbf{g}(\mathbf{x}_k^{(i)}). \tag{11.36}$$

Performance of the SIR algorithm depends on the quality of the importance distribution $\pi(\cdot)$. The importance distribution should be in such a functional form that we can easily draw samples from it and that it is possible to evaluate the probability densities of the sample points. *The optimal importance distribution* in terms of variance (see, e.g., Doucet et al., 2001; Ristic et al., 2004) is

$$\pi(\mathbf{x}_k \mid \mathbf{x}_{0:k-1}, \mathbf{y}_{1:k}) = p(\mathbf{x}_k \mid \mathbf{x}_{k-1}, \mathbf{y}_k). \tag{11.37}$$

If the optimal importance distribution cannot be directly used, importance distributions can sometimes be obtained by *local linearization*, where a mixture of extended Kalman filters (EKF), unscented Kalman filters (UKF), or other types of non-linear Kalman filters are used for forming the importance distribution (Doucet et al., 2000; Van der Merwe et al., 2001). It is also possible to use a Metropolis–Hastings step after (or in place of) the resampling step to smooth the resulting distribution (Van der Merwe et al., 2001). A particle filter with UKF importance distribution is also referred to as the *unscented particle filter* (UPF). Similarly, we could call a particle filter with the Gauss–Hermite Kalman filter importance distribution the *Gauss–Hermite particle filter* (GHPF) and one with the cubature Kalman filter importance distribution the *cubature particle filter* (CPF). The use of iterated posterior linearization filters to form the proposal has also been investigated in Hostettler et al. (2020). However, the use of this kind of importance distribution can also lead to the failure of particle filter convergence, and hence they should be used with extreme care. Instead of forming the importance distribution directly by using the Gaussian approximation provided by the EKF, UKF, or other Gaussian filter, it may be advisable to artificially increase the covariance of the distribution or to replace the Gaussian distribution with a Student's t distribution with a suitable number of degrees of freedom (see Cappé et al., 2005).

By tuning the resampling algorithm to specific estimation problems and possibly changing the order of weight computation and sampling, accuracy and computational efficiency of the algorithm can be improved (Fearnhead and Clifford, 2003). An important issue is that sampling is more efficient without replacement, such that duplicate samples are not stored. There is also evidence that in some situations it is more efficient to use a simple deterministic algorithm for preserving the N most likely particles. In the article by Punskaya et al. (2002), it is shown that in digital demodulation, where the sampled space is discrete and the optimization criterion is the minimum error, the deterministic algorithm performs better.

The bootstrap filter (Gordon et al., 1993) is a variation of SIR where the dynamic model $p(\mathbf{x}_k \mid \mathbf{x}_{k-1})$ is used as the importance distribution. This makes the implementation of the algorithm very easy, but due to the inefficiency of the importance distribution it may require a very large number of Monte Carlo samples for accurate estimation results. In the bootstrap filter the resampling is normally done at each time step.

Algorithm 11.9 (Bootstrap filter). *The bootstrap filter algorithm is as follows.*

1. *Draw a new point* $\mathbf{x}_k^{(i)}$ *for each point in the sample set* $\{\mathbf{x}_{k-1}^{(i)} : i = 1, \ldots, N\}$ *from the dynamic model:*

$$\mathbf{x}_k^{(i)} \sim p(\mathbf{x}_k \mid \mathbf{x}_{k-1}^{(i)}), \qquad i = 1, \ldots, N. \tag{11.38}$$

2. *Calculate the weights*

$$w_k^{(i)} \propto p(\mathbf{y}_k \mid \mathbf{x}_k^{(i)}), \qquad i = 1, \ldots, N, \tag{11.39}$$

and normalize them to sum to unity.

3. *Do resampling.*

One problem encountered in particle filtering, even when using a resampling procedure, is called *sample impoverishment* (see, e.g., Ristic et al., 2004). It refers to the effect that when the noise in the dynamic model is very small, many of the particles in the particle set will turn out to have exactly the same value. That is, the resampling step simply multiplies a few (or one) particle(s), and thus we end up having a set of identical copies of certain high weighted particles. This problem can be diminished by using, for example, the resample-move algorithm, regularization, or MCMC steps (Ristic et al., 2004).

Because low noise in the dynamic model causes sample impoverishment, it also implies that pure recursive estimation with particle filters is challenging. This is because in pure recursive estimation the process noise is formally zero and thus a basic SIR-based particle filter is likely to perform very badly. Pure recursive estimation, such as recursive estimation of static parameters, can sometimes be done by applying a Rao–Blackwellized particle filter instead of the basic SIR particle filter (see

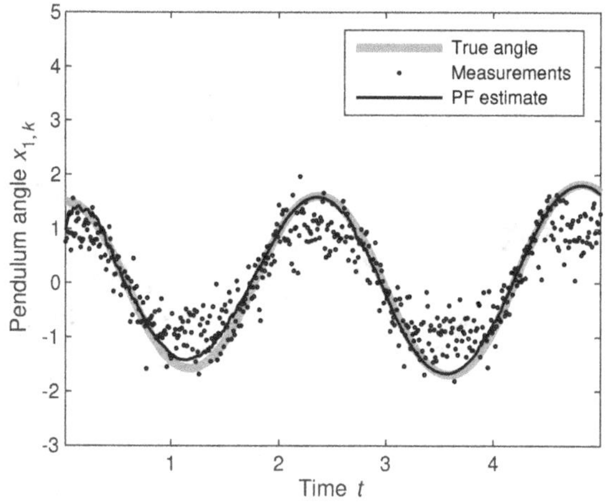

Figure 11.3 Simulated pendulum data and the result of tracking the pendulum described in Example 7.6 with a bootstrap type of particle filter (PF) with 10,000 particles. The resulting RMSE was 0.10, which is approximately the same as the best Gaussian filters.

Section 16.3.6). However, the more common use of Rao–Blackwellization is in the conditionally linear Gaussian state space models, which we will discuss in Section 11.7.

Example 11.10 (Pendulum tracking with a particle filter). *The result of the bootstrap filter with 10,000 particles in the pendulum model (Example 7.6) is shown in Figure 11.3. The RMSE of* 0.10 *is approximately the same as that of the best Gaussian filters. This implies that in this case the filtering distribution is indeed quite well approximated by a Gaussian distribution, and thus using a particle filter is not particularly beneficial.*

In the above example the model is of the type that is suitable for Gaussian approximation-based filters, and thus the particle filter produces much the same result as they do. But often the noises in the system are not Gaussian, or there might be clutter (outlier) measurements that do not fit into the Gaussian non-linear state space modeling framework at all. In these kinds of models, the particle filter still produces good results, whereas Gaussian filters do not work at all. The next example illustrates this kind of situation.

Example 11.11 (Cluttered pendulum tracking with a particle filter). *In this scenario, the pendulum sensor is broken such that at each time instant it produces clutter (a random number in the range* $[-2, 2]$*) with probability*

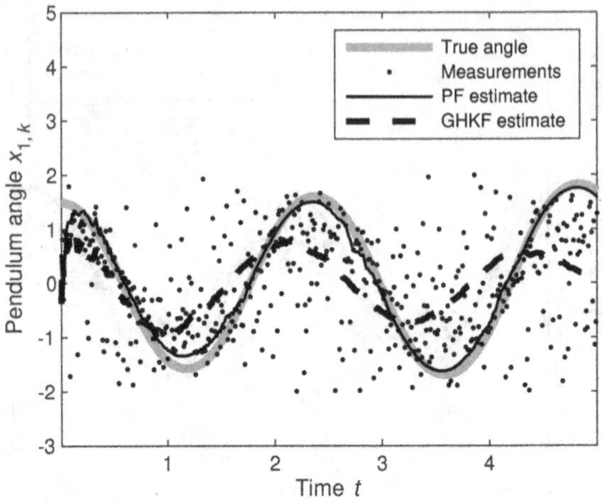

Figure 11.4 Simulated pendulum data in the presence of 50% clutter (outlier) measurements and the result of tracking with a bootstrap type of particle filter (PF) with 10,000 particles (see Example 11.11). The resulting RMSE was 0.20. The corresponding error of GHKF was 0.89.

50%, which is a situation that can be modeled with the model given in Example 5.6. The result of the bootstrap filter using that model is shown in Figure 11.4. The RMSE of 0.20 *is higher than in the clutter-free case, but still moderate. In Gaussian filters a clutter model cannot be easily implemented. In Figure 11.4 we also show the result of a Gauss–Hermite Kalman filter (GHKF), with no clutter model at all, applied to the data. The resulting RMSE is 0.89, which is significantly higher than that of the bootstrap filter.*

11.6 Auxiliary Particle Filter

The idea of an auxiliary particle filter (APF, Pitt and Shephard, 1999; Arulampalam et al., 2002; Johansen and Doucet, 2008; Doucet and Johansen, 2011) is to introduce an auxiliary variable κ to enable us to use $\mathbf{y}_k$ to adjust the probability by which we sample different particles. The construction results in a filter that resembles a particle filter that uses the optimal importance distribution.

If our weighted particle set at time step $k-1$ is $\{\mathbf{x}_{k-1}^{(i)}, w_{k-1}^{(i)}\}$ for $i = 1, \ldots, N$, then the predicted density at time step k can be approximated as

$$p(\mathbf{x}_k \mid \mathbf{y}_{1:k-1}) \approx \sum_i p(\mathbf{x}_k \mid \mathbf{x}_{k-1}^{(i)})\, w_{k-1}^{(i)}, \tag{11.40}$$

and an approximate filtering distribution can be formed as

$$\hat{p}(\mathbf{x}_k \mid \mathbf{y}_{1:k}) \propto p(\mathbf{y}_k \mid \mathbf{x}_k) \sum_i p(\mathbf{x}_k \mid \mathbf{x}_{k-1}^{(i)})\, w_{k-1}^{(i)}. \tag{11.41}$$

A key challenge is that many particles may explain $\mathbf{y}_k$ poorly, and if we sample particles at time step k as described in earlier sections, those particles are likely to obtain a small weight, which means that they do not contribute much to our representation of the posterior.

Let us instead introduce the augmented target distribution

$$\hat{p}(\mathbf{x}_k, \kappa \mid \mathbf{y}_{1:k}) \propto p(\mathbf{y}_k \mid \mathbf{x}_k)\, p(\mathbf{x}_k \mid \mathbf{x}_{k-1}^{(\kappa)})\, w_{k-1}^{(\kappa)}, \tag{11.42}$$

where $\kappa = 1, \ldots, N$ is the auxiliary variable. By sampling from this joint distribution and then discarding κ, we can generate a sample from the distribution in Equation (11.41). To do that, we can use sequential importance sampling, which amounts to sampling from an importance distribution $\mathbf{x}_k^{(i)}, \kappa^{(i)} \sim \pi_k(\mathbf{x}_k, \kappa \mid \mathbf{y}_{1:k})$ and then computing weights as

$$w_k^{(i)} \propto \frac{p(\mathbf{y}_k \mid \mathbf{x}_k^{(i)})\, p(\mathbf{x}_k^{(i)} \mid \mathbf{x}_{k-1}^{(\kappa^{(i)})})\, w_{k-1}^{(\kappa^{(i)})}}{\pi_k(\mathbf{x}_k^{(i)}, \kappa^{(i)} \mid \mathbf{y}_{1:k})}. \tag{11.43}$$

A convenient choice of importance distribution (Pitt and Shephard, 1999) is

$$\pi_k(\mathbf{x}_k, \kappa \mid \mathbf{y}_{1:k}) \propto p(\mathbf{y}_k \mid \boldsymbol{\mu}_k^{(\kappa)})\, p(\mathbf{x}_k \mid \mathbf{x}_{k-1}^{(\kappa)})\, w_{k-1}^{(\kappa)}, \tag{11.44}$$

where $\boldsymbol{\mu}_k^{(\kappa)}$ is, for example, the mean, the mode, a draw, or some other likely value associated with $p(\mathbf{x}_k \mid \mathbf{x}_{k-1}^{(\kappa)})$. By construction, the importance distribution has the marginal and conditional distributions as follows:

$$\begin{aligned} \pi_k(\kappa \mid \mathbf{y}_{1:k}) &\propto w_{k-1}^{(\kappa)}\, p(\mathbf{y}_k \mid \boldsymbol{\mu}_k^{(\kappa)}), \\ \pi_k(\mathbf{x}_k \mid \kappa, \mathbf{y}_{1:k}) &= p(\mathbf{x}_k \mid \mathbf{x}_{k-1}^{(\kappa)}), \end{aligned} \tag{11.45}$$

which implies that we can sample from the importance distribution by first generating an auxiliary variable with probability $\lambda^{(\kappa)} \propto \pi_k(\kappa \mid \mathbf{y}_{1:k})$ and then sampling from $p(\mathbf{x}_k \mid \mathbf{x}_{k-1}^{(\kappa)})$. Note that we have managed to incorporate knowledge about $\mathbf{y}_k$ to adjust the probability by which we sample different particle indexes. The probabilities $\lambda^{(\kappa)}$ are often called the first-stage weights in the algorithm (Pitt and Shephard, 1999).

Using the importance distribution in Equation (11.44), the weights in Equation (11.43) reduce to

$$w_k^{(i)} \propto \frac{p(\mathbf{y}_k \mid \mathbf{x}_k^{(i)})}{p(\mathbf{y}_k \mid \boldsymbol{\mu}_k^{(\kappa^{(i)})})}. \tag{11.46}$$

These weights could now be used as in a standard SIR algorithm to do (adaptive) resampling. However, it has been shown (see, e.g., Johansen and Doucet, 2008) that this additional resampling step is not required or beneficial.

The resulting algorithm is the following.

Algorithm 11.12 (Auxiliary particle filter). *The steps of* auxiliary particle filter (APF) *are:*

- *Draw samples* $\mathbf{x}_0^{(i)}$ *from the prior and set* $w_0^{(i)} = 1/N$ *for* $i = 1, \ldots, N$.
- *For each* $k = 1, \ldots, T$ *do the following:*
  1. *Generate* $\boldsymbol{\mu}_k^{(i)}$ *from* $p(\mathbf{x}_k \mid \mathbf{x}_{k-1}^{(i)})$.
  2. *Compute the first-stage weights*
  $$\lambda^{(i)} \propto w_{k-1}^{(i)}\, p(\mathbf{y}_k \mid \boldsymbol{\mu}_k^{(i)}), \tag{11.47}$$
  and normalize them to sum to unity.
  3. *Resample* $\{(\cdot, \lambda^{(i)})\}$ *to get a new set of indices* $\kappa^{(i)}$.
  4. *Draw new samples from the dynamic model*
  $$\mathbf{x}_k^{(i)} \sim p(\mathbf{x}_k^{(i)} \mid \mathbf{x}_{k-1}^{(\kappa^{(i)})}). \tag{11.48}$$
  5. *Compute new weights*
  $$w_k^{(i)} \propto \frac{p(\mathbf{y}_k \mid \mathbf{x}_k^{(i)})}{p(\mathbf{y}_k \mid \boldsymbol{\mu}_k^{(\kappa^{(i)})})}, \tag{11.49}$$
  and normalize them to sum to unity.

In step 3 we have used resampling in a slightly different way than in the previous particle filtering algorithms. Recall that resampling is usually implemented by drawing new indices from the discrete distribution defined by the probabilities $\lambda^{(i)}$ (or $w_k^{(i)}$ in conventional particle filters), and then these indices are used to select the resampled particles. In step 3 of Algorithm 11.12, we instead store the indices themselves to use them in steps

4 and 5. For a discussion on the relationship and differences between auxiliary particle filters and conventional particle filters, please see Johansen and Doucet (2008).

11.7 Rao–Blackwellized Particle Filter

One way of improving the efficiency of SIR is to use Rao–Blackwellization. The idea of the *Rao–Blackwellized particle filter* (RBPF) (Akashi and Kumamoto, 1977; Doucet et al., 2001; Ristic et al., 2004), which is also called the *mixture Kalman filter* (MKF) (Chen and Liu, 2000), is that sometimes it is possible to evaluate some of the filtering equations analytically and the others with Monte Carlo sampling instead of computing everything with pure sampling. According to the *Rao–Blackwell theorem* (see, e.g., Berger, 1985), this leads to estimators with less variance than could be obtained with pure Monte Carlo sampling. An intuitive way of understanding this is that the marginalization replaces the finite Monte Carlo particle set representation with an infinite closed form particle set, which is always more accurate than any finite set.

Most commonly Rao–Blackwellized particle filtering refers to marginalized filtering of conditionally linear Gaussian models of the form

$$\begin{aligned} p(\mathbf{x}_k \mid \mathbf{x}_{k-1}, \mathbf{u}_{k-1}) &= \mathrm{N}(\mathbf{x}_k \mid \mathbf{A}_{k-1}(\mathbf{u}_{k-1})\,\mathbf{x}_{k-1}, \mathbf{Q}_{k-1}(\mathbf{u}_{k-1})), \\ p(\mathbf{y}_k \mid \mathbf{x}_k, \mathbf{u}_k) &= \mathrm{N}(\mathbf{y}_k \mid \mathbf{H}_k(\mathbf{u}_k)\,\mathbf{x}_k, \mathbf{R}_k(\mathbf{u}_k)), \\ p(\mathbf{u}_k \mid \mathbf{u}_{k-1}) &= (\text{any given form}), \end{aligned} \tag{11.50}$$

where $\mathbf{x}_k$ is the state, $\mathbf{y}_k$ is the measurement, and $\mathbf{u}_k$ is an arbitrary latent variable. If in addition the prior of $\mathbf{x}_k$ is Gaussian, then due to the conditionally linear Gaussian structure of the model, the state variables $\mathbf{x}_k$ can be integrated out analytically, and only the latent variables $\mathbf{u}_k$ need to be sampled. The Rao–Blackwellized particle filter uses SIR for the latent variables and computes the conditionally Gaussian part in closed form.

To derive the filtering algorithm, first note that the full posterior distribution at step k can be factored as

$$p(\mathbf{u}_{0:k}, \mathbf{x}_{0:k} \mid \mathbf{y}_{1:k}) = p(\mathbf{x}_{0:k} \mid \mathbf{u}_{0:k}, \mathbf{y}_{1:k})\, p(\mathbf{u}_{0:k} \mid \mathbf{y}_{1:k}), \tag{11.51}$$

where the first term is Gaussian and computable with the Kalman filter and RTS smoother. For the second term we get the following recursion analogously to Equation (11.17):

$$\begin{aligned}
&p(\mathbf{u}_{0:k} \mid \mathbf{y}_{1:k}) \\
&\propto p(\mathbf{y}_k \mid \mathbf{u}_{0:k}, \mathbf{y}_{1:k-1}) \, p(\mathbf{u}_{0:k} \mid \mathbf{y}_{1:k-1}) \\
&= p(\mathbf{y}_k \mid \mathbf{u}_{0:k}, \mathbf{y}_{1:k-1}) \, p(\mathbf{u}_k \mid \mathbf{u}_{0:k-1}, \mathbf{y}_{1:k-1}) \, p(\mathbf{u}_{0:k-1} \mid \mathbf{y}_{1:k-1}) \\
&= p(\mathbf{y}_k \mid \mathbf{u}_{0:k}, \mathbf{y}_{1:k-1}) \, p(\mathbf{u}_k \mid \mathbf{u}_{k-1}) \, p(\mathbf{u}_{0:k-1} \mid \mathbf{y}_{1:k-1}),
\end{aligned} \tag{11.52}$$

where we have used the Markovianity of $\mathbf{u}_k$. Now the measurements are not conditionally independent given $\mathbf{u}_k$, and thus the first term differs from the corresponding term in Equation (11.17). The first term can be computed by running the Kalman filter with fixed $\mathbf{u}_{0:k}$ over the measurement sequence. The second term is just the dynamic model, and the third term is the posterior from the previous step.

If we form the importance distribution recursively as follows:

$$\pi(\mathbf{u}_{0:k} \mid \mathbf{y}_{1:k}) = \pi(\mathbf{u}_k \mid \mathbf{u}_{0:k-1}, \mathbf{y}_{1:k}) \, \pi(\mathbf{u}_{0:k-1} \mid \mathbf{y}_{1:k-1}), \tag{11.53}$$

then by following the same derivation as in Section 11.3, we get the following recursion for the weights:

$$w_k^{(i)} \propto \frac{p(\mathbf{y}_k \mid \mathbf{u}_{0:k}^{(i)}, \mathbf{y}_{1:k-1}) \, p(\mathbf{u}_k^{(i)} \mid \mathbf{u}_{k-1}^{(i)})}{\pi(\mathbf{u}_k^{(i)} \mid \mathbf{u}_{0:k-1}^{(i)}, \mathbf{y}_{1:k})} \, w_{k-1}^{(i)}, \tag{11.54}$$

which corresponds to Equation (11.22). Thus via the above recursion we can form an importance sampling-based approximation to the marginal distribution $p(\mathbf{u}_{0:k} \mid \mathbf{y}_{1:k})$. But because, given $\mathbf{u}_{0:k}$, the distribution $p(\mathbf{x}_{0:k} \mid \mathbf{u}_{0:k}, \mathbf{y}_{1:k})$ is Gaussian, we can form the full posterior distribution by using Equation (11.51). Computing the distribution jointly for the full history $\mathbf{x}_{0:k}$ would require running both the Kalman filter and the RTS smoother over the sequences $\mathbf{u}_{0:k}$ and $\mathbf{y}_{1:k}$, but if we are only interested in the posterior of the last time step $\mathbf{x}_k$, we only need to run the Kalman filter. The resulting algorithm is the following.

Algorithm 11.13 (Rao–Blackwellized particle filter). *Given a sequence of importance distributions* $\pi(\mathbf{u}_k \mid \mathbf{u}_{0:k-1}^{(i)}, \mathbf{y}_{1:k})$ *and a set of weighted samples* $\{w_{k-1}^{(i)}, \mathbf{u}_{k-1}^{(i)}, \mathbf{m}_{k-1}^{(i)}, \mathbf{P}_{k-1}^{(i)} : i = 1, \ldots, N\}$, *the Rao–Blackwellized particle filter (RBPF) processes the measurement* $\mathbf{y}_k$ *as follows (Doucet et al., 2001).*

1. *Perform Kalman filter predictions for each of the Kalman filter means and covariances in the particles $i = 1, \ldots, N$ conditional on the previously drawn latent variable values $\mathbf{u}_{k-1}^{(i)}$:*

$$\begin{aligned} \mathbf{m}_k^{-(i)} &= \mathbf{A}_{k-1}(\mathbf{u}_{k-1}^{(i)})\,\mathbf{m}_{k-1}^{(i)}, \\ \mathbf{P}_k^{-(i)} &= \mathbf{A}_{k-1}(\mathbf{u}_{k-1}^{(i)})\,\mathbf{P}_{k-1}^{(i)}\,\mathbf{A}_{k-1}^{\mathsf{T}}(\mathbf{u}_{k-1}^{(i)}) + \mathbf{Q}_{k-1}(\mathbf{u}_{k-1}^{(i)}). \end{aligned} \tag{11.55}$$

2. *Draw new latent variables $\mathbf{u}_k^{(i)}$ for each particle in $i = 1, \ldots, N$ from the corresponding importance distributions:*

$$\mathbf{u}_k^{(i)} \sim \pi(\mathbf{u}_k \mid \mathbf{u}_{0:k-1}^{(i)}, \mathbf{y}_{1:k}). \tag{11.56}$$

3. *Calculate new weights as follows:*

$$w_k^{(i)} \propto w_{k-1}^{(i)} \frac{p(\mathbf{y}_k \mid \mathbf{u}_{0:k}^{(i)}, \mathbf{y}_{1:k-1})\, p(\mathbf{u}_k^{(i)} \mid \mathbf{u}_{k-1}^{(i)})}{\pi(\mathbf{u}_k^{(i)} \mid \mathbf{u}_{0:k-1}^{(i)}, \mathbf{y}_{1:k})}, \tag{11.57}$$

where the likelihood term is the marginal measurement likelihood of the Kalman filter

$$\begin{aligned} &p(\mathbf{y}_k \mid \mathbf{u}_{0:k}^{(i)}, \mathbf{y}_{1:k-1}) \\ &= \mathrm{N}\left(\mathbf{y}_k \,\middle|\, \mathbf{H}_k(\mathbf{u}_k^{(i)})\,\mathbf{m}_k^{-(i)}, \mathbf{H}_k(\mathbf{u}_k^{(i)})\,\mathbf{P}_k^{-(i)}\,\mathbf{H}_k^{\mathsf{T}}(\mathbf{u}_k^{(i)}) + \mathbf{R}_k(\mathbf{u}_k^{(i)})\right) \end{aligned} \tag{11.58}$$

such that the model parameters in the Kalman filter are conditioned on the drawn latent variable value $\mathbf{u}_k^{(i)}$. Then normalize the weights to sum to unity.

4. *Perform Kalman filter updates for each of the particles conditional on the drawn latent variables $\mathbf{u}_k^{(i)}$:*

$$\begin{aligned} \mathbf{v}_k^{(i)} &= \mathbf{y}_k - \mathbf{H}_k(\mathbf{u}_k^{(i)})\,\mathbf{m}_k^{-(i)}, \\ \mathbf{S}_k^{(i)} &= \mathbf{H}_k(\mathbf{u}_k^{(i)})\,\mathbf{P}_k^{-(i)}\,\mathbf{H}_k^{\mathsf{T}}(\mathbf{u}_k^{(i)}) + \mathbf{R}_k(\mathbf{u}_k^{(i)}), \\ \mathbf{K}_k^{(i)} &= \mathbf{P}_k^{-(i)}\,\mathbf{H}_k^{\mathsf{T}}(\mathbf{u}_k^{(i)}) \left[\mathbf{S}_k^{(i)}\right]^{-1}, \\ \mathbf{m}_k^{(i)} &= \mathbf{m}_k^{-(i)} + \mathbf{K}_k^{(i)}\,\mathbf{v}_k^{(i)}, \\ \mathbf{P}_k^{(i)} &= \mathbf{P}_k^{-(i)} - \mathbf{K}_k^{(i)}\,\mathbf{S}_k^{(i)}\,[\mathbf{K}_k^{(i)}]^{\mathsf{T}}. \end{aligned} \tag{11.59}$$

5. *If the effective number of particles* (11.31) *is too low, perform* resampling.

The Rao–Blackwellized particle filter produces for each time step k a set of weighted samples $\{w_k^{(i)}, \mathbf{u}_k^{(i)}, \mathbf{m}_k^{(i)}, \mathbf{P}_k^{(i)} : i = 1, \ldots, N\}$ such that the expectation of a function $\mathbf{g}(\cdot)$ can be approximated as

$$\mathrm{E}[\mathbf{g}(\mathbf{x}_k, \mathbf{u}_k) \mid \mathbf{y}_{1:k}] \approx \sum_{i=1}^{N} w_k^{(i)} \int \mathbf{g}(\mathbf{x}_k, \mathbf{u}_k^{(i)}) \, \mathrm{N}(\mathbf{x}_k \mid \mathbf{m}_k^{(i)}, \mathbf{P}_k^{(i)}) \, \mathrm{d}\mathbf{x}_k. \tag{11.60}$$

Equivalently, the RBPF can be interpreted to form an approximation to the filtering distribution as

$$p(\mathbf{x}_k, \mathbf{u}_k \mid \mathbf{y}_{1:k}) \approx \sum_{i=1}^{N} w_k^{(i)} \, \delta(\mathbf{u}_k - \mathbf{u}_k^{(i)}) \, \mathrm{N}(\mathbf{x}_k \mid \mathbf{m}_k^{(i)}, \mathbf{P}_k^{(i)}). \tag{11.61}$$

The optimal importance distribution, that is, the importance distribution that minimizes the variance of the importance weights in the RBPF case, is

$$\begin{aligned} \pi(\mathbf{u}_k \mid \mathbf{y}_{1:k}, \mathbf{u}_{0:k-1}^{(i)}) &= p(\mathbf{u}_k \mid \mathbf{y}_{1:k}, \mathbf{u}_{0:k-1}^{(i)}) \\ &\propto p(\mathbf{y}_k \mid \mathbf{u}_k, \mathbf{u}_{0:k-1}^{(i)}) \, p(\mathbf{u}_k \mid \mathbf{u}_{0:k-1}^{(i)}, \mathbf{y}_{1:k-1}). \end{aligned} \tag{11.62}$$

In general, normalizing this distribution or drawing samples from this distribution directly is not possible. But, if the latent variables $\mathbf{u}_k$ are discrete, we can normalize this distribution and use this optimal importance distribution directly.

The class of models where Rao–Blackwellization of some linear state components can be carried out can be extended beyond the conditionally linear Gaussian models presented here. We can, for example, include additional latent-variable-dependent non-linear terms into the dynamic and measurement models (Schön et al., 2005). In some cases, when the filtering model is not strictly Gaussian due to slight non-linearities in either the dynamic or measurement models, it is possible to replace the exact Kalman filter update and prediction steps in RBPF with the extended Kalman filter (EKF), the unscented Kalman filter (UKF) prediction and update steps, or with any other Gaussian approximation-based filters (Särkkä et al., 2007b).

11.8 Exercises

11.1 Consider the following linear Gaussian state space model:

$$\begin{aligned} \mathbf{x}_k &= \begin{pmatrix} 1 & 1 \\ 0 & 1 \end{pmatrix} \mathbf{x}_{k-1} + \mathbf{q}_{k-1}, \\ y_k &= \begin{pmatrix} 1 & 0 \end{pmatrix} \mathbf{x}_k + r_k, \end{aligned} \tag{11.63}$$

where $\mathbf{x}_k = (x_k, \dot{x}_k)$ is the state, y_k is the measurement, and $\mathbf{q}_k \sim \mathrm{N}(\mathbf{0}, \mathrm{diag}(1/10^2, 1^2))$ and $r_k \sim \mathrm{N}(0, 10^2)$ are white Gaussian noise processes.

(a) Write down the Kalman filter equations for this model.

(b) Derive an expression for the optimal importance distribution for the model:

$$\pi(\mathbf{x}_k) = p(\mathbf{x}_k \mid \mathbf{x}_{k-1}, \mathbf{y}_{1:k}). \tag{11.64}$$

(c) Write pseudo-code for the corresponding particle filter algorithm (sequential importance resampling algorithm). Also write down the equations for the weight update.

(d) Compare the number of CPU steps (multiplications and additions) needed by the particle filter and Kalman filter. Which implementation would you choose for a real implementation?

11.2 Consider the following partially observed random-walk model:

$$\begin{aligned} p(x_k \mid x_{k-1}) &= \mathrm{N}(x_k \mid x_{k-1}, 1), \\ p(y_k \mid x_k) &= \mathrm{LogNormal}(y_k \mid \log|x_k| - 1/2, 1), \end{aligned} \tag{11.65}$$

where $\mathrm{LogNormal}(a, b)$ has the density

$$\mathrm{LogNormal}(x \mid a, b) \propto \exp\left(-\frac{(\log(x) - a)^2}{2b^2}\right). \tag{11.66}$$

Hint: $\mathrm{LogNormal}(\ln|x_k| - 1/2, 1)$ is defined such that $\mathbb{E}[y_k \mid x_k] = |x_k|$. Generate synthetic data from this model and apply

(a) one of the Gaussian approximation-based filters

(b) a bootstrap particle filter using 1,000 particles.

Compare the posterior distributions given by the two methods above: for the Gaussian filter, plot the mean and uncertainty (0.95 standard deviation), and for the bootstrap filter, show a scatter plot of the filtering distribution at each time step.

11.3 Implement the bootstrap filter for the model in Exercise 7.1, and test its performance against the non-linear Kalman filters.

11.4 Implement the auxiliary particle filter for the model in Exercise 7.1, and test its performance against the bootstrap filter.

11.5 Implement a sequential importance resampling filter with an EKF-, UKF-, GHKF-, or CKF-based importance distribution for the model in Exercise 7.1. Note that you might want to use a small non-zero covariance as the prior of the previous step instead of plain zero to get the filters to work better.

11.6 Implement the bootstrap filter for the bearings-only target tracking model in Exercise 7.2. Plot the results, and compare the RMSE values to those of the non-linear Kalman filters.

11.7 Implement the bootstrap filter for the Cartesian coordinated turn model in Equations (4.61) and (4.90) with the linear position measurement model. Also compare its performance with the EKF and UKF.

11.8 Implement the bootstrap filter for the bicycle model considered in Exercise 9.11, and compare its performance with a non-linear Kalman filter.

11.9 Implement a Rao–Blackwellized particle filter for the following clutter model (outlier model) :

$$x_k = x_{k-1} + q_{k-1},$$
$$y_k = \begin{cases} x_k + r_k, & \text{if } u_k = 0, \\ r_k^{\mathrm{c}}, & \text{if } u_k = 1, \end{cases} \tag{11.67}$$

where $q_{k-1} \sim \mathrm{N}(0, 1)$, $r_k \sim \mathrm{N}(0, 1)$, and $r_k^{\mathrm{c}} \sim \mathrm{N}(0, 10^2)$. The indicator variables u_k are modeled as independent random variables that take the value $u_k = 0$ with prior probability 0.9 and the value $u_k = 1$ with prior probability 0.1. Test the performance of the filter with simulated data, and compare the performance to a Kalman filter, where the clutter r_k^{c} is ignored. What is the optimal importance distribution for this model?

12

Bayesian Smoothing Equations and Exact Solutions

So far in this book we have only considered filtering algorithms that use the measurements obtained before and at the current step for computing the best possible estimate of the current state (and possibly future states). However, sometimes it is also of interest to estimate states for each time step conditional on all the measurements that we have obtained. This problem can be solved with Bayesian smoothing. In this chapter, we present the Bayesian theory of smoothing. After that we derive the Rauch–Tung–Striebel (RTS) smoother, which is the closed form smoothing solution to linear Gaussian models, as well as its affine extension. We also specialize the equations to discrete state spaces and present the Viterbi algorithm for computing maximum a posteriori (MAP) state trajectories.

12.1 Bayesian Smoothing Equations

The purpose of *Bayesian smoothing*[1] is to compute the marginal posterior distribution of the state $\mathbf{x}_k$ at the time step k after receiving the measurements up to a time step T, where $T > k$:

$$p(\mathbf{x}_k \mid \mathbf{y}_{1:T}). \tag{12.1}$$

The difference between filters and smoothers is that *the Bayesian filter* computes its estimates using only the measurements obtained before and at the time step k, but *the Bayesian smoother* also uses the future measurements for computing its estimates. After obtaining the filtering posterior state distributions, the following theorem gives the equations for computing the marginal posterior distributions for each time step conditionally on all the measurements up to the time step T.

[1] This definition actually applies to the fixed-interval type of smoothing.

Theorem 12.1 (Bayesian smoothing equations). *The backward recursive equations (the* Bayesian smoother*) for computing the* smoothed distributions $p(\mathbf{x}_k \mid \mathbf{y}_{1:T})$ *for any* $k < T$ *are given by the following* Bayesian (fixed-interval) smoothing equations *(Kitagawa, 1987):*

$$p(\mathbf{x}_{k+1} \mid \mathbf{y}_{1:k}) = \int p(\mathbf{x}_{k+1} \mid \mathbf{x}_k)\, p(\mathbf{x}_k \mid \mathbf{y}_{1:k})\, \mathrm{d}\mathbf{x}_k,$$
$$p(\mathbf{x}_k \mid \mathbf{y}_{1:T}) = p(\mathbf{x}_k \mid \mathbf{y}_{1:k}) \int \left[\frac{p(\mathbf{x}_{k+1} \mid \mathbf{x}_k)\, p(\mathbf{x}_{k+1} \mid \mathbf{y}_{1:T})}{p(\mathbf{x}_{k+1} \mid \mathbf{y}_{1:k})} \right] \mathrm{d}\mathbf{x}_{k+1}, \tag{12.2}$$

where $p(\mathbf{x}_k \mid \mathbf{y}_{1:k})$ *is the filtering distribution of the time step* k*. Note that the term* $p(\mathbf{x}_{k+1} \mid \mathbf{y}_{1:k})$ *is simply the predicted distribution of time step* $k+1$*. The integrations are replaced by summations if some of the state components are discrete. For more details on this, see Section 12.4.*

Proof Due to the Markov properties, the state $\mathbf{x}_k$ is independent of $\mathbf{y}_{k+1:T}$ given $\mathbf{x}_{k+1}$, which gives $p(\mathbf{x}_k \mid \mathbf{x}_{k+1}, \mathbf{y}_{1:T}) = p(\mathbf{x}_k \mid \mathbf{x}_{k+1}, \mathbf{y}_{1:k})$. By using *Bayes' rule*, the distribution of $\mathbf{x}_k$ given $\mathbf{x}_{k+1}$ and $\mathbf{y}_{1:T}$ can be expressed as

$$\begin{aligned} p(\mathbf{x}_k \mid \mathbf{x}_{k+1}, \mathbf{y}_{1:T}) &= p(\mathbf{x}_k \mid \mathbf{x}_{k+1}, \mathbf{y}_{1:k}) \\ &= \frac{p(\mathbf{x}_k, \mathbf{x}_{k+1} \mid \mathbf{y}_{1:k})}{p(\mathbf{x}_{k+1} \mid \mathbf{y}_{1:k})} \\ &= \frac{p(\mathbf{x}_{k+1} \mid \mathbf{x}_k, \mathbf{y}_{1:k})\, p(\mathbf{x}_k \mid \mathbf{y}_{1:k})}{p(\mathbf{x}_{k+1} \mid \mathbf{y}_{1:k})} \\ &= \frac{p(\mathbf{x}_{k+1} \mid \mathbf{x}_k)\, p(\mathbf{x}_k \mid \mathbf{y}_{1:k})}{p(\mathbf{x}_{k+1} \mid \mathbf{y}_{1:k})}. \end{aligned} \tag{12.3}$$

The joint distribution of $\mathbf{x}_k$ and $\mathbf{x}_{k+1}$ given $\mathbf{y}_{1:T}$ can be now computed as

$$\begin{aligned} p(\mathbf{x}_k, \mathbf{x}_{k+1} \mid \mathbf{y}_{1:T}) &= p(\mathbf{x}_k \mid \mathbf{x}_{k+1}, \mathbf{y}_{1:T})\, p(\mathbf{x}_{k+1} \mid \mathbf{y}_{1:T}) \\ &= p(\mathbf{x}_k \mid \mathbf{x}_{k+1}, \mathbf{y}_{1:k})\, p(\mathbf{x}_{k+1} \mid \mathbf{y}_{1:T}) \\ &= \frac{p(\mathbf{x}_{k+1} \mid \mathbf{x}_k)\, p(\mathbf{x}_k \mid \mathbf{y}_{1:k})\, p(\mathbf{x}_{k+1} \mid \mathbf{y}_{1:T})}{p(\mathbf{x}_{k+1} \mid \mathbf{y}_{1:k})}, \end{aligned} \tag{12.4}$$

where $p(\mathbf{x}_{k+1} \mid \mathbf{y}_{1:T})$ is the smoothed distribution of the time step $k+1$. The marginal distribution of $\mathbf{x}_k$ given $\mathbf{y}_{1:T}$ is given by integration (or summation) over $\mathbf{x}_{k+1}$ in Equation (12.4), which gives the desired result. □

A more precise name for the above smoother could be *Bayesian forward-backward smoother*, because it assumes that we first run a filter to compute $p(\mathbf{x}_k \mid \mathbf{y}_{1:k})$ and then use Equations (12.2) to compute the smoothing density $p(\mathbf{x}_k \mid \mathbf{y}_{1:T})$ from $p(\mathbf{x}_{k+1} \mid \mathbf{y}_{1:T})$. That is, we first run a filter forward in time, recursively computing $p(\mathbf{x}_k \mid \mathbf{y}_{1:k})$ from $p(\mathbf{x}_{k-1} \mid \mathbf{y}_{1:k-1})$, and we then run a smoother backward in time, recursively computing $p(\mathbf{x}_k \mid \mathbf{y}_{1:T})$ from $p(\mathbf{x}_{k+1} \mid \mathbf{y}_{1:T})$.

An alternative approach to the above forward-backward approach is the two-filter smoother (Fraser and Potter, 1969; Kitagawa, 1994). That approach is based on the factorization $p(\mathbf{x}_k \mid \mathbf{y}_{1:T}) \propto p(\mathbf{x}_k \mid \mathbf{y}_{1:k-1})\, p(\mathbf{y}_{k:T} \mid \mathbf{x}_k)$, where the first factor is given by the prediction step in the Bayesian filter, and the second one can be obtained with a backward recursion. The backward recursion for the latter term resembles a filter that runs backward, hence the term two-filter smoother. Two-filter smoothing is particularly useful in linear Gaussian and discrete-state models, and the so-called belief propagation algorithm for state space models (see, e.g., Koller and Friedman, 2009) can be seen as an instance of two-filter smoothing. However, because two-filter smoothing is less useful in constructing approximate smoothers for non-linear models, in this book we focus on the forward-backward type of smoothing formulations.

12.2 Rauch–Tung–Striebel Smoother

The *Rauch–Tung–Striebel smoother* (RTSS, Rauch et al., 1965), which is also called the *Kalman smoother*, can be used for computing the closed form smoothing solution

$$p(\mathbf{x}_k \mid \mathbf{y}_{1:T}) = \mathrm{N}(\mathbf{x}_k \mid \mathbf{m}_k^{\mathrm{s}}, \mathbf{P}_k^{\mathrm{s}}) \tag{12.5}$$

to the linear filtering model (6.17). The difference to the solution computed by the *Kalman filter* is that the smoothed solution is conditional on the whole measurement data $\mathbf{y}_{1:T}$, while the filtering solution is conditional only on the measurements obtained before and at the time step k, that is, on the measurements $\mathbf{y}_{1:k}$.

Theorem 12.2 (RTS smoother). *The backward recursion equations for the (fixed interval) Rauch–Tung–Striebel smoother are given as*

$$\begin{aligned}
\mathbf{m}_{k+1}^- &= \mathbf{A}_k\, \mathbf{m}_k, \\
\mathbf{P}_{k+1}^- &= \mathbf{A}_k\, \mathbf{P}_k\, \mathbf{A}_k^\mathsf{T} + \mathbf{Q}_k, \\
\mathbf{G}_k &= \mathbf{P}_k\, \mathbf{A}_k^\mathsf{T}\, [\mathbf{P}_{k+1}^-]^{-1}, \\
\mathbf{m}_k^{\mathrm{s}} &= \mathbf{m}_k + \mathbf{G}_k\, [\mathbf{m}_{k+1}^{\mathrm{s}} - \mathbf{m}_{k+1}^-], \\
\mathbf{P}_k^{\mathrm{s}} &= \mathbf{P}_k + \mathbf{G}_k\, [\mathbf{P}_{k+1}^{\mathrm{s}} - \mathbf{P}_{k+1}^-]\, \mathbf{G}_k^\mathsf{T},
\end{aligned} \tag{12.6}$$

where $\mathbf{m}_k$ *and* $\mathbf{P}_k$ *are the mean and covariance computed by the Kalman filter. Note that the first two of the equations are simply the Kalman filter prediction equations.*

The above procedure is a recursion that can be used for computing the smoothing distribution of time step k *from the smoothing distribution of time step* $k+1$*. Because the smoothing distribution and filtering distribution of the last time step* T *are the same, we have* $\mathbf{m}_T^{\mathrm{s}} = \mathbf{m}_T$*,* $\mathbf{P}_T^{\mathrm{s}} = \mathbf{P}_T$*, and thus the recursion can be used for computing the smoothing distributions of all time steps by starting from the last step* $k = T$ *and proceeding backward to the initial step* $k = 0$*.*

Proof Similarly to the Kalman filter case, by Lemma A.2, the joint distribution of $\mathbf{x}_k$ and $\mathbf{x}_{k+1}$ given $\mathbf{y}_{1:k}$ is

$$\begin{aligned} p(\mathbf{x}_k, \mathbf{x}_{k+1} \mid \mathbf{y}_{1:k}) &= p(\mathbf{x}_{k+1} \mid \mathbf{x}_k)\, p(\mathbf{x}_k \mid \mathbf{y}_{1:k}) \\ &= \mathrm{N}(\mathbf{x}_{k+1} \mid \mathbf{A}_k\, \mathbf{x}_k, \mathbf{Q}_k)\, \mathrm{N}(\mathbf{x}_k \mid \mathbf{m}_k, \mathbf{P}_k) \\ &= \mathrm{N}\left(\begin{pmatrix} \mathbf{x}_k \\ \mathbf{x}_{k+1} \end{pmatrix} \,\middle|\, \tilde{\mathbf{m}}_1, \tilde{\mathbf{P}}_1 \right), \end{aligned} \tag{12.7}$$

where

$$\tilde{\mathbf{m}}_1 = \begin{pmatrix} \mathbf{m}_k \\ \mathbf{A}_k\, \mathbf{m}_k \end{pmatrix}, \qquad \tilde{\mathbf{P}}_1 = \begin{pmatrix} \mathbf{P}_k & \mathbf{P}_k\, \mathbf{A}_k^{\mathsf{T}} \\ \mathbf{A}_k\, \mathbf{P}_k & \mathbf{A}_k\, \mathbf{P}_k\, \mathbf{A}_k^{\mathsf{T}} + \mathbf{Q}_k \end{pmatrix}. \tag{12.8}$$

Due to the Markov property of the states, we have

$$p(\mathbf{x}_k \mid \mathbf{x}_{k+1}, \mathbf{y}_{1:T}) = p(\mathbf{x}_k \mid \mathbf{x}_{k+1}, \mathbf{y}_{1:k}), \tag{12.9}$$

and thus by Lemma A.3 we get the conditional distribution

$$\begin{aligned} p(\mathbf{x}_k \mid \mathbf{x}_{k+1}, \mathbf{y}_{1:T}) &= p(\mathbf{x}_k \mid \mathbf{x}_{k+1}, \mathbf{y}_{1:k}) \\ &= \mathrm{N}(\mathbf{x}_k \mid \tilde{\mathbf{m}}_2, \tilde{\mathbf{P}}_2), \end{aligned} \tag{12.10}$$

where

$$\begin{aligned} \mathbf{G}_k &= \mathbf{P}_k\, \mathbf{A}_k^{\mathsf{T}}\, (\mathbf{A}_k\, \mathbf{P}_k\, \mathbf{A}_k^{\mathsf{T}} + \mathbf{Q}_k)^{-1} \\ \tilde{\mathbf{m}}_2 &= \mathbf{m}_k + \mathbf{G}_k\, (\mathbf{x}_{k+1} - \mathbf{A}_k\, \mathbf{m}_k) \\ \tilde{\mathbf{P}}_2 &= \mathbf{P}_k - \mathbf{G}_k\, (\mathbf{A}_k\, \mathbf{P}_k\, \mathbf{A}_k^{\mathsf{T}} + \mathbf{Q}_k)\, \mathbf{G}_k^{\mathsf{T}}. \end{aligned} \tag{12.11}$$

The joint distribution of $\mathbf{x}_k$ and $\mathbf{x}_{k+1}$ given all the data is

$$\begin{aligned} p(\mathbf{x}_{k+1}, \mathbf{x}_k \mid \mathbf{y}_{1:T}) &= p(\mathbf{x}_k \mid \mathbf{x}_{k+1}, \mathbf{y}_{1:T})\, p(\mathbf{x}_{k+1} \mid \mathbf{y}_{1:T}) \\ &= \mathrm{N}(\mathbf{x}_k \mid \tilde{\mathbf{m}}_2, \tilde{\mathbf{P}}_2)\, \mathrm{N}(\mathbf{x}_{k+1} \mid \mathbf{m}_{k+1}^{\mathrm{s}}, \mathbf{P}_{k+1}^{\mathrm{s}}) \\ &= \mathrm{N}\left(\begin{pmatrix} \mathbf{x}_{k+1} \\ \mathbf{x}_k \end{pmatrix} \,\middle|\, \tilde{\mathbf{m}}_3, \tilde{\mathbf{P}}_3 \right), \end{aligned} \tag{12.12}$$

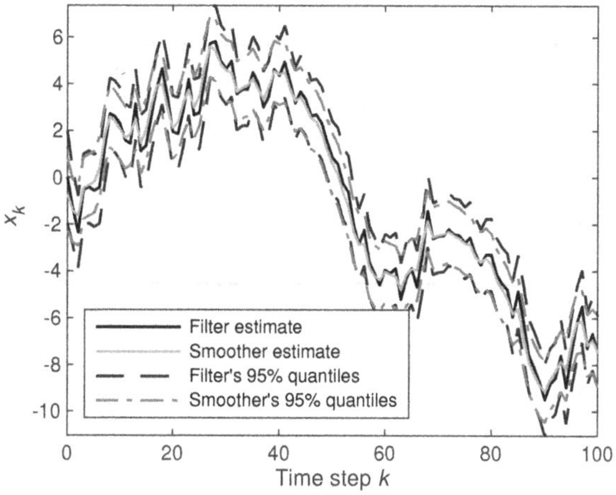

Figure 12.1 Filter and smoother estimates in the Gaussian random walk smoothing example (Example 12.3).

where

$$\begin{aligned} \tilde{\mathbf{m}}_3 &= \begin{pmatrix} \mathbf{m}^s_{k+1} \\ \mathbf{m}_k + \mathbf{G}_k\,(\mathbf{m}^s_{k+1} - \mathbf{A}_k\,\mathbf{m}_k) \end{pmatrix}, \\ \tilde{\mathbf{P}}_3 &= \begin{pmatrix} \mathbf{P}^s_{k+1} & \mathbf{P}^s_{k+1}\,\mathbf{G}_k^\mathsf{T} \\ \mathbf{G}_k\,\mathbf{P}^s_{k+1} & \mathbf{G}_k\,\mathbf{P}^s_{k+1}\,\mathbf{G}_k^\mathsf{T} + \tilde{\mathbf{P}}_2 \end{pmatrix}. \end{aligned} \tag{12.13}$$

Thus by Lemma A.3, the marginal distribution of $\mathbf{x}_k$ is given as

$$p(\mathbf{x}_k \mid \mathbf{y}_{1:T}) = \mathrm{N}(\mathbf{x}_k \mid \mathbf{m}^s_k, \mathbf{P}^s_k), \tag{12.14}$$

where

$$\begin{aligned} \mathbf{m}^s_k &= \mathbf{m}_k + \mathbf{G}_k\,(\mathbf{m}^s_{k+1} - \mathbf{A}_k\,\mathbf{m}_k), \\ \mathbf{P}^s_k &= \mathbf{P}_k + \mathbf{G}_k\,(\mathbf{P}^s_{k+1} - \mathbf{A}_k\,\mathbf{P}_k\,\mathbf{A}_k^\mathsf{T} - \mathbf{Q}_k)\,\mathbf{G}_k^\mathsf{T}. \end{aligned} \tag{12.15}$$

□

The RTS smoother described in Theorem (12.2) assumes that we first run a Kalman filter, and store $\mathbf{m}_k$ and $\mathbf{P}_k$ for $k = 1, 2, \ldots, T$, before performing smoothing. However, the Kalman filter actually also computes $\mathbf{m}_k^-$ and $\mathbf{P}_k^-$, which are also computed in Equations (12.6). An alternative version of the RTS smoother is therefore to store the values of $\mathbf{m}_k^-$ and $\mathbf{P}_k^-$ computed by the Kalman filter, and to only compute $\mathbf{G}_k$, $\mathbf{m}^s_k$, and $\mathbf{P}^s_k$ in Equations (12.6) during smoothing.

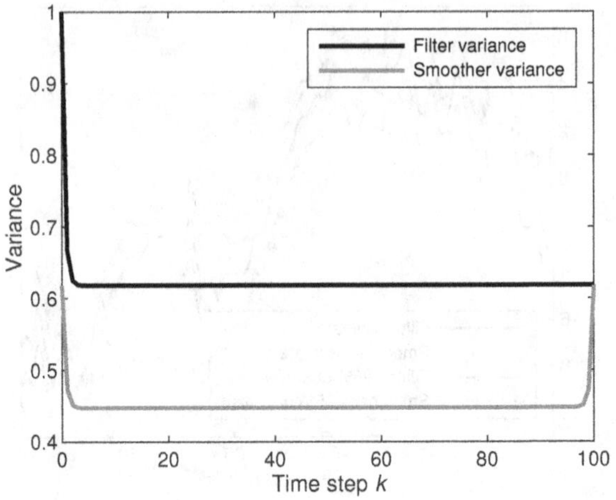

Figure 12.2 Filter and smoother variances in the Gaussian random walk smoothing example (Example 12.3). The variance of the smoother is always smaller than that of the filter. The only exception is at the final step, where the variances are the same.

Example 12.3 (RTS smoother for Gaussian random walk). *The RTS smoother for the random walk model given in Example 6.4 is given by the equations*

$$
\begin{aligned}
m_{k+1}^- &= m_k, \\
P_{k+1}^- &= P_k + Q, \\
m_k^s &= m_k + \frac{P_k}{P_{k+1}^-}(m_{k+1}^s - m_{k+1}^-), \\
P_k^s &= P_k + \left(\frac{P_k}{P_{k+1}^-}\right)^2 [P_{k+1}^s - P_{k+1}^-],
\end{aligned}
\tag{12.16}
$$

where m_k and P_k are the updated mean and covariance from the Kalman filter in Example 6.7. The result of applying the smoother to simulated data is shown in Figure 12.1. The evolution of the filter and smoother variances is illustrated in Figure 12.2.

Example 12.4 (RTS smoother for car tracking). *The backward recursion equations required for implementing the RTS smoother for the car tracking* *problem in Example 6.8 are the following:*

$$\mathbf{m}_{k+1}^{-} = \begin{pmatrix} 1 & 0 & \Delta t & 0 \\ 0 & 1 & 0 & \Delta t \\ 0 & 0 & 1 & 0 \\ 0 & 0 & 0 & 1 \end{pmatrix} \mathbf{m}_k,$$

$$\mathbf{P}_{k+1}^{-} = \begin{pmatrix} 1 & 0 & \Delta t & 0 \\ 0 & 1 & 0 & \Delta t \\ 0 & 0 & 1 & 0 \\ 0 & 0 & 0 & 1 \end{pmatrix} \mathbf{P}_k \begin{pmatrix} 1 & 0 & \Delta t & 0 \\ 0 & 1 & 0 & \Delta t \\ 0 & 0 & 1 & 0 \\ 0 & 0 & 0 & 1 \end{pmatrix}^{\mathsf{T}} + \begin{pmatrix} \frac{q_1^c\, \Delta t^3}{3} & 0 & \frac{q_1^c\, \Delta t^2}{2} & 0 \\ 0 & \frac{q_2^c\, \Delta t^3}{3} & 0 & \frac{q_2^c\, \Delta t^2}{2} \\ \frac{q_1^c\, \Delta t^2}{2} & 0 & q_1^c\, \Delta t & 0 \\ 0 & \frac{q_2^c\, \Delta t^2}{2} & 0 & q_2^c\, \Delta t \end{pmatrix},$$

$$\begin{aligned} \mathbf{G}_k &= \mathbf{P}_k \begin{pmatrix} 1 & 0 & \Delta t & 0 \\ 0 & 1 & 0 & \Delta t \\ 0 & 0 & 1 & 0 \\ 0 & 0 & 0 & 1 \end{pmatrix}^{\mathsf{T}} [\mathbf{P}_{k+1}^{-}]^{-1}, \\ \mathbf{m}_k^{s} &= \mathbf{m}_k + \mathbf{G}_k\, [\mathbf{m}_{k+1}^{s} - \mathbf{m}_{k+1}^{-}], \\ \mathbf{P}_k^{s} &= \mathbf{P}_k + \mathbf{G}_k\, [\mathbf{P}_{k+1}^{s} - \mathbf{P}_{k+1}^{-}]\, \mathbf{G}_k^{\mathsf{T}}. \end{aligned}$$

The terms $\mathbf{m}_k$ *and* $\mathbf{P}_k$ *are the Kalman filter means and covariances computed with the equations given in Example 6.8. It would also be possible to store the values* $\mathbf{m}_{k+1}^{-}$ *and* $\mathbf{P}_{k+1}^{-}$ *during Kalman filtering to avoid recomputation of them in the first two equations above. The gains* $\mathbf{G}_k$ *could be computed already during the Kalman filtering as well. The result of applying the RTS smoother to simulated data is shown in Figure 12.3.*

12.3 Affine Rauch–Tung–Striebel Smoother

As in the case of filtering, the smoothing equations also have closed form expressions for affine models of the form in Equation (6.34). The affine Kalman filtering equations for affine models are given in Theorem 6.9. The RTS smoother for affine models is almost identical to the RTS smoother for linear models but is included here for completeness as it is useful later in constructing non-linear smoothers. Note that the measurement model does

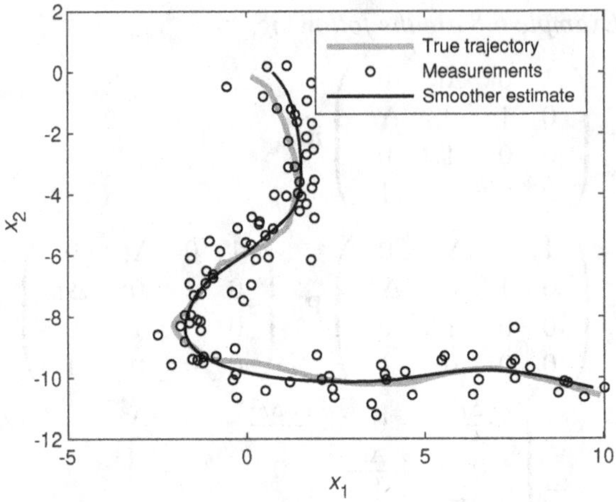

Figure 12.3 Simulated trajectory, measurements, and result of RTS smoother-based car tracking in Example 6.8. The starting point is at the top of the trajectory. The RMSE position error based on the measurements only is 0.77, the position RMSE of the Kalman filter estimate is 0.43, and the error of the RTS smoother is 0.27. It can be seen that the estimate is much "smoother" than the result of the Kalman filter in Figure 6.5.

not influence the RTS smoother, and therefore the only difference compared to Theorem 12.2 is that the constant $\mathbf{a}_k$ leads to a slight modification to the expression for $\mathbf{m}_{k+1}^-$.

Theorem 12.5 (Affine RTS smoother). *The backward recursion equations for the affine (fixed interval) Rauch–Tung–Striebel smoother are given as*

$$
\begin{aligned}
\mathbf{m}_{k+1}^- &= \mathbf{A}_k\,\mathbf{m}_k + \mathbf{a}_k, \\
\mathbf{P}_{k+1}^- &= \mathbf{A}_k\,\mathbf{P}_k\,\mathbf{A}_k^\mathsf{T} + \mathbf{Q}_k, \\
\mathbf{G}_k &= \mathbf{P}_k\,\mathbf{A}_k^\mathsf{T}\,[\mathbf{P}_{k+1}^-]^{-1}, \\
\mathbf{m}_k^\mathrm{s} &= \mathbf{m}_k + \mathbf{G}_k\,[\mathbf{m}_{k+1}^\mathrm{s} - \mathbf{m}_{k+1}^-], \\
\mathbf{P}_k^\mathrm{s} &= \mathbf{P}_k + \mathbf{G}_k\,[\mathbf{P}_{k+1}^\mathrm{s} - \mathbf{P}_{k+1}^-]\,\mathbf{G}_k^\mathsf{T},
\end{aligned}
\tag{12.17}
$$

where $\mathbf{m}_k$ and $\mathbf{P}_k$ are the mean and covariance computed by the affine Kalman filter. The recursion is started from the last time step T, with $\mathbf{m}_T^\mathrm{s} = \mathbf{m}_T$ and $\mathbf{P}_T^\mathrm{s} = \mathbf{P}_T$. Note that the first two of the equations are simply the Kalman filter prediction equations for affine models.

12.4 Bayesian Smoother for Discrete State Spaces

In Section 6.5 we saw how Bayesian filtering equations can be specialized to the case where the states take values in a discrete space, $\mathbf{x}_k \in \mathbb{X}$. Similarly to the Bayesian filter, the Bayesian smoother can also be formulated for this kind of model. The result is the following.

Corollary 12.6 (Discrete Bayesian smoothing equations). *For a discrete state-space* $\mathbf{x}_k \in \mathbb{X}$, *the equations for computing the* smoothed distributions $p(\mathbf{x}_k \mid \mathbf{y}_{1:T})$ *are given as follows:*

$$p(\mathbf{x}_{k+1} \mid \mathbf{y}_{1:k}) = \sum_{\mathbf{x}_k \in \mathbb{X}} p(\mathbf{x}_{k+1} \mid \mathbf{x}_k)\, p(\mathbf{x}_k \mid \mathbf{y}_{1:k}),$$

$$p(\mathbf{x}_k \mid \mathbf{y}_{1:T}) = p(\mathbf{x}_k \mid \mathbf{y}_{1:k}) \sum_{\mathbf{x}_{k+1} \in \mathbb{X}} \left[\frac{p(\mathbf{x}_{k+1} \mid \mathbf{x}_k)\, p(\mathbf{x}_{k+1} \mid \mathbf{y}_{1:T})}{p(\mathbf{x}_{k+1} \mid \mathbf{y}_{1:k})} \right], \tag{12.18}$$

where $p(\mathbf{x}_k \mid \mathbf{y}_{1:k})$ *is the filtering distribution at time step* k.

As in the case of filtering, for models with finite numbers of possible states and measurements, it is convenient to define state transition and emission matrices $\boldsymbol{\Pi}$ and $\mathbf{O}$, respectively, via Equations (6.41). If we further denote the filtering and prediction distributions as in Equations (6.42) and denote

$$p^{s}_{j,k} = p(x_k = j \mid y_{1:T}), \tag{12.19}$$

then the smoothing equations can be written as follows.

Corollary 12.7 (Bayesian smoother for discrete HMMs). *The Bayesian smoothing equations for hidden Markov models (HMMs) with finite state and measurement spaces can be written as*

$$\begin{aligned} p^{-}_{j,k+1} &= \sum_i \Pi_{i,j}\, p_{i,k}, \\ p^{s}_{j,k} &= p_{i,k} \sum_j \Pi_{i,j} \frac{p^{s}_{j,k+1}}{p^{-}_{j,k+1}}. \end{aligned} \tag{12.20}$$

Example 12.8 (Gilbert–Elliot channel smoother). *Figure 12.4 shows the result of applying the Bayesian smoother in Corollary 12.7 to the Gilbert–Elliot model that we already considered in Examples 5.14 and 6.12.*

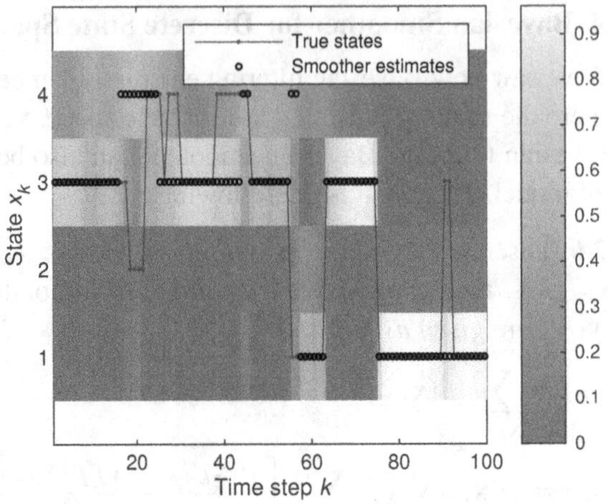

Figure 12.4 Smoothing result for the Gilbert–Elliot channel model (Example 12.8). The probabilities of the states are denoted by shades of gray, and the smoother estimates are the most probable states computed from the smoothing marginals. The error rate of the smoother when the most probable state is guessed is 8% while the error rate of the filter reported in Example 6.12 was 11%.

12.5 Viterbi Algorithm

The Viterbi algorithm (Larson and Peschon, 1966; Viterbi, 1967; Cappé et al., 2005) is an algorithm that can be used to compute the maximum a posterior (MAP) estimates of state trajectories. The algorithm can be applied to models with discrete or continuous state or measurement spaces, but the most common practical use is when both of them are discrete.

The aim of the Viterbi algorithm is to find the MAP estimate, which corresponds to maximization of

$$p(\mathbf{x}_{0:T} \mid \mathbf{y}_{1:T}) = \frac{p(\mathbf{y}_{1:T}, \mathbf{x}_{0:T})}{p(\mathbf{y}_{1:T})} \propto p(\mathbf{y}_{0:T}, \mathbf{x}_{1:T}), \tag{12.21}$$

which is equivalent to maximizing

$$p(\mathbf{y}_{1:T}, \mathbf{x}_{0:T}) = p(\mathbf{x}_0) \prod_{i=1}^{T} [p(\mathbf{y}_i \mid \mathbf{x}_i)\, p(\mathbf{x}_i \mid \mathbf{x}_{i-1})], \tag{12.22}$$

that is,

$$\mathbf{x}^*_{0:T} = \arg\max_{\mathbf{x}_{0:T}} p(\mathbf{y}_{1:T}, \mathbf{x}_{0:T}). \tag{12.23}$$

It is worth noting that the computation of the MAP estimate of the state trajectory is different from the problem of estimating the smoothing marginals $p(\mathbf{x}_k \mid \mathbf{y}_{1:T})$, and the maxima of these marginals are not necessarily the same as the intermediate states along the MAP trajectory. For discrete state-spaces, it is therefore relevant to develop both Bayesian smoothers, as presented in Section 12.4, and the MAP estimators presented here, whereas one algorithm is sufficient for linear Gaussian models since the solutions to both objectives always coincide for such models.

Theorem 12.9 (Viterbi algorithm for computing the MAP path). *The following procedure can be used to compute a MAP path* $\mathbf{x}^*_{0:T}$. *On the* forward pass *we start from* $V_0(\mathbf{x}_0) = p(\mathbf{x}_0)$ *and compute the following for* $k = 1, \ldots, T$*:*

$$V_k(\mathbf{x}_k) = \max_{\mathbf{x}_{k-1}} [p(\mathbf{y}_k \mid \mathbf{x}_k)\, p(\mathbf{x}_k \mid \mathbf{x}_{k-1})\, V_{k-1}(\mathbf{x}_{k-1})]. \tag{12.24}$$

In the backward pass *we first take*

$$\mathbf{x}^*_T = \arg\max_{\mathbf{x}_T} V_T(\mathbf{x}_T) \tag{12.25}$$

and then compute

$$\mathbf{x}^*_k = \arg\max_{\mathbf{x}_k} [p(\mathbf{x}^*_{k+1} \mid \mathbf{x}_k)\, V_k(\mathbf{x}_k)] \tag{12.26}$$

for $k = T-1, \ldots, 0$, *where* arg max[$\cdot$] *can return any of the maximizing states if there are more than one of them.*

Proof By definition, we have $V_0(\mathbf{x}_0) = p(\mathbf{x}_0)$. We first show that

$$V_k(\mathbf{x}_k) = \max_{\mathbf{x}_{0:k-1}} p(\mathbf{x}_0) \prod_{i=1}^{k} [p(\mathbf{y}_i \mid \mathbf{x}_i)\, p(\mathbf{x}_i \mid \mathbf{x}_{i-1})], \tag{12.27}$$

using an inductive proof. We note that this holds trivially for $k = 0$ and we can use this as our base case. Furthermore, to complete the inductive proof, let us assume that the relation holds for $k-1$ and prove that it then must hold for k. To this end, we note that

$$\begin{aligned} V_k(\mathbf{x}_k) &= \max_{\mathbf{x}_{k-1}} \Bigg[p(\mathbf{y}_k \mid \mathbf{x}_k)\, p(\mathbf{x}_k \mid \mathbf{x}_{k-1}) \\ &\qquad \times \max_{\mathbf{x}_{0:k-2}} p(\mathbf{x}_0) \prod_{i=1}^{k-1} [p(\mathbf{y}_i \mid \mathbf{x}_i)\, p(\mathbf{x}_i \mid \mathbf{x}_{i-1})] \Bigg] \\ &= \max_{\mathbf{x}_{k-1}} [p(\mathbf{y}_k \mid \mathbf{x}_k)\, p(\mathbf{x}_k \mid \mathbf{x}_{k-1})\, V_{k-1}(\mathbf{x}_{k-1})], \end{aligned} \tag{12.28}$$

which proves that $V_k(\mathbf{x}_k)$ satisfies Equation (12.24).

To complete the proof, we will now use a second inductive proof to show that the backward pass gives the MAP sequence $\mathbf{x}^*_{0:T}$. In this case, we first note that the relation $\mathbf{x}^*_T = \arg\max V_T(\mathbf{x}_T)$ follows from Equation (12.27), and we can use this as our base case. Finally, we assume that we know the tail of a MAP path $\mathbf{x}^*_{k+1:T}$ and show that Equation (12.26) must give the MAP estimate at time step k for $k = T-1, \ldots, 0$.

When $\mathbf{x}^*_{k+1:T}$ is given, the MAP estimation problem simplifies to finding

$$\begin{aligned} \mathbf{x}^*_{0:k} &= \arg\max_{\mathbf{x}_{0:k}} p(\mathbf{y}_{1:T}, \mathbf{x}_{0:k}, \mathbf{x}^*_{k+1:T}) \\ &= \arg\max_{\mathbf{x}_{0:k}} p(\mathbf{x}^*_{k+1} \mid \mathbf{x}_k) p(\mathbf{x}_0) \prod_{i=1}^{k} [p(\mathbf{y}_i \mid \mathbf{x}_i)\, p(\mathbf{x}_i \mid \mathbf{x}_{i-1})]. \end{aligned} \tag{12.29}$$

We can hence express the MAP estimate at time k as

$$\begin{aligned} \mathbf{x}^*_k &= \arg\max_{\mathbf{x}_k} p(\mathbf{x}^*_{k+1} \mid \mathbf{x}_k) \max_{\mathbf{x}_{0:k-1}} p(\mathbf{x}_0) \prod_{i=1}^{k} [p(\mathbf{y}_i \mid \mathbf{x}_i)\, p(\mathbf{x}_i \mid \mathbf{x}_{i-1})], \\ &= \arg\max_{\mathbf{x}_k} p(\mathbf{x}^*_{k+1} \mid \mathbf{x}_k)\, V_k(\mathbf{x}_k), \end{aligned} \tag{12.30}$$

which proves that Equation (12.26) yields the MAP sequence $\mathbf{x}^*_{0:T}$, by induction.

□

The Viterbi algorithm has interesting connections to several other algorithms. For instance, the forward pass resembles an operation that jointly performs prediction and update, but in contrast to our filtering distributions, $V_k(\mathbf{x}_k)$ is not normalized, and $\mathbf{x}_{k-1}$ is not marginalized by summation (as in the Chapman–Kolmogorov equation) but instead selects the value of $\mathbf{x}_{k-1}$ that maximizes $V_k(\mathbf{x}_k)$, for each value of $\mathbf{x}_k$.

The Viterbi algorithm also has a close connection with the so-called max-product algorithm, which is often used in the context of Bayesian graphical models (Koller and Friedman, 2009). The max-product algorithm can roughly be seen as a two-filter formulation of the Viterbi algorithm using an analogy with forward-backward versus two-filter smoothing.

As with discrete-state filters and smoothers, the Viterbi algorithm can also be specialized to the case where the state and measurement spaces are finite, in which case the model is defined by the state transition and

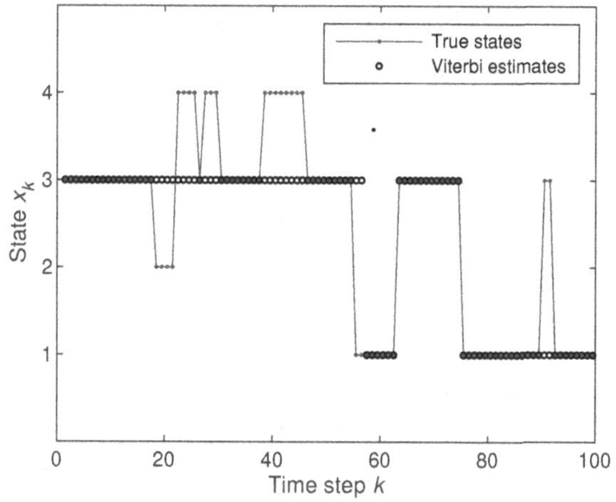

Figure 12.5 Result of applying the Viterbi algorithm to the Gilbert–Elliot channel model (Example 12.11). The error rate of the algorithm was 8% while the filter rate in Example 6.12 was 11%.

emission matrices $\mathbf{\Pi}$ and $\mathbf{O}$ given by Equations (6.41). We can then also denote

$$V_{i,k} = V_k(x_k = i). \tag{12.31}$$

With this notation, the Viterbi algorithm becomes the following.

Corollary 12.10 (Viterbi algorithm for discrete HMMs). *On the forward pass we start from* $V_{i,0} = p_{i,0}$ *and compute the following for* $k = 1, \ldots, T$*:*

$$V_{i,k} = \max_j \left[O_{i,y_k} \, \Pi_{j,i} \, V_{j,k-1} \right]. \tag{12.32}$$

On the backward pass we first put $x_T^* = \arg\max_i V_{i,T}$ *and then proceed as*

$$x_k^* = \arg\max_j \left[\Pi_{j,x_{k+1}^*} \, V_{j,k} \right]. \tag{12.33}$$

Example 12.11 (Viterbi algorithm for Gilbert–Elliot channel). *The result of applying the Viterbi algorithm in Corollary 12.10 to the Gilbert–Elliot model introduced in Example 5.14 is shown in Figure 12.5.*

12.6 Exercises

12.1 Derive the linear RTS smoother for the non-zero-mean noise model in Exercise 6.1.

12.2 Implement the Gaussian random walk model smoother in Example 12.3, and compare its performance to the corresponding Kalman filter. Plot the evolution of the smoothing distribution.

12.3 The Gaussian random walk model considered in Example 6.4 also defines a joint Gaussian prior distribution $p(x_0, \ldots, x_T)$. The measurement model $p(y_1, \ldots, y_T \mid x_0, \ldots, x_T)$ is Gaussian as well. Construct these distributions, and compute the posterior distribution $p(x_0, \ldots, x_T \mid y_1, \ldots, y_T)$. Check numerically that the mean and the diagonal covariance entries of this distribution exactly match the smoother means and variances.

12.4 Implement the RTS smoother for the resonator model in Exercise 6.4. Compare its RMSE performance to the filtering and baseline solutions, and plot the results.

12.5 Implement an HMM smoother to the model considered in Exercise 6.5, and compute the error rate of the smoother when the highest probability state in the smoothing distribution is used as the estimate. Also compare the performance with the HMM filter.

12.6 Implement Viterbi algorithm to the model in Exercise 6.5, and compare its performance with the filter and the smoother in Exercise 12.5.

12.7 Form a grid-based approximation to the Gaussian random walk model smoother in the same way as was done for the filtering equations in Exercise 6.6. Verify that the result is practically the same as in the RTS smoother above.

12.8 Implement Viterbi algorithm for the problem in Exercise 12.7.

12.9 Show that in the linear Gaussian case, the Viterbi algorithm in Theorem 12.9 reduces to the RTS smoother.

13

Extended Rauch–Tung–Striebel Smoothing

In this chapter, we present Taylor series-based extended Rauch–Tung–Striebel (i.e., Kalman) smoothers along with their iterated versions. These smoothers use similar approximations as the extended Kalman filters that we encountered in Chapter 7. The iterated extended Rauch–Tung–Striebel smoother also turns out to be an instance of the Gauss–Newton method, but in contrast to the iterated extended Kalman filter, it computes the maximum a posteriori (MAP) estimate of the whole trajectory instead of just a single state.

13.1 Extended Rauch–Tung–Striebel Smoother

The first order (i.e., linearized) extended Rauch–Tung–Striebel smoother (ERTSS) (Cox, 1964; Sage and Melsa, 1971) can be obtained from the basic RTS smoother equations by replacing the prediction equations with first order approximations. Various kinds of higher order extended Kalman smoothers are also possible (see, e.g., Cox, 1964; Sage and Melsa, 1971), and we present a second order smoother in Section 13.2. The ERTSS forms (or assumes) a Gaussian approximation to the smoothing distribution as follows:

$$p(\mathbf{x}_k \mid \mathbf{y}_{1:T}) \simeq \mathrm{N}(\mathbf{x}_k \mid \mathbf{m}_k^{\mathrm{s}}, \mathbf{P}_k^{\mathrm{s}}). \tag{13.1}$$

For the additive model in Equation (7.24), the extended Rauch–Tung–Striebel smoother algorithm is the following.

Algorithm 13.1 (Extended RTS smoother). *The equations for the extended RTS smoother are*

$$\begin{aligned}
\mathbf{m}_{k+1}^{-} &= \mathbf{f}(\mathbf{m}_k), \\
\mathbf{P}_{k+1}^{-} &= \mathbf{F}_{\mathbf{x}}(\mathbf{m}_k)\, \mathbf{P}_k\, \mathbf{F}_{\mathbf{x}}^{\mathsf{T}}(\mathbf{m}_k) + \mathbf{Q}_k, \\
\mathbf{G}_k &= \mathbf{P}_k\, \mathbf{F}_{\mathbf{x}}^{\mathsf{T}}(\mathbf{m}_k)\, [\mathbf{P}_{k+1}^{-}]^{-1}, \\
\mathbf{m}_k^{\mathrm{s}} &= \mathbf{m}_k + \mathbf{G}_k\, [\mathbf{m}_{k+1}^{\mathrm{s}} - \mathbf{m}_{k+1}^{-}], \\
\mathbf{P}_k^{\mathrm{s}} &= \mathbf{P}_k + \mathbf{G}_k\, [\mathbf{P}_{k+1}^{\mathrm{s}} - \mathbf{P}_{k+1}^{-}]\, \mathbf{G}_k^{\mathsf{T}},
\end{aligned} \tag{13.2}$$

where the matrix $\mathbf{F}_{\mathbf{x}}(\mathbf{m}_k)$ *is the Jacobian matrix of* $\mathbf{f}(\mathbf{x})$ *evaluated at* $\mathbf{m}_k$. *The recursion is started from the last time step* T, *with* $\mathbf{m}_T^{\mathrm{s}} = \mathbf{m}_T$ *and* $\mathbf{P}_T^{\mathrm{s}} = \mathbf{P}_T$.

Derivation Assume that the filtering distributions for the model (7.24) are approximately Gaussian:

$$p(\mathbf{x}_k \mid \mathbf{y}_{1:k}) \simeq \mathrm{N}(\mathbf{x}_k \mid \mathbf{m}_k, \mathbf{P}_k),$$

and we have already computed the means and covariance using the extended Kalman filter or a similar method. Further assume that the smoothing distribution of time step $k+1$ is known and approximately Gaussian

$$p(\mathbf{x}_{k+1} \mid \mathbf{y}_{1:T}) \simeq \mathrm{N}(\mathbf{x}_{k+1} \mid \mathbf{m}_{k+1}^{\mathrm{s}}, \mathbf{P}_{k+1}^{\mathrm{s}}).$$

As in the derivation of the prediction step of the EKF in Section 7.2, the approximate joint distribution of $\mathbf{x}_k$ and $\mathbf{x}_{k+1}$ given $\mathbf{y}_{1:k}$ is

$$p(\mathbf{x}_k, \mathbf{x}_{k+1} \mid \mathbf{y}_{1:k}) = \mathrm{N}\left(\begin{pmatrix} \mathbf{x}_k \\ \mathbf{x}_{k+1} \end{pmatrix} \middle|\, \tilde{\mathbf{m}}_1, \tilde{\mathbf{P}}_1 \right), \tag{13.3}$$

where

$$\begin{aligned}
\tilde{\mathbf{m}}_1 &= \begin{pmatrix} \mathbf{m}_k \\ \mathbf{f}(\mathbf{m}_k) \end{pmatrix}, \\
\tilde{\mathbf{P}}_1 &= \begin{pmatrix} \mathbf{P}_k & \mathbf{P}_k\, \mathbf{F}_x^{\mathsf{T}} \\ \mathbf{F}_x\, \mathbf{P}_k & \mathbf{F}_x\, \mathbf{P}_k\, \mathbf{F}_x^{\mathsf{T}} + \mathbf{Q}_k \end{pmatrix},
\end{aligned} \tag{13.4}$$

and the Jacobian matrix $\mathbf{F}_x$ of $\mathbf{f}(\mathbf{x})$ is evaluated at $\mathbf{x} = \mathbf{m}_k$. By conditioning on $\mathbf{x}_{k+1}$ as in the RTS derivation in Section 12.2, we get

$$\begin{aligned}
p(\mathbf{x}_k \mid \mathbf{x}_{k+1}, \mathbf{y}_{1:T}) &= p(\mathbf{x}_k \mid \mathbf{x}_{k+1}, \mathbf{y}_{1:k}) \\
&= \mathrm{N}(\mathbf{x}_k \mid \tilde{\mathbf{m}}_2, \tilde{\mathbf{P}}_2),
\end{aligned} \tag{13.5}$$

where

$$\begin{aligned}
\mathbf{G}_k &= \mathbf{P}_k\,\mathbf{F}_x^\mathsf{T}\,(\mathbf{F}_x\,\mathbf{P}_k\,\mathbf{F}_x^\mathsf{T} + \mathbf{Q}_k)^{-1}, \\
\tilde{\mathbf{m}}_2 &= \mathbf{m}_k + \mathbf{G}_k\,(\mathbf{x}_{k+1} - \mathbf{f}(\mathbf{m}_k)), \\
\tilde{\mathbf{P}}_2 &= \mathbf{P}_k - \mathbf{G}_k\,(\mathbf{F}_x\,\mathbf{P}_k\,\mathbf{F}_x^\mathsf{T} + \mathbf{Q}_k)\,\mathbf{G}_k^\mathsf{T}.
\end{aligned} \tag{13.6}$$

The joint distribution of $\mathbf{x}_k$ and $\mathbf{x}_{k+1}$ given all the data is now

$$\begin{aligned}
p(\mathbf{x}_{k+1}, \mathbf{x}_k \mid \mathbf{y}_{1:T}) &= p(\mathbf{x}_k \mid \mathbf{x}_{k+1}, \mathbf{y}_{1:T})\,p(\mathbf{x}_{k+1} \mid \mathbf{y}_{1:T}) \\
&= \mathrm{N}\left(\begin{pmatrix} \mathbf{x}_{k+1} \\ \mathbf{x}_k \end{pmatrix} \,\middle|\, \tilde{\mathbf{m}}_3, \tilde{\mathbf{P}}_3 \right),
\end{aligned} \tag{13.7}$$

where

$$\begin{aligned}
\tilde{\mathbf{m}}_3 &= \begin{pmatrix} \mathbf{m}_{k+1}^\mathrm{s} \\ \mathbf{m}_k + \mathbf{G}_k\,(\mathbf{m}_{k+1}^\mathrm{s} - \mathbf{f}(\mathbf{m}_k)) \end{pmatrix}, \\
\tilde{\mathbf{P}}_3 &= \begin{pmatrix} \mathbf{P}_{k+1}^\mathrm{s} & \mathbf{P}_{k+1}^\mathrm{s}\,\mathbf{G}_k^\mathsf{T} \\ \mathbf{G}_k\,\mathbf{P}_{k+1}^\mathrm{s} & \mathbf{G}_k\,\mathbf{P}_{k+1}^\mathrm{s}\,\mathbf{G}_k^\mathsf{T} + \tilde{\mathbf{P}}_2 \end{pmatrix}.
\end{aligned} \tag{13.8}$$

The marginal distribution of $\mathbf{x}_k$ is then

$$p(\mathbf{x}_k \mid \mathbf{y}_{1:T}) = \mathrm{N}(\mathbf{x}_k \mid \mathbf{m}_k^\mathrm{s}, \mathbf{P}_k^\mathrm{s}), \tag{13.9}$$

where

$$\begin{aligned}
\mathbf{m}_k^\mathrm{s} &= \mathbf{m}_k + \mathbf{G}_k\,(\mathbf{m}_{k+1}^\mathrm{s} - \mathbf{f}(\mathbf{m}_k)), \\
\mathbf{P}_k^\mathrm{s} &= \mathbf{P}_k + \mathbf{G}_k\,(\mathbf{P}_{k+1}^\mathrm{s} - \mathbf{F}_x\,\mathbf{P}_k\,\mathbf{F}_x^\mathsf{T} - \mathbf{Q}_k)\,\mathbf{G}_k^\mathsf{T}.
\end{aligned} \tag{13.10}$$

□

The generalization to the non-additive model (7.39) is analogous to the filtering case – we just need to replace the first two of Equations (13.2) with their non-additive versions as in Algorithm 7.5.

Example 13.2 (Pendulum tracking with ERTSS). *The result of applying the ERTSS to the pendulum model in Example 7.6 is shown in Figure 13.1. The resulting RMSE was* 0.06, *which is much lower than the error of the EKF, which was* 0.17. *It is also lower than the errors of any other filters, which were in the range* 0.10–0.17.

Example 13.3 (Coordinated turn model with ERTSS). *Figure 13.2 shows the result of ERTSS in the coordinated turn model problem considered in Example 7.7. The resulting error was* 0.016 *(as opposed to* 0.31 *for EKF).*

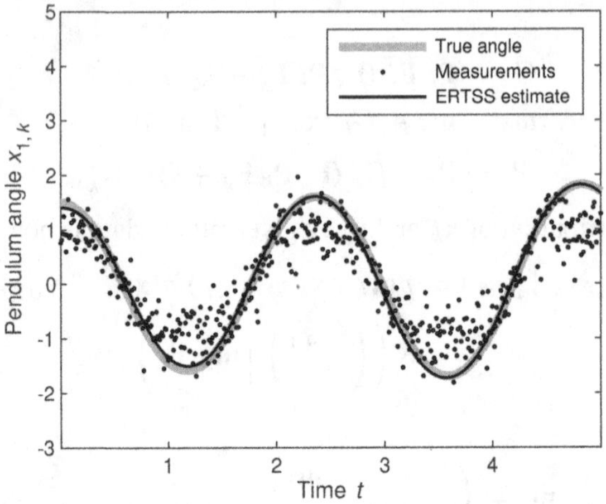

Figure 13.1 Simulated pendulum data and the result of tracking the pendulum described in Example 7.6 with the ERTSS (see Example 13.2). The resulting RMSE is 0.06 (recall that the RMSE of the EKF was 0.17).

13.2 Higher Order Extended Rauch–Tung–Striebel Smoothers

We can also form a second order ERTSS by replacing the prediction equations in the ERTSS with their second order versions appearing in Algorithm 7.8, which results in the following.

Algorithm 13.4 (Second order extended RTS smoother). *The equations for the second order extended RTS smoother are*

$$
\begin{aligned}
\mathbf{m}_{k+1}^{-} &= \mathbf{f}(\mathbf{m}_k) + \frac{1}{2} \sum_i \mathbf{e}_i \operatorname{tr}\left\{ \mathbf{F}_{\mathbf{xx}}^{(i)}(\mathbf{m}_k)\, \mathbf{P}_k \right\}, \\
\mathbf{P}_{k+1}^{-} &= \mathbf{F}_{\mathbf{x}}(\mathbf{m}_k)\, \mathbf{P}_k\, \mathbf{F}_{\mathbf{x}}^{\mathsf{T}}(\mathbf{m}_k) \\
&\quad + \frac{1}{2} \sum_{i,i'} \mathbf{e}_i\, \mathbf{e}_{i'}^{\mathsf{T}} \operatorname{tr}\left\{ \mathbf{F}_{\mathbf{xx}}^{(i)}(\mathbf{m}_k) \mathbf{P}_k \mathbf{F}_{\mathbf{xx}}^{(i')}(\mathbf{m}_k) \mathbf{P}_k \right\} + \mathbf{Q}_k, \qquad (13.11) \\
\mathbf{G}_k &= \mathbf{P}_k\, \mathbf{F}_{\mathbf{x}}^{\mathsf{T}}(\mathbf{m}_k)\, [\mathbf{P}_{k+1}^{-}]^{-1}, \\
\mathbf{m}_k^{\mathrm{s}} &= \mathbf{m}_k + \mathbf{G}_k\, [\mathbf{m}_{k+1}^{\mathrm{s}} - \mathbf{m}_{k+1}^{-}], \\
\mathbf{P}_k^{\mathrm{s}} &= \mathbf{P}_k + \mathbf{G}_k\, [\mathbf{P}_{k+1}^{\mathrm{s}} - \mathbf{P}_{k+1}^{-}]\, \mathbf{G}_k^{\mathsf{T}},
\end{aligned}
$$

where $\mathbf{F}_{\mathbf{x}}(\mathbf{m}_k)$ *is the Jacobian matrix of* $\mathbf{f}(\mathbf{x})$ *evaluated at* $\mathbf{m}_k$*, and* $\mathbf{F}_{\mathbf{xx}}^{(i)}(\mathbf{m}_k)$ *is its Hessian as defined in Equation* (7.54).

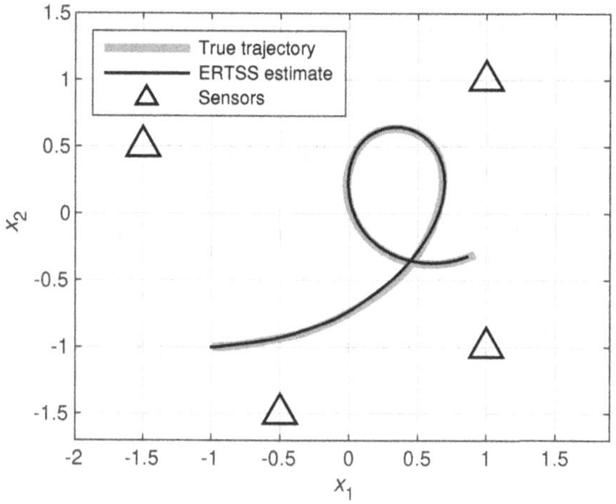

Figure 13.2 The result of applying ERTSS to simulated data from the (polar) coordinated turn model in Example 7.7. The resulting error was 0.016, whereas for the EKF it was 0.31.

13.3 Iterated Extended Rauch–Tung–Striebel Smoother

Similarly to the IEKF presented in Section 7.4, it is also possible to formulate an iterated extended Rauch–Tung–Striebel smoother (IERTSS), which is often called the iterated extended Kalman smoother (IEKS) or just iterated Kalman smoother (IKS, Bell, 1994), as a Gauss–Newton method that aims to compute the maximum-a-posteriori (MAP) estimate of the whole trajectory $\mathbf{x}_{0:T}$ given the measurements. The intuitive idea is that once we have computed an ERTSS estimate of the trajectory, we can relinearize the model at all the time steps simultaneously along this smoother estimate (conditioned on all the data) and run the smoother again. Iterating this will result in the IERTSS algorithm. Please note that in the IEKF we only relinearized the measurement model at the previous estimate produced by the EKF, but in the IERTSS we relinearize both dynamic and measurement models at all time steps before doing the next iteration. Let us now derive this procedure from the Gauss–Newton point of view.

In order to do that, first recall that the joint posterior distribution of the trajectory can be written as

$$p(\mathbf{x}_{0:T} \mid \mathbf{y}_{1:T}) \propto p(\mathbf{x}_0) \prod_k p(\mathbf{y}_k \mid \mathbf{x}_k)\, p(\mathbf{x}_k \mid \mathbf{x}_{k-1}), \tag{13.12}$$

and the MAP estimate of the trajectory can be thus computed by

$$\begin{aligned} \mathbf{x}_{0:T}^{\mathrm{MAP}} &= \arg\max_{\mathbf{x}_{0:T}} p(\mathbf{x}_{0:T} \mid \mathbf{y}_{1:T}) \\ &= \arg\max_{\mathbf{x}_{0:T}} p(\mathbf{x}_0) \prod_k p(\mathbf{y}_k \mid \mathbf{x}_k)\, p(\mathbf{x}_k \mid \mathbf{x}_{k-1}). \end{aligned} \tag{13.13}$$

Similarly to Section 7.4, it is convenient to work with the negative logarithm

$$\begin{aligned} L(\mathbf{x}_{0:T}) &= -\log\left[p(\mathbf{x}_0) \prod_k p(\mathbf{y}_k \mid \mathbf{x}_k)\, p(\mathbf{x}_k \mid \mathbf{x}_{k-1}) \right] \\ &= -\log p(\mathbf{x}_0) - \sum_k \log p(\mathbf{y}_k \mid \mathbf{x}_k) - \sum_k \log p(\mathbf{x}_k \mid \mathbf{x}_{k-1}), \end{aligned} \tag{13.14}$$

in which case the MAP estimation problem becomes

$$\mathbf{x}_{0:T}^{\mathrm{MAP}} = \arg\min_{\mathbf{x}_{0:T}} L(\mathbf{x}_{0:T}). \tag{13.15}$$

When the dynamic and measurement models have the form

$$\begin{aligned} p(\mathbf{x}_k \mid \mathbf{x}_{k-1}) &= \mathrm{N}(\mathbf{x}_k \mid \mathbf{f}(\mathbf{x}_{k-1}), \mathbf{Q}_{k-1}), \\ p(\mathbf{y}_k \mid \mathbf{x}_k) &= \mathrm{N}(\mathbf{y}_k \mid \mathbf{h}(\mathbf{x}_k), \mathbf{R}_k), \end{aligned} \tag{13.16}$$

and the initial distribution is $p(\mathbf{x}_0) = \mathrm{N}(\mathbf{x}_0 \mid \mathbf{m}_0, \mathbf{P}_0)$, then the negative logarithm is

$$\begin{aligned} L(\mathbf{x}_{0:T}) &= C + \frac{1}{2}(\mathbf{x}_0 - \mathbf{m}_0)^{\mathsf{T}} \mathbf{P}_0^{-1} (\mathbf{x}_0 - \mathbf{m}_0) \\ &\quad + \frac{1}{2}\sum_k (\mathbf{y}_k - \mathbf{h}(\mathbf{x}_k))^{\mathsf{T}} \mathbf{R}_k^{-1} (\mathbf{y}_k - \mathbf{h}(\mathbf{x}_k)) \\ &\quad + \frac{1}{2}\sum_k (\mathbf{x}_k - \mathbf{f}(\mathbf{x}_{k-1}))^{\mathsf{T}} \mathbf{Q}_{k-1}^{-1} (\mathbf{x}_k - \mathbf{f}(\mathbf{x}_{k-1})), \end{aligned} \tag{13.17}$$

where C is a constant, independent of the states. The MAP estimate of the trajectory can now be obtained as the minimum of the above cost function.

In Section 7.4, we constructed the IEKF by noticing that the cost function optimized by the IEKF in Equation (7.61) is a special case of cost functions that the Gauss–Newton method can handle, which have the form of Equation (7.63). We can now observe that the cost function in Equation (13.17) is also a special case of (7.63), and therefore we can use Gauss–Newton to compute the MAP estimate of the whole trajectory by iteratively minimizing (13.17).

The Gauss–Newton iterations are started from some initial trajectory $\mathbf{x}_{0:T}^{(0)}$. That can be computed, for example, by using a non-iterated ERTSS. Let us then assume that we have already computed an estimate of the whole trajectory $\mathbf{x}_{0:T}^{(i-1)}$ for iteration step $i-1$. We can now simultaneously linearize all the dynamic and measurement model functions by using Taylor series:

$$\begin{aligned}
\mathbf{f}(\mathbf{x}_{k-1}) &\simeq \mathbf{f}(\mathbf{x}_{k-1}^{(i-1)}) + \mathbf{F}_{\mathbf{x}}(\mathbf{x}_{k-1}^{(i-1)})\,(\mathbf{x}_{k-1} - \mathbf{x}_{k-1}^{(i-1)}) \\
&= \mathbf{F}_{\mathbf{x}}(\mathbf{x}_{k-1}^{(i-1)})\,\mathbf{x}_{k-1} + \mathbf{f}(\mathbf{x}_{k-1}^{(i-1)}) - \mathbf{F}_{\mathbf{x}}(\mathbf{x}_{k-1}^{(i-1)})\,\mathbf{x}_{k-1}^{(i-1)}, \\
\mathbf{h}(\mathbf{x}_{k}) &\simeq \mathbf{h}(\mathbf{x}_{k}^{(i-1)}) + \mathbf{H}_{\mathbf{x}}(\mathbf{x}_{k}^{(i-1)})\,(\mathbf{x}_{k} - \mathbf{x}_{k}^{(i-1)}) \\
&= \mathbf{H}_{\mathbf{x}}(\mathbf{x}_{k}^{(i-1)})\,\mathbf{x}_{k} + \mathbf{h}(\mathbf{x}_{k}^{(i-1)}) - \mathbf{H}_{\mathbf{x}}(\mathbf{x}_{k}^{(i-1)})\,\mathbf{x}_{k}^{(i-1)},
\end{aligned} \tag{13.18}$$

which is done for all $k = 1, \ldots, T$. Let us now temporarily denote

$$\begin{aligned}
\mathbf{u}_{k-1} &= \mathbf{f}(\mathbf{x}_{k-1}^{(i-1)}) - \mathbf{F}_{\mathbf{x}}(\mathbf{x}_{k-1}^{(i-1)})\,\mathbf{x}_{k-1}^{(i-1)}, \\
\mathbf{A}_{k-1} &= \mathbf{F}_{\mathbf{x}}(\mathbf{x}_{k-1}^{(i-1)}), \\
\mathbf{d}_{k} &= \mathbf{h}(\mathbf{x}_{k}^{(i-1)}) - \mathbf{H}_{\mathbf{x}}(\mathbf{x}_{k}^{(i-1)})\,\mathbf{x}_{k}^{(i-1)}, \\
\mathbf{H}_{k} &= \mathbf{H}_{\mathbf{x}}(\mathbf{x}_{k}^{(i-1)}),
\end{aligned} \tag{13.19}$$

which then simplifies the notation of the linearizations to

$$\begin{aligned}
\mathbf{f}(\mathbf{x}_{k-1}) &\simeq \mathbf{A}_{k-1}\,\mathbf{x}_{k-1} + \mathbf{u}_{k-1}, \\
\mathbf{h}(\mathbf{x}_{k}) &\simeq \mathbf{H}_{k}\,\mathbf{x}_{k} + \mathbf{d}_{k}.
\end{aligned} \tag{13.20}$$

Substituting the above approximations to $L(\mathbf{x}_{0:T})$ in (13.17) then results in

$$\begin{aligned}
L(\mathbf{x}_{0:T}) &\simeq C + \frac{1}{2}(\mathbf{x}_0 - \mathbf{m}_0)^\top \mathbf{P}_0^{-1}\,(\mathbf{x}_0 - \mathbf{m}_0) \\
&+ \frac{1}{2}\sum_k (\mathbf{y}_k - \mathbf{H}_k\,\mathbf{x}_k - \mathbf{d}_k)^\top \mathbf{R}_k^{-1}\,(\mathbf{y}_k - \mathbf{H}_k\,\mathbf{x}_k - \mathbf{d}_k) \\
&+ \frac{1}{2}\sum_k (\mathbf{x}_k - \mathbf{A}_{k-1}\,\mathbf{x}_{k-1} - \mathbf{u}_{k-1})^\top \mathbf{Q}_{k-1}^{-1}\,(\mathbf{x}_k - \mathbf{A}_{k-1}\,\mathbf{x}_{k-1} - \mathbf{u}_{k-1}),
\end{aligned} \tag{13.21}$$

which can be recognized to correspond to the affine Gaussian model

$$\begin{aligned}
p(\mathbf{x}_k \mid \mathbf{x}_{k-1}) &\simeq \mathrm{N}(\mathbf{x}_k \mid \mathbf{A}_{k-1}\,\mathbf{x}_{k-1} + \mathbf{u}_{k-1}, \mathbf{Q}_{k-1}), \\
p(\mathbf{y}_k \mid \mathbf{x}_k) &\simeq \mathrm{N}(\mathbf{y}_k \mid \mathbf{H}_k\,\mathbf{x}_k + \mathbf{d}_k, \mathbf{R}_k).
\end{aligned} \tag{13.22}$$

For a Gaussian distribution, the mean matches the MAP estimate, and hence we can compute the MAP estimate for this affine Gaussian model – which is the minimum of (13.21) – by running the affine Kalman filter in Theorem 6.9 and the affine Rauch–Tung–Striebel smoother in Theorem 12.5 for the above model. The smoother mean sequence $\mathbf{m}^{\mathrm{s}}_{0:T}$ then gives the MAP estimate. The MAP estimate $\mathbf{m}^{\mathrm{s}}_{0:T}$ computed for the linearized model can then be used as the next iterate $\mathbf{x}^{(i)}_{0:T} = \mathbf{m}^{\mathrm{s}}_{0:T}$. This leads to an iterative algorithm given in the following.

Algorithm 13.5 (Iterated extended RTS smoother). *The iterated extended RTS smoother is started from an initial guess for the trajectory $\mathbf{x}^{(0)}_{0:T}$, which can be computed, for example, using the (non-iterated) extended Kalman filter and smoother. For each $i = 1, 2, \ldots, I_{\max}$ we then do the following.*

1. *Run the following filter by starting from $\mathbf{m}^{(i)}_0 = \mathbf{m}_0$ and $\mathbf{P}^{(i)}_0 = \mathbf{P}_0$ and by performing the following predictions and updates for $k = 1, \ldots, T$:*

$$
\begin{aligned}
\mathbf{m}^{-(i)}_k &= \mathbf{f}(\mathbf{x}^{(i-1)}_{k-1}) + \mathbf{F}_{\mathbf{x}}(\mathbf{x}^{(i-1)}_{k-1})\,(\mathbf{m}^{(i)}_{k-1} - \mathbf{x}^{(i-1)}_{k-1}), \\
\mathbf{P}^{-(i)}_k &= \mathbf{F}_{\mathbf{x}}(\mathbf{x}^{(i-1)}_{k-1})\,\mathbf{P}^{(i)}_{k-1}\,\mathbf{F}^{\mathsf{T}}_{\mathbf{x}}(\mathbf{x}^{(i-1)}_{k-1}) + \mathbf{Q}_{k-1}, \\
\mathbf{v}^{(i)}_k &= \mathbf{y}_k - \mathbf{h}(\mathbf{x}^{(i-1)}_k) - \mathbf{H}_{\mathbf{x}}(\mathbf{x}^{(i-1)}_k)\,(\mathbf{m}^{-(i)}_k - \mathbf{x}^{(i-1)}_k), \\
\mathbf{S}^{(i)}_k &= \mathbf{H}_{\mathbf{x}}(\mathbf{x}^{(i-1)}_k)\,\mathbf{P}^{-(i)}_k\,\mathbf{H}^{\mathsf{T}}_{\mathbf{x}}(\mathbf{x}^{(i-1)}_k) + \mathbf{R}_k, \\
\mathbf{K}^{(i)}_k &= \mathbf{P}^{-(i)}_k\,\mathbf{H}^{\mathsf{T}}_{\mathbf{x}}(\mathbf{x}^{(i-1)}_k)\,[\mathbf{S}^{(i)}_k]^{-1}, \\
\mathbf{m}^{(i)}_k &= \mathbf{m}^{-(i)}_k + \mathbf{K}^{(i)}_k\,\mathbf{v}^{(i)}_k, \\
\mathbf{P}^{(i)}_k &= \mathbf{P}^{-(i)}_k - \mathbf{K}^{(i)}_k\,\mathbf{S}^{(i)}_k\,[\mathbf{K}^{(i)}_k]^{\mathsf{T}}.
\end{aligned}
\tag{13.23}
$$

2. *Run the following smoother by starting from $\mathbf{m}^{\mathrm{s},(i)}_T = \mathbf{m}_T$ and $\mathbf{P}^{\mathrm{s},(i)}_T = \mathbf{P}_T$ and performing the following smoothing steps for $k = T-1, \ldots, 0$:*

$$
\begin{aligned}
\mathbf{m}^{-(i)}_{k+1} &= \mathbf{f}(\mathbf{x}^{(i-1)}_k) + \mathbf{F}_{\mathbf{x}}(\mathbf{x}^{(i-1)}_k)\,(\mathbf{m}^{(i)}_k - \mathbf{x}^{(i-1)}_k), \\
\mathbf{P}^{-(i)}_{k+1} &= \mathbf{F}_{\mathbf{x}}(\mathbf{x}^{(i-1)}_k)\,\mathbf{P}^{(i)}_k\,\mathbf{F}^{\mathsf{T}}_{\mathbf{x}}(\mathbf{x}^{(i-1)}_k) + \mathbf{Q}_k, \\
\mathbf{G}^{(i)}_k &= \mathbf{P}^{(i)}_k\,\mathbf{F}^{\mathsf{T}}_{\mathbf{x}}(\mathbf{x}^{(i-1)}_k)\,[\mathbf{P}^{-(i)}_{k+1}]^{-1}, \\
\mathbf{m}^{\mathrm{s},(i)}_k &= \mathbf{m}^{(i)}_k + \mathbf{G}^{(i)}_k\,[\mathbf{m}^{\mathrm{s},(i)}_{k+1} - \mathbf{m}^{-(i)}_{k+1}], \\
\mathbf{P}^{\mathrm{s},(i)}_k &= \mathbf{P}^{(i)}_k + \mathbf{G}^{(i)}_k\,[\mathbf{P}^{\mathrm{s},(i)}_{k+1} - \mathbf{P}^{-(i)}_{k+1}]\,[\mathbf{G}^{(i)}_k]^{\mathsf{T}}.
\end{aligned}
\tag{13.24}
$$

3. *Form the next iterate as $\mathbf{x}^{(i)}_{0:T} = \mathbf{m}^{\mathrm{s},(i)}_{0:T}$ and go to step 1.*

Above, $\mathbf{F}_{\mathbf{x}}$ and $\mathbf{H}_{\mathbf{x}}$ denote the Jacobians of $\mathbf{f}$ and $\mathbf{h}$, respectively, defined as in Equations (7.28) and (7.29). After the iterations, the final smoother means and covariances can be approximated as

$$
\begin{aligned}
\mathbf{m}^{\mathrm{s}}_{0:T} &= \mathbf{m}^{\mathrm{s},(I_{\max})}_{0:T}, \\
\mathbf{P}^{\mathrm{s}}_{0:T} &= \mathbf{P}^{\mathrm{s},(I_{\max})}_{0:T}.
\end{aligned}
\tag{13.25}
$$

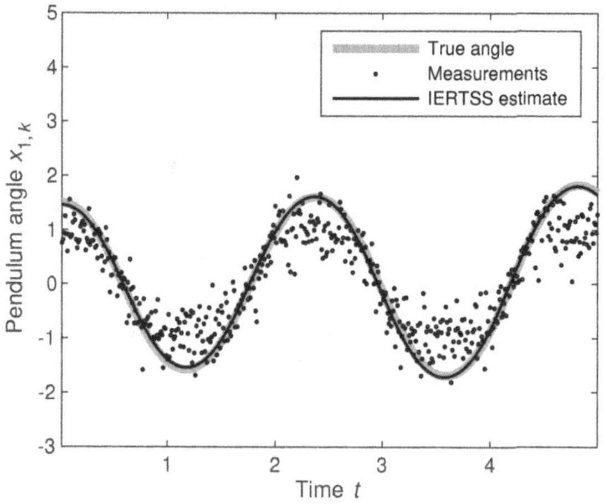

Figure 13.3 Simulated pendulum data and the result of tracking the pendulum described in Example 7.6 with the IERTSS (see Example 13.6). The resulting RMSE is 0.03 (recall that the RMSE of ERTSS was 0.06, and the RMSE of EKF was 0.17).

Because the IERTSS is an instance of the Gauss–Newton algorithm, it can be shown to converge in well-defined conditions (Bell, 1994), and the convergence is to the MAP estimate of the trajectory. This is in contrast to the IEKF (Algorithm 7.9), which only does local Gauss–Newton iteration at the update step and hence does not converge to the global MAP estimate. Furthermore, as mentioned in the IERTSS algorithm description, one way to initialize the iterations is by using an ERTSS estimate from Algorithm 13.1. However, the initialization only affects the convergence speed of the method, and the point of convergence (if the algorithm convergences) will be the MAP estimate regardless of the initial trajectory guess.

Example 13.6 (Pendulum tracking with IERTSS). *The result of applying the IERTSS to the pendulum model in Example 7.6 is shown in Figure 13.3. The resulting RMSE was* 0.03*, which is much lower than the error of the EKF, which was* 0.17*. It is also lower than the errors of any other filters, which were in the range* 0.10–0.17 *and lower than the ERTSS, which had error of* 0.06.

Example 13.7 (Coordinated turn model with IERTSS). *Figure 13.4 shows the result of the IERTSS on the coordinated turn model problem considered*

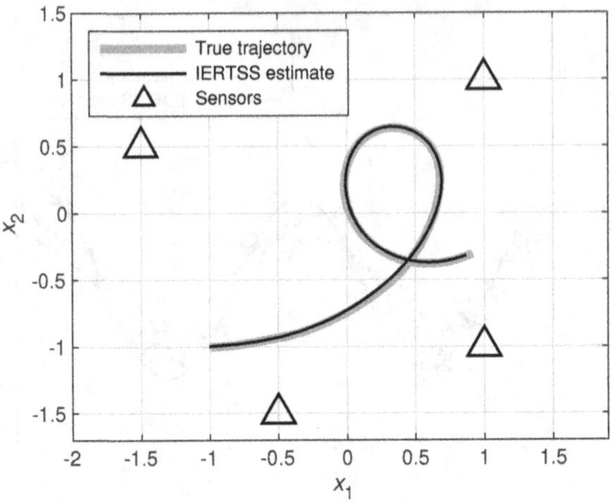

Figure 13.4 The result of applying IERTSS to simulated data from the (polar) coordinated turn model in Example 7.11 when using an inaccurate initial guess (see Example 13.7). The resulting RMSE was 0.015, while for the plain ERTSS it is 0.017.

in Example 7.11. The iteration reduces the error from 0.017 *of the ERTSS to* 0.015*. We also ran the IERTSS on the more accurate initial guess data in Example 13.3, and the resulting error was the same* 0.015*.*

13.4 Levenberg–Marquardt and Line-Search IERTSSs

Similarly to the IEKF case in Section 7.5 it is also possible to improve the convergence of the Gauss–Newton method in IERTSS by replacing it with the Levenberg–Marquardt method or by including a line-search procedure into it. These kinds of extensions have been presented in Särkkä and Svensson (2020). Although we are not giving details here, the basic ideas are:

- *The Levenberg–Marquardt* method can be implemented by modifying the filter portion in the IERTSS iteration to have a (pseudo-) measurement of the previous state iterate $\mathbf{x}_k^{(i-1)}$ at each time step k separately with a suitably defined covariance matrix.
- *Line-search* can be implemented by modifying the computation of the new iterate $\mathbf{x}_{0:T}^{(i)}$ into the following:

$$\mathbf{x}_{0:T}^{(i)} = \mathbf{x}_{0:T}^{(i-1)} + \gamma \, (\mathbf{m}_{0:T}^{\mathrm{s},(i)} - \mathbf{x}_{0:T}^{(i-1)}), \tag{13.26}$$

where the step size γ is chosen to minimize $L(\mathbf{x}_{0:T}^{(i-1)} + \gamma\,(\mathbf{m}_{0:T}^{\mathrm{s},(i)} - \mathbf{x}_{0:T}^{(i-1)}))$ where L is as defined in Equation (13.17). This minimization does not necessarily need to be exact (Särkkä and Svensson, 2020).

13.5 Exercises

13.1 Derive and implement the extended RTS smoother to the model in Exercise 7.1, and compare the errors of the filters and smoothers.

13.2 Derive and implement the second order extended RTS smoother to the model in Exercise 7.1, and compare the errors of the filters and smoothers.

13.3 Implement the extended RTS smoother to the bearings-only target tracking problem in Exercise 7.2. Note that even though the full model is non-linear, due to the linear dynamic model the smoother is linear.

13.4 Implement extended RTS for the Cartesian coordinated turn model in Equations (4.61) and (4.90) with the linear position measurement model.

13.5 Show that single iteration of the iterated extended RTS smoother corresponds to the non-iterated extended RTS smoother if the initial guess is the EKF result.

13.6 Derive and implement the iterated extended RTS smoother to the model in Exercise 7.1, and compare the errors of the filters and smoothers.

13.7 Implement the iterated extended RTS smoother to the bearings-only target tracking problem in Exercise 7.2.

13.8 Implement the iterated extended RTS for the Cartesian coordinated turn model in Equations (4.61) and (4.90) with the linear position measurement model.

14

General Gaussian Smoothing

In Chapter 12 we encountered the general Bayesian smoothing equations as well as the Rauch–Tung–Striebel (RTS) smoother for linear Gaussian (and affine) models. In Chapter 13 we then extended the RTS smoothers to non-linear models by using Taylor series expansions. These extensions followed the path from Kalman filters to extended Kalman filters, which we saw in Chapters 6 and 7. In the filtering case we then proceeded to Gaussian filters (Chapter 8), enabling approximations (Chapter 9), and posterior linearization (Chapter 10).

The principles used to develop Gaussian filters can also be used to obtain Gaussian smoothers. The aim of this chapter is to directly proceed from the Taylor series expansion-based smoothers in Chapter 13 to all general Gaussian smoothers, including the enabling approximation and posterior linearization smoothers. In particular, the moment matching approach used in Chapter 8, the statistical linear regression described in Chapter 9, and the posterior linearization presented in Chapter 10 can all be used to develop Gaussian forward-backward smoothers by leveraging the RTS equations. In this chapter, we introduce these Gaussian smoothers and explain how they can be combined with sigma-point methods to obtain practical algorithms.

14.1 General Gaussian Rauch–Tung–Striebel Smoother

The Gaussian moment matching described in Section 8.1 can be used in smoothers in an analogous manner to the Gaussian filters in Section 8.2. If we follow the extended RTS smoother derivation in Section 13.1, we get the following algorithm, which originally appears in Särkkä and Hartikainen (2010a).

Algorithm 14.1 (Gaussian RTS smoother I). *The equations of the additive form* Gaussian RTS smoother *are the following:*

$$
\begin{aligned}
\mathbf{m}_{k+1}^- &= \int \mathbf{f}(\mathbf{x}_k)\, \mathrm{N}(\mathbf{x}_k \mid \mathbf{m}_k, \mathbf{P}_k)\, \mathrm{d}\mathbf{x}_k, \\
\mathbf{P}_{k+1}^- &= \int [\mathbf{f}(\mathbf{x}_k) - \mathbf{m}_{k+1}^-]\, [\mathbf{f}(\mathbf{x}_k) - \mathbf{m}_{k+1}^-]^\mathsf{T}\, \mathrm{N}(\mathbf{x}_k \mid \mathbf{m}_k, \mathbf{P}_k)\, \mathrm{d}\mathbf{x}_k + \mathbf{Q}_k, \\
\mathbf{D}_{k+1} &= \int [\mathbf{x}_k - \mathbf{m}_k]\, [\mathbf{f}(\mathbf{x}_k) - \mathbf{m}_{k+1}^-]^\mathsf{T}\, \mathrm{N}(\mathbf{x}_k \mid \mathbf{m}_k, \mathbf{P}_k)\, \mathrm{d}\mathbf{x}_k, \\
\mathbf{G}_k &= \mathbf{D}_{k+1}\, [\mathbf{P}_{k+1}^-]^{-1}, \\
\mathbf{m}_k^\mathrm{s} &= \mathbf{m}_k + \mathbf{G}_k\, (\mathbf{m}_{k+1}^\mathrm{s} - \mathbf{m}_{k+1}^-), \\
\mathbf{P}_k^\mathrm{s} &= \mathbf{P}_k + \mathbf{G}_k\, (\mathbf{P}_{k+1}^\mathrm{s} - \mathbf{P}_{k+1}^-)\, \mathbf{G}_k^\mathsf{T}.
\end{aligned}
\tag{14.1}
$$

Derivation Assume that the approximate means and covariances of the filtering distributions are available:

$$
p(\mathbf{x}_k \mid \mathbf{y}_{1:k}) \simeq \mathrm{N}(\mathbf{x}_k \mid \mathbf{m}_k, \mathbf{P}_k),
$$

and the smoothing distribution of time step $k+1$ is known and approximately Gaussian:

$$
p(\mathbf{x}_{k+1} \mid \mathbf{y}_{1:T}) \simeq \mathrm{N}(\mathbf{x}_{k+1} \mid \mathbf{m}_{k+1}^\mathrm{s}, \mathbf{P}_{k+1}^\mathrm{s}).
$$

We can now derive moment matching-based Gaussian approximation to the smoothing distribution at step k as follows.

1. In similar way to the derivation of the Gaussian filter in Section 8.2, we apply the moment matching in Algorithm 8.1 to $\mathbf{x}_{k+1} = \mathbf{f}(\mathbf{x}_k) + \mathbf{q}_k$ with $\mathbf{x}_k \sim \mathrm{N}(\mathbf{m}_k, \mathbf{P}_k)$, to get
$$
p(\mathbf{x}_k, \mathbf{x}_{k+1} \mid \mathbf{y}_{1:k}) \simeq \mathrm{N}\left(\begin{pmatrix} \mathbf{x}_k \\ \mathbf{x}_{k+1} \end{pmatrix} \,\middle|\, \begin{pmatrix} \mathbf{m}_k \\ \mathbf{m}_{k+1}^- \end{pmatrix}, \begin{pmatrix} \mathbf{P}_k & \mathbf{D}_{k+1} \\ \mathbf{D}_{k+1}^\mathsf{T} & \mathbf{P}_{k+1}^- \end{pmatrix} \right),
\tag{14.2}
$$
where $\mathbf{m}_{k+1}^-$, $\mathbf{P}_{k+1}^-$, and $\mathbf{D}_{k+1}$ are as defined in Equation (14.1).
2. Because the distribution (14.2) is Gaussian, by the computation rules of Gaussian distributions, the conditional distribution of $\mathbf{x}_k$ is given as
$$
p(\mathbf{x}_k \mid \mathbf{x}_{k+1}, \mathbf{y}_{1:T}) \simeq \mathrm{N}(\mathbf{x}_k \mid \tilde{\mathbf{m}}_2, \tilde{\mathbf{P}}_2),
$$
where
$$
\begin{aligned}
\mathbf{G}_k &= \mathbf{D}_{k+1}\, [\mathbf{P}_{k+1}^-]^{-1}, \\
\tilde{\mathbf{m}}_2 &= \mathbf{m}_k + \mathbf{G}_k (\mathbf{x}_{k+1} - \mathbf{m}_{k+1}^-), \\
\tilde{\mathbf{P}}_2 &= \mathbf{P}_k - \mathbf{G}_k\, \mathbf{P}_{k+1}^-\, \mathbf{G}_k^\mathsf{T}.
\end{aligned}
$$

3. The rest of the derivation is completely analogous to the derivation of the ERTSS in Section 13.1.

□

The integrals above can be approximated using analogous numerical integration or analytical approximation schemes as in the filtering case, that is, with Gauss–Hermite cubatures (Ito and Xiong, 2000; Wu et al., 2006), spherical cubature rules (McNamee and Stenger, 1967; Arasaratnam and Haykin, 2009), or with many other numerical integration schemes. In the non-additive case, the Gaussian smoother becomes the following (Särkkä and Hartikainen, 2010a).

Algorithm 14.2 (Gaussian RTS smoother II). *The equations of the non-additive form* Gaussian RTS smoother *are the following:*

$$\begin{aligned}
\mathbf{m}_{k+1}^{-} &= \int \mathbf{f}(\mathbf{x}_k, \mathbf{q}_k)\, \mathrm{N}(\mathbf{x}_k \mid \mathbf{m}_k, \mathbf{P}_k)\, \mathrm{N}(\mathbf{q}_k \mid \mathbf{0}, \mathbf{Q}_k)\, \mathrm{d}\mathbf{x}_k\, \mathrm{d}\mathbf{q}_k, \\
\mathbf{P}_{k+1}^{-} &= \int [\mathbf{f}(\mathbf{x}_k, \mathbf{q}_k) - \mathbf{m}_{k+1}^{-}]\, [\mathbf{f}(\mathbf{x}_k, \mathbf{q}_k) - \mathbf{m}_{k+1}^{-}]^{\mathsf{T}} \\
&\qquad \times \mathrm{N}(\mathbf{x}_k \mid \mathbf{m}_k, \mathbf{P}_k)\, \mathrm{N}(\mathbf{q}_k \mid \mathbf{0}, \mathbf{Q}_k)\, \mathrm{d}\mathbf{x}_k\, \mathrm{d}\mathbf{q}_k, \\
\mathbf{D}_{k+1} &= \int [\mathbf{x}_k - \mathbf{m}_k]\, [\mathbf{f}(\mathbf{x}_k, \mathbf{q}_k) - \mathbf{m}_{k+1}^{-}]^{\mathsf{T}} \\
&\qquad \times \mathrm{N}(\mathbf{x}_k \mid \mathbf{m}_k, \mathbf{P}_k)\, \mathrm{N}(\mathbf{q}_k \mid \mathbf{0}, \mathbf{Q}_k)\, \mathrm{d}\mathbf{x}_k\, \mathrm{d}\mathbf{q}_k, \\
\mathbf{G}_k &= \mathbf{D}_{k+1}\, [\mathbf{P}_{k+1}^{-}]^{-1}, \\
\mathbf{m}_k^{\mathrm{s}} &= \mathbf{m}_k + \mathbf{G}_k\, (\mathbf{m}_{k+1}^{\mathrm{s}} - \mathbf{m}_{k+1}^{-}), \\
\mathbf{P}_k^{\mathrm{s}} &= \mathbf{P}_k + \mathbf{G}_k\, (\mathbf{P}_{k+1}^{\mathrm{s}} - \mathbf{P}_{k+1}^{-})\, \mathbf{G}_k^{\mathsf{T}}.
\end{aligned} \tag{14.3}$$

As in the Gaussian filtering case, the above algorithms are mainly theoretical, because the integrals can be solved in closed form only in special cases. Fixed-point and fixed-lag (as opposed to fixed-interval) versions of the algorithms above can also be found in Särkkä and Hartikainen (2010a).

14.2 Gauss–Hermite Rauch–Tung–Striebel Smoother

By using the Gauss–Hermite cubature integration approximation from Section 8.3 in the additive form Gaussian RTS smoother, we get the following Gauss–Hermite Rauch–Tung–Striebel smoother (GHRTSS) algorithm.

Algorithm 14.3 (Gauss–Hermite Rauch–Tung–Striebel smoother). *The* additive form Gauss–Hermite RTS smoother algorithm *is the following.*

1. *Form the sigma points as*

$$\mathcal{X}_k^{(i_1,\ldots,i_n)} = \mathbf{m}_k + \sqrt{\mathbf{P}_k}\,\boldsymbol{\xi}^{(i_1,\ldots,i_n)}, \qquad i_1,\ldots,i_n = 1,\ldots,p, \quad (14.4)$$

where the unit sigma points $\boldsymbol{\xi}^{(i_1,\ldots,i_n)}$ *were defined in Equation* (8.30).

2. *Propagate the sigma points through the dynamic model:*

$$\hat{\mathcal{X}}_{k+1}^{(i_1,\ldots,i_n)} = \mathbf{f}(\mathcal{X}_k^{(i_1,\ldots,i_n)}), \quad i_1,\ldots,i_n = 1,\ldots,p. \quad (14.5)$$

3. *Compute the predicted mean* $\mathbf{m}_{k+1}^-$, *the predicted covariance* $\mathbf{P}_{k+1}^-$, *and the cross-covariance* $\mathbf{D}_{k+1}$*:*

$$\begin{aligned}
\mathbf{m}_{k+1}^- &= \sum_{i_1,\ldots,i_n} W_{i_1,\ldots,i_n} \hat{\mathcal{X}}_{k+1}^{(i_1,\ldots,i_n)}, \\
\mathbf{P}_{k+1}^- &= \sum_{i_1,\ldots,i_n} W_{i_1,\ldots,i_n} (\hat{\mathcal{X}}_{k+1}^{(i_1,\ldots,i_n)} - \mathbf{m}_{k+1}^-)\,(\hat{\mathcal{X}}_{k+1}^{(i_1,\ldots,i_n)} - \mathbf{m}_{k+1}^-)^\mathsf{T} + \mathbf{Q}_k, \\
\mathbf{D}_{k+1} &= \sum_{i_1,\ldots,i_n} W_{i_1,\ldots,i_n} (\mathcal{X}_k^{(i_1,\ldots,i_n)} - \mathbf{m}_k)\,(\hat{\mathcal{X}}_{k+1}^{(i_1,\ldots,i_n)} - \mathbf{m}_{k+1}^-)^\mathsf{T},
\end{aligned} \quad (14.6)$$

where the weights $W_{i_1,\ldots,i_n}$ *were defined in Equation* (8.29).

4. *Compute the gain* $\mathbf{G}_k$, *mean* $\mathbf{m}_k^s$ *and covariance* $\mathbf{P}_k^s$ *as follows:*

$$\begin{aligned}
\mathbf{G}_k &= \mathbf{D}_{k+1}\,[\mathbf{P}_{k+1}^-]^{-1}, \\
\mathbf{m}_k^s &= \mathbf{m}_k + \mathbf{G}_k\,(\mathbf{m}_{k+1}^s - \mathbf{m}_{k+1}^-), \\
\mathbf{P}_k^s &= \mathbf{P}_k + \mathbf{G}_k\,(\mathbf{P}_{k+1}^s - \mathbf{P}_{k+1}^-)\,\mathbf{G}_k^\mathsf{T}.
\end{aligned} \quad (14.7)$$

It would also be possible to formulate a non-additive version of the above smoother analogously, but due to unpleasant exponential computational scaling of the Gauss–Hermite cubature method in the state dimension, that extension is not very useful in practice. Recall that the state dimension doubles when using the non-additive transform, because we need to integrate over the state and process noise jointly.

Example 14.4 (Pendulum tracking with GHRTSS). *The result of applying GHRTSS to the pendulum model in Example 7.6 is shown in Figure 14.1. The resulting RMSE error is* 0.04, *which is between the ERTSS and IERTSS errors (which were* 0.06 *and* 0.03) *and significantly lower than the GHKF error of* 0.10.

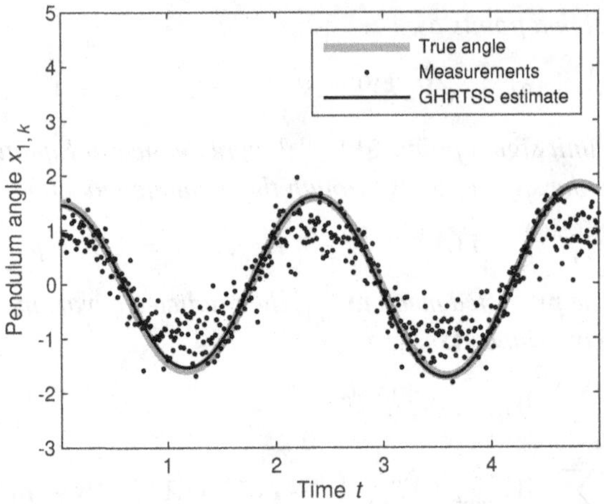

Figure 14.1 Simulated pendulum data and the result of tracking the pendulum described in Example 7.6 with the GHRTSS. The resulting RMSE is 0.04, which is lower than that of ERTSS (0.06) but higher than that of IERTSS (0.03). Recall that GHKF had an RMSE of 0.10.

14.3 Cubature Rauch–Tung–Striebel Smoother

By using the third order spherical cubature approximation (Section 8.5) to the additive form Gaussian RTS smoother, we get the following cubature Rauch–Tung–Striebel smoother (CRTSS) algorithm (see Arasaratnam and Haykin, 2011).

Algorithm 14.5 (Cubature Rauch–Tung–Striebel smoother I). *The* additive form cubature RTS smoother algorithm *is the following.*

1. *Form the sigma points:*

$$\mathcal{X}_k^{(i)} = \mathbf{m}_k + \sqrt{\mathbf{P}_k}\, \boldsymbol{\xi}^{(i)}, \qquad i = 1, \ldots, 2n, \tag{14.8}$$

where the unit sigma points are defined as

$$\boldsymbol{\xi}^{(i)} = \begin{cases} \sqrt{n}\, \mathbf{e}_i, & i = 1, \ldots, n, \\ -\sqrt{n}\, \mathbf{e}_{i-n}, & i = n+1, \ldots, 2n, \end{cases} \tag{14.9}$$

where $\mathbf{e}_i$ *denotes a unit vector in the direction of the coordinate axis* i*.*

2. *Propagate the sigma points through the dynamic model:*

$$\hat{\mathcal{X}}_{k+1}^{(i)} = \mathbf{f}(\mathcal{X}_k^{(i)}), \quad i = 1, \ldots, 2n.$$

3. *Compute the predicted mean* $\mathbf{m}_{k+1}^-$, *the predicted covariance* $\mathbf{P}_{k+1}^-$, *and the cross-covariance* $\mathbf{D}_{k+1}$*:*

$$\begin{aligned}
\mathbf{m}_{k+1}^- &= \frac{1}{2n} \sum_{i=1}^{2n} \hat{\mathcal{X}}_{k+1}^{(i)}, \\
\mathbf{P}_{k+1}^- &= \frac{1}{2n} \sum_{i=1}^{2n} (\hat{\mathcal{X}}_{k+1}^{(i)} - \mathbf{m}_{k+1}^-) (\hat{\mathcal{X}}_{k+1}^{(i)} - \mathbf{m}_{k+1}^-)^\mathsf{T} + \mathbf{Q}_k, \\
\mathbf{D}_{k+1} &= \frac{1}{2n} \sum_{i=1}^{2n} (\mathcal{X}_k^{(i)} - \mathbf{m}_k) (\hat{\mathcal{X}}_{k+1}^{(i)} - \mathbf{m}_{k+1}^-)^\mathsf{T}.
\end{aligned} \tag{14.10}$$

4. *Compute the gain* $\mathbf{G}_k$, *mean* $\mathbf{m}_k^\mathrm{s}$, *and covariance* $\mathbf{P}_k^\mathrm{s}$ *as follows:*

$$\begin{aligned}
\mathbf{G}_k &= \mathbf{D}_{k+1} \, [\mathbf{P}_{k+1}^-]^{-1}, \\
\mathbf{m}_k^\mathrm{s} &= \mathbf{m}_k + \mathbf{G}_k \, (\mathbf{m}_{k+1}^\mathrm{s} - \mathbf{m}_{k+1}^-), \\
\mathbf{P}_k^\mathrm{s} &= \mathbf{P}_k + \mathbf{G}_k \, (\mathbf{P}_{k+1}^\mathrm{s} - \mathbf{P}_{k+1}^-) \, \mathbf{G}_k^\mathsf{T}.
\end{aligned} \tag{14.11}$$

By using the third order spherical cubature approximation to the non-additive form Gaussian RTS smoother, we get the following algorithm.

Algorithm 14.6 (Cubature Rauch–Tung–Striebel smoother II)**.** *A single step of the* non-additive augmented form cubature RTS smoother *is as follows.*

1. *Form the sigma points for the* $n' = n + n_q$*-dimensional augmented random variable* $(\mathbf{x}_k, \mathbf{q}_k)$*:*

$$\tilde{\mathcal{X}}_k^{(i)} = \tilde{\mathbf{m}}_k + \sqrt{\tilde{\mathbf{P}}_k} \, \boldsymbol{\xi}^{(i)'}, \qquad i = 1, \ldots, 2n', \tag{14.12}$$

where

$$\tilde{\mathbf{m}}_k = \begin{pmatrix} \mathbf{m}_k \\ \mathbf{0} \end{pmatrix}, \qquad \tilde{\mathbf{P}}_k = \begin{pmatrix} \mathbf{P}_k & \mathbf{0} \\ \mathbf{0} & \mathbf{Q}_k \end{pmatrix}.$$

2. *Propagate the sigma points through the dynamic model:*

$$\hat{\mathcal{X}}_{k+1}^{(i)} = \mathbf{f}\,(\tilde{\mathcal{X}}_k^{(i),x}, \tilde{\mathcal{X}}_k^{(i),q}), \quad i = 1, \ldots, 2n',$$

where $\tilde{\mathcal{X}}_k^{(i),x}$ *and* $\tilde{\mathbf{X}}_k^{(i),q}$ *denote the parts of the augmented sigma point* i *that correspond to* $\mathbf{x}_k$ *and* $\mathbf{q}_k$*, respectively.*

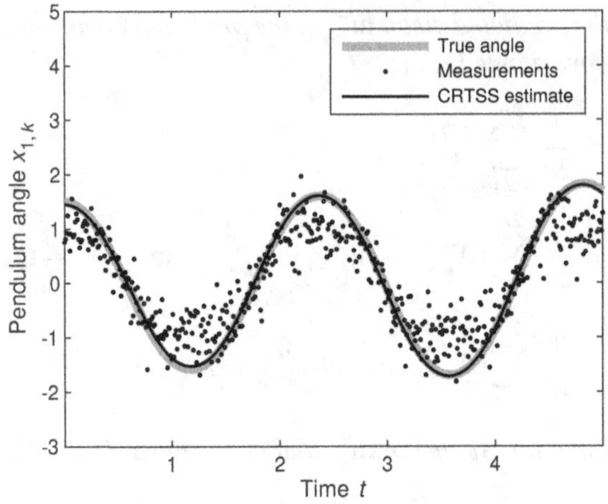

Figure 14.2 Simulated pendulum data and the result of tracking the pendulum described in Example 7.6 with the CRTSS. The resulting RMSE is 0.04, which is similar as that of GHRTSS.

3. *Compute the predicted mean* $\mathbf{m}_{k+1}^-$*, the predicted covariance* $\mathbf{P}_{k+1}^-$*, and the cross-covariance* $\mathbf{D}_{k+1}$*:*

$$
\begin{aligned}
\mathbf{m}_{k+1}^- &= \frac{1}{2n'} \sum_{i=1}^{2n'} \hat{\mathcal{X}}_{k+1}^{(i)}, \\
\mathbf{P}_{k+1}^- &= \frac{1}{2n'} \sum_{i=1}^{2n'} (\hat{\mathcal{X}}_{k+1}^{(i)} - \mathbf{m}_{k+1}^-)\,(\hat{\mathcal{X}}_{k+1}^{(i)} - \mathbf{m}_{k+1}^-)^\mathsf{T}, \\
\mathbf{D}_{k+1} &= \frac{1}{2n'} \sum_{i=1}^{2n'} (\tilde{\mathcal{X}}_{k}^{(i),x} - \mathbf{m}_k)\,(\hat{\mathcal{X}}_{k+1}^{(i)} - \mathbf{m}_{k+1}^-)^\mathsf{T}.
\end{aligned}
\tag{14.13}
$$

4. *Compute the gain* $\mathbf{G}_k$*, mean* $\mathbf{m}_k^s$*, and covariance* $\mathbf{P}_k^s$*:*

$$
\begin{aligned}
\mathbf{G}_k &= \mathbf{D}_{k+1}\,[\mathbf{P}_{k+1}^-]^{-1}, \\
\mathbf{m}_k^s &= \mathbf{m}_k + \mathbf{G}_k\left(\mathbf{m}_{k+1}^s - \mathbf{m}_{k+1}^-\right), \\
\mathbf{P}_k^s &= \mathbf{P}_k + \mathbf{G}_k\left(\mathbf{P}_{k+1}^s - \mathbf{P}_{k+1}^-\right)\mathbf{G}_k^\mathsf{T}.
\end{aligned}
\tag{14.14}
$$

Example 14.7 (Pendulum tracking with CRTSS). *The result of applying the CRTSS for the pendulum model in Example 7.6 is shown in Figure 14.2. The error* 0.04 *and the overall result are similar to GHRTSS.*

14.4 Unscented Rauch–Tung–Striebel Smoother

The *unscented Rauch–Tung–Striebel smoother* (URTSS, Särkkä, 2006; Šimandl and Duník, 2006; Särkkä, 2008) is a Gaussian approximation-based smoother where the non-linearity is approximated using the unscented transform, which is a numerical integration solution to the Gaussian moment matching integrals (recall Sections 8.7 and 8.8). The smoother equations for the additive model (7.39) are given as follows.

Algorithm 14.8 (Unscented Rauch–Tung–Striebel smoother I)**.** *The* additive form unscented RTS smoother algorithm *is the following.*

1. *Form the sigma points:*

$$\begin{aligned} \mathcal{X}_k^{(0)} &= \mathbf{m}_k, \\ \mathcal{X}_k^{(i)} &= \mathbf{m}_k + \sqrt{n+\lambda}\,\left[\sqrt{\mathbf{P}_k}\right]_i, \\ \mathcal{X}_k^{(i+n)} &= \mathbf{m}_k - \sqrt{n+\lambda}\,\left[\sqrt{\mathbf{P}_k}\right]_i, \quad i = 1,\dots,n, \end{aligned} \tag{14.15}$$

where the parameter λ *was defined in Equation* (8.74).

2. *Propagate the sigma points through the dynamic model:*

$$\hat{\mathcal{X}}_{k+1}^{(i)} = \mathbf{f}(\mathcal{X}_k^{(i)}), \quad i = 0,\dots,2n.$$

3. *Compute the predicted mean* $\mathbf{m}_{k+1}^-$*, the predicted covariance* $\mathbf{P}_{k+1}^-$*, and the cross-covariance* $\mathbf{D}_{k+1}$*:*

$$\begin{aligned} \mathbf{m}_{k+1}^- &= \sum_{i=0}^{2n} W_i^{(\mathrm{m})}\,\hat{\mathcal{X}}_{k+1}^{(i)}, \\ \mathbf{P}_{k+1}^- &= \sum_{i=0}^{2n} W_i^{(\mathrm{c})}\,(\hat{\mathcal{X}}_{k+1}^{(i)} - \mathbf{m}_{k+1}^-)\,(\hat{\mathcal{X}}_{k+1}^{(i)} - \mathbf{m}_{k+1}^-)^\mathsf{T} + \mathbf{Q}_k, \\ \mathbf{D}_{k+1} &= \sum_{i=0}^{2n} W_i^{(\mathrm{c})}\,(\mathcal{X}_k^{(i)} - \mathbf{m}_k)\,(\hat{\mathcal{X}}_{k+1}^{(i)} - \mathbf{m}_{k+1}^-)^\mathsf{T}, \end{aligned} \tag{14.16}$$

where the weights were defined in Equation (8.75).

4. *Compute the smoother gain* $\mathbf{G}_k$*, the smoothed mean* $\mathbf{m}_k^\mathrm{s}$*, and the covariance* $\mathbf{P}_k^\mathrm{s}$ *as follows:*

$$\begin{aligned} \mathbf{G}_k &= \mathbf{D}_{k+1}\,[\mathbf{P}_{k+1}^-]^{-1}, \\ \mathbf{m}_k^\mathrm{s} &= \mathbf{m}_k + \mathbf{G}_k\,(\mathbf{m}_{k+1}^\mathrm{s} - \mathbf{m}_{k+1}^-), \\ \mathbf{P}_k^\mathrm{s} &= \mathbf{P}_k + \mathbf{G}_k\,(\mathbf{P}_{k+1}^\mathrm{s} - \mathbf{P}_{k+1}^-)\,\mathbf{G}_k^\mathsf{T}. \end{aligned} \tag{14.17}$$

The above computations are started from the filtering result of the last time step $\mathbf{m}_T^{\mathrm{s}} = \mathbf{m}_T$, $\mathbf{P}_T^{\mathrm{s}} = \mathbf{P}_T$, *and the recursion runs backward for* $k = T-1,\ldots,0$.

Derivation The derivation is the same as of the general Gaussian smoother in Algorithm 14.1 except that the moment matching approximation is replaced with the unscented transform. □

It is easy to see that the CRTSS in Algorithm 14.5 is indeed a special case of the URTSS method with parameters $\alpha = 1$, $\beta = 0$, $\kappa = 0$. However, that particular selection of parameters tends to work well in practice, and due to the simplicity of the sigma-point and weight selection rules, the method is very simple to implement.

The corresponding augmented version of the smoother for non-additive models of the form (7.39) is almost the same, except that the augmented UT in Algorithm 8.16 is used instead of the additive UT in Algorithm 8.15. The smoother can be formulated as follows (Särkkä, 2008).

Algorithm 14.9 (Unscented Rauch–Tung–Striebel smoother II). *A single step of the* non-additive augmented form unscented RTS smoother *for non-additive models is as follows.*

1. *Form the sigma points for the* $n' = n + n_q$*-dimensional augmented random variable* $(\mathbf{x}_k, \mathbf{q}_k)$:

$$\begin{aligned} \tilde{\mathcal{X}}_k^{(0)} &= \tilde{\mathbf{m}}_k, \\ \tilde{\mathcal{X}}_k^{(i)} &= \tilde{\mathbf{m}}_k + \sqrt{n' + \lambda'} \left[\sqrt{\tilde{\mathbf{P}}_k} \right]_i, \\ \tilde{\mathcal{X}}_k^{(i+n')} &= \tilde{\mathbf{m}}_k - \sqrt{n' + \lambda'} \left[\sqrt{\tilde{\mathbf{P}}_k} \right]_i, \quad i = 1,\ldots,n', \end{aligned} \tag{14.18}$$

where

$$\tilde{\mathbf{m}}_k = \begin{pmatrix} \mathbf{m}_k \\ \mathbf{0} \end{pmatrix}, \qquad \tilde{\mathbf{P}}_k = \begin{pmatrix} \mathbf{P}_k & \mathbf{0} \\ \mathbf{0} & \mathbf{Q}_k \end{pmatrix}.$$

2. *Propagate the sigma points through the dynamic model:*

$$\hat{\mathcal{X}}_{k+1}^{(i)} = \mathbf{f}(\tilde{\mathcal{X}}_k^{(i),x}, \tilde{\mathcal{X}}_k^{(i),q}), \quad i = 0,\ldots,2n',$$

where $\tilde{\mathcal{X}}_k^{(i),x}$ *and* $\tilde{\mathcal{X}}_k^{(i),q}$ *denote the parts of the augmented sigma point* i *that correspond to* $\mathbf{x}_k$ *and* $\mathbf{q}_k$*, respectively.*

3. *Compute the predicted mean* $\mathbf{m}_{k+1}^-$, *the predicted covariance* $\mathbf{P}_{k+1}^-$, *and the cross-covariance* $\mathbf{D}_{k+1}$:

$$
\begin{aligned}
\mathbf{m}_{k+1}^- &= \sum_{i=0}^{2n'} W_i^{(\mathrm{m})'} \, \hat{\mathcal{X}}_{k+1}^{(i)}, \\
\mathbf{P}_{k+1}^- &= \sum_{i=0}^{2n'} W_i^{(\mathrm{c})'} \, (\hat{\mathcal{X}}_{k+1}^{(i)} - \mathbf{m}_{k+1}^-) \, (\hat{\mathcal{X}}_{k+1}^{(i)} - \mathbf{m}_{k+1}^-)^\mathsf{T}, \\
\mathbf{D}_{k+1} &= \sum_{i=0}^{2n'} W_i^{(\mathrm{c})'} \, (\tilde{\mathcal{X}}_k^{(i),x} - \mathbf{m}_k) \, (\hat{\mathcal{X}}_{k+1}^{(i)} - \mathbf{m}_{k+1}^-)^\mathsf{T},
\end{aligned}
\tag{14.19}
$$

where the definitions of the parameter λ' *and the weights* $W_i^{(\mathrm{m})'}$ *and* $W_i^{(\mathrm{c})'}$ *are the same as in Section 8.7.*

4. *Compute the smoother gain* $\mathbf{G}_k$, *the smoothed mean* $\mathbf{m}_k^\mathrm{s}$, *and the covariance* $\mathbf{P}_k^\mathrm{s}$:

$$
\begin{aligned}
\mathbf{G}_k &= \mathbf{D}_{k+1} \, [\mathbf{P}_{k+1}^-]^{-1}, \\
\mathbf{m}_k^\mathrm{s} &= \mathbf{m}_k + \mathbf{G}_k \left[\mathbf{m}_{k+1}^\mathrm{s} - \mathbf{m}_{k+1}^-\right], \\
\mathbf{P}_k^\mathrm{s} &= \mathbf{P}_k + \mathbf{G}_k \left[\mathbf{P}_{k+1}^\mathrm{s} - \mathbf{P}_{k+1}^-\right] \mathbf{G}_k^\mathsf{T}.
\end{aligned}
\tag{14.20}
$$

Example 14.10 (Pendulum tracking with URTSS). *The result of applying the URTSS to the pendulum model in Example 7.6 is shown in Figure 14.3. The resulting RMSE is* 0.04, *which is the same as for the GHRTSS and CRTSS.*

14.5 Higher Order Cubature/Unscented RTS Smoothers

Similarly to the filtering case discussed in Section 8.9, we can use higher order cubature or unscented rules (e.g., McNamee and Stenger, 1967; Wu et al., 2006) to construct approximate Gaussian smoothers. For example, using the fifth order rule that we used in Algorithm 8.22, we get the following smoother.

Algorithm 14.11 (Fifth order cubature/unscented RTS smoother I). *The* additive form of the fifth order cubature/unscented RTS smoother algorithm (URTSS5) *is the following.*

1. *Form the sigma points:*

$$
\mathcal{X}_k^{(i)} = \mathbf{m}_k + \sqrt{\mathbf{P}_k} \, \boldsymbol{\xi}^{(i)}, \qquad i = 0, 1, 2, \ldots, \tag{14.21}
$$

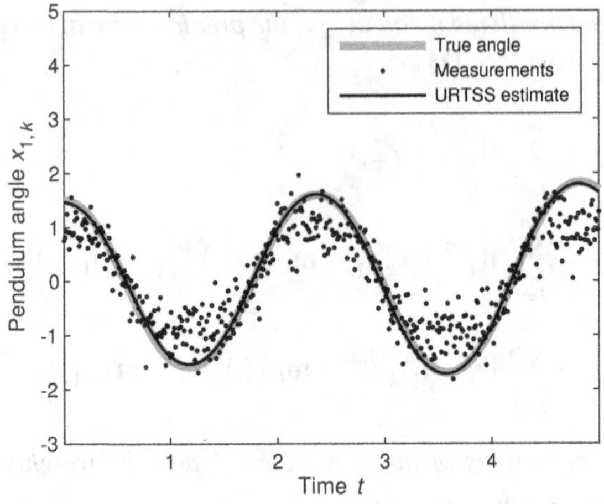

Figure 14.3 Simulated pendulum data and the result of tracking the pendulum described in Example 7.6 with the URTSS. The resulting RMSE is 0.04, which is practically the same as with the GHRTSS and CRTSS. Recall that the RMSE of the UKF was 0.10, which is significantly higher.

where the unit sigma points are defined as in Equation (8.102).

2. *Propagate the sigma points through the dynamic model:*

$$\hat{\mathcal{X}}_{k+1}^{(i)} = \mathbf{f}(\mathcal{X}_k^{(i)}), \quad i = 0, 1, 2, \ldots.$$

3. *Compute the predicted mean* $\mathbf{m}_{k+1}^-$, *the predicted covariance* $\mathbf{P}_{k+1}^-$, *and the cross-covariance* $\mathbf{D}_{k+1}$*:*

$$\begin{aligned}
\mathbf{m}_{k+1}^- &= \sum_i W_i \, \hat{\mathcal{X}}_{k+1}^{(i)}, \\
\mathbf{P}_{k+1}^- &= \sum_i W_i \, (\hat{\mathcal{X}}_{k+1}^{(i)} - \mathbf{m}_{k+1}^-) \, (\hat{\mathcal{X}}_{k+1}^{(i)} - \mathbf{m}_{k+1}^-)^\mathsf{T} + \mathbf{Q}_k, \\
\mathbf{D}_{k+1} &= \sum_i W_i \, (\mathcal{X}_k^{(i)} - \mathbf{m}_k) \, (\hat{\mathcal{X}}_{k+1}^{(i)} - \mathbf{m}_{k+1}^-)^\mathsf{T},
\end{aligned} \tag{14.22}$$

where the weights W_i *are defined as in Equation* (8.103).

4. *Compute the gain* $\mathbf{G}_k$, *mean* $\mathbf{m}_k^\mathrm{s}$, *and covariance* $\mathbf{P}_k^\mathrm{s}$ *as follows:*

$$\begin{aligned}
\mathbf{G}_k &= \mathbf{D}_{k+1} \, [\mathbf{P}_{k+1}^-]^{-1}, \\
\mathbf{m}_k^\mathrm{s} &= \mathbf{m}_k + \mathbf{G}_k \, (\mathbf{m}_{k+1}^\mathrm{s} - \mathbf{m}_{k+1}^-), \\
\mathbf{P}_k^\mathrm{s} &= \mathbf{P}_k + \mathbf{G}_k \, (\mathbf{P}_{k+1}^\mathrm{s} - \mathbf{P}_{k+1}^-) \, \mathbf{G}_k^\mathsf{T}.
\end{aligned} \tag{14.23}$$

The corresponding non-additive form is the following.

Algorithm 14.12 (Fifth order cubature/unscented RTS smoother II). *A single step of the* augmented form the fifth order cubature/unscented RTS smoother *is as follows.*

1. *Form the sigma points for the* $n' = n + n_q$*-dimensional augmented random variable* $(\mathbf{x}_k, \mathbf{q}_k)$*:*

$$\tilde{\mathcal{X}}_k^{(i)} = \tilde{\mathbf{m}}_k + \sqrt{\tilde{\mathbf{P}}_k}\, \boldsymbol{\xi}^{(i)'}, \qquad i = 0, 1, 2, \ldots, \tag{14.24}$$

where

$$\tilde{\mathbf{m}}_k = \begin{pmatrix} \mathbf{m}_k \\ \mathbf{0} \end{pmatrix}, \qquad \tilde{\mathbf{P}}_k = \begin{pmatrix} \mathbf{P}_k & \mathbf{0} \\ \mathbf{0} & \mathbf{Q}_k \end{pmatrix},$$

and the unit sigma points $\boldsymbol{\xi}^{(i)'}$ *are defined as in Equation* (8.102) *(with* $n \leftarrow n + n_q$*).*

2. *Propagate the sigma points through the dynamic model:*

$$\hat{\mathcal{X}}_{k+1}^{(i)} = \mathbf{f}(\tilde{\mathcal{X}}_k^{(i),x}, \tilde{\mathcal{X}}_k^{(i),q}), \quad i = 0, 1, 2, \ldots,$$

where $\tilde{\mathcal{X}}_k^{(i),x}$ *and* $\tilde{\mathbf{X}}_k^{(i),q}$ *denote the parts of the augmented sigma point* i *that correspond to* $\mathbf{x}_k$ *and* $\mathbf{q}_k$*, respectively.*

3. *Compute the predicted mean* $\mathbf{m}_{k+1}^-$*, the predicted covariance* $\mathbf{P}_{k+1}^-$*, and the cross-covariance* $\mathbf{D}_{k+1}$*:*

$$\begin{aligned}
\mathbf{m}_{k+1}^- &= \sum_i W_i' \, \hat{\mathcal{X}}_{k+1}^{(i)}, \\
\mathbf{P}_{k+1}^- &= \sum_i W_i' \, (\hat{\mathcal{X}}_{k+1}^{(i)} - \mathbf{m}_{k+1}^-) \, (\hat{\mathcal{X}}_{k+1}^{(i)} - \mathbf{m}_{k+1}^-)^\mathsf{T}, \\
\mathbf{D}_{k+1} &= \sum_i W_i' \, (\tilde{\mathcal{X}}_k^{(i),x} - \mathbf{m}_k) \, (\hat{\mathcal{X}}_{k+1}^{(i)} - \mathbf{m}_{k+1}^-)^\mathsf{T},
\end{aligned} \tag{14.25}$$

where the weights W_i' *are defined as in Equation* (8.103) *(with* $n \leftarrow n + n_q$*).*

4. *Compute the gain* $\mathbf{G}_k$*, mean* $\mathbf{m}_k^\mathrm{s}$*, and covariance* $\mathbf{P}_k^\mathrm{s}$*:*

$$\begin{aligned}
\mathbf{G}_k &= \mathbf{D}_{k+1} \, [\mathbf{P}_{k+1}^-]^{-1}, \\
\mathbf{m}_k^\mathrm{s} &= \mathbf{m}_k + \mathbf{G}_k \, \left(\mathbf{m}_{k+1}^\mathrm{s} - \mathbf{m}_{k+1}^-\right), \\
\mathbf{P}_k^\mathrm{s} &= \mathbf{P}_k + \mathbf{G}_k \, \left(\mathbf{P}_{k+1}^\mathrm{s} - \mathbf{P}_{k+1}^-\right) \mathbf{G}_k^\mathsf{T}.
\end{aligned} \tag{14.26}$$

Please note that the above two algorithms can be easily modified to use any (sigma-point) integration method (e.g., McNamee and Stenger, 1967; Ito and Xiong, 2000; Julier et al., 2000; Wu et al., 2006; Arasaratnam

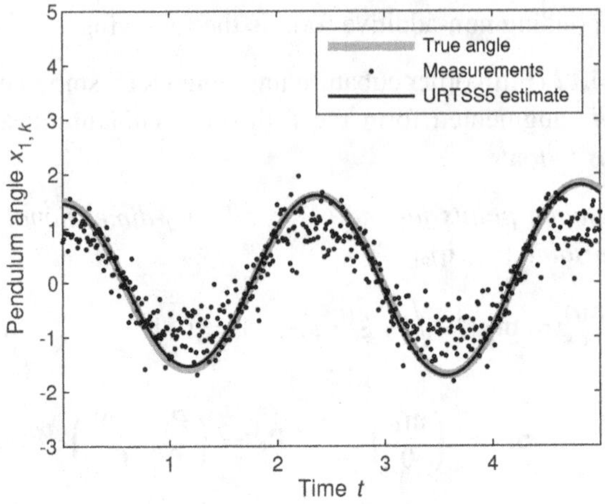

Figure 14.4 Simulated pendulum data and the result of tracking the pendulum described in Example 7.6 with the URTSS5. The resulting RMSE is 0.04, which is similar as those of the other Gaussian smoothers.

and Haykin, 2009; Särkkä et al., 2016; Karvonen et al., 2019) simply by changing the unit sigma points and weights.

Example 14.13 (Pendulum tracking with URTSS5). *The result of applying the URTSS5 to the pendulum model in Example 7.6 is shown in Figure 14.4. The error* 0.04 *and the overall result are similar to the other Gaussian integration-based smoothers that we have encountered so far.*

Example 14.14 (Coordinated turn model with Gaussian smoothers). *We applied numerical integration-based Gaussian smoothers to the coordinated turn model introduced in Example 7.7 using Gauss–Hermite, spherical cubature, unscented, and fifth order cubature/unscented integration. The resulting RMSE was* 0.016 *for all of the methods (for ERTSS it was* 0.017*). Figure 14.5 shows the result of the URTSS, which looks similar to the other smoothers, together with the ERTSS result.*

14.6 Statistical Linear Regression Smoothers

Similarly to Gaussian filters, the Gaussian smoothers presented here can also be understood as algorithms that make use of enabling approximations (see Chapter 9) – that is, as algorithms that first linearize the models

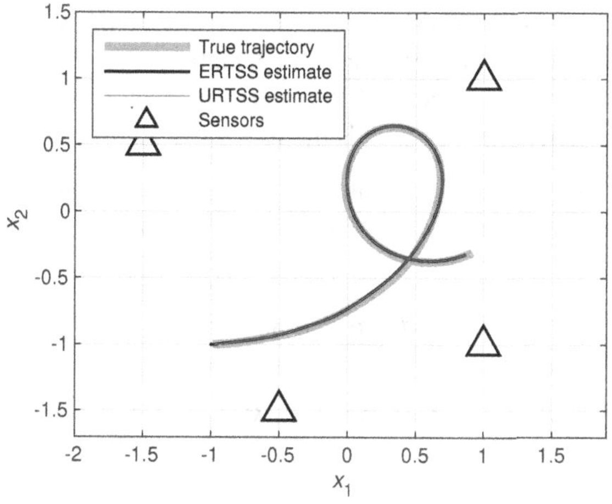

Figure 14.5 The result of applying URTSS to the coordinated turn model. The result is practically the same as with the other numerical integration-based smoothers with RMSE of 0.016.

and then make use of closed-form solutions provided by the RTS smoother to compute the posterior. For example, the extended RTS smoother in Algorithm 13.1 makes use of the same first-order Taylor expansions as an EKF and linearizes the dynamic model about the filtering mean, $\mathbf{m}_k$. Similarly, the iterated ERTSS in Algorithm 13.5 instead uses a first order Taylor expansion about the current estimate of the smoothed posterior mean, $\mathbf{m}_k^{s,(i-1)}$.

The general Gaussian smoothers can be understood as methods that linearize the models using statistical linear regression (SLR). The affine RTS smoother in Theorem 12.5 does not use the measurement model, and, to use the RTS smoother, it is sufficient to perform SLR on the dynamic model such that we obtain a linearized model

$$\mathbf{x}_{k+1} = \mathbf{A}_k \, \mathbf{x}_k + \mathbf{a}_k + \tilde{\mathbf{e}}_k, \tag{14.27}$$

where $\tilde{\mathbf{e}}_k \sim \mathrm{N}(\mathbf{0}, \mathbf{\Lambda}_k)$. As discussed in Chapter 10 (see Definition 10.1 and Theorem 10.2), it is possible to perform SLR with respect to different choices of distributions. Depending on the distribution that we use to perform SLR, we obtain different smoothers. It turns out that if we use the filtering density $\mathrm{N}(\mathbf{x}_k \mid \mathbf{m}_k, \mathbf{P}_k)$ to perform SLR on the dynamic model, we obtain Gaussian RTS smoothers that yield the same result as Algorithms 14.1 and 14.2.

Let us now see how to do general Gaussian smoothing using SLR for additive noise models (as given, e.g., in Equation (9.1)). Combining SLR of the dynamic model, where SLR is performed with respect to $\mathrm{N}(\mathbf{x}_k \mid \mathbf{m}_k, \mathbf{P}_k)$, with Theorem 12.5 yields the following algorithm.

Algorithm 14.15 (Statistical linear regression smoother)**.** *The additive noise statistical linear regression RTS smoother (SLRRTSS) is the following.*

- *Linearize the dynamic model at time step k:*
  Find

$$\begin{aligned}\mathbf{A}_k &= (\mathbf{P}_{k+1}^{xx})^\mathsf{T}\, \mathbf{P}_k^{-1}, \\ \mathbf{a}_k &= \boldsymbol{\mu}_{k+1}^- - \mathbf{A}_k\, \mathbf{m}_k, \\ \boldsymbol{\Lambda}_k &= \mathbf{P}_{k+1}^{x} - \mathbf{A}_k\, \mathbf{P}_k\, \mathbf{A}_k^\mathsf{T},\end{aligned} \tag{14.28}$$

  where

$$\begin{aligned}\boldsymbol{\mu}_{k+1}^- &= \int \mathbf{f}(\mathbf{x})\, \mathrm{N}(\mathbf{x} \mid \mathbf{m}_k, \mathbf{P}_k)\, \mathrm{d}\mathbf{x}, \\ \mathbf{P}_{k+1}^{xx} &= \int (\mathbf{x} - \mathbf{m}_k)\, (\mathbf{f}(\mathbf{x}) - \boldsymbol{\mu}_{k+1}^-)^\mathsf{T}\, \mathrm{N}(\mathbf{x} \mid \mathbf{m}_k, \mathbf{P}_k)\, \mathrm{d}\mathbf{x}, \\ \mathbf{P}_{k+1}^{x} &= \int (\mathbf{f}(\mathbf{x}) - \boldsymbol{\mu}_{k+1}^-)\, (\mathbf{f}(\mathbf{x}) - \boldsymbol{\mu}_{k+1}^-)^\mathsf{T}\, \mathrm{N}(\mathbf{x} \mid \mathbf{m}_k, \mathbf{P}_k)\, \mathrm{d}\mathbf{x} + \mathbf{Q}_k.\end{aligned} \tag{14.29}$$

- *Perform RTS smoothing using the affine model*

$$\begin{aligned}\mathbf{m}_{k+1}^- &= \mathbf{A}_k\, \mathbf{m}_k + \mathbf{a}_{k-1}, \\ \mathbf{P}_{k+1}^- &= \mathbf{A}_k\, \mathbf{P}_k\, \mathbf{A}_k^\mathsf{T} + \boldsymbol{\Lambda}_k, \\ \mathbf{G}_k &= \mathbf{P}_k\, \mathbf{A}_k^\mathsf{T}\, [\mathbf{P}_{k+1}^-]^{-1}, \\ \mathbf{m}_k^\mathrm{s} &= \mathbf{m}_k + \mathbf{G}_k\, [\mathbf{m}_{k+1}^\mathrm{s} - \mathbf{m}_{k+1}^-], \\ \mathbf{P}_k^\mathrm{s} &= \mathbf{P}_k + \mathbf{G}_k\, [\mathbf{P}_{k+1}^\mathrm{s} - \mathbf{P}_{k+1}^-]\, \mathbf{G}_k^\mathsf{T}.\end{aligned} \tag{14.30}$$

In Algorithm 14.15 we explicitly formulate the linearized model and then use that to perform RTS smoothing. However, if we compare (14.28) with (14.30), we notice that $\mathbf{m}_{k+1}^- = \boldsymbol{\mu}_{k+1}^-$ and $\mathbf{P}_{k+1}^- = \mathbf{P}_{k+1}^{x}$. If we also introduce the notation $\mathbf{D}_{k+1} = \mathbf{P}_k\, \mathbf{A}_k^\mathsf{T}$, we get the relation $\mathbf{D}_{k+1} = \mathbf{P}_{k+1}^{x}$. Having established these relations, we note that (14.29) precisely matches the first three lines in (14.1). We can conclude that some of the equations in Algorithm 14.15 are redundant and that the algorithm can be simplified into Algorithm 14.1.

Similarly, if we formulate an SLR smoother for non-additive models, by performing SLR with respect to $\mathrm{N}(\mathbf{x}_k \mid \mathbf{m}_k, \mathbf{P}_k)$, that algorithm can be simplified into Algorithm 14.2. More generally, in SLR smoothing, if we linearize the dynamic model with respect to the filtering density $\mathrm{N}(\mathbf{x}_k \mid \mathbf{m}_k, \mathbf{P}_k)$, then it yields the same result as the corresponding Gaussian RTS smoother.

In general, to use a Gaussian smoother we need to compute the moments $\mathbf{m}_{k+1}^- = \mathbb{E}[\mathbf{x}_{k+1} \mid \mathbf{y}_{1:k}]$, $\mathbf{P}_{k+1}^- = \mathrm{Cov}[\mathbf{x}_{k+1} \mid \mathbf{y}_{1:k}]$, and $\mathbf{D}_{k+1} = \mathrm{Cov}[\mathbf{x}_k, \mathbf{x}_{k+1} \mid \mathbf{y}_{1:k}]$. Based on the theory presented in Chapter 9, it is possible to present five different expressions for these moments: one for additive noise models (9.1), and two each for non-additive (9.9) and conditional distribution models (9.10). Two of those expressions were used to formulate Algorithms 8.3 and 8.4, and below we use a third expression to present a Gaussian RTS smoother for conditional distribution models. For the remaining two expressions, which make use of the conditional moments formulations, see Exercise 14.3.

Algorithm 14.16 (Gaussian RTS smoother III). *The conditional distribution* Gaussian RTS smoother *is the following:*

$$
\begin{aligned}
\mathbf{m}_{k+1}^- &= \int \mathbf{x}_{k+1}\, p(\mathbf{x}_{k+1} \mid \mathbf{x}_k)\, \mathrm{N}(\mathbf{x}_k \mid \mathbf{m}_k, \mathbf{P}_k)\, \mathrm{d}\mathbf{x}_{k+1}\, \mathrm{d}\mathbf{x}_k, \\
\mathbf{P}_{k+1}^- &= \int (\mathbf{x}_{k+1} - \mathbf{m}_{k+1}^-)\, (\mathbf{x}_{k+1} - \mathbf{m}_{k+1}^-)^\mathsf{T} \\
&\qquad \times p(\mathbf{x}_{k+1} \mid \mathbf{x}_k)\, \mathrm{N}(\mathbf{x}_k \mid \mathbf{m}_k, \mathbf{P}_k)\, \mathrm{d}\mathbf{x}_{k+1}\, \mathrm{d}\mathbf{x}_k, \\
\mathbf{D}_{k+1} &= \int (\mathbf{x}_k - \mathbf{m}_k)\, (\mathbf{x}_{k+1} - \mathbf{m}_{k+1}^-)^\mathsf{T} \\
&\qquad \times p(\mathbf{x}_{k+1} \mid \mathbf{x}_k)\, \mathrm{N}(\mathbf{x}_k \mid \mathbf{m}_k, \mathbf{P}_k)\, \mathrm{d}\mathbf{x}_{k+1}\, \mathrm{d}\mathbf{x}_k, \\
\mathbf{G}_k &= \mathbf{D}_{k+1}\, [\mathbf{P}_{k+1}^-]^{-1}, \\
\mathbf{m}_k^\mathrm{s} &= \mathbf{m}_k + \mathbf{G}_k\, (\mathbf{m}_{k+1}^\mathrm{s} - \mathbf{m}_{k+1}^-), \\
\mathbf{P}_k^\mathrm{s} &= \mathbf{P}_k + \mathbf{G}_k\, (\mathbf{P}_{k+1}^\mathrm{s} - \mathbf{P}_{k+1}^-)\, \mathbf{G}_k^\mathsf{T}.
\end{aligned}
\tag{14.31}
$$

The three integrals in (14.31) might not have closed form solutions. In that case, we may need to approximate them numerically, for instance, using Monte Carlo sampling, as in the following algorithm.

Algorithm 14.17 (Monte Carlo RTS smoother). *The conditional distribution form of the Monte Carlo RTS smoother (MCRTSS) is the following.*

1. *Generate samples of* $\mathbf{x}_k$:

$$\mathbf{x}_k^{(i)} \sim \mathrm{N}(\mathbf{m}_k, \mathbf{P}_k), \qquad i = 1, \ldots, N. \tag{14.32}$$

2. *Propagate the samples through the dynamic model:*

$$\mathbf{x}_{k+1}^{(i)} \sim p(\mathbf{x}_{k+1} \mid \mathbf{x}_k^{(i)}), \quad i = 1, \ldots, N. \tag{14.33}$$

3. *Compute the predicted mean* $\mathbf{m}_{k+1}^{-}$, *the predicted covariance* $\mathbf{P}_{k+1}^{-}$, *and the cross-covariance* $\mathbf{D}_{k+1}$:

$$\begin{aligned}
\mathbf{m}_{k+1}^{-} &= \frac{1}{N}\sum_{i=1}^{N} \mathbf{x}_{k+1}^{(i)}, \\
\mathbf{P}_{k+1}^{-} &= \frac{1}{N}\sum_{i=1}^{N} (\mathbf{x}_{k+1}^{(i)} - \mathbf{m}_{k+1}^{-})\,(\mathbf{x}_{k+1}^{(i)} - \mathbf{m}_{k+1}^{-})^{\mathsf{T}}, \\
\mathbf{D}_{k+1} &= \frac{1}{N}\sum_{i=1}^{N} (\mathbf{x}_{k}^{(i)} - \mathbf{m}_{k})\,(\mathbf{x}_{k+1}^{(i)} - \mathbf{m}_{k+1}^{-})^{\mathsf{T}}.
\end{aligned} \tag{14.34}$$

4. *Perform the RTS backward recursion:*

$$\begin{aligned}
\mathbf{G}_k &= \mathbf{D}_{k+1}\,[\mathbf{P}_{k+1}^{-}]^{-1}, \\
\mathbf{m}_k^{\mathrm{s}} &= \mathbf{m}_k + \mathbf{G}_k\,(\mathbf{m}_{k+1}^{\mathrm{s}} - \mathbf{m}_{k+1}^{-}), \\
\mathbf{P}_k^{\mathrm{s}} &= \mathbf{P}_k + \mathbf{G}_k\,(\mathbf{P}_{k+1}^{\mathrm{s}} - \mathbf{P}_{k+1}^{-})\,\mathbf{G}_k^{\mathsf{T}}.
\end{aligned} \tag{14.35}$$

An interesting aspect of the Gaussian smoothers that we have seen so far in this chapter is that they rely on the same linearizations as the Gaussian filters. That is, the Gaussian filters implicitly linearize the dynamic model by performing SLR with respect to the filtering distribution, and so do the Gaussian smoothers. Consequently, if we use Algorithm 9.20 to perform filtering such that we explicitly linearize the dynamic model at each time step, we could directly use those linearizations to perform RTS smoothing using Theorem 12.5, and this would yield the same result as running Algorithm 8.3 followed by Algorithm 14.1.

14.7 Posterior Linearization Smoothers

As in the filtering setting (see Chapter 10), there is no inherent reason to linearize our models with respect to the filtering distributions, and we can use the generalized SLR introduced in Definition 10.1 to linearize the models also for smoothing. Consequently, for each of the dynamic and measurement models, we are free to select any linearization distribution when linearizing the models. Let $\pi_k^{\mathbf{f}}(\mathbf{x}_k)$ and $\pi_k^{\mathbf{h}}(\mathbf{x}_k)$ denote our linearization distributions at time step k for the dynamic and measurement models, respectively, and let $\mathbf{m}_k^{\mathbf{f}} = \mathrm{E}_{\pi_k^{\mathbf{f}}}[\mathbf{x}_k]$, $\mathbf{P}_k^{\mathbf{f}} = \mathrm{Cov}_{\pi_k^{\mathbf{f}}}[\mathbf{x}_k]$, $\mathbf{m}_k^{\mathbf{h}} = \mathrm{E}_{\pi_k^{\mathbf{h}}}[\mathbf{x}_k]$, and $\mathbf{P}_k^{\mathbf{h}} = \mathrm{Cov}_{\pi_k^{\mathbf{h}}}[\mathbf{x}_k]$ be the corresponding means and covariances for $k = 0, 1, 2, \ldots, T$. Given the linearized models, we can perform (affine) Kalman filtering and RTS smoothing to compute the smoothing distributions.

Algorithm 14.18 (Generalized SLR smoother). *The generalized SLR smoother for additive noise models is the following.*

1. *For $k = 0, 1, \ldots, T-1$, perform SLR of the dynamic model with respect to $\pi_k^{\mathbf{f}}(\mathbf{x}_k)$.*
   (i) Compute moments with respect to $\mathbf{x}_k \sim \pi_k^{\mathbf{f}}(\mathbf{x}_k)$:

$$\begin{aligned}
\boldsymbol{\mu}_{k+1}^- &= \int \mathbf{f}(\mathbf{x})\, \pi_k^{\mathbf{f}}(\mathbf{x})\, \mathrm{d}\mathbf{x}, \\
\mathbf{P}_{k+1}^{\mathrm{xx}} &= \int (\mathbf{x} - \mathbf{m}_k^{\mathbf{f}})\, (\mathbf{f}(\mathbf{x}) - \boldsymbol{\mu}_{k+1}^-)^\mathsf{T}\, \pi_k^{\mathbf{f}}(\mathbf{x})\, \mathrm{d}\mathbf{x}, \\
\mathbf{P}_{k+1}^{\mathrm{x}} &= \int (\mathbf{f}(\mathbf{x}) - \boldsymbol{\mu}_{k+1}^-)\, (\mathbf{f}(\mathbf{x}) - \boldsymbol{\mu}_{k+1}^-)^\mathsf{T}\, \pi_k^{\mathbf{f}}(\mathbf{x})\, \mathrm{d}\mathbf{x} + \mathbf{Q}_k.
\end{aligned} \tag{14.36}$$

   (ii) Linearize the dynamic model:

$$\begin{aligned}
\mathbf{A}_k &= (\mathbf{P}_{k+1}^{\mathrm{xx}})^\mathsf{T}\, (\mathbf{P}_k^{\mathbf{f}})^{-1}, \\
\mathbf{a}_k &= \boldsymbol{\mu}_{k+1}^- - \mathbf{A}_k\, \mathbf{m}_k^{\mathbf{f}}, \\
\boldsymbol{\Lambda}_k &= \mathbf{P}_{k+1}^{\mathrm{x}} - \mathbf{A}_k\, \mathbf{P}_k^{\mathbf{f}}\, \mathbf{A}_k^\mathsf{T}.
\end{aligned} \tag{14.37}$$

2. *For $k = 1, 2, \ldots, T$, perform SLR of the measurement model with respect to $\pi_k^{\mathbf{h}}(\mathbf{x}_k)$.*
   (i) Compute moments with respect to $\mathbf{x}_k \sim \pi_k^{\mathbf{h}}(\mathbf{x}_k)$:

$$\begin{aligned}
\boldsymbol{\mu}_k^+ &= \int \mathbf{h}(\mathbf{x})\, \pi_k^{\mathbf{h}}(\mathbf{x})\, \mathrm{d}\mathbf{x}, \\
\mathbf{P}_k^{\mathrm{xy}} &= \int (\mathbf{x} - \mathbf{m}_k^{\mathbf{h}})\, (\mathbf{h}(\mathbf{x}) - \boldsymbol{\mu}_k^+)^\mathsf{T}\, \pi_k^{\mathbf{h}}(\mathbf{x})\, \mathrm{d}\mathbf{x}, \\
\mathbf{P}_k^{\mathrm{y}} &= \int (\mathbf{h}(\mathbf{x}) - \boldsymbol{\mu}_k^+)\, (\mathbf{h}(\mathbf{x}) - \boldsymbol{\mu}_k^+)^\mathsf{T}\, \pi_k^{\mathbf{h}}(\mathbf{x})\, \mathrm{d}\mathbf{x} + \mathbf{R}_k.
\end{aligned} \tag{14.38}$$

(ii) Linearize the measurement model:

$$
\begin{aligned}
\mathbf{H}_k &= (\mathbf{P}_k^{\mathrm{xy}})^{\mathsf{T}} \, (\mathbf{P}_k^{\mathbf{h}})^{-1}, \\
\mathbf{b}_k &= \boldsymbol{\mu}_k^{+} - \mathbf{H}_k \, \mathbf{m}_k^{\mathbf{h}}, \\
\boldsymbol{\Omega}_k &= \mathbf{P}_k^{\mathrm{y}} - \mathbf{H}_k \, \mathbf{P}_k^{\mathbf{h}} \, \mathbf{H}_k^{\mathsf{T}}.
\end{aligned}
\tag{14.39}
$$

3. *Run an affine Kalman filter on the linearized model. For* $k = 1, 2, \ldots, T$*:*

   (i) Prediction

$$
\begin{aligned}
\mathbf{m}_k^{-} &= \mathbf{A}_{k-1} \, \mathbf{m}_{k-1} + \mathbf{a}_{k-1}, \\
\mathbf{P}_k^{-} &= \mathbf{A}_{k-1} \, \mathbf{P}_{k-1} \, \mathbf{A}_{k-1}^{\mathsf{T}} + \boldsymbol{\Lambda}_{k-1}.
\end{aligned}
\tag{14.40}
$$

   (ii) Update

$$
\begin{aligned}
\boldsymbol{\mu}_k &= \mathbf{H}_k \, \mathbf{m}_k^{-} + \mathbf{b}_k, \\
\mathbf{S}_k &= \mathbf{H}_k \, \mathbf{P}_k^{-} \, \mathbf{H}_k^{\mathsf{T}} + \boldsymbol{\Omega}_k, \\
\mathbf{K}_k &= \mathbf{P}_k \, \mathbf{H}_k^{\mathsf{T}} \, \mathbf{S}_k^{-1}, \\
\mathbf{m}_k &= \mathbf{m}_k^{-} + \mathbf{K}_k \, (\mathbf{y}_k - \boldsymbol{\mu}_k), \\
\mathbf{P}_k &= \mathbf{P}_k^{-} - \mathbf{K}_k \, \mathbf{S}_k \, \mathbf{K}_k^{\mathsf{T}}.
\end{aligned}
\tag{14.41}
$$

4. *Run an affine RTS smoother on the linearized model. Set* $\mathbf{m}_T^{\mathrm{s}} = \mathbf{m}_T$ *and* $\mathbf{P}_T^{s} = \mathbf{P}_T$*. For* $k = T-1, T-2, \ldots, 0$*:*

$$
\begin{aligned}
\mathbf{m}_{k+1}^{-} &= \mathbf{A}_k \, \mathbf{m}_k + \mathbf{a}_k, \\
\mathbf{P}_{k+1}^{-} &= \mathbf{A}_k \, \mathbf{P}_k \, \mathbf{A}_k^{\mathsf{T}} + \boldsymbol{\Lambda}_k, \\
\mathbf{G}_k &= \mathbf{P}_k \, \mathbf{A}_k^{\mathsf{T}} \, (\mathbf{P}_{k+1}^{-})^{-1}, \\
\mathbf{m}_k^{\mathrm{s}} &= \mathbf{m}_k + \mathbf{G}_k \, [\mathbf{m}_{k+1}^{\mathrm{s}} - \mathbf{m}_{k+1}^{-}], \\
\mathbf{P}_k^{\mathrm{s}} &= \mathbf{P}_k + \mathbf{G}_k \, [\mathbf{P}_{k+1}^{\mathrm{s}} - \mathbf{P}_{k+1}^{-}] \, \mathbf{G}_k^{\mathsf{T}}.
\end{aligned}
\tag{14.42}
$$

The posterior linearization filters introduced in Chapter 10 were based on the philosophy that we should construct linearizations that are accurate on average across our posterior density of the state $\mathbf{x}_k$. We can extend this principle to smoothing and perform SLR with respect to the smoothing distribution $p(\mathbf{x}_k \mid \mathbf{y}_{1:T}) \simeq \mathrm{N}(\mathbf{x}_k \mid \mathbf{m}_k^{\mathrm{s}}, \mathbf{P}_k^{\mathrm{s}})$. In other words, we would like to perform generalized SLR smoothing with $\pi_k^{\mathbf{f}}(\mathbf{x}_k) = \pi_k^{\mathbf{h}}(\mathbf{x}_k) = p(\mathbf{x}_k \mid \mathbf{y}_{1:T})$ or at least with respect to $p(\mathbf{x}_k \mid \mathbf{y}_{1:T}) \simeq \mathrm{N}(\mathbf{x}_k \mid \mathbf{m}_k^{\mathrm{s}}, \mathbf{P}_k^{\mathrm{s}})$. In practice, the moments $(\mathbf{m}_k^{\mathrm{s}}, \mathbf{P}_k^{\mathrm{s}})$ are unknown to us as they are the end-products that we wish to compute. Fortunately, as for the posterior linearization filters, we can construct an iterative algorithm where we use our current best approximations of the posterior moments at each step. The resulting algorithm is called the iterated posterior linearization smoother (IPLS, García-Fernández et al., 2017; Tronarp et al., 2018).

Algorithm 14.19 (Iterated posterior linearization smoother I). *The iterated posterior linearization smoother for additive noise models is started from an initial guess for the trajectory moments* $\mathbf{m}_{0:T}^{\mathrm{s},(0)}$, $\mathbf{P}_{0:T}^{\mathrm{s},(0)}$. *For* $i = 1, 2, \ldots$ *we do the following.*

1. *For* $k = 0, 1, \ldots, T-1$, *perform SLR of the dynamic model with respect to* $\mathrm{N}(\mathbf{x}_k \mid \mathbf{m}_k^{\mathrm{s},(i-1)}, \mathbf{P}_k^{\mathrm{s},(i-1)})$.

   (i) Compute moments with respect to $\mathbf{x}_k \sim \mathrm{N}(\mathbf{x}_k \mid \mathbf{m}_k^{\mathrm{s},(i-1)}, \mathbf{P}_k^{\mathrm{s},(i-1)})$*:*

$$
\begin{aligned}
\boldsymbol{\mu}_{k+1}^{-,(i-1)} &= \int \mathbf{f}(\mathbf{x})\, \mathrm{N}(\mathbf{x} \mid \mathbf{m}_k^{\mathrm{s},(i-1)}, \mathbf{P}_k^{\mathrm{s},(i-1)})\, \mathrm{d}\mathbf{x}, \\
\mathbf{P}_{k+1}^{\mathrm{xx},(i-1)} &= \int (\mathbf{x} - \mathbf{m}_k^{\mathrm{s},(i-1)})\, (\mathbf{f}(\mathbf{x}) - \boldsymbol{\mu}_{k+1}^{-,(i-1)})^{\mathsf{T}} \\
&\qquad \times \mathrm{N}(\mathbf{x} \mid \mathbf{m}_k^{\mathrm{s},(i-1)}, \mathbf{P}_k^{\mathrm{s},(i-1)})\, \mathrm{d}\mathbf{x}, \\
\mathbf{P}_{k+1}^{\mathrm{x},(i-1)} &= \int (\mathbf{f}(\mathbf{x}) - \boldsymbol{\mu}_{k+1}^{-,(i-1)})\, (\mathbf{f}(\mathbf{x}) - \boldsymbol{\mu}_{k+1}^{-,(i-1)})^{\mathsf{T}} \\
&\qquad \times \mathrm{N}(\mathbf{x} \mid \mathbf{m}_k^{\mathrm{s},(i-1)}, \mathbf{P}_k^{\mathrm{s},(i-1)})\, \mathrm{d}\mathbf{x} + \mathbf{Q}_k.
\end{aligned}
\tag{14.43}
$$

   (ii) Linearize the dynamic model:

$$
\begin{aligned}
\mathbf{A}_k^{(i)} &= (\mathbf{P}_{k+1}^{\mathrm{xx},(i-1)})^{\mathsf{T}}\, (\mathbf{P}_k^{\mathrm{s},(i-1)})^{-1}, \\
\mathbf{a}_k^{(i)} &= \boldsymbol{\mu}_{k+1}^{-,(i-1)} - \mathbf{A}_k^{(i)}\, \mathbf{m}_k^{\mathrm{s},(i-1)}, \\
\boldsymbol{\Lambda}_k^{(i)} &= \mathbf{P}_{k+1}^{\mathrm{x},(i-1)} - \mathbf{A}_k^{(i)}\, \mathbf{P}_k^{\mathrm{s},(i-1)}\, (\mathbf{A}_k^{(i)})^{\mathsf{T}}.
\end{aligned}
\tag{14.44}
$$

2. *For* $k = 1, 2, \ldots, T$, *perform SLR of the measurement model with respect to* $\mathrm{N}(\mathbf{x}_k \mid \mathbf{m}_k^{\mathrm{s},(i-1)}, \mathbf{P}_k^{\mathrm{s},(i-1)})$.

   (i) Compute moments with respect to $\mathbf{x}_k \sim \mathrm{N}(\mathbf{x}_k \mid \mathbf{m}_k^{\mathrm{s},(i-1)}, \mathbf{P}_k^{\mathrm{s},(i-1)})$*:*

$$
\begin{aligned}
\boldsymbol{\mu}_{k}^{+,(i-1)} &= \int \mathbf{h}(\mathbf{x})\, \mathrm{N}(\mathbf{x} \mid \mathbf{m}_k^{\mathrm{s},(i-1)}, \mathbf{P}_k^{\mathrm{s},(i-1)})\, \mathrm{d}\mathbf{x}, \\
\mathbf{P}_{k}^{\mathrm{xy},(i-1)} &= \int (\mathbf{x} - \mathbf{m}_k^{\mathrm{s},(i-1)})\, (\mathbf{h}(\mathbf{x}) - \boldsymbol{\mu}_{k}^{+,(i-1)})^{\mathsf{T}} \\
&\qquad \times \mathrm{N}(\mathbf{x} \mid \mathbf{m}_k^{\mathrm{s},(i-1)}, \mathbf{P}_k^{\mathrm{s},(i-1)})\, \mathrm{d}\mathbf{x}, \\
\mathbf{P}_{k}^{\mathrm{y},(i-1)} &= \int (\mathbf{h}(\mathbf{x}) - \boldsymbol{\mu}_{k}^{+,(i-1)})\, (\mathbf{h}(\mathbf{x}) - \boldsymbol{\mu}_{k}^{+,(i-1)})^{\mathsf{T}} \\
&\qquad \times \mathrm{N}(\mathbf{x} \mid \mathbf{m}_k^{\mathrm{s},(i-1)}, \mathbf{P}_k^{\mathrm{s},(i-1)})\, \mathrm{d}\mathbf{x} + \mathbf{R}_k.
\end{aligned}
\tag{14.45}
$$

(ii) Linearize the measurement model:

$$
\begin{aligned}
\mathbf{H}_k^{(i)} &= (\mathbf{P}_k^{\mathrm{xy},(i-1)})^\mathsf{T} \, (\mathbf{P}_k^{\mathrm{s},(i-1)})^{-1}, \\
\mathbf{b}_k^{(i)} &= \boldsymbol{\mu}_k^{+,(i-1)} - \mathbf{H}_k^{(i)} \, \mathbf{m}_k^{\mathrm{s},(i-1)}, \\
\boldsymbol{\Omega}_k^{(i)} &= \mathbf{P}_k^{\mathrm{y},(i-1)} - \mathbf{H}_k^{(i)} \, \mathbf{P}_k^{\mathrm{s},(i-1)} \, (\mathbf{H}_k^{(i)})^\mathsf{T}.
\end{aligned}
\tag{14.46}
$$

3. *Run a Kalman filter on the linearized model. To initiate the algorithm we set* $\mathbf{m}_0^{(i)} = \mathbf{m}_0$ *and* $\mathbf{P}_0^{(i)} = \mathbf{P}_0$. *For* $k = 1, 2, \ldots, T$:

   (i) Prediction

$$
\begin{aligned}
\mathbf{m}_k^{-(i)} &= \mathbf{A}_{k-1}^{(i)} \, \mathbf{m}_{k-1}^{(i)} + \mathbf{a}_{k-1}^{(i)}, \\
\mathbf{P}_k^{-(i)} &= \mathbf{A}_{k-1}^{(i)} \, \mathbf{P}_{k-1}^{(i)} \, (\mathbf{A}_{k-1}^{(i)})^\mathsf{T} + \boldsymbol{\Lambda}_{k-1}^{(i)}.
\end{aligned}
\tag{14.47}
$$

   (ii) Update

$$
\begin{aligned}
\boldsymbol{\mu}_k^{(i)} &= \mathbf{H}_k^{(i)} \, \mathbf{m}_k^{-(i)} + \mathbf{b}_k^{(i)}, \\
\mathbf{S}_k^{(i)} &= \mathbf{H}_k^{(i)} \, \mathbf{P}_k^{-(i)} \, (\mathbf{H}_k^{(i)})^\mathsf{T} + \boldsymbol{\Omega}_k^{(i)}, \\
\mathbf{K}_k^{(i)} &= \mathbf{P}_k^{-(i)} \, (\mathbf{H}_k^{(i)})^\mathsf{T} \, (\mathbf{S}_k^{(i)})^{-1}, \\
\mathbf{m}_k^{(i)} &= \mathbf{m}_k^{-(i)} + \mathbf{K}_k^{(i)} \, (\mathbf{y}_k - \boldsymbol{\mu}_k^{(i)}), \\
\mathbf{P}_k^{(i)} &= \mathbf{P}_k^{-(i)} - \mathbf{K}_k^{(i)} \, \mathbf{S}_k^{(i)} \, (\mathbf{K}_k^{(i)})^\mathsf{T}.
\end{aligned}
\tag{14.48}
$$

4. *Run an RTS smoother on the linearized model. Set* $\mathbf{m}_T^{\mathrm{s},(i)} = \mathbf{m}_T^{(i)}$ *and* $\mathbf{P}_T^{s(i)} = \mathbf{P}_T^{(i)}$. *For* $k = T-1, T-2, \ldots, 0$:

$$
\begin{aligned}
\mathbf{m}_{k+1}^{-(i)} &= \mathbf{A}_k^{(i)} \, \mathbf{m}_k^{(i)} + \mathbf{a}_k^{(i)}, \\
\mathbf{P}_{k+1}^{-(i)} &= \mathbf{A}_k^{(i)} \, \mathbf{P}_k^{(i)} \, (\mathbf{A}_k^{(i)})^\mathsf{T} + \boldsymbol{\Lambda}_k^{(i)}, \\
\mathbf{G}_k^{(i)} &= \mathbf{P}_k^{(i)} \, (\mathbf{A}_k^{(i)})^\mathsf{T} \, (\mathbf{P}_{k+1}^{-(i)})^{-1}, \\
\mathbf{m}_k^{\mathrm{s},(i)} &= \mathbf{m}_k^{(i)} + \mathbf{G}_k^{(i)} \, [\mathbf{m}_{k+1}^{\mathrm{s},(i)} - \mathbf{m}_{k+1}^{-(i)}], \\
\mathbf{P}_k^{\mathrm{s},(i)} &= \mathbf{P}_k^{(i)} + \mathbf{G}_k^{(i)} \, [\mathbf{P}_{k+1}^{\mathrm{s},(i)} - \mathbf{P}_{k+1}^{-(i)}] \, (\mathbf{G}_k^{(i)})^\mathsf{T}.
\end{aligned}
\tag{14.49}
$$

At convergence, set $\mathbf{m}_{1:T}^{\mathrm{s},} = \mathbf{m}_{1:T}^{\mathrm{s},(i)}$ *and* $\mathbf{P}_{1:T}^{\mathrm{s},} = \mathbf{P}_{1:T}^{\mathrm{s},(i)}$.

Similarly to the iterated posterior linearization filter (IPLF), the IPLS seeks to iteratively improve the linearizations, but there are also important differences. First, the IPLF performs SLR with respect to the filtering density, whereas the the IPLS performs SLR with respect to the smoothing density. The IPLS thus makes use of all measurements $\mathbf{y}_{1:T}$ when selecting the linearizations, which has the potential to yield linearizations that are much more accurate in the relevant regions. Second, the IPLF only tries to improve on how we linearize the measurement model (and only at one time step at a time), whereas the IPLS iteratively updates the linearizations of both dynamic and measurement models, at all time steps.

The IPLS also has important connections to the IERTSS presented in Section 13.3. Both algorithms use the current approximation to the smoothed posterior to linearize the models. Given the linearized models, new smoothed posteriors are found in closed form by running an RTS smoother. The difference between the two algorithms is that the IEKS makes use of first order Taylor expansions about the mean of the smoothed posterior, whereas the IPLS performs SLR with respect to the smoothed posterior.

Algorithm 14.19 is designed for additive noise models, but we can also develop iterated posterior linearization smoothers for non-additive and conditional distribution models. As for the SLR smoothers, the IPLSs rely on certain moments to compute the SLR; in Algorithm 14.19 these moments are computed in (14.43) and (14.45). What changes when we use the IPLS for the other model classes is that the expressions for these moments are modified. As we saw in Section 9.4, we can express these moments in two different forms for both non-additive and conditional distribution models. Here we present an IPLS that makes use of the conditional moments form. This algorithm assumes that we are able to compute (numerically or analytically) the moments $\boldsymbol{\mu}_k^-(\mathbf{x}_{k-1}) = \mathrm{E}[\mathbf{x}_k \mid \mathbf{x}_{k-1}]$, $\mathbf{P}_k^x(\mathbf{x}_{k-1}) = \mathrm{Cov}[\mathbf{x}_k \mid \mathbf{x}_{k-1}]$, $\boldsymbol{\mu}_k(\mathbf{x}_k) = \mathrm{E}[\mathbf{y}_k \mid \mathbf{x}_k]$, and $\mathbf{P}_k^y(\mathbf{x}_k) = \mathrm{Cov}[\mathbf{y}_k \mid \mathbf{x}_k]$, and the algorithm can be used for additive, non-additive, and conditional distribution models as long as we can compute these moments.

Algorithm 14.20 (Iterated posterior linearization smoother II). *The conditional moments iterated posterior linearization smoother starts from an initial guess for the trajectory moments* $\mathbf{m}_{0:T}^{s,(0)}$, $\mathbf{P}_{0:T}^{s,(0)}$. *For* $i = 1, 2, \ldots$ *we do the following.*

1. *For* $k = 0, 1, \ldots, T-1$, *perform SLR of the dynamic model with respect to* $\mathrm{N}(\mathbf{x}_k \mid \mathbf{m}_k^{s,(i-1)}, \mathbf{P}_k^{s,(i-1)})$.

(i) Compute moments with respect to $\mathbf{x}_k \sim \mathrm{N}(\mathbf{x}_k \mid \mathbf{m}_k^{\mathrm{s},(i-1)}, \mathbf{P}_k^{\mathrm{s},(i-1)})$:

$$\begin{aligned}
\boldsymbol{\mu}_{k+1}^{-,(i-1)} &= \int \boldsymbol{\mu}_k^{-}(\mathbf{x})\, \mathrm{N}(\mathbf{x} \mid \mathbf{m}_k^{\mathrm{s},(i-1)}, \mathbf{P}_k^{\mathrm{s},(i-1)})\, \mathrm{d}\mathbf{x}, \\
\mathbf{P}_{k+1}^{\mathrm{xx},(i-1)} &= \int (\mathbf{x} - \mathbf{m}_k^{\mathrm{s},(i-1)})\, (\boldsymbol{\mu}_k^{-}(\mathbf{x}) - \boldsymbol{\mu}_{k+1}^{-,(i-1)})^{\mathsf{T}} \\
&\qquad \times \mathrm{N}(\mathbf{x} \mid \mathbf{m}_k^{\mathrm{s},(i-1)}, \mathbf{P}_k^{\mathrm{s},(i-1)})\, \mathrm{d}\mathbf{x}, \\
\mathbf{P}_{k+1}^{\mathrm{x},(i-1)} &= \int (\boldsymbol{\mu}_k^{-}(\mathbf{x}) - \boldsymbol{\mu}_{k+1}^{-,(i-1)})\, (\boldsymbol{\mu}_k^{-}(\mathbf{x}) - \boldsymbol{\mu}_{k+1}^{-,(i-1)})^{\mathsf{T}} \\
&\qquad \times \mathrm{N}(\mathbf{x} \mid \mathbf{m}_k^{\mathrm{s},(i-1)}, \mathbf{P}_k^{\mathrm{s},(i-1)})\, \mathrm{d}\mathbf{x} \\
&\qquad + \int \mathbf{P}_k^{x}(\mathbf{x})\, \mathrm{N}(\mathbf{x} \mid \mathbf{m}_k^{\mathrm{s},(i-1)}, \mathbf{P}_k^{\mathrm{s},(i-1)})\, \mathrm{d}\mathbf{x}.
\end{aligned} \tag{14.50}$$

(ii) Linearize the dynamic model using Equation (14.44).

2. *For* $k = 1, 2, \ldots, T$*, perform SLR of the measurement model with respect to* $\mathrm{N}(\mathbf{x}_k \mid \mathbf{m}_k^{\mathrm{s},(i-1)}, \mathbf{P}_k^{\mathrm{s},(i-1)})$.

(i) Compute moments with respect to $\mathbf{x}_k \sim \mathrm{N}(\mathbf{x}_k \mid \mathbf{m}_k^{\mathrm{s},(i-1)}, \mathbf{P}_k^{\mathrm{s},(i-1)})$:

$$\begin{aligned}
\boldsymbol{\mu}_{k}^{+,(i-1)} &= \int \boldsymbol{\mu}_k(\mathbf{x})\, \mathrm{N}(\mathbf{x} \mid \mathbf{m}_k^{\mathrm{s},(i-1)}, \mathbf{P}_k^{\mathrm{s},(i-1)})\, \mathrm{d}\mathbf{x}, \\
\mathbf{P}_{k}^{\mathrm{xy},(i-1)} &= \int (\mathbf{x} - \mathbf{m}_k^{\mathrm{s},(i-1)})\, (\boldsymbol{\mu}_k(\mathbf{x}) - \boldsymbol{\mu}_{k}^{+,(i-1)})^{\mathsf{T}} \\
&\qquad \times \mathrm{N}(\mathbf{x} \mid \mathbf{m}_k^{\mathrm{s},(i-1)}, \mathbf{P}_k^{\mathrm{s},(i-1)})\, \mathrm{d}\mathbf{x}, \\
\mathbf{P}_{k}^{\mathrm{y},(i-1)} &= \int (\boldsymbol{\mu}_k(\mathbf{x}) - \boldsymbol{\mu}_{k}^{+,(i-1)})\, (\boldsymbol{\mu}_k(\mathbf{x}) - \boldsymbol{\mu}_{k}^{+,(i-1)})^{\mathsf{T}} \\
&\qquad \times \mathrm{N}(\mathbf{x} \mid \mathbf{m}_k^{\mathrm{s},(i-1)}, \mathbf{P}_k^{\mathrm{s},(i-1)})\, \mathrm{d}\mathbf{x} \\
&\qquad + \int \mathbf{P}_k^{y}(\mathbf{x})\, \mathrm{N}(\mathbf{x} \mid \mathbf{m}_k^{\mathrm{s},(i-1)}, \mathbf{P}_k^{\mathrm{s},(i-1)})\, \mathrm{d}\mathbf{x}.
\end{aligned} \tag{14.51}$$

(ii) Linearize the measurement model using Equation (14.46).

3. *Run a Kalman filter on the linearized model. To initiate the algorithm we set* $\mathbf{m}_0^{(i)} = \mathbf{m}_0$ *and* $\mathbf{P}_0^{(i)} = \mathbf{P}_0$. *For* $k = 1, 2, \ldots, T$*, perform prediction and update using Equations* (14.47) *and* (14.48).
4. *Run an RTS smoother on the linearized model. Set* $\mathbf{m}_T^{\mathrm{s},(i)} = \mathbf{m}_T^{(i)}$ *and* $\mathbf{P}_T^{s(i)} = \mathbf{P}_T^{(i)}$. *For* $k = T-1, T-2, \ldots, 0$*, perform RTS backward recursion using Equation* (14.49).

At convergence, set $\mathbf{m}_{1:T}^{\mathrm{s}} = \mathbf{m}_{1:T}^{\mathrm{s},(i)}$ *and* $\mathbf{P}_{1:T}^{\mathrm{s}} = \mathbf{P}_{1:T}^{\mathrm{s},(i)}$.

To convert Algorithm 14.19 into a practical algorithm, we need a strategy to approximate the integrals in (14.43) and (14.45). Here we present a general sigma point version of the IPLS for additive noise models.

Algorithm 14.21 (Sigma-point iterated posterior linearization smoother I). *The sigma-point iterated posterior linearization smoother for additive noise is started from an initial guess for the trajectory moments $\mathbf{m}^{s,(0)}_{0:T}$ and $\mathbf{P}^{s,(0)}_{0:T}$. For $i = 1, 2, \ldots$ we do the following.*

1. *For $k = 0, 1, \ldots, T-1$, perform SLR of the dynamic model with respect to $\mathrm{N}(\mathbf{x}_k \mid \mathbf{m}^{s,(i-1)}_k, \mathbf{P}^{s,(i-1)}_k)$.*
   (i) *Form the sigma points as:*
   $$\mathcal{X}^{(j)}_k = \mathbf{m}^{s,(i-1)}_k + \sqrt{\mathbf{P}^{s,(i-1)}_k}\,\boldsymbol{\xi}^{(j)}, \qquad j = 1, \ldots, m, \tag{14.52}$$
   and select the weights $W_1, \ldots, W_m$.
   (ii) *Propagate the sigma points through the dynamic model:*
   $$\mathcal{X}^{-(j)}_{k+1} = \mathbf{f}(\mathcal{X}^{(j)}_k), \qquad j = 1, \ldots, m. \tag{14.53}$$
   (iii) *Compute moments with respect to $\mathbf{x}_k \sim \mathrm{N}(\mathbf{x}_k \mid \mathbf{m}^{s,(i-1)}_k, \mathbf{P}^{s,(i-1)}_k)$:*
   $$\begin{aligned}
   \boldsymbol{\mu}^{-,(i-1)}_{k+1} &= \sum_{j=1}^{m} W_j\, \mathcal{X}^{-(j)}_{k+1}, \\
   \mathbf{P}^{xx,(i-1)}_{k+1} &= \sum_{j=1}^{m} W_j\, (\mathcal{X}^{(j)}_k - \mathbf{m}^{s,(i-1)}_k)\,(\mathcal{X}^{-(j)}_{k+1} - \boldsymbol{\mu}^{-,(i-1)}_{k+1})^\mathsf{T}, \\
   \mathbf{P}^{x,(i-1)}_{k+1} &= \sum_{j=1}^{m} W_j\, (\mathcal{X}^{-(j)}_{k+1} - \boldsymbol{\mu}^{-,(i-1)}_{k+1})\,(\mathcal{X}^{-(j)}_{k+1} - \boldsymbol{\mu}^{-,(i-1)}_{k+1})^\mathsf{T} + \mathbf{Q}_k.
   \end{aligned} \tag{14.54}$$
   (iv) *Linearize the dynamic model using* (14.44).
2. *For $k = 1, 2, \ldots, T$, perform SLR of the measurement model with respect to $\mathrm{N}(\mathbf{x}_k \mid \mathbf{m}^{s,(i-1)}_k, \mathbf{P}^{s,(i-1)}_k)$.*
   (i) *Form the sigma points as:*
   $$\mathcal{X}^{(j)}_k = \mathbf{m}^{s,(i-1)}_k + \sqrt{\mathbf{P}^{s,(i-1)}_k}\,\boldsymbol{\xi}^{(j)}, \qquad j = 1, \ldots, m, \tag{14.55}$$
   and select the weights $W_1, \ldots, W_m$.
   (ii) *Propagate the sigma-points through the measurement model:*
   $$\mathcal{Y}^{(j)}_k = \mathbf{h}(\mathcal{X}^{(j)}_k), \quad j = 1, \ldots, m. \tag{14.56}$$

(iii) Compute moments with respect to $\mathbf{x}_k \sim \mathrm{N}(\mathbf{x}_k \mid \mathbf{m}_k^{\mathrm{s},(i-1)}, \mathbf{P}_k^{\mathrm{s},(i-1)})$*:*

$$\begin{aligned}
\boldsymbol{\mu}_k^{+,(i-1)} &= \sum_{j=1}^{m} W_j \, \mathcal{Y}_k^{(j)}, \\
\mathbf{P}_k^{\mathrm{xy},(i-1)} &= \sum_{j=1}^{m} W_j \, (\mathcal{X}_k^{(j)} - \mathbf{m}_k^{\mathrm{s},(i-1)}) \, (\mathcal{Y}_k^{(j)} - \boldsymbol{\mu}_k^{+,(i-1)})^{\mathsf{T}}, \\
\mathbf{P}_k^{\mathrm{y},(i-1)} &= \sum_{j=1}^{m} W_j \, (\mathcal{Y}_k^{(j)} - \boldsymbol{\mu}_k^{+,(i-1)}) \, (\mathcal{Y}_k^{(j)} - \boldsymbol{\mu}_k^{+,(i-1)})^{\mathsf{T}} + \mathbf{R}_k.
\end{aligned} \tag{14.57}$$

(iv) Linearize the measurement model using (14.46).

3. *Run an (affine) Kalman filter on the linearized model. To initiate the algorithm we set* $\mathbf{m}_0^{(i)} = \mathbf{m}_0, \mathbf{P}_0^{(i)} = \mathbf{P}_0$*. For* $k = 1, 2, \ldots, T$*, perform prediction using* (14.47) *and update using* (14.48).
4. *Run an (affine) RTS smoother on the linearized model. Set* $\mathbf{m}_T^{\mathrm{s},(i)} = \mathbf{m}_T^{(i)}$ *and* $\mathbf{P}_T^{s(i)} = \mathbf{P}_T^{(i)}$*. For* $k = T-1, T-2, \ldots, 0$*, perform the backward recursion using* (14.49).

At convergence, set $\mathbf{m}_{1:T}^{\mathrm{s}} = \mathbf{m}_{1:T}^{\mathrm{s},(i)}$ *and* $\mathbf{P}_{1:T}^{\mathrm{s}} = \mathbf{P}_{1:T}^{\mathrm{s},(i)}$.

To create the initial guess for the trajectory moments $\mathbf{m}_{0:T}^{\mathrm{s},(0)}$ and $\mathbf{P}_{0:T}^{\mathrm{s},(0)}$, it is common to use a sigma-point RTS smoother. We can also create a sigma-point implementation of Algorithm 14.20.

Algorithm 14.22 (Sigma-point iterated posterior linearization smoother II)**.** *The sigma-point conditional moment iterated posterior linearization smoother is started from an initial guess for the trajectory moments* $\mathbf{m}_{0:T}^{\mathrm{s},(0)}$ *and* $\mathbf{P}_{0:T}^{\mathrm{s},(0)}$*. For* $i = 1, 2, \ldots$ *we do the following.*

1. *For* $k = 0, 1, \ldots, T-1$*, perform SLR of the dynamic model with respect to* $\mathrm{N}(\mathbf{x}_k \mid \mathbf{m}_k^{\mathrm{s},(i-1)}, \mathbf{P}_k^{\mathrm{s},(i-1)})$.

   (i) Form the sigma points as:

$$\mathcal{X}_k^{(j)} = \mathbf{m}_k^{\mathrm{s},(i-1)} + \sqrt{\mathbf{P}_k^{\mathrm{s},(i-1)}} \, \boldsymbol{\xi}^{(j)}, \qquad j = 1, \ldots, m, \tag{14.58}$$

   and select the weights $W_1, \ldots, W_m$.

   (ii) Propagate the sigma points through the conditional mean and covariance functions of the dynamic model:

$$\begin{aligned}
\boldsymbol{\mu}_{k+1}^{-,(j)} &= \boldsymbol{\mu}_k^{-}(\mathcal{X}_k^{(j)}), \quad j = 1, \ldots, m, \\
\mathbf{P}_{k+1}^{x,(j)} &= \mathbf{P}_k^{x}(\mathcal{X}_k^{(j)}), \quad j = 1, \ldots, m.
\end{aligned} \tag{14.59}$$

(iii) *Compute moments with respect to* $\mathbf{x}_k \sim \mathrm{N}(\mathbf{x}_k \mid \mathbf{m}_k^{\mathrm{s},(i-1)}, \mathbf{P}_k^{\mathrm{s},(i-1)})$:

$$\begin{aligned}
\boldsymbol{\mu}_{k+1}^{-,(i-1)} &= \sum_{j=1}^{m} W_j \, \boldsymbol{\mu}_{k+1}^{-,(j)}, \\
\mathbf{P}_{k+1}^{\mathrm{xx},(i-1)} &= \sum_{j=1}^{m} W_j \, (\mathcal{X}_k^{(j)} - \mathbf{m}_k^{\mathrm{s},(i-1)}) \, (\boldsymbol{\mu}_{k+1}^{-,(j)} - \boldsymbol{\mu}_{k+1}^{-,(i-1)})^{\mathsf{T}}, \\
\mathbf{P}_{k+1}^{\mathrm{x},(i-1)} &= \sum_{j=1}^{m} W_j \, (\boldsymbol{\mu}_{k+1}^{-,(j)} - \boldsymbol{\mu}_{k+1}^{-,(i-1)}) \, (\boldsymbol{\mu}_{k+1}^{-,(j)} - \boldsymbol{\mu}_{k+1}^{-,(i-1)})^{\mathsf{T}} \\
&\quad + \sum_{j=1}^{m} W_j \, \mathbf{P}_{k+1}^{x,(j)}.
\end{aligned} \tag{14.60}$$

(iv) *Linearize the dynamic model using* (14.44).

2. *For* $k = 1, 2, \ldots, T$, *perform SLR of the measurement model with respect to* $\mathrm{N}(\mathbf{x}_k \mid \mathbf{m}_k^{\mathrm{s},(i-1)}, \mathbf{P}_k^{\mathrm{s},(i-1)})$.

(i) *Form the sigma points as:*

$$\mathcal{X}_k^{(j)} = \mathbf{m}_k^{\mathrm{s},(i-1)} + \sqrt{\mathbf{P}_k^{\mathrm{s},(i-1)}} \, \boldsymbol{\xi}^{(j)}, \qquad j = 1, \ldots, m, \tag{14.61}$$

and select the weights $W_1, \ldots, W_m$.

(ii) *Propagate the sigma points through the conditional mean and covariance functions of the measurement model:*

$$\begin{aligned}
\boldsymbol{\mu}_k^{+,(j)} &= \boldsymbol{\mu}_k(\mathcal{X}_k^{(j)}), \quad j = 1, \ldots, m, \\
\mathbf{P}_k^{y,(j)} &= \mathbf{P}_k^{y}(\mathcal{X}_k^{(j)}), \quad j = 1, \ldots, m.
\end{aligned} \tag{14.62}$$

(iii) *Compute moments with respect to* $\mathbf{x}_k \sim \mathrm{N}(\mathbf{x}_k \mid \mathbf{m}_k^{\mathrm{s},(i-1)}, \mathbf{P}_k^{\mathrm{s},(i-1)})$:

$$\begin{aligned}
\boldsymbol{\mu}_k^{+,(i-1)} &= \sum_{j=1}^{m} W_j \, \boldsymbol{\mu}_k^{+,(j)}, \\
\mathbf{P}_k^{\mathrm{xy},(i-1)} &= \sum_{j=1}^{m} W_j \, (\mathcal{X}_k^{(j)} - \mathbf{m}_k^{\mathrm{s},(i-1)}) \, (\boldsymbol{\mu}_k^{+,(j)} - \boldsymbol{\mu}_k^{+,(i-1)})^{\mathsf{T}}, \\
\mathbf{P}_k^{\mathrm{y},(i-1)} &= \sum_{j=1}^{m} W_j \, (\boldsymbol{\mu}_k^{+,(j)} - \boldsymbol{\mu}_k^{+,(i-1)}) \, (\boldsymbol{\mu}_k^{+,(j)} - \boldsymbol{\mu}_k^{+,(i-1)})^{\mathsf{T}} \\
&\quad + \sum_{j=1}^{m} W_j \, \mathbf{P}_k^{y,(j)}.
\end{aligned} \tag{14.63}$$

(iv) *Linearize the measurement model using* (14.46).

3. *Run an (affine) Kalman filter on the linearized model. To initiate the algorithm we set* $\mathbf{m}_0^{(i)} = \mathbf{m}_0, \mathbf{P}_0^{(i)} = \mathbf{P}_0$. *For* $k = 1, 2, \ldots, T$, *perform prediction using* (14.47) *and update using* (14.48).
4. *Run an (affine) RTS smoother on the linearized model. Set* $\mathbf{m}_T^{\mathrm{s},(i)} = \mathbf{m}_T^{(i)}$ *and* $\mathbf{P}_T^{s(i)} = \mathbf{P}_T^{(i)}$. *For* $k = T-1, T-2, \ldots, 0$, *perform the backward recursion using* (14.49).

At convergence, set $\mathbf{m}_{1:T}^{\mathrm{s}} = \mathbf{m}_{1:T}^{\mathrm{s},(i)}$ *and* $\mathbf{P}_{1:T}^{\mathrm{s}} = \mathbf{P}_{1:T}^{\mathrm{s},(i)}$.

In the above algorithms, we are free to choose any of the numerical integration methods that we first saw in the context of filtering in Chapter 8 are then later in the start of this chapter.

Example 14.23 (Pendulum tracking with IPLS). *The result of applying IPLS with Gauss–Hermite integration to the pendulum model in Example 7.6 is shown in Figure 14.6. The resulting RMSE error is* 0.03, *which matches that of the IERTSS.*

Example 14.24 (Coordinated turn model with Gaussian smoothers). *We applied iterated posterior linearization smoothers to the coordinated turn model introduced in Example 7.11 using Gauss–Hermite, spherical cubature, unscented, and fifth order cubature integration. The resulting RMSE was* 0.015 *for all of the methods (for IERTSS it was also* 0.015*). Figure 14.7 shows the result of the IURTSS, which looks similar to the other smoothers, together with the IERTSS result.*

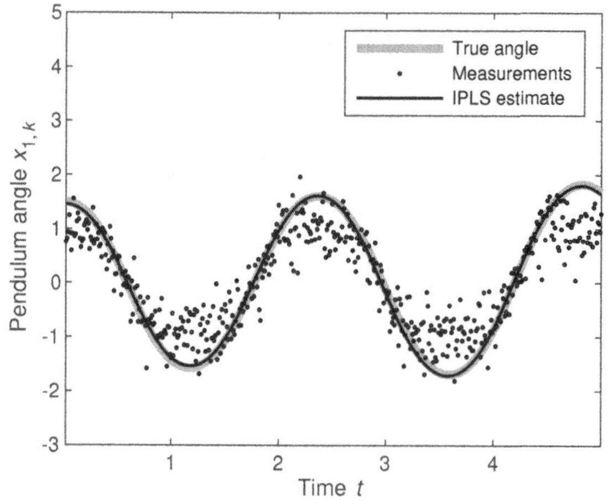

Figure 14.6 Simulated pendulum data and the result of tracking the pendulum described in Example 7.6 with the IPLS with Gauss–Hermite integration. The resulting RMSE is 0.03, which is practically the same as for IERTSS.

14.8 Exercises

14.1 Implement the Gauss–Hermite-based RTS smoother to the model in Exercise 7.1, and compare the errors of the filters and smoothers.

14.2 Formulate a Gaussian RTS smoother for non-additive noise models using the conditional moments formulation of the involved moments. *Hint:* Modify Algorithm 14.2 by using expressions for $\mathbf{m}_{k+1}^{-}$, $\mathbf{P}_{k+1}^{-}$, and $\mathbf{D}_{k+1}$ on the conditional form described in (9.51) and (9.52). See also Appendix A.9.

14.3 Formulate a Gaussian RTS smoother that uses conditional moments instead of the conditional distributions analogously to Algorithm 14.16.

14.4 Write down the detailed derivation of the (additive form) statistically linearized RTS smoother that uses statistical linearization from Section 9.2 instead of Taylor series expansions. You can follow the same steps as in the derivation of the extended RTS smoother.

14.5 Derive and implement the statistically linearized RTS smoother above to the model in Exercise 7.1 and compare the errors of the filters and smoothers.

14.6 Derive and implement an analytical SLR-based RTS smoother to the model in Exercise 7.1 (cf. Exercise 9.7), and compare the errors of the filters and smoothers.

14.7 Implement a cubature integration-based smoother for the bearings-only target tracking problem in Exercise 7.2. Compare the performance with the filters.

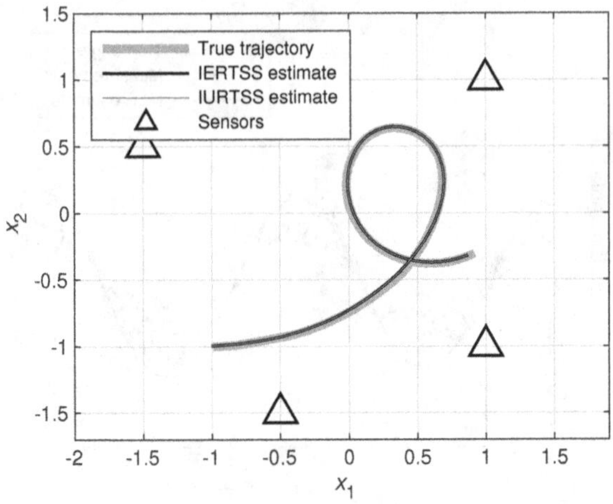

Figure 14.7 The result of applying IURTSS to the coordinated turn model. The result is practically the same as IERTSS, with an RMSE of 0.015.

14.8 Implement an unscented transform-based smoother for the Cartesian coordinated turn model in Equations (4.61) and (4.90) with the linear position measurement model. Also compare its performance with filters.

14.9 Implement the conditional moment-based smoother from Exercise 14.3 to the bicycle model considered in Exercise 9.11, and compare its performance with the Gaussian approximation-based filters.

14.10 Implement the Gauss–Hermite-based iterated posterior linearization smoother to the model in Exercise 7.1, and compare the errors of the filters and smoothers.

14.11 Formulate an iterated posterior linearization smoother (IPLS) for the non-additive state space model in (9.9). *Hint:* The algorithm is identical to the IPLS for additive noise models described in Algorithm 14.19 except for the expressions for the moments in (14.43) and (14.45). Similarly to Algorithm 14.21, you can therefore refer to equations in Algorithm 14.19 for the other parts.

14.12 Implement a cubature integration-based iterated posterior linearization smoother for the bearings-only target tracking problem in Exercise 7.2. Compare the performance with other smoothers.

14.13 Implement an unscented transform-based iterated posterior linearization smoother for the Cartesian coordinated turn model in Equations (4.61) and (4.90) with linear position measurement model. Also compare its performance with other smoothers.

14.14 Implement an iterated posterior linearization smoother to the bicycle model considered in Exercise 9.11 and compare its performance with other smoothers.

15

Particle Smoothing

When smoothing solutions to non-linear/non-Gaussian problems are sought, Gaussian approximations might not lead to accurate enough approximations. In that case it is better to use Monte Carlo (particle) approximations, which in principle can be used for approximating arbitrary smoothing distributions. Although the same SIR algorithm that is used for particle filtering provides an approximation to the smoothing distribution as a by-product, it does not yet solve the problem of particle smoothing. The challenge is that the resulting approximation tends to be degenerate. For this reason other types of particle smoothers have been developed, and here we present the most commonly used ones, the forward-filtering backward-sampling (FFBS) smoother and the reweighting-based (marginal) particle smoother.

15.1 SIR Particle Smoother

The *SIR particle smoother* (SIR-PS) of Kitagawa (1996) is based on direct usage of the SIR for smoothing. Recall that in Section 11.3 we derived the sequential importance sampling (SIS) method to approximate the full posterior distribution, not just the filtering distributions. We then discarded the sample histories $\mathbf{x}_{0:k-1}^{(i)}$ and only kept the current states $\mathbf{x}_k^{(i)}$, because we were interested in the filtering distributions. But we can get an approximation to the smoothing distribution by keeping the full histories. To get the smoothing solution from sequential importance resampling (SIR), we also need to resample the state histories, not only the current states, to prevent the resampling from breaking the state histories. The resulting algorithm is the following.

Algorithm 15.1 (SIR particle smoother)**.** *The direct sequential importance resampling (SIR)-based smoother algorithm is the following.*

- *Draw N samples $\mathbf{x}_0^{(i)}$ from the prior:*

$$\mathbf{x}_0^{(i)} \sim p(\mathbf{x}_0), \qquad i = 1, \ldots, N, \tag{15.1}$$

and set $w_0^{(i)} = 1/N$, for all $i = 1, \ldots, N$. Initialize the state histories to contain the prior samples $\mathbf{x}_0^{(i)}$.
- *For each $k = 1, \ldots, T$ do the following:*
  1. *Draw N new samples $\mathbf{x}_k^{(i)}$ from the importance distributions:*

$$\mathbf{x}_k^{(i)} \sim \pi(\mathbf{x}_k \mid \mathbf{x}_{k-1}^{(i)}, \mathbf{y}_{1:k}), \qquad i = 1, \ldots, N, \tag{15.2}$$

     where $\mathbf{x}_{k-1}^{(i)}$ is the $k-1$th (last) element in the sample history $\mathbf{x}_{0:k-1}^{(i)}$.
  2. *Calculate the new weights according to*

$$w_k^{(i)} \propto w_{k-1}^{(i)} \frac{p(\mathbf{y}_k \mid \mathbf{x}_k^{(i)})\, p(\mathbf{x}_k^{(i)} \mid \mathbf{x}_{k-1}^{(i)})}{\pi(\mathbf{x}_k^{(i)} \mid \mathbf{x}_{k-1}^{(i)}, \mathbf{y}_{1:k})}, \tag{15.3}$$

     and normalize them to sum to unity.
  3. *Append the samples to the state histories:*

$$\mathbf{x}_{0:k}^{(i)} = (\mathbf{x}_{0:k-1}^{(i)}, \mathbf{x}_k^{(i)}). \tag{15.4}$$

  4. *If the effective number of particles* (11.31) *is too low, perform resampling on the state histories.*

The approximation to the full posterior (smoothed) distribution is (Kitagawa, 1996; Doucet et al., 2000)

$$p(\mathbf{x}_{0:T} \mid \mathbf{y}_{1:T}) \approx \sum_{i=1}^{N} w_T^{(i)}\, \delta(\mathbf{x}_{0:T} - \mathbf{x}_{0:T}^{(i)}). \tag{15.5}$$

The approximation to the smoothed posterior distribution at time step k, given the measurements up to time step $T > k$, is

$$p(\mathbf{x}_k \mid \mathbf{y}_{1:T}) \approx \sum_{i=1}^{N} w_T^{(i)}\, \delta(\mathbf{x}_k - \mathbf{x}_k^{(i)}), \tag{15.6}$$

where $\mathbf{x}_k^{(i)}$ is the kth component in $\mathbf{x}_{0:T}^{(i)}$. However, if $T \gg k$, the direct SIR smoother algorithm is known to produce degenerate approximations (Kitagawa, 1996; Doucet et al., 2000).

15.2 Backward-Simulation Particle Smoother

A less degenerate particle smoother than the SIR particle smoother can be obtained by reusing the filtering results instead of simply storing the full particle histories in SIR. The *backward-simulation particle smoother* (BS-PS, Godsill et al., 2004; Lindsten and Schön, 2013), also called the *forward-filtering backward-sampling (FFBS) smoother*, is based on simulation of individual trajectories backward, starting from the last step and proceeding to the first. The algorithm is the following.

Algorithm 15.2 (Backward-simulation particle smoother). *Given the weighted set of particles* $\{w_k^{(i)}, \mathbf{x}_k^{(i)} \; : \; i = 1, \ldots, N, \; k = 1, \ldots, T\}$ *representing the filtering distributions:*

- *Choose* $\tilde{\mathbf{x}}_T = \mathbf{x}_T^{(i)}$ *with probability* $w_T^{(i)}$.
- *For* $k = T - 1, \ldots, 0$*:*

1. *Compute new weights by*

$$w_{k|k+1}^{(i)} \propto w_k^{(i)} \, p(\tilde{\mathbf{x}}_{k+1} \mid \mathbf{x}_k^{(i)}). \tag{15.7}$$

2. *Choose* $\tilde{\mathbf{x}}_k = \mathbf{x}_k^{(i)}$ *with probability* $w_{k|k+1}^{(i)}$.

Derivation Assume that we have already simulated a trajectory $\tilde{\mathbf{x}}_{k+1:T}$ from the smoothing distribution. By using Equation (12.3) we get

$$\begin{aligned} p(\mathbf{x}_k \mid \tilde{\mathbf{x}}_{k+1}, \mathbf{y}_{1:T}) &= \frac{p(\tilde{\mathbf{x}}_{k+1} \mid \mathbf{x}_k) \, p(\mathbf{x}_k \mid \mathbf{y}_{1:k})}{p(\tilde{\mathbf{x}}_{k+1} \mid \mathbf{y}_{1:k})} \\ &= Z \, p(\tilde{\mathbf{x}}_{k+1} \mid \mathbf{x}_k) \, p(\mathbf{x}_k \mid \mathbf{y}_{1:k}), \end{aligned} \tag{15.8}$$

where Z is a normalization constant. By substituting the SIR filter approximation in Equation (11.35) we get

$$p(\mathbf{x}_k \mid \tilde{\mathbf{x}}_{k+1}, \mathbf{y}_{1:T}) \approx Z \sum_i w_k^{(i)} \, p(\tilde{\mathbf{x}}_{k+1} \mid \mathbf{x}_k) \, \delta(\mathbf{x}_k - \mathbf{x}_k^{(i)}). \tag{15.9}$$

We can now draw a sample from this distribution by sampling $\mathbf{x}_k^{(i)}$ with probability $\propto w_k^{(i)} \, p(\tilde{\mathbf{x}}_{k+1} \mid \mathbf{x}_k^{(i)})$. □

Given S iterations of Algorithm 15.2, resulting in sample trajectories $\tilde{\mathbf{x}}_{0:T}^{(j)}$ for $j = 1, \ldots, S$, the smoothing distribution can now be approximated as

$$p(\mathbf{x}_{0:T} \mid \mathbf{y}_{1:T}) \approx \frac{1}{S} \sum_j \delta(\mathbf{x}_{0:T} - \tilde{\mathbf{x}}_{0:T}^{(j)}). \tag{15.10}$$

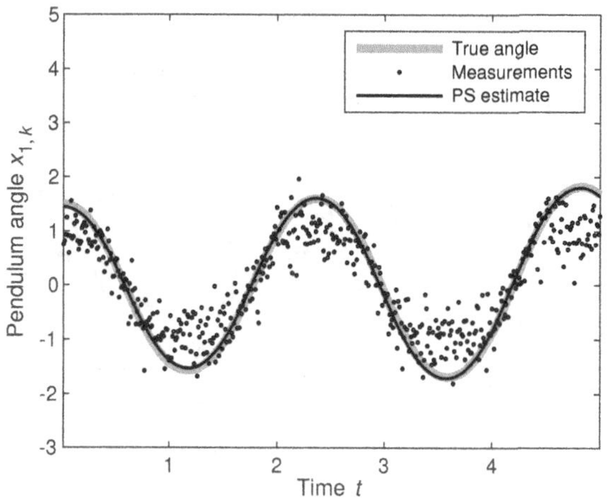

Figure 15.1 Simulated pendulum data, and the result of tracking the pendulum described in Example 7.6 with the backward-simulation particle smoother. The resulting RMSE is 0.04, which is similar to the Gaussian smoother errors. Recall that the RMSE of the bootstrap filter was 0.10, and thus the smoother reduces the error significantly.

The marginal distribution samples for a step k can be obtained by extracting the kth components from the above trajectories. The computational complexity of the method is $O(S\,T\,N)$. However, the result is much less degenerate than that of the particle smoother of Kitagawa (1996). Under suitable conditions, it is also possible to reduce the number of computations to linear in the number of particles by implementing the backward simulation using rejection sampling (see next section).

Example 15.3 (Pendulum tracking with BS-PS). *The result of applying the backward-simulation particle smoother with 100 samples (with a bootstrap filter with 10,000 samples as the filter part) to the pendulum model in Example 7.6 is shown in Figure 15.1. The RMSE error* 0.04 *is smaller than the RMSE of the ERTSS, which was* 0.06*, but slightly higher than that of the IERTSS, which was* 0.03*, while being similar to the errors of Gaussian smoothers. As already concluded in Example 11.10, the use of particle approximations is not really beneficial in this particular model.*

Example 15.4 (Cluttered pendulum tracking with BS-PS). *The result of applying the backward-simulation particle smoother with 100 samples*

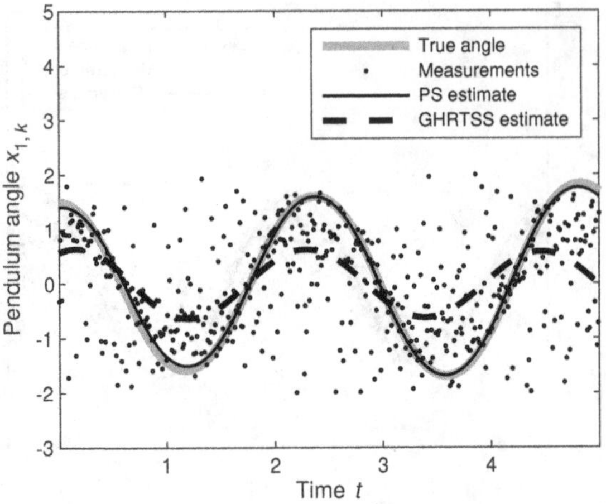

Figure 15.2 Smoothing results for cluttered pendulum tracking with particle smoother and GHRTSS. The RMSE errors of the methods were 0.05 and 0.83, respectively.

(with a bootstrap filter with 10,000 samples as the filter part) to the cluttered pendulum model in Example 11.11 is shown in Figure 15.2. The RMSE error of the particle smoother was 0.05*, while the RMSE of the particle filter was* 0.20*. The RMSE of a GHRTSS without a clutter model in this case was* 0.83*. Thus in this case the particle smoother gives a significant improvement over the Gaussian approximation-based smoothers.*

15.3 Backward-Simulation with Rejection Sampling

A challenge in the backward-simulation-based smoother introduced in the previous section is that generating $S = N$ smoothed trajectories from a particle filter result with N particles requires a computational complexity of $O(T\, N^2)$, where the quadratic dependence on N can be prohibitive when N is large. The reason for this computational complexity is the normalization of weights required for sampling $\mathbf{x}_k^{(i)}$ from the distribution $\propto w_k^{(i)}\, p(\tilde{\mathbf{x}}_{k+1} \mid \mathbf{x}_k^{(i)})$. However, it is also possible to use rejection sampling to avoid this computational complexity (Douc et al., 2011; Chopin and Papaspiliopoulos, 2020).

Rejection sampling (see, e.g., Luengo et al., 2020) is a method of drawing samples from a distribution $p(\mathbf{x})$ by using proposals from another distribution $q(\mathbf{x})$ that is selected such that $p(\mathbf{x}) \leq M\, q(\mathbf{x})$ for some constant M. The algorithm is as follows.

Algorithm 15.5 (Rejection sampling). *The rejection sampling algorithm has the following steps:*

- *Repeat the following until a candidate is accepted:*
  - *Draw a candidate* $\mathbf{x}' \sim q(\mathbf{x}')$ *and* $u \sim \mathrm{U}(0, 1)$.
  - *If*

$$u \leq \frac{p(\mathbf{x}')}{M\, q(\mathbf{x}')}, \tag{15.11}$$

    accept the candidate $\mathbf{x}'$, *otherwise continue with the loop.*
- *The value* $\mathbf{x} = \mathbf{x}'$ *is now a sample from the distribution* $p(\mathbf{x})$.

We can now use the above algorithm to sample from the following distribution, which is the one we sample from in the backward-simulation particle smoothers:

$$p(i) = \frac{w_k^{(i)}\, p(\tilde{\mathbf{x}}_{k+1} \mid \mathbf{x}_k^{(i)})}{\sum_j w_k^{(j)}\, p(\tilde{\mathbf{x}}_{k+1} \mid \mathbf{x}_k^{(j)})} = \frac{1}{Z_p}\, w_k^{(i)}\, p(\tilde{\mathbf{x}}_{k+1} \mid \mathbf{x}_k^{(i)}), \tag{15.12}$$

where $Z_p = \sum_j w_k^{(j)}\, p(\tilde{\mathbf{x}}_{k+1} \mid \mathbf{x}_k^{(j)})$. We can use rejection sampling for it provided that there exists a function $C(\mathbf{x}_k)$ such that

$$p(\tilde{\mathbf{x}}_{k+1} \mid \mathbf{x}_k) \leq C(\mathbf{x}_k). \tag{15.13}$$

What now follows is that if we use the proposal distribution

$$q(i) = \frac{w_k^{(i)}\, C(\mathbf{x}_k^{(i)})}{\sum_j w_k^{(j)}\, C(\mathbf{x}_k^{(j)})} = \frac{1}{Z_q}\, w_k^{(i)}\, C(\mathbf{x}_k^{(i)}), \tag{15.14}$$

where $Z_q = \sum_j w_k^{(j)}\, C(\mathbf{x}_k^{(j)})$, then we have:

$$p(i) = \frac{1}{Z_p}\, w_k^{(i)}\, p(\tilde{\mathbf{x}}_{k+1} \mid \mathbf{x}_k^{(i)}) \leq \frac{1}{Z_p}\, w_k^{(i)}\, C(\mathbf{x}_k^{(i)}) = \frac{Z_q}{Z_p}\, q(i), \tag{15.15}$$

where the constant for rejection sampling is thus $M = Z_q / Z_p$. The ratio needed in the rejection sampling then simplifies to:

$$\frac{p(i)}{(Z_q/Z_p)\, q(i)} = \frac{p(\tilde{\mathbf{x}}_{k+1} \mid \mathbf{x}_k^{(i)})}{C(\mathbf{x}_k^{(i)})}. \tag{15.16}$$

We get the following algorithm.

Algorithm 15.6 (Backward-simulation particle smoother with rejection sampling). *Given the weighted set of particles* $\{w_k^{(i)}, \mathbf{x}_k^{(i)} : i = 1, \ldots, N, k = 1, \ldots, T\}$ *representing the filtering distributions and a function* $C(\mathbf{x}_k)$ *such that* $p(\tilde{\mathbf{x}}_{k+1} \mid \mathbf{x}_k) \le C(\mathbf{x}_k)$*:*

- *Choose* $\tilde{\mathbf{x}}_T = \mathbf{x}_T^{(i)}$ *with probabilities given by the weights* $w_T^{(i)}$.
- *For* $k = T - 1, \ldots, 0$*:*

  1. *While sample not accepted:*
     (a) *Draw proposal index i from distribution* $q(i) \propto w_k^{(i)}\, C(\mathbf{x}_k^{(i)})$.
     (b) *Draw* $u \sim \mathrm{U}(0, 1)$*, and accept* $\mathbf{x}_k^{(i)}$ *if* $u \le \frac{p(\tilde{\mathbf{x}}_{k+1} \mid \mathbf{x}_k^{(i)})}{C(\mathbf{x}_k^{(i)})}$.
  2. *Choose* $\tilde{\mathbf{x}}_k = \mathbf{x}_k^{(i)}$.

In practice, to obtain an algorithm with complexity close to $O(TN)$ for obtaining N trajectories from the smoothing distribution, we need to sample them jointly, not one-by-one as the above simple algorithm description implies. For more details, please see Douc et al. (2011) and Chopin and Papaspiliopoulos (2020).

One practical challenge with Algorithm 15.6 is that the acceptance probabilities for some particles can be very low, which can cause the run-time of the algorithm to be extremely long. For that reason, Taghavi et al. (2013) propose a hybrid strategy which falls back to the backward-simulation particle smoother when the rejection sampling step takes too long.

15.4 Reweighting Particle Smoother

The *reweighting particle smoother* of Hürzeler and Kunsch (1998) and Doucet et al. (2000), which is also called the *marginal particle smoother*, is based on computing new weights for the SIR filtering particles such that we get an approximation to the marginal smoothing distribution.

Algorithm 15.7 (Reweighting particle smoother). *Given the weighted set of particles* $\{w_k^{(i)}, \mathbf{x}_k^{(i)} : i = 1, \ldots, N\}$ *representing the filtering distribution, we can form approximations to the marginal smoothing distributions as follows.*

- *Start by setting* $w_{T|T}^{(i)} = w_T^{(i)}$ *for* $i = 1, \ldots, N$.

- *For each $k = T-1, \ldots, 0$, compute new weights by*

$$w_{k|T}^{(i)} = \sum_j w_{k+1|T}^{(j)} \frac{w_k^{(i)} \, p(\mathbf{x}_{k+1}^{(j)} \mid \mathbf{x}_k^{(i)})}{\left[\sum_l w_k^{(l)} \, p(\mathbf{x}_{k+1}^{(j)} \mid \mathbf{x}_k^{(l)})\right]}. \tag{15.17}$$

At each step k, the marginal smoothing distribution can be approximated as

$$p(\mathbf{x}_k \mid \mathbf{y}_{1:T}) \approx \sum_i w_{k|T}^{(i)} \, \delta(\mathbf{x}_k - \mathbf{x}_k^{(i)}). \tag{15.18}$$

Derivation Assume that we have already computed the weights for the following approximation, where the particles $\mathbf{x}_{k+1}^{(i)}$ are from the SIR filter:

$$p(\mathbf{x}_{k+1} \mid \mathbf{y}_{1:T}) \approx \sum_i w_{k+1|T}^{(i)} \, \delta(\mathbf{x}_{k+1} - \mathbf{x}_{k+1}^{(i)}). \tag{15.19}$$

The integral in the second of Equations (12.2) can be now approximated as

$$\begin{aligned}
&\int \frac{p(\mathbf{x}_{k+1} \mid \mathbf{x}_k) \, p(\mathbf{x}_{k+1} \mid \mathbf{y}_{1:T})}{p(\mathbf{x}_{k+1} \mid \mathbf{y}_{1:k})} \, \mathrm{d}\mathbf{x}_{k+1} \\
&\quad \approx \int \frac{p(\mathbf{x}_{k+1} \mid \mathbf{x}_k)}{p(\mathbf{x}_{k+1} \mid \mathbf{y}_{1:k})} \sum_i \left[w_{k+1|T}^{(i)} \, \delta(\mathbf{x}_{k+1} - \mathbf{x}_{k+1}^{(i)}) \right] \mathrm{d}\mathbf{x}_{k+1} \\
&\quad = \sum_i w_{k+1|T}^{(i)} \frac{p(\mathbf{x}_{k+1}^{(i)} \mid \mathbf{x}_k)}{p(\mathbf{x}_{k+1}^{(i)} \mid \mathbf{y}_{1:k})}.
\end{aligned} \tag{15.20}$$

By using SIR filter approximation in Equation (11.35), we get the following approximation for the predicted distribution in the denominator:

$$p(\mathbf{x}_{k+1} \mid \mathbf{y}_{1:k}) \approx \sum_j w_k^{(j)} \, p(\mathbf{x}_{k+1} \mid \mathbf{x}_k^{(j)}), \tag{15.21}$$

which gives

$$\begin{aligned}
&\int \frac{p(\mathbf{x}_{k+1} \mid \mathbf{y}_{1:T}) \, p(\mathbf{x}_{k+1} \mid \mathbf{x}_k)}{p(\mathbf{x}_{k+1} \mid \mathbf{y}_{1:k})} \, \mathrm{d}\mathbf{x}_{k+1} \\
&\quad \approx \sum_i w_{k+1|T}^{(i)} \frac{p(\mathbf{x}_{k+1}^{(i)} \mid \mathbf{x}_k)}{\left[\sum_j w_k^{(j)} \, p(\mathbf{x}_{k+1}^{(i)} \mid \mathbf{x}_k^{(j)})\right]}.
\end{aligned} \tag{15.22}$$

By substituting the SIR filter approximation and the approximation above into the Bayesian optimal smoothing equation, we get

$$
\begin{aligned}
&p(\mathbf{x}_k \mid \mathbf{y}_{1:T}) \\
&\quad = p(\mathbf{x}_k \mid \mathbf{y}_{1:k}) \int \left[\frac{p(\mathbf{x}_{k+1} \mid \mathbf{x}_k)\, p(\mathbf{x}_{k+1} \mid \mathbf{y}_{1:T})}{p(\mathbf{x}_{k+1} \mid \mathbf{y}_{1:k})} \right] \mathrm{d}\mathbf{x}_{k+1} \\
&\quad \approx \sum_l w_k^{(l)}\, \delta(\mathbf{x}_k - \mathbf{x}_k^{(l)}) \sum_i w_{k+1|T}^{(i)} \frac{p(\mathbf{x}_{k+1}^{(i)} \mid \mathbf{x}_k^{(l)})}{\left[\sum_j w_k^{(j)}\, p(\mathbf{x}_{k+1}^{(i)} \mid \mathbf{x}_k^{(j)}) \right]},
\end{aligned}
\tag{15.23}
$$

where we can identify the weights as

$$
w_{k|T}^{(l)} = \sum_i w_{k+1|T}^{(i)} \frac{w_k^{(l)}\, p(\mathbf{x}_{k+1}^{(i)} \mid \mathbf{x}_k^{(l)})}{\left[\sum_j w_k^{(j)}\, p(\mathbf{x}_{k+1}^{(i)} \mid \mathbf{x}_k^{(j)}) \right]}. \tag{15.24}
$$

□

The computational complexity of the method is $O(T\,N^2)$, that is, the same as of the backward-simulation smoother with $S = N$ simulated trajectories.

15.5 Rao–Blackwellized Particle Smoothers

Rao–Blackwellized particle smoothers (RBPS) can be used for computing approximate smoothing solutions to the conditionally Gaussian models defined in Equation (11.50). A simple way to implement an RBPS is to store the histories instead of the single states in the RBPF, as in the case of the SIR particle smoother (Algorithm 15.1). The corresponding histories of the means and the covariances are then conditional on the *latent variable histories* $\mathbf{u}_{0:T}$. However, the means and covariances at time step k are only conditional on the *measurement histories* up to k, not on the later measurements. In order to correct this, RTS smoothers have to be applied to each history of the means and the covariances.

Algorithm 15.8 (Rao–Blackwellized SIR particle smoother). *A set of weighted samples* $\{w_T^{\mathrm{s},(i)}, \mathbf{u}_{0:T}^{\mathrm{s},(i)}, \mathbf{m}_{0:T}^{\mathrm{s},(i)}, \mathbf{P}_{0:T}^{\mathrm{s},(i)} : i = 1, \ldots, N\}$ *representing the smoothed distribution can be computed as follows.*

1. *Compute the weighted set of Rao–Blackwellized state histories*

$$
\{w_T^{(i)}, \mathbf{u}_{0:T}^{(i)}, \mathbf{m}_{0:T}^{(i)}, \mathbf{P}_{0:T}^{(i)} : i = 1, \ldots, N\} \tag{15.25}
$$

by storing histories in the Rao–Blackwellized particle filter analogously to the SIR particle smoother in Algorithm 15.1.

2. *Set*

$$\begin{aligned} w_T^{\mathrm{s},(i)} &= w_T^{(i)}, \\ \mathbf{u}_{0:T}^{\mathrm{s},(i)} &= \mathbf{u}_{0:T}^{(i)}. \end{aligned} \tag{15.26}$$

3. *Apply the RTS smoother to each of the mean and covariance histories* $\mathbf{m}_{0:T}^{(i)}, \mathbf{P}_{0:T}^{(i)}$ *for* $i = 1, \ldots, N$ *to produce the smoothed mean and covariance histories* $\mathbf{m}_{0:T}^{\mathrm{s},(i)}, \mathbf{P}_{0:T}^{\mathrm{s},(i)}$.

The Rao–Blackwellized particle smoother in this simple form also has the same disadvantage as the SIR particle smoother, that is, the smoothed estimate of $\mathbf{u}_k$ can be quite degenerate if $T \gg k$. Fortunately, the smoothed estimates of the actual states $\mathbf{x}_k$ can still be relatively good, because their degeneracy is avoided by the Rao–Blackwellization.

To avoid the degeneracy in estimates of $\mathbf{u}_k$, it is possible to use better sampling procedures for generating samples from the smoothing distributions analogously to the plain particle smoothing. The backward-simulation has indeed been generalized to the Rao–Blackwellized case, but the implementation of the Rao–Blackwellized reweighting smoother seems to be quite problematic. The Rao–Blackwellized backward-simulation smoother (see Särkkä et al., 2012a; Lindsten et al., 2016) can be used for drawing backward trajectories from the marginal posterior of the latent variables $\mathbf{u}_k$, and the posterior of the conditionally Gaussian part is obtained via Kalman filtering and RTS smoothing. Another option is to simulate backward trajectories from the joint distribution of $(\mathbf{x}_k, \mathbf{u}_k)$ (Fong et al., 2002; Lindsten, 2011). However, this approach does not really lead to Rao–Blackwellized estimates of the smoothing distribution, because the Gaussian part of the state is sampled as well.

It is also possible to construct approximate Rao–Blackwellized backward-simulation smoothers by using Kim's approximation (cf. Kim, 1994; Barber, 2006; Särkkä et al., 2012a)

$$\begin{aligned} &\int p(\mathbf{u}_k \mid \mathbf{u}_{k+1}, \mathbf{x}_{k+1}, \mathbf{y}_{1:k})\, p(\mathbf{x}_{k+1} \mid \mathbf{u}_{k+1}, \mathbf{y}_{1:T})\, \mathrm{d}\mathbf{x}_{k+1} \\ &\qquad \simeq p(\mathbf{u}_k \mid \mathbf{u}_{k+1}, \mathbf{y}_{1:k}). \end{aligned} \tag{15.27}$$

The result is an algorithm where we first apply the backward-simulation smoother in the Algorithm 15.2 to the marginal samples of $\mathbf{u}_k$ alone, that is, we simply use $p(\mathbf{u}_{k+1} \mid \mathbf{u}_k)$ instead of $p(\mathbf{x}_{k+1} \mid \mathbf{x}_k)$ in the algorithm. Given a trajectory of the non-Gaussian variable, the linear Gaussian part may be recovered with a Kalman filter and RTS smoother. However, this procedure is only an approximation and does not lead to a true Rao–Blackwellized Monte Carlo representation of the smoothing distribution.

15.6 Exercises

15.1 Implement the backward-simulation particle smoother for the model in Exercise 7.1, and compare its performance to the Gaussian approximation-based smoothers.

15.2 Implement the backward-simulation particle smoother with rejection sampling for the model in Exercise 7.1, and compare its performance to the other smoothers.

15.3 Implement the reweighting smoother for the model in Exercise 7.1 and compare its performance to the other smoothers.

15.4 Show that the latent variable sequence in conditionally Gaussian models is not Markovian in general in the sense that

$$p(\mathbf{u}_k \mid \mathbf{u}_{k+1}, \mathbf{y}_{1:T}) \neq p(\mathbf{u}_k \mid \mathbf{u}_{k+1}, \mathbf{y}_{1:k}) \tag{15.28}$$

when $T > k$, and thus simple backward smoothing in $\mathbf{u}_k$ does not lead to the correct result.

15.5 Implement the Rao–Blackwellized SIR particle smoother for the clutter model in Exercise 11.9.

15.6 Let's again consider the clutter model in Exercise 11.9. Assume that you have implemented the filter as a Rao–Blackwellized particle filter with resampling at every step (thus the weights are all equal). Write down the algorithm for Kim's approximation-based backward simulation smoother for the model. What peculiar property does the smoother have? Does this have something to do with the property in Equation (15.28)?

16

Parameter Estimation

In the previous chapters we have assumed that the parameters of the state space model are known, and only the state needs to be estimated. However, in practical models, the parameters are unknown as well. In this chapter we concentrate on three types of method for parameter estimation: optimization-based methods for computing maximum a posteriori (MAP) or maximum likelihood (ML) estimates, expectation-maximization (EM) algorithms for computing the MAP or ML estimates, and Markov chain Monte Carlo (MCMC) methods for generating Monte Carlo approximations of the posterior distributions. We also show how Kalman filters and RTS smoothers, Gaussian filters and smoothers, and particle filters and smoothers can be used for approximating the marginal likelihoods, parameter posteriors, and other quantities needed by the methods.

16.1 Bayesian Estimation of Parameters in State Space Models

The Bayesian way of treating unknown parameters $\boldsymbol{\theta} \in \mathbb{R}^d$ is to model them as random variables with a certain prior distribution $p(\boldsymbol{\theta})$. A state space model with unknown parameters can be written in the form

$$\begin{aligned}
\boldsymbol{\theta} &\sim p(\boldsymbol{\theta}), \\
\mathbf{x}_0 &\sim p(\mathbf{x}_0 \mid \boldsymbol{\theta}), \\
\mathbf{x}_k &\sim p(\mathbf{x}_k \mid \mathbf{x}_{k-1}, \boldsymbol{\theta}), \\
\mathbf{y}_k &\sim p(\mathbf{y}_k \mid \mathbf{x}_k, \boldsymbol{\theta}).
\end{aligned} \tag{16.1}$$

A straightforward way to proceed would now be to form the full posterior distribution via Bayes' rule:

$$p(\mathbf{x}_{0:T}, \boldsymbol{\theta} \mid \mathbf{y}_{1:T}) = \frac{p(\mathbf{y}_{1:T} \mid \mathbf{x}_{0:T}, \boldsymbol{\theta})\, p(\mathbf{x}_{0:T} \mid \boldsymbol{\theta})\, p(\boldsymbol{\theta})}{p(\mathbf{y}_{1:T})}, \tag{16.2}$$

where the terms $p(\mathbf{x}_{0:T} \mid \boldsymbol{\theta})$ and $p(\mathbf{y}_{1:T} \mid \mathbf{x}_{0:T}, \boldsymbol{\theta})$ can be computed as

$$p(\mathbf{x}_{0:T} \mid \boldsymbol{\theta}) = p(\mathbf{x}_0 \mid \boldsymbol{\theta}) \prod_{k=1}^{T} p(\mathbf{x}_k \mid \mathbf{x}_{k-1}, \boldsymbol{\theta}),$$

$$p(\mathbf{y}_{1:T} \mid \mathbf{x}_{0:T}, \boldsymbol{\theta}) = \prod_{k=1}^{T} p(\mathbf{y}_k \mid \mathbf{x}_k, \boldsymbol{\theta}).$$

If we are only interested in the parameters $\boldsymbol{\theta}$, the proper Bayesian way to proceed is to integrate the states out, which gives the marginal posterior of parameters:

$$p(\boldsymbol{\theta} \mid \mathbf{y}_{1:T}) = \int p(\mathbf{x}_{0:T}, \boldsymbol{\theta} \mid \mathbf{y}_{1:T}) \, \mathrm{d}\mathbf{x}_{0:T}. \tag{16.3}$$

Unfortunately, computation of this high-dimensional integral is hard and becomes even harder as we obtain more measurements. In this approach we encounter the same computational problem as was discussed in Sections 1.3 and 6.1, which led us to consider optimal filtering and smoothing instead of the straightforward Bayesian approach. Thus it is again advantageous to look at recursive, filtering, and smoothing kinds of solutions.

In this chapter we present methods for parameter estimation that are based on approximating the marginal posterior distribution

$$p(\boldsymbol{\theta} \mid \mathbf{y}_{1:T}) \propto p(\mathbf{y}_{1:T} \mid \boldsymbol{\theta}) \, p(\boldsymbol{\theta}), \tag{16.4}$$

without explicitly forming the joint posterior distribution of the states and parameters as in Equation (16.2). Instead, we present recursive algorithms for direct computation of the above distribution. For linear state space models, this can be done exactly, and in non-linear and non-Gaussian models we can use Gaussian filtering or particle filtering-based approximations. Once we know how to evaluate the above distribution, we can estimate the parameters, for example, by finding their maximum a posteriori (MAP) estimate or by sampling from the posterior by Markov chain Monte Carlo (MCMC) methods. If direct evaluation of the distribution is not feasible, we can use the expectation-maximization (EM) algorithm to iteratively find the ML or MAP estimate.

16.1.1 Parameter Posterior and Energy Function

The difficult part in Equation (16.4) is the evaluation of the marginal likelihood $p(\mathbf{y}_{1:T} \mid \boldsymbol{\theta})$. The prior distribution can usually be selected such that it is easy to evaluate. Although evaluation of the normalization constant for the posterior distribution is a difficult problem, its evaluation is usually avoided by Bayesian computational methods, and thus we do not need to worry about it.

The key to recursive computation of the parameter posterior in state space models is the following factorization (often called prediction error decomposition):

$$p(\mathbf{y}_{1:T} \mid \boldsymbol{\theta}) = \prod_{k=1}^{T} p(\mathbf{y}_k \mid \mathbf{y}_{1:k-1}, \boldsymbol{\theta}), \tag{16.5}$$

where we have denoted $p(\mathbf{y}_1 \mid \mathbf{y}_{1:0}, \boldsymbol{\theta}) \triangleq p(\mathbf{y}_1 \mid \boldsymbol{\theta})$ for notational convenience. Because each of the terms in the above product can be computed recursively, the whole marginal likelihood can be computed recursively as follows.

Theorem 16.1 (Recursion for marginal likelihood of parameters). *The marginal likelihood of parameters is given by Equation* (16.5)*, where the terms in the product can be computed recursively as*

$$p(\mathbf{y}_k \mid \mathbf{y}_{1:k-1}, \boldsymbol{\theta}) = \int p(\mathbf{y}_k \mid \mathbf{x}_k, \boldsymbol{\theta})\, p(\mathbf{x}_k \mid \mathbf{y}_{1:k-1}, \boldsymbol{\theta})\, \mathrm{d}\mathbf{x}_k, \tag{16.6}$$

where $p(\mathbf{y}_k \mid \mathbf{x}_k, \boldsymbol{\theta})$ is the measurement model, and $p(\mathbf{x}_k \mid \mathbf{y}_{1:k-1}, \boldsymbol{\theta})$ is the predictive distribution of the state, which obeys the recursion

$$\begin{aligned} p(\mathbf{x}_k \mid \mathbf{y}_{1:k-1}, \boldsymbol{\theta}) &= \int p(\mathbf{x}_k \mid \mathbf{x}_{k-1}, \boldsymbol{\theta})\, p(\mathbf{x}_{k-1} \mid \mathbf{y}_{1:k-1}, \boldsymbol{\theta})\, \mathrm{d}\mathbf{x}_{k-1}, \\ p(\mathbf{x}_k \mid \mathbf{y}_{1:k}, \boldsymbol{\theta}) &= \frac{p(\mathbf{y}_k \mid \mathbf{x}_k, \boldsymbol{\theta})\, p(\mathbf{x}_k \mid \mathbf{y}_{1:k-1}, \boldsymbol{\theta})}{p(\mathbf{y}_k \mid \mathbf{y}_{1:k-1}, \boldsymbol{\theta})}. \end{aligned} \tag{16.7}$$

Note that the latter equations are just the Bayesian filtering Equations (6.11) *and* (6.12)*, where we have explicitly written down the parameter dependence.*

Proof Due to the conditional independence of the measurements (Property 6.3), we have

$$\begin{aligned} p(\mathbf{y}_k, \mathbf{x}_k \mid \mathbf{y}_{1:k-1}, \boldsymbol{\theta}) &= p(\mathbf{y}_k \mid \mathbf{x}_k, \mathbf{y}_{1:k-1}, \boldsymbol{\theta})\, p(\mathbf{x}_k \mid \mathbf{y}_{1:k-1}, \boldsymbol{\theta}) \\ &= p(\mathbf{y}_k \mid \mathbf{x}_k, \boldsymbol{\theta})\, p(\mathbf{x}_k \mid \mathbf{y}_{1:k-1}, \boldsymbol{\theta}). \end{aligned} \tag{16.8}$$

Integrating over $\mathbf{x}_k$ gives Equation (16.6). □

The marginal likelihood obtained via Theorem 16.1 can then be substituted into Equation (16.4) to give the marginal posterior distribution of the parameters. However, instead of working with marginal likelihood or marginal posterior explicitly, in parameter estimation it is often convenient to define the unnormalized negative log-posterior or *energy function* as follows.

Definition 16.2 (Energy function).

$$\varphi_T(\boldsymbol{\theta}) = -\log p(\mathbf{y}_{1:T} \mid \boldsymbol{\theta}) - \log p(\boldsymbol{\theta}). \tag{16.9}$$

Remark 16.3. *The definition of energy function thus implies*

$$p(\boldsymbol{\theta} \mid \mathbf{y}_{1:T}) \propto \exp(-\varphi_T(\boldsymbol{\theta})). \tag{16.10}$$

The energy function can be seen to obey the following simple recursion.

Theorem 16.4 (Recursion for energy function). *The energy function defined in Equation* (16.9) *can be evaluated recursively as follows.*

- *Start from* $\varphi_0(\theta) = -\log p(\boldsymbol{\theta})$.
- *At each step* $k = 1, 2, \ldots, T$, *compute the following:*

$$\varphi_k(\boldsymbol{\theta}) = \varphi_{k-1}(\boldsymbol{\theta}) - \log p(\mathbf{y}_k \mid \mathbf{y}_{1:k-1}, \boldsymbol{\theta}), \tag{16.11}$$

where the terms $p(\mathbf{y}_k \mid \mathbf{y}_{1:k-1}, \boldsymbol{\theta})$ *can be computed recursively by Theorem 16.1.*

Proof The result follows from substituting Equation (16.5) into the definition of the energy function in Equation (16.9) and identifying the terms corresponding to $\varphi_{k-1}(\boldsymbol{\theta})$. □

16.2 Computational Methods for Parameter Estimation

In this section we briefly go through the underlying ideas in ML- and MAP-based parameter estimation and implementation by direct optimization and by the EM algorithm, as well as the basics of Markov chain Monte Carlo (MCMC) methods. There exist many other parameter estimation methods for state space models and for more general statistical models, but here we concentrate on these, because these approaches are the most widely used (probabilistic methods) in the state space context.

16.2.1 Maximum A Posteriori and Laplace Approximations

The *maximum a posteriori (MAP)* estimate is obtained by determining the location of the maximum of the posterior distribution and using it as the point estimate:

$$\hat{\boldsymbol{\theta}}^{\mathrm{MAP}} = \arg\max_{\boldsymbol{\theta}} \left[p(\boldsymbol{\theta} \mid \mathbf{y}_{1:T}) \right]. \tag{16.12}$$

The MAP estimate can be equivalently computed as the minimum of the error function defined in Equation (16.9):

$$\hat{\boldsymbol{\theta}}^{\mathrm{MAP}} = \arg\min_{\boldsymbol{\theta}} \left[\varphi_T(\boldsymbol{\theta}) \right], \tag{16.13}$$

which is usually numerically more stable and easier to compute. The maximum likelihood (ML) estimate of the parameter is a MAP estimate with a formally uniform prior $p(\boldsymbol{\theta}) \propto 1$.

The minimum of the energy function can be computed by using various gradient-free or gradient-based general optimization algorithms (see, e.g., Luenberger and Ye, 2008). However, to be able to use gradient-based optimization, we will need to evaluate the derivatives of the energy function as well. It is possible to find the derivatives in basically two ways (see, e.g., Cappé et al., 2005).

1. By formally differentiating the energy function recursion equations for a particular method. This results in so-called *sensitivity equations*, which can be implemented as additional recursion equations computed along with the filtering.
2. Using *Fisher's identity*, which expresses the gradient of the energy function as an expectation of the derivative of the complete data log likelihood over the smoothing distribution. The advantage of this approach over direct differentiation is that there is no need for an additional recursion.

The disadvantage of the MAP estimate is that it essentially approximates the posterior distribution with the Dirac delta function

$$p(\boldsymbol{\theta} \mid \mathbf{y}_{1:T}) \simeq \delta(\boldsymbol{\theta} - \hat{\boldsymbol{\theta}}^{\mathrm{MAP}}) \tag{16.14}$$

and thus ignores the spread of the distribution completely.

It is also possible to use a Laplace approximation (Gelman et al., 2013), which uses the second derivative (Hessian) of the energy function to form a Gaussian approximation to the posterior distribution:

$$p(\boldsymbol{\theta} \mid \mathbf{y}_{1:T}) \simeq \mathrm{N}(\boldsymbol{\theta} \mid \hat{\boldsymbol{\theta}}^{\mathrm{MAP}}, [\mathbf{H}(\hat{\boldsymbol{\theta}}^{\mathrm{MAP}})]^{-1}), \tag{16.15}$$

where $\mathbf{H}(\hat{\boldsymbol{\theta}}^{\text{MAP}})$ is the Hessian matrix evaluated at the MAP estimate. However, to implement the Laplace approximation, we need to have a method to compute (or approximate) the second order derivatives of the energy function.

16.2.2 Parameter Inference via Markov Chain Monte Carlo

Markov chain Monte Carlo (MCMC) methods (see, e.g., Liu, 2001; Brooks et al., 2011) are a class of algorithms for drawing random variables from a given distribution by simulating a Markov chain that has the desired distribution as its stationary distribution. The methods are particularly suited for simulating samples from Bayesian posterior distributions $p(\boldsymbol{\theta} \mid \mathbf{y}_{1:T})$, because to implement the methods, we only need to know the unnormalized posterior distribution $\tilde{p}(\boldsymbol{\theta} \mid \mathbf{y}_{1:T}) = p(\mathbf{y}_{1:T} \mid \boldsymbol{\theta})\, p(\boldsymbol{\theta})$ or equivalently the energy function in Equation (16.9), and knowledge of the normalization constant of the posterior distribution is not required. The usage of MCMC methods in the state space context has been discussed, for example, by Ninness and Henriksen (2010) and Andrieu et al. (2010).

The *Metropolis–Hastings* (MH) algorithm is the most common type of MCMC method. MH uses a *proposal density* $q(\boldsymbol{\theta}^{(i)} \mid \boldsymbol{\theta}^{(i-1)})$ for suggesting new samples $\boldsymbol{\theta}^{(i)}$ given the previous ones $\boldsymbol{\theta}^{(i-1)}$. The algorithm is the following.

Algorithm 16.5 (Metropolis–Hastings). *The Metropolis–Hastings (MH) algorithm consists of the following steps.*

- *Draw the starting point $\boldsymbol{\theta}^{(0)}$ from an arbitrary initial distribution.*
- *For $i = 1, 2, \ldots, N$, do*

1. *Sample a candidate point $\boldsymbol{\theta}^*$ from the proposal distribution:*

$$\boldsymbol{\theta}^* \sim q(\boldsymbol{\theta}^* \mid \boldsymbol{\theta}^{(i-1)}). \tag{16.16}$$

2. *Evaluate the acceptance probability*

$$\alpha_i = \min\left\{1, \exp(\varphi_T(\boldsymbol{\theta}^{(i-1)}) - \varphi_T(\boldsymbol{\theta}^*))\,\frac{q(\boldsymbol{\theta}^{(i-1)} \mid \boldsymbol{\theta}^*)}{q(\boldsymbol{\theta}^* \mid \boldsymbol{\theta}^{(i-1)})}\right\}. \tag{16.17}$$

3. *Generate a uniform random variable $u \sim \mathrm{U}(0, 1)$ and set*

$$\boldsymbol{\theta}^{(i)} = \begin{cases} \boldsymbol{\theta}^*, & \text{if } u \le \alpha_i, \\ \boldsymbol{\theta}^{(i-1)}, & \text{otherwise}. \end{cases} \tag{16.18}$$

The Metropolis algorithm is a commonly used special case of Metropolis–Hastings, where the proposal distribution is symmetric, $q(\boldsymbol{\theta}^{(i-1)} \mid \boldsymbol{\theta}^{(i)}) = q(\boldsymbol{\theta}^{(i)} \mid \boldsymbol{\theta}^{(i-1)})$. In this case the acceptance probability reduces to

$$\alpha_i = \min\left\{1, \exp(\varphi_T(\boldsymbol{\theta}^{(i-1)}) - \varphi_T(\boldsymbol{\theta}^*))\right\}. \tag{16.19}$$

The choice of the proposal distribution is crucial for performance of the Metropolis–Hastings method, and finding a good one is a hard task (see, e.g., Liu, 2001; Brooks et al., 2011). Some choices will lead to Metropolis–Hastings methods where the samples are highly correlated, whereas with some choices the rejection rate becomes too high.

One commonly used choice is to use a Gaussian distribution as the proposal distribution,

$$q(\boldsymbol{\theta}^{(i)} \mid \boldsymbol{\theta}^{(i-1)}) = \mathrm{N}(\boldsymbol{\theta}^{(i)} \mid \boldsymbol{\theta}^{(i-1)}, \boldsymbol{\Sigma}_{i-1}), \tag{16.20}$$

where $\boldsymbol{\Sigma}_{i-1}$ is some suitable covariance matrix. The resulting algorithm is called the random walk Metropolis algorithm, because the transition distribution above defines a Gaussian random walk in parameter space. With this selection of proposal distribution, the challenge is now to find a suitable covariance matrix for the random walk.

One approach to the problem of selection of the covariance matrix is to use *adaptive Markov chain Monte Carlo (AMCMC)* methods where the covariance of the Gaussian proposal in the Metropolis algorithm is automatically adapted during the MCMC run (see, e.g., Haario et al., 1999, 2001; Andrieu and Thoms, 2008; Vihola, 2012). A typical idea in AMCMC methods is to use the covariance of the previously generated samples as an estimate of the actual covariance of the posterior distribution. Given the covariance, it is possible to compute the covariance of the proposal distribution such that it causes an acceptance rate $\bar{\alpha}_*$ that is optimal in some suitable sense. For the random walk Metropolis algorithm, the optimal acceptance rate in certain ideal conditions is $\bar{\alpha}_* = 0.234$ (Roberts and Rosenthal, 2001).

For example, the *robust adaptive Metropolis (RAM)* algorithm of Vihola (2012) is similar to the adaptive Metropolis (AM) algorithm of Haario et al. (2001) except that the adaptation of the covariance $\boldsymbol{\Sigma}_i$ is done in a slightly different way. The algorithm is the following.

Algorithm 16.6 (RAM algorithm)**.** *The RAM algorithm consists of the following steps.*

1. *Draw $\boldsymbol{\theta}^{(0)}$ from an initial distribution $p_0(\boldsymbol{\theta})$, and initialize $\mathbf{S}_0$ to be the lower-triangular Cholesky factor of the initial covariance $\boldsymbol{\Sigma}_0$.*
2. *Sample a candidate point by $\boldsymbol{\theta}^* = \boldsymbol{\theta}_{i-1} + \mathbf{S}_{i-1}\,\mathbf{r}_i$, where $\mathbf{r}_i \sim \mathrm{N}(\mathbf{0}, \mathbf{I})$.*
3. *Compute the acceptance probability*

$$\alpha_i = \min\{1, \exp(\varphi_T(\boldsymbol{\theta}_{i-1}) - \varphi_T(\boldsymbol{\theta}^*))\}. \tag{16.21}$$

4. *Sample a uniform random variable u from the uniform distribution* $\mathrm{U}(0, 1)$.
5. *If $u \le \alpha_i$, set $\boldsymbol{\theta}^{(i)} = \boldsymbol{\theta}^*$. Otherwise set $\boldsymbol{\theta}^{(i)} = \boldsymbol{\theta}^{(i-1)}$.*
6. *Compute a lower-triangular matrix $\mathbf{S}_i$ with positive diagonal elements satisfying the equation*

$$\mathbf{S}_i\,\mathbf{S}_i^{\mathsf{T}} = \mathbf{S}_{i-1}\left(\mathbf{I} + \eta_i\,(\alpha_i - \bar{\alpha}_*)\,\frac{\mathbf{r}_i\,\mathbf{r}_i^{\mathsf{T}}}{||\mathbf{r}_i||^2}\right)\mathbf{S}_{i-1}^{\mathsf{T}}, \tag{16.22}$$

where $\{\eta_i\}_{i\ge1} \subset (0, 1]$ is an adaptation step size sequence decaying to zero. Although any such sequence will do, Vihola (2012) suggests $\eta_i = i^{-\gamma}$ with a suitable exponent $\gamma \in (1/2, 1]$.
7. *Set $i \leftarrow i + 1$, and go to step 2 until the desired number of samples has been generated.*

Instead of the random walk Metropolis algorithm with covariance adaptation it is also possible to use the gradient information in the construction of the proposal distribution. This is the idea used in the Hamiltonian Monte Carlo (HMC) or hybrid Monte Carlo (HMC) method (Duane et al., 1987; Neal, 2011). In HMC, the proposal distribution is constructed by simulating a physical system consisting of particles moving under the influence of a potential (the energy function) and heat bath. The gradient of the energy function enters the equations as the force caused by the potential. The HMC method has been applied in the state space context by Mbalawata et al. (2013).

Another commonly used MCMC method is Gibbs' sampling (see, e.g., Liu, 2001; Brooks et al., 2011), which samples components of the parameters one at a time from their conditional distributions given the other parameters. The advantage of this method is that no rejections are needed: the acceptance probability is identically one. However, in order to implement the method it is necessary to be able to generate samples from the conditional distributions of parameters, which is possible only in a restricted class of models. For various other methods the reader is referred to Brooks et al. (2011).

16.2.3 Expectation Maximization

The expectation-maximization (EM) algorithm is a method to iteratively find an ML estimate of the parameters when direct optimization of the posterior distribution (or equivalently, energy function) is not feasible. The algorithm was originally introduced by Dempster et al. (1977), and applications to state space models have been discussed, for example, in Shumway and Stoffer (1982), Roweis and Ghahramani (2001), and Schön et al. (2011). Gaussian smoothing and sigma point-based approximations in an EM context have also been discussed in Väänänen (2012) and Kokkala et al. (2016). Although the EM algorithm was originally an algorithm for computing ML estimates, it can also be easily modified for computation of MAP estimates, as discussed below.

The EM algorithm is based on the result that even when we cannot evaluate the marginal likelihood as such, we are still often able to compute a lower bound for it as follows. Let $q(\mathbf{x}_{0:T})$ be an arbitrary probability density over the states; then we have

$$\log p(\mathbf{y}_{1:T} \mid \boldsymbol{\theta}) \geq F[q(\mathbf{x}_{0:T}), \boldsymbol{\theta}], \tag{16.23}$$

where the functional F is defined as

$$F[q(\mathbf{x}_{0:T}), \boldsymbol{\theta}] = \int q(\mathbf{x}_{0:T}) \log \frac{p(\mathbf{x}_{0:T}, \mathbf{y}_{1:T} \mid \boldsymbol{\theta})}{q(\mathbf{x}_{0:T})} \, \mathrm{d}\mathbf{x}_{0:T}. \tag{16.24}$$

The key idea behind the EM algorithm is that it is possible to maximize the left-hand side of Equation (16.23) by iteratively maximizing the lower bound $F[q(\mathbf{x}_{0:T}), \boldsymbol{\theta}]$. A simple way to do that is the following iteration (Neal and Hinton, 1999).

Algorithm 16.7 (Abstract EM). *The maximization of the lower bound in Equation* (16.24) *can be done by coordinate ascent as follows.*

- *Start from initial guesses* $q^{(0)}$, $\boldsymbol{\theta}^{(0)}$.
- *For* $n = 0, 1, 2, \ldots$ *do the following steps:*
  1. E-step: *Find* $q^{(n+1)} = \arg\max_q F[q, \boldsymbol{\theta}^{(n)}]$.
  2. M-step: *Find* $\boldsymbol{\theta}^{(n+1)} = \arg\max_{\boldsymbol{\theta}} F[q^{(n+1)}, \boldsymbol{\theta}]$.

In order to implement the EM algorithm, we need to be able to do the above maximizations in practice. Fortunately, it can be shown (see, e.g., Neal and Hinton, 1999) that the result of the maximization at the E-step is

$$q^{(n+1)}(\mathbf{x}_{0:T}) = p(\mathbf{x}_{0:T} \mid \mathbf{y}_{1:T}, \boldsymbol{\theta}^{(n)}). \tag{16.25}$$

Plugging this into the expression of $F[q^{(n+1)}(\mathbf{x}_{0:T}), \boldsymbol{\theta}]$ gives

$$\begin{aligned} F[q^{(n+1)}(\mathbf{x}_{0:T}), \boldsymbol{\theta}] \\ &= \int p(\mathbf{x}_{0:T} \mid \mathbf{y}_{1:T}, \boldsymbol{\theta}^{(n)}) \log p(\mathbf{x}_{0:T}, \mathbf{y}_{1:T} \mid \boldsymbol{\theta}) \, \mathrm{d}\mathbf{x}_{0:T} \\ &\quad - \int p(\mathbf{x}_{0:T} \mid \mathbf{y}_{1:T}, \boldsymbol{\theta}^{(n)}) \log p(\mathbf{x}_{0:T} \mid \mathbf{y}_{1:T}, \boldsymbol{\theta}^{(n)}) \, \mathrm{d}\mathbf{x}_{0:T}. \end{aligned} \tag{16.26}$$

Because the latter term does not depend on $\boldsymbol{\theta}$, maximizing $F[q^{(n+1)}, \boldsymbol{\theta}]$ is equivalent to maximizing the first term above, which in the EM context is commonly denoted as

$$\mathcal{Q}(\boldsymbol{\theta}, \boldsymbol{\theta}^{(n)}) = \int p(\mathbf{x}_{0:T} \mid \mathbf{y}_{1:T}, \boldsymbol{\theta}^{(n)}) \log p(\mathbf{x}_{0:T}, \mathbf{y}_{1:T} \mid \boldsymbol{\theta}) \, \mathrm{d}\mathbf{x}_{0:T}, \tag{16.27}$$

which is thus the expectation of the logarithm of the complete data likelihood $p(\mathbf{x}_{0:T}, \mathbf{y}_{1:T} \mid \boldsymbol{\theta})$ over the full joint posterior of the states given the parameters $\boldsymbol{\theta}^{(n)}$. The resulting algorithm is the following.

Algorithm 16.8 (EM algorithm). *The EM algorithm consists of the following steps.*

- *Start from an initial guess* $\boldsymbol{\theta}^{(0)}$.
- *For* $n = 0, 1, 2, \ldots$ *do the following steps:*
  1. E-step: *compute* $\mathcal{Q}(\boldsymbol{\theta}, \boldsymbol{\theta}^{(n)})$.
  2. M-step: *compute* $\boldsymbol{\theta}^{(n+1)} = \arg\max_{\boldsymbol{\theta}} \mathcal{Q}(\boldsymbol{\theta}, \boldsymbol{\theta}^{(n)})$.

Due to the Markovian structure of the state space model in Equation (16.1), the complete data log-likelihood has the form

$$\begin{aligned} \log p(\mathbf{x}_{0:T}, \mathbf{y}_{1:T} \mid \boldsymbol{\theta}) \\ &= \log p(\mathbf{x}_0 \mid \boldsymbol{\theta}) + \sum_{k=1}^{T} \log p(\mathbf{x}_k \mid \mathbf{x}_{k-1}, \boldsymbol{\theta}) + \sum_{k=1}^{T} \log p(\mathbf{y}_k \mid \mathbf{x}_k, \boldsymbol{\theta}). \end{aligned} \tag{16.28}$$

The expression for $\mathcal{Q}$ in Equation (16.27) and thus the E-step in Algorithm 16.8 now reduces to computation of (see Schön et al., 2011)

$$\mathcal{Q}(\boldsymbol{\theta}, \boldsymbol{\theta}^{(n)}) = I_1(\boldsymbol{\theta}, \boldsymbol{\theta}^{(n)}) + I_2(\boldsymbol{\theta}, \boldsymbol{\theta}^{(n)}) + I_3(\boldsymbol{\theta}, \boldsymbol{\theta}^{(n)}), \tag{16.29}$$

where

$$
\begin{aligned}
I_1(\boldsymbol{\theta}, \boldsymbol{\theta}^{(n)}) &= \int p(\mathbf{x}_0 \mid \mathbf{y}_{1:T}, \boldsymbol{\theta}^{(n)}) \log p(\mathbf{x}_0 \mid \boldsymbol{\theta}) \, \mathrm{d}\mathbf{x}_0, \\
I_2(\boldsymbol{\theta}, \boldsymbol{\theta}^{(n)}) &= \sum_{k=1}^{T} \int p(\mathbf{x}_k, \mathbf{x}_{k-1} \mid \mathbf{y}_{1:T}, \boldsymbol{\theta}^{(n)}) \\
&\qquad \times \log p(\mathbf{x}_k \mid \mathbf{x}_{k-1}, \boldsymbol{\theta}) \, \mathrm{d}\mathbf{x}_k \, \mathrm{d}\mathbf{x}_{k-1}, \\
I_3(\boldsymbol{\theta}, \boldsymbol{\theta}^{(n)}) &= \sum_{k=1}^{T} \int p(\mathbf{x}_k \mid \mathbf{y}_{1:T}, \boldsymbol{\theta}^{(n)}) \log p(\mathbf{y}_k \mid \mathbf{x}_k, \boldsymbol{\theta}) \, \mathrm{d}\mathbf{x}_k.
\end{aligned}
\tag{16.30}
$$

The above expectations are over the smoothing distribution, and the key thing to observe is that we do not need to compute expectations over the full posterior, but only over the smoothing distributions $p(\mathbf{x}_k \mid \mathbf{y}_{1:T}, \boldsymbol{\theta}^{(n)})$ and pairwise smoothing distributions $p(\mathbf{x}_k, \mathbf{x}_{k-1} \mid \mathbf{y}_{1:T}, \boldsymbol{\theta}^{(n)})$. It turns out that the required expectations can be easily (approximately) evaluated using smoother results. In the case of linear state space models, we can find a closed form expression for the above integrals in terms of RTS smoother results. In the non-linear case, we can approximate the integrals by using non-linear smoothers such as Gaussian smoothers. In the more general probabilistic state space model, we can use particle smoothers to approximate them.

On the E-step of Algorithm 16.8, we need to maximize the expression for $\mathcal{Q}$ in Equation (16.29) with respect to the parameters $\boldsymbol{\theta}$. In principle, we can utilize various gradient-free and gradient-based optimization methods (see, e.g., Luenberger and Ye, 2008) for doing that, but the most useful case occurs when we can do the maximization analytically via setting the gradient to zero:

$$
\frac{\partial \mathcal{Q}(\boldsymbol{\theta}, \boldsymbol{\theta}^{(n)})}{\partial \boldsymbol{\theta}} = 0. \tag{16.31}
$$

This happens, for example, when estimating the parameters of linear state space models and in certain classes of non-linear state space models.

It turns out that we can calculate MAP estimates using the EM algorithm by replacing $p(\mathbf{x}_{0:T}, \mathbf{y}_{1:T} \mid \boldsymbol{\theta})$ in Equation (16.27) with $p(\mathbf{x}_{0:T}, \mathbf{y}_{1:T}, \boldsymbol{\theta})$. In practice, this can be implemented by maximizing $\mathcal{Q}(\boldsymbol{\theta}, \boldsymbol{\theta}^{(n)}) + \log p(\boldsymbol{\theta})$ at the M-step instead of the plain $\mathcal{Q}$.

As a side product of the EM formulation above, we also get a method to compute the gradient of the energy function needed in gradient-based optimization for finding the MAP or ML estimates. *Fisher's identity* (see, e.g., Cappé et al., 2005) states that if we evaluate the gradient of $\mathcal{Q}$ at $\boldsymbol{\theta}^{(n)} = \boldsymbol{\theta}$, we get exactly the gradient of the marginal log-likelihood. This implies that the gradient of the energy function can be evaluated as

$$\frac{\partial \varphi_T(\boldsymbol{\theta})}{\partial \boldsymbol{\theta}} = -\frac{\partial \log p(\boldsymbol{\theta})}{\partial \boldsymbol{\theta}} - \left.\frac{\partial \mathcal{Q}(\boldsymbol{\theta}, \boldsymbol{\theta}^{(n)})}{\partial \boldsymbol{\theta}}\right|_{\boldsymbol{\theta}^{(n)}=\boldsymbol{\theta}}. \tag{16.32}$$

Note that here we refer to the above identity as Fisher's identity although the original identity is the relationship with the log marginal likelihood and $\mathcal{Q}$, not with the log posterior and $\mathcal{Q}$. In any case this identity is useful in linear state space models, because it is often easier to compute and computationally lighter (Segal and Weinstein, 1989; Olsson et al., 2007). However, in non-linear state space models it is not as useful, because the approximations involved in computation of the filtering and smoothing solutions often cause the gradient to have different approximations from the energy function approximation implied by the same method. That is, the gradient approximation computed with Fisher's identity and Gaussian smoothing might not exactly match the gradient of the energy function approximation computed with the corresponding Gaussian filter. However, in the case of particle filters, Fisher's identity provides a feasible way to approximate the gradient of the energy function.

16.3 Practical Parameter Estimation in State Space Models

In this section we discuss practical parameter estimation methods for state space models using linear Kalman filters and RTS smoothers, Gaussian approximation-based non-linear Kalman filters and RTS smoothers, and particle filters and smoothers. But before going to them, we outline the simple but sometimes effective state augmentation approach.

16.3.1 State Augmentation Approach

Before going to more elaborate parameter estimation methods for state space models, we recall that already in Chapter 3 we used the Kalman filter for estimating static parameters in a regression model. The same approach can be generalized to the *state augmentation approach*, which simply means that we augment the parameter as part of the state. For example, let us say that we have a non-linear model with unknown parameters $\boldsymbol{\theta}$:

$$\begin{aligned} \mathbf{x}_k &= \mathbf{f}(\mathbf{x}_{k-1}, \boldsymbol{\theta}) + \mathbf{q}_{k-1}, \\ \mathbf{y}_k &= \mathbf{h}(\mathbf{x}_k, \boldsymbol{\theta}) + \mathbf{r}_k. \end{aligned} \tag{16.33}$$

We can now rewrite the model as

$$\begin{aligned}\boldsymbol{\theta}_k &= \boldsymbol{\theta}_{k-1},\\ \mathbf{x}_k &= \mathbf{f}(\mathbf{x}_{k-1}, \boldsymbol{\theta}_{k-1}) + \mathbf{q}_{k-1},\\ \mathbf{y}_k &= \mathbf{h}(\mathbf{x}_k, \boldsymbol{\theta}_k) + \mathbf{r}_k,\end{aligned} \tag{16.34}$$

where the dynamic model for the parameter essentially says that it is constant. If we now redefine the state as $\tilde{\mathbf{x}}_k = (\mathbf{x}_k, \boldsymbol{\theta}_k)$, we get a state space model of the form

$$\begin{aligned}\tilde{\mathbf{x}}_k &= \tilde{\mathbf{f}}(\tilde{\mathbf{x}}_{k-1}) + \tilde{\mathbf{q}}_{k-1},\\ \mathbf{y}_k &= \mathbf{h}(\tilde{\mathbf{x}}_k) + \mathbf{r}_k,\end{aligned} \tag{16.35}$$

which does not contain any unknown parameter anymore. The problem in this *state augmentation* is the singularity of the dynamic model for the parameter. It works well when the whole system is linear and we do not have any approximation errors in the estimator. If the parameters appear linearly in the system, it sometimes is a good idea to include the parameters as part of the state. However, this can also fail sometimes.

With approximate non-linear filters, the singularity of the parameter dynamic model indeed causes problems. With non-linear Kalman filters, the Gaussian approximation tends to become singular, which causes the filter to diverge. As discussed in Section 11.5, particle filters have a problem with small amounts of noise in the dynamic model because this causes sample impoverishment. As the noise in the dynamic model above is exactly zero, this case is particularly problematic for particle filters.

A common way to circumvent the problem is to introduce artificial noise to the dynamic model of the parameter, that is, replace $\boldsymbol{\theta}_k = \boldsymbol{\theta}_{k-1}$ with

$$\boldsymbol{\theta}_k = \boldsymbol{\theta}_{k-1} + \boldsymbol{\varepsilon}_{k-1}, \tag{16.36}$$

where $\boldsymbol{\varepsilon}_{k-1}$ is a "small" noise process. But the problem is that we are no longer solving the original parameter estimation problem but another one with a time-varying parameter. Anyway, this approach is sometimes applicable and should be considered before jumping into more complicated parameter estimation methods.

There is also a form of Rao–Blackwellization that sometimes helps. This approach is discussed in Section 16.3.6, and the idea is to use a closed form solution for the static parameter ("Rao–Blackwellize" the parameter) and sample only the original state part. This works if the parameter appears in the model in a suitable conjugate form.

16.3.2 Parameter Estimation in Linear State Space Models

Consider the following linear Gaussian state space model with unknown parameters $\boldsymbol{\theta}$:

$$\begin{aligned} \mathbf{x}_k &= \mathbf{A}(\boldsymbol{\theta})\, \mathbf{x}_{k-1} + \mathbf{q}_{k-1}, \\ \mathbf{y}_k &= \mathbf{H}(\boldsymbol{\theta})\, \mathbf{x}_k + \mathbf{r}_k, \end{aligned} \tag{16.37}$$

where $\mathbf{q}_{k-1} \sim \mathrm{N}(\mathbf{0}, \mathbf{Q}(\boldsymbol{\theta}))$, $\mathbf{r}_k \sim \mathrm{N}(\mathbf{0}, \mathbf{R}(\boldsymbol{\theta}))$, and $\mathbf{x}_0 \sim \mathrm{N}(\mathbf{m}_0(\boldsymbol{\theta}), \mathbf{P}_0(\boldsymbol{\theta}))$. In the above model, for notational convenience, we have assumed that the model matrices do not explicitly depend on time. The energy function and thus the marginal parameter posterior for the linear Gaussian model above can be obtained as follows.

Theorem 16.9 (Energy function for linear Gaussian model). *The recursion for the energy function is given as*

$$\varphi_k(\boldsymbol{\theta}) = \varphi_{k-1}(\boldsymbol{\theta}) + \frac{1}{2} \log |2\pi\, \mathbf{S}_k(\boldsymbol{\theta})| + \frac{1}{2} \mathbf{v}_k^\mathsf{T}(\boldsymbol{\theta})\, \mathbf{S}_k^{-1}(\boldsymbol{\theta})\, \mathbf{v}_k(\boldsymbol{\theta}), \tag{16.38}$$

where the terms $\mathbf{v}_k(\boldsymbol{\theta})$ and $\mathbf{S}_k(\boldsymbol{\theta})$ are given by the Kalman filter with the parameters fixed to $\boldsymbol{\theta}$.

- Prediction:

$$\begin{aligned} \mathbf{m}_k^-(\boldsymbol{\theta}) &= \mathbf{A}(\boldsymbol{\theta})\, \mathbf{m}_{k-1}(\boldsymbol{\theta}), \\ \mathbf{P}_k^-(\boldsymbol{\theta}) &= \mathbf{A}(\boldsymbol{\theta})\, \mathbf{P}_{k-1}(\boldsymbol{\theta})\, \mathbf{A}^\mathsf{T}(\boldsymbol{\theta}) + \mathbf{Q}(\boldsymbol{\theta}). \end{aligned} \tag{16.39}$$

- Update:

$$\begin{aligned} \mathbf{v}_k(\boldsymbol{\theta}) &= \mathbf{y}_k - \mathbf{H}(\boldsymbol{\theta})\, \mathbf{m}_k^-(\boldsymbol{\theta}), \\ \mathbf{S}_k(\boldsymbol{\theta}) &= \mathbf{H}(\boldsymbol{\theta})\, \mathbf{P}_k^-(\boldsymbol{\theta})\, \mathbf{H}^\mathsf{T}(\boldsymbol{\theta}) + \mathbf{R}(\boldsymbol{\theta}), \\ \mathbf{K}_k(\boldsymbol{\theta}) &= \mathbf{P}_k^-(\boldsymbol{\theta})\, \mathbf{H}^\mathsf{T}(\boldsymbol{\theta})\, \mathbf{S}_k^{-1}(\boldsymbol{\theta}), \\ \mathbf{m}_k(\boldsymbol{\theta}) &= \mathbf{m}_k^-(\boldsymbol{\theta}) + \mathbf{K}_k(\boldsymbol{\theta})\, \mathbf{v}_k(\boldsymbol{\theta}), \\ \mathbf{P}_k(\boldsymbol{\theta}) &= \mathbf{P}_k^-(\boldsymbol{\theta}) - \mathbf{K}_k(\boldsymbol{\theta})\, \mathbf{S}_k(\boldsymbol{\theta})\, \mathbf{K}_k^\mathsf{T}(\boldsymbol{\theta}). \end{aligned} \tag{16.40}$$

Proof The Kalman filter gives us the Gaussian predictive distribution $p(\mathbf{x}_k \mid \mathbf{y}_{1:k-1}, \boldsymbol{\theta}) = \mathrm{N}(\mathbf{x}_k \mid \mathbf{m}_k^-(\boldsymbol{\theta}), \mathbf{P}_k^-(\boldsymbol{\theta}))$, which via Theorem 16.1 thus gives

$$\begin{aligned} & p(\mathbf{y}_k \mid \mathbf{y}_{1:k-1}, \boldsymbol{\theta}) \\ &= \int \mathrm{N}(\mathbf{y}_k \mid \mathbf{H}(\boldsymbol{\theta})\, \mathbf{x}_k, \mathbf{R}(\boldsymbol{\theta}))\, \mathrm{N}(\mathbf{x}_k \mid \mathbf{m}_k^-(\boldsymbol{\theta}), \mathbf{P}_k^-(\boldsymbol{\theta}))\, \mathrm{d}\mathbf{x}_k \\ &= \mathrm{N}(\mathbf{y}_k \mid \mathbf{H}(\boldsymbol{\theta})\, \mathbf{m}_k^-(\boldsymbol{\theta}), \mathbf{H}(\boldsymbol{\theta})\, \mathbf{P}_k^-(\boldsymbol{\theta})\, \mathbf{H}^\mathsf{T}(\boldsymbol{\theta}) + \mathbf{R}(\boldsymbol{\theta})). \end{aligned} \tag{16.41}$$

The rest follows from Theorem 16.4. □

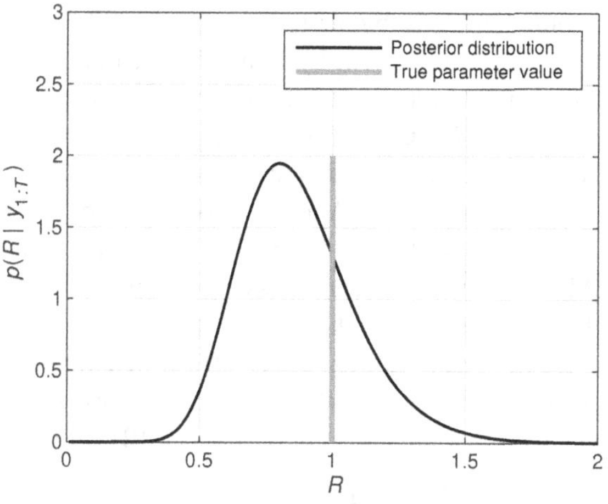

Figure 16.1 Posterior distribution of noise variance R in the Gaussian random walk model (see Example 6.4).

Thus if we fix $\boldsymbol{\theta}$ and run the above algorithm from $\varphi_0(\boldsymbol{\theta}) = -\log p(\boldsymbol{\theta})$ at $k = 0$ to the step $k = T$, then the full energy function is $\varphi_T(\boldsymbol{\theta})$. That is, the marginal posterior density at the point $\boldsymbol{\theta}$ can be evaluated up to a normalization constant by Equation (16.10) as

$$p(\boldsymbol{\theta} \mid \mathbf{y}_{1:T}) \propto \exp(-\varphi_T(\boldsymbol{\theta})).$$

Given the energy function, it is now easy to implement, for example, a Metropolis–Hastings-based MCMC sampler for generating a Monte Carlo approximation of the posterior distribution or to use the energy function in a gradient-free optimization for finding the ML or MAP estimates of the parameters.

Example 16.10 (Parameter posterior for Gaussian random walk). *The posterior distribution of the noise variance $p(R \mid y_{1:T})$ for the Gaussian random walk model in Example 6.4 is shown in Figure 16.1. A formally uniform prior $p(R) \propto 1$ was assumed. The true value used in the simulation is indeed well within the high density area of the posterior distribution. However, it can also been seen that if we computed the MAP (or equivalently ML) estimate of the parameter, we would get a smaller value than the true one.*

In order to implement a gradient-based optimization method, we need to have the gradient of the energy function as well. One way to implement the gradient computation is by first differentiating the energy function expression in Theorem 16.9 term-by-term and then each of the Kalman filter equations. This results in a recursion called *sensitivity equations* (Gupta and Mehra, 1974; Cappé et al., 2005), which can be evaluated along with the Kalman filtering computations. The equations are given in Theorem A.10 in Section A.10. Another way to compute the gradient is by using Fisher's identity (16.32), but before going into that, let us take a look at the EM algorithm for linear Gaussian models.

Recall that for implementing the EM algorithm we need to be able to compute the expectations in Equations (16.30), which in terms requires the knowledge of the smoothing distributions and pairwise smoothing distributions. Fortunately, by Equations (12.5) and (12.12) we know that

$$\begin{aligned} p(\mathbf{x}_k \mid \mathbf{y}_{1:T}, \boldsymbol{\theta}^{(n)}) &= \mathrm{N}(\mathbf{x}_k \mid \mathbf{m}_k^{\mathrm{s}}, \mathbf{P}_k^{\mathrm{s}}), \\ p(\mathbf{x}_k, \mathbf{x}_{k-1} \mid \mathbf{y}_{1:T}, \boldsymbol{\theta}^{(n)}) & \\ = \mathrm{N}\left(\begin{pmatrix} \mathbf{x}_k \\ \mathbf{x}_{k-1} \end{pmatrix} \middle| \begin{pmatrix} \mathbf{m}_k^{\mathrm{s}} \\ \mathbf{m}_{k-1}^{\mathrm{s}} \end{pmatrix}, \right. & \left. \begin{pmatrix} \mathbf{P}_k^{\mathrm{s}} & \mathbf{P}_k^{\mathrm{s}} \mathbf{G}_{k-1}^{\mathsf{T}} \\ \mathbf{G}_{k-1} \mathbf{P}_k^{\mathrm{s}} & \mathbf{P}_{k-1}^{\mathrm{s}} \end{pmatrix} \right), \end{aligned} \tag{16.42}$$

where the means, covariances, and gains are computed with an RTS smoother with the model parameters fixed to $\boldsymbol{\theta}^{(n)}$. Note that in the EM algorithms appearing in the literature, the cross term $\mathbf{P}_k^{\mathrm{s}} \mathbf{G}_{k-1}^{\mathsf{T}}$ is sometimes computed with a separate recursion (see, e.g., Shumway and Stoffer, 1982), but in fact this is unnecessary due to the above. The required expectations for EM in Equations (16.30) can now be computed in closed form, and the result is the following (cf. Shumway and Stoffer, 1982).

Theorem 16.11 (Evaluation of $\mathcal{Q}$ for linear Gaussian model). *The expression for $\mathcal{Q}$ for the linear Gaussian model in Equation* (16.37) *can be written as*

$$\begin{aligned}
&\mathcal{Q}(\boldsymbol{\theta}, \boldsymbol{\theta}^{(n)}) \\
&= -\frac{1}{2} \log |2\pi \, \mathbf{P}_0(\boldsymbol{\theta})| - \frac{T}{2} \log |2\pi \, \mathbf{Q}(\boldsymbol{\theta})| - \frac{T}{2} \log |2\pi \, \mathbf{R}(\boldsymbol{\theta})| \\
&- \frac{1}{2} \operatorname{tr} \left\{ \mathbf{P}_0^{-1}(\boldsymbol{\theta}) \left[\mathbf{P}_0^{\mathrm{s}} + (\mathbf{m}_0^{\mathrm{s}} - \mathbf{m}_0(\boldsymbol{\theta})) \, (\mathbf{m}_0^{\mathrm{s}} - \mathbf{m}_0(\boldsymbol{\theta}))^{\mathsf{T}} \right] \right\} \\
&- \frac{T}{2} \operatorname{tr} \left\{ \mathbf{Q}^{-1}(\boldsymbol{\theta}) \left[\boldsymbol{\Sigma} - \mathbf{C} \, \mathbf{A}^{\mathsf{T}}(\boldsymbol{\theta}) - \mathbf{A}(\boldsymbol{\theta}) \, \mathbf{C}^{\mathsf{T}} + \mathbf{A}(\boldsymbol{\theta}) \, \boldsymbol{\Phi} \, \mathbf{A}^{\mathsf{T}}(\boldsymbol{\theta}) \right] \right\} \\
&- \frac{T}{2} \operatorname{tr} \left\{ \mathbf{R}^{-1}(\boldsymbol{\theta}) \left[\mathbf{D} - \mathbf{B} \, \mathbf{H}^{\mathsf{T}}(\boldsymbol{\theta}) - \mathbf{H}(\boldsymbol{\theta}) \, \mathbf{B}^{\mathsf{T}} + \mathbf{H}(\boldsymbol{\theta}) \, \boldsymbol{\Sigma} \, \mathbf{H}^{\mathsf{T}}(\boldsymbol{\theta}) \right] \right\},
\end{aligned} \tag{16.43}$$

where the following quantities are computed from the results of RTS smoothers run with parameter values $\boldsymbol{\theta}^{(n)}$:

$$\begin{aligned}
\boldsymbol{\Sigma} &= \frac{1}{T} \sum_{k=1}^{T} \mathbf{P}_k^{\mathrm{s}} + \mathbf{m}_k^{\mathrm{s}} \, [\mathbf{m}_k^{\mathrm{s}}]^{\mathsf{T}}, \\
\boldsymbol{\Phi} &= \frac{1}{T} \sum_{k=1}^{T} \mathbf{P}_{k-1}^{\mathrm{s}} + \mathbf{m}_{k-1}^{\mathrm{s}} \, [\mathbf{m}_{k-1}^{\mathrm{s}}]^{\mathsf{T}}, \\
\mathbf{B} &= \frac{1}{T} \sum_{k=1}^{T} \mathbf{y}_k \, [\mathbf{m}_k^{\mathrm{s}}]^{\mathsf{T}}, \\
\mathbf{C} &= \frac{1}{T} \sum_{k=1}^{T} \mathbf{P}_k^{\mathrm{s}} \, \mathbf{G}_{k-1}^{\mathsf{T}} + \mathbf{m}_k^{\mathrm{s}} \, [\mathbf{m}_{k-1}^{\mathrm{s}}]^{\mathsf{T}}, \\
\mathbf{D} &= \frac{1}{T} \sum_{k=1}^{T} \mathbf{y}_k \, \mathbf{y}_k^{\mathsf{T}}.
\end{aligned} \tag{16.44}$$

The usefulness of the EM algorithm for linear state space models stems from the fact that if the parameters are selected to be some of the full model matrices (or initial mean), we can indeed perform the M-step of the EM algorithm in closed form. The same thing happens if the parameters appear linearly in one of the model matrices (e.g., are some subcomponents of the matrices), but application of the EM algorithm to the estimation of the full matrices is the classical result. By setting the gradients of $\partial \mathcal{Q}(\boldsymbol{\theta}, \boldsymbol{\theta}^{(n)}) / \partial \boldsymbol{\theta}$

to zero for each $\boldsymbol{\theta} = \{\mathbf{A}, \mathbf{H}, \mathbf{Q}, \mathbf{R}, \mathbf{P}_0, \mathbf{m}_0\}$ separately, we get the following result.

Theorem 16.12 (Maximization of $\mathcal{Q}$ for linear model parameters). *The maximum $\boldsymbol{\theta}^* = \arg\max_{\boldsymbol{\theta}} \mathcal{Q}(\boldsymbol{\theta}, \boldsymbol{\theta}^{(n)})$, when the parameters are selected to be one of the model parameters $\boldsymbol{\theta} \in \{\mathbf{A}, \mathbf{H}, \mathbf{Q}, \mathbf{R}, \mathbf{P}_0, \mathbf{m}_0\}$, can be computed as follows.*

- *When $\boldsymbol{\theta} = \mathbf{P}_0$ we get*

$$\mathbf{P}_0^* = \mathbf{P}_0^{\mathrm{s}} + (\mathbf{m}_0^{\mathrm{s}} - \mathbf{m}_0)\,(\mathbf{m}_0^{\mathrm{s}} - \mathbf{m}_0)^{\mathsf{T}}. \tag{16.45}$$

- *When $\boldsymbol{\theta} = \mathbf{A}$ we get*

$$\mathbf{A}^* = \mathbf{C}\,\boldsymbol{\Phi}^{-1}. \tag{16.46}$$

- *When $\boldsymbol{\theta} = \mathbf{Q}$ we get*

$$\mathbf{Q}^* = \boldsymbol{\Sigma} - \mathbf{C}\,\mathbf{A}^{\mathsf{T}} - \mathbf{A}\,\mathbf{C}^{\mathsf{T}} + \mathbf{A}\,\boldsymbol{\Phi}\,\mathbf{A}^{\mathsf{T}}. \tag{16.47}$$

- *When $\boldsymbol{\theta} = \mathbf{H}$ we get*

$$\mathbf{H}^* = \mathbf{B}\,\boldsymbol{\Sigma}^{-1}. \tag{16.48}$$

- *When $\boldsymbol{\theta} = \mathbf{R}$ we get*

$$\mathbf{R}^* = \mathbf{D} - \mathbf{H}\,\mathbf{B}^{\mathsf{T}} - \mathbf{B}\,\mathbf{H}^{\mathsf{T}} + \mathbf{H}\,\boldsymbol{\Sigma}\,\mathbf{H}^{\mathsf{T}}. \tag{16.49}$$

- *Finally, the maximum with respect to the initial mean $\boldsymbol{\theta} = \mathbf{m}_0$ is*

$$\mathbf{m}_0^* = \mathbf{m}_0^{\mathrm{s}}. \tag{16.50}$$

Obviously the above theorem can also be used for solving the maximum of $\mathcal{Q}$ with respect to any subset of model matrices by solving the resulting equations jointly. The EM algorithm for finding the maximum likelihood estimates of the linear state space model parameters is now the following.

Algorithm 16.13 (EM algorithm for linear state space models). *Let $\boldsymbol{\theta}$ contain some subset of the model parameters $\{\mathbf{A}, \mathbf{H}, \mathbf{Q}, \mathbf{R}, \mathbf{P}_0, \mathbf{m}_0\}$. We can find maximum likelihood estimates of them via the following iteration.*

- *Start from some initial guess $\boldsymbol{\theta}^{(0)}$.*
- *For $n = 0, 1, 2, \ldots$ do the following steps.*
  1. E-step: *Run the RTS smoother using the current parameter values in $\boldsymbol{\theta}^{(n)}$, and compute the quantities in Equation* (16.44) *from the smoother results.*
  2. M-step: *Find new parameter values by using Equations* (16.45) – (16.50)*, and store them in $\boldsymbol{\theta}^{(n+1)}$.*

The expression for $\mathcal{Q}(\boldsymbol{\theta}, \boldsymbol{\theta}^{(n)})$ also provides an "easy gradient recipe" (Olsson et al., 2007) for computation of the energy function gradient via Fisher's identity (Equation (16.32)), as it enables the computation of the gradient without an additional recursion (sensitivity equations). The resulting expression is given in Theorem A.11 in Section A.10.

16.3.3 Parameter Estimation with Gaussian Filtering and Smoothing

One way to implement parameter estimation in non-linear models is by replacing the Kalman filters and RTS smoothers used in the linear case with their non-linear counterparts. Let's consider parameter estimation in models of the form

$$\begin{aligned} \mathbf{x}_k &= \mathbf{f}(\mathbf{x}_{k-1}, \boldsymbol{\theta}) + \mathbf{q}_{k-1}, \\ \mathbf{y}_k &= \mathbf{h}(\mathbf{x}_k, \boldsymbol{\theta}) + \mathbf{r}_k, \end{aligned} \tag{16.51}$$

where $\mathbf{q}_{k-1} \sim \mathrm{N}(\mathbf{0}, \mathbf{Q}(\boldsymbol{\theta}))$, $\mathbf{r}_k \sim \mathrm{N}(\mathbf{0}, \mathbf{R}(\boldsymbol{\theta}))$, and $\mathbf{x}_0 \sim \mathrm{N}(\mathbf{m}_0(\boldsymbol{\theta}), \mathbf{P}_0(\boldsymbol{\theta}))$. The energy function can now be approximated with the following Gaussian filtering-based algorithm.

Algorithm 16.14 (Gaussian filtering-based energy function). *The recursion for the approximate energy function is*

$$\varphi_k(\boldsymbol{\theta}) \simeq \varphi_{k-1}(\boldsymbol{\theta}) + \frac{1}{2} \log |2\pi\, \mathbf{S}_k(\boldsymbol{\theta})| + \frac{1}{2} \mathbf{v}_k^\mathsf{T}(\boldsymbol{\theta})\, \mathbf{S}_k^{-1}(\boldsymbol{\theta})\, \mathbf{v}_k(\boldsymbol{\theta}), \tag{16.52}$$

where the terms $\mathbf{v}_k(\boldsymbol{\theta})$ and $\mathbf{S}_k(\boldsymbol{\theta})$ are given by the Gaussian filter with the parameters fixed to $\boldsymbol{\theta}$.

- *Prediction:*

$$\begin{aligned} \mathbf{m}_k^-(\boldsymbol{\theta}) &= \int \mathbf{f}(\mathbf{x}_{k-1}, \boldsymbol{\theta})\, \mathrm{N}(\mathbf{x}_{k-1} \mid \mathbf{m}_{k-1}(\boldsymbol{\theta}), \mathbf{P}_{k-1}(\boldsymbol{\theta}))\, \mathrm{d}\mathbf{x}_{k-1}, \\ \mathbf{P}_k^-(\boldsymbol{\theta}) &= \int (\mathbf{f}(\mathbf{x}_{k-1}, \boldsymbol{\theta}) - \mathbf{m}_k^-(\boldsymbol{\theta}))\, (\mathbf{f}(\mathbf{x}_{k-1}, \boldsymbol{\theta}) - \mathbf{m}_k^-(\boldsymbol{\theta}))^\mathsf{T} \\ &\qquad \times \mathrm{N}(\mathbf{x}_{k-1} \mid \mathbf{m}_{k-1}(\boldsymbol{\theta}), \mathbf{P}_{k-1}(\boldsymbol{\theta}))\, \mathrm{d}\mathbf{x}_{k-1} + \mathbf{Q}_{k-1}(\boldsymbol{\theta}). \end{aligned} \tag{16.53}$$

- *Update:*

$$
\begin{aligned}
\boldsymbol{\mu}_k(\boldsymbol{\theta}) &= \int \mathbf{h}(\mathbf{x}_k, \boldsymbol{\theta})\, \mathrm{N}(\mathbf{x}_k \mid \mathbf{m}_k^-(\boldsymbol{\theta}), \mathbf{P}_k^-(\boldsymbol{\theta}))\, \mathrm{d}\mathbf{x}_k, \\
\mathbf{v}_k(\boldsymbol{\theta}) &= \mathbf{y}_k - \boldsymbol{\mu}_k(\boldsymbol{\theta}), \\
\mathbf{S}_k(\boldsymbol{\theta}) &= \int (\mathbf{h}(\mathbf{x}_k, \boldsymbol{\theta}) - \boldsymbol{\mu}_k(\boldsymbol{\theta}))\, (\mathbf{h}(\mathbf{x}_k, \boldsymbol{\theta}) - \boldsymbol{\mu}_k(\boldsymbol{\theta}))^\mathsf{T} \\
&\qquad \times \mathrm{N}(\mathbf{x}_k \mid \mathbf{m}_k^-(\boldsymbol{\theta}), \mathbf{P}_k^-(\boldsymbol{\theta}))\, \mathrm{d}\mathbf{x}_k + \mathbf{R}_k(\boldsymbol{\theta}), \\
\mathbf{C}_k(\boldsymbol{\theta}) &= \int (\mathbf{x}_k - \mathbf{m}_k^-(\boldsymbol{\theta}))\, (\mathbf{h}(\mathbf{x}_k, \boldsymbol{\theta}) - \boldsymbol{\mu}_k(\boldsymbol{\theta}))^\mathsf{T} \\
&\qquad \times \mathrm{N}(\mathbf{x}_k \mid \mathbf{m}_k^-(\boldsymbol{\theta}), \mathbf{P}_k^-(\boldsymbol{\theta}))\, \mathrm{d}\mathbf{x}_k, \\
\mathbf{K}_k(\boldsymbol{\theta}) &= \mathbf{C}_k(\boldsymbol{\theta})\, \mathbf{S}_k^{-1}(\boldsymbol{\theta}), \\
\mathbf{m}_k(\boldsymbol{\theta}) &= \mathbf{m}_k^-(\boldsymbol{\theta}) + \mathbf{K}_k(\boldsymbol{\theta})\, \mathbf{v}_k(\boldsymbol{\theta}), \\
\mathbf{P}_k(\boldsymbol{\theta}) &= \mathbf{P}_k^-(\boldsymbol{\theta}) - \mathbf{K}_k(\boldsymbol{\theta})\, \mathbf{S}_k(\boldsymbol{\theta})\, \mathbf{K}_k^\mathsf{T}(\boldsymbol{\theta}).
\end{aligned}
\tag{16.54}
$$

Derivation This approximation can be derived in the same way as the linear case in Theorem 16.9 except that Gaussian moment matching-based approximations are used instead of the true Gaussian distributions. □

The above energy function can now be directly used in MCMC sampling or in gradient-free optimization algorithms for computing ML or MAP estimates. However, because the energy function is based on a Gaussian approximation, the implied posterior distribution is an approximation as well, and thus the parameter estimates will be biased. The posterior distribution approximation is also typically thinner than the true posterior distribution, and thus the uncertainty in the parameter is underestimated. This issue is illustrated in Example 16.17.

It is also possible to compute the derivatives of the above energy function analogously to the linear case. In the case of the extended Kalman filter (EKF), the derivatives can be easily derived by formally differentiating the EKF equations (see Mbalawata et al., 2013). When sigma point filters are used, a similar approach works, but additional care is needed for computation of the derivative of the square root matrix $\partial \sqrt{\mathbf{P}(\boldsymbol{\theta})}/\partial \theta_i$ arising in the equations. The equations for computing the derivatives of the energy function are given in Algorithm A.12.

To compute the expectations required for implementing the EM algorithm, we can approximate the integrals in Equations (16.30) using the Gaussian assumed density approximation (i.e., moment matching). The resulting expression for $\mathcal{Q}$ is the following.

Algorithm 16.15 (Evaluation of $\mathcal{Q}$ via Gaussian smoothing). *The expression for $\mathcal{Q}$ for the non-linear state space model in Equation* (16.51) *can be written as*

$$\begin{aligned}
&\mathcal{Q}(\boldsymbol{\theta}, \boldsymbol{\theta}^{(n)}) \\
&\simeq -\frac{1}{2} \log |2\pi \, \mathbf{P}_0(\boldsymbol{\theta})| - \frac{T}{2} \log |2\pi \, \mathbf{Q}(\boldsymbol{\theta})| - \frac{T}{2} \log |2\pi \, \mathbf{R}(\boldsymbol{\theta})| \\
&- \frac{1}{2} \operatorname{tr} \left\{ \mathbf{P}_0^{-1}(\boldsymbol{\theta}) \left[\mathbf{P}_0^{\mathrm{s}} + (\mathbf{m}_0^{\mathrm{s}} - \mathbf{m}_0(\boldsymbol{\theta})) \, (\mathbf{m}_0^{\mathrm{s}} - \mathbf{m}_0(\boldsymbol{\theta}))^{\mathsf{T}} \right] \right\} \\
&- \frac{1}{2} \sum_{k=1}^{T} \operatorname{tr} \left\{ \mathbf{Q}^{-1}(\boldsymbol{\theta}) \, \mathrm{E}\left[(\mathbf{x}_k - \mathbf{f}(\mathbf{x}_{k-1}, \boldsymbol{\theta})) \, (\mathbf{x}_k - \mathbf{f}(\mathbf{x}_{k-1}, \boldsymbol{\theta}))^{\mathsf{T}} \mid \mathbf{y}_{1:T} \right] \right\} \\
&- \frac{1}{2} \sum_{k=1}^{T} \operatorname{tr} \left\{ \mathbf{R}^{-1}(\boldsymbol{\theta}) \, \mathrm{E}\left[(\mathbf{y}_k - \mathbf{h}(\mathbf{x}_k, \boldsymbol{\theta})) \, (\mathbf{y}_k - \mathbf{h}(\mathbf{x}_k, \boldsymbol{\theta}))^{\mathsf{T}} \mid \mathbf{y}_{1:T} \right] \right\},
\end{aligned} \tag{16.55}$$

where the expectations are over the counterparts of the distributions in Equations (16.42) *obtained from the Gaussian smoother.*

In practice, we can approximate the Gaussian smoother and Gaussian integrals above with Taylor series approximations (EKF/ERTSS) or by sigma point methods such as Gauss–Hermite or spherical cubature integration or the unscented transform. The M-step for the noise parameters can indeed be implemented analogously to the linear case in Theorem 16.12, because the maxima of the above $\mathcal{Q}$ with respect to the noise covariance are simply

$$\begin{aligned}
\mathbf{Q}^* &= \frac{1}{T} \sum_{k=1}^{T} \mathrm{E}\left[(\mathbf{x}_k - \mathbf{f}(\mathbf{x}_{k-1}, \boldsymbol{\theta})) \, (\mathbf{x}_k - \mathbf{f}(\mathbf{x}_{k-1}, \boldsymbol{\theta}))^{\mathsf{T}} \mid \mathbf{y}_{1:T} \right], \\
\mathbf{R}^* &= \frac{1}{T} \sum_{k=1}^{T} \mathrm{E}\left[(\mathbf{y}_k - \mathbf{h}(\mathbf{x}_k, \boldsymbol{\theta})) \, (\mathbf{y}_k - \mathbf{h}(\mathbf{x}_k, \boldsymbol{\theta}))^{\mathsf{T}} \mid \mathbf{y}_{1:T} \right].
\end{aligned} \tag{16.56}$$

The details of implementation of the M-step for other kinds of parameter depend on the actual functional form of $\mathbf{f}$ and $\mathbf{h}$. If the parameters appear linearly in the functions, it is possible to find closed form solutions for the maxima analogously to the linear case (Theorem 16.12). Obviously, even when analytical solutions cannot be found, it would be possible to use iterative optimization methods inside EM. But if iterative methods need to be used anyway, then with the same effort we can try to find the minimum of the energy function directly (recall that this is what EM tries to find as well).

Also in the non-linear case, Fisher's identity (Equation (16.32)), in principle, gives an easy way to evaluate the gradients of the energy function. The problem is that both the energy function and the gradient given by Fisher's identity are approximations, and there is no guarantee that the approximations involved are the same. That is, the derivative given by Fisher's identity might not be exactly the derivative of the approximate energy function given by the Gaussian filter. The derivation of the Fisher's identity-based derivative expression is left as an exercise to the reader (Exercise 16.6).

16.3.4 Parameter Estimation with Iterated Filtering and Smoothing

In Chapters 7, 10, 13, and 14 we saw a collection of iterated filters and smoothers that improve the filter and smoother estimates by using iteration either on the update step or over the whole linearization trajectory. In principle, iterated filters can be used to approximate the energy function in the same way as Gaussian filters simply by taking the final result of iteration as the final filter approximation. We can also use iterated smoothers to find the best linearization trajectory along with the corresponding linearization and then use the energy function provided by the (affine) Kalman filter for parameter estimation, or alternatively, use the corresponding $\mathcal{Q}$ approximation.

Indeed, the aforementioned approaches work well for parameter estimation in principle, but there is one catch: The linearization resulting from the iteration will depend on the model parameters, and hence the iterated linearization result will be a complicated function of the parameters. For evaluation of the energy function this is not a problem, but it causes problems when computing the energy gradient. The gradient of the approximation will no longer be given by, for example, the sensitivity equations, but instead it will also depend on the derivative of the iterated linear linearization result with respect to the parameters. This issue is addressed, for example, in Christianson (1994).

16.3.5 Parameter Estimation via Particle Filtering and Smoothing

Particle filtering can also be used for approximate evaluation of the marginal likelihood and also of the energy function needed in parameter estimation. In the particle filtering approach, we can consider generic models of the form

$$\begin{aligned}\boldsymbol{\theta} &\sim p(\boldsymbol{\theta}),\\ \mathbf{x}_0 &\sim p(\mathbf{x}_0 \mid \boldsymbol{\theta}),\\ \mathbf{x}_k &\sim p(\mathbf{x}_k \mid \mathbf{x}_{k-1}, \boldsymbol{\theta}),\\ \mathbf{y}_k &\sim p(\mathbf{y}_k \mid \mathbf{x}_k, \boldsymbol{\theta}),\end{aligned} \tag{16.57}$$

where $\boldsymbol{\theta} \in \mathbb{R}^d$ is the unknown parameter to be estimated. The approximate evaluation of the marginal likelihood can be done with the following modification of the SIR particle filter (see, e.g., Andrieu et al., 2004; Creal, 2012).

Algorithm 16.16 (SIR-based energy function approximation)**.** *An approximation to the marginal likelihood of the parameters can be evaluated during the sequential importance resampling (SIR) algorithm (particle filter), as follows.*

- *Draw N samples $\mathbf{x}_0^{(i)}$ from the prior:*

$$\mathbf{x}_0^{(i)} \sim p(\mathbf{x}_0 \mid \boldsymbol{\theta}), \qquad i = 1, \ldots, N, \tag{16.58}$$

  and set $w_0^{(i)} = 1/N$ for all $i = 1, \ldots, N$.
- *For each $k = 1, \ldots, T$, do the following.*

  1. *Draw samples $\mathbf{x}_k^{(i)}$ from the importance distributions:*

$$\mathbf{x}_k^{(i)} \sim \pi(\mathbf{x}_k \mid \mathbf{x}_{k-1}^{(i)}, \mathbf{y}_{1:k}, \boldsymbol{\theta}), \qquad i = 1, \ldots, N. \tag{16.59}$$

  2. *Compute the following weights:*

$$v_k^{(i)} = \frac{p(\mathbf{y}_k \mid \mathbf{x}_k^{(i)}, \boldsymbol{\theta})\, p(\mathbf{x}_k^{(i)} \mid \mathbf{x}_{k-1}^{(i)}, \boldsymbol{\theta})}{\pi(\mathbf{x}_k^{(i)} \mid \mathbf{x}_{k-1}^{(i)}, \mathbf{y}_{1:k}, \boldsymbol{\theta})}, \tag{16.60}$$

     and compute the estimate of $p(\mathbf{y}_k \mid \mathbf{y}_{1:k-1}, \boldsymbol{\theta})$ as

$$\hat{p}(\mathbf{y}_k \mid \mathbf{y}_{1:k-1}, \boldsymbol{\theta}) = \sum_i w_{k-1}^{(i)}\, v_k^{(i)}. \tag{16.61}$$

  3. *Compute the normalized weights as*

$$w_k^{(i)} \propto w_{k-1}^{(i)}\, v_k^{(i)}. \tag{16.62}$$

  4. *If the effective number of particles* (11.31) *is too low, perform resampling.*

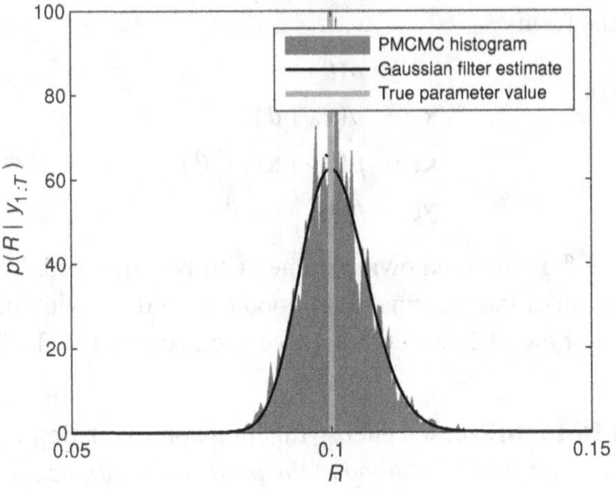

Figure 16.2 Posterior distribution of the noise variance R in the pendulum model (see Example 16.17). The approximation computed with the the Gaussian filter is very close to the reference result computed with the PMCMC.

The approximation of the marginal likelihood of the parameters is

$$p(\mathbf{y}_{1:T} \mid \boldsymbol{\theta}) \approx \prod_k \hat{p}(\mathbf{y}_k \mid \mathbf{y}_{1:k-1}, \boldsymbol{\theta}), \tag{16.63}$$

and the corresponding energy function approximation is

$$\varphi_T(\boldsymbol{\theta}) \approx -\log p(\boldsymbol{\theta}) - \sum_{k=1}^{T} \log \hat{p}(\mathbf{y}_k \mid \mathbf{y}_{1:k-1}, \boldsymbol{\theta}). \tag{16.64}$$

The energy function approximation could also be computed recursively during the SIR algorithm above.

The particle filter-based energy function approximation can now be used, for example, in the Metropolis–Hastings-based MCMC algorithm. The result is the particle Markov chain Monte Carlo (PMCMC) method (Andrieu et al., 2010) and, more specifically, the particle marginal Metropolis–Hastings (PMMH) algorithm variant of it. It turns out that the resulting MCMC algorithm is exact in the sense that it samples from the right distribution Andrieu et al. (2010). Thus the result is (asymptotically) the same as if we had used the exact energy function instead of the particle filter-based approximation in the MCMC method.

Example 16.17 (Estimation of noise variance in the pendulum model). *Figure 16.2 shows the posterior distribution approximation for the noise variance R computed with a Gaussian filter for the pendulum model in Example 7.6. The figure also shows the histogram of samples from the PM-CMC method, which thus should approach the true posterior of the parameter. As can be seen, the posterior distribution estimate of the Gaussian filter (fifth order Gauss–Hermite filter) is a good approximation to the true posterior distribution.*

In principle, it would also be possible to use the likelihood or energy function approximation in gradient-free optimization methods for finding MAP or ML estimates. However, this might turn out to be hard, because even if we fixed the random number generator sequence in the particle filter, the likelihood function would not be continuous in $\boldsymbol{\theta}$ (see, e.g., Kantas et al., 2009). This also renders the use of gradient-based optimization methods impossible.

By comparing to the Rao–Blackwellized particle filter in Algorithm 11.13, it is easy to see that the corresponding likelihood approximation can be obtained by setting

$$v_k^{(i)} = \frac{p(\mathbf{y}_k \mid \mathbf{u}_{0:k}^{(i)}, \mathbf{y}_{1:k-1}, \boldsymbol{\theta})\, p(\mathbf{u}_k^{(i)} \mid \mathbf{u}_{k-1}^{(i)}, \boldsymbol{\theta})}{\pi(\mathbf{u}_k^{(i)} \mid \mathbf{u}_{0:k-1}^{(i)}, \mathbf{y}_{1:k}, \boldsymbol{\theta})}. \tag{16.65}$$

The likelihood approximation itself remains the same as in Equation (16.61):

$$\hat{p}(\mathbf{y}_k \mid \mathbf{y}_{1:k-1}, \boldsymbol{\theta}) = \sum_i w_{k-1}^{(i)}\, v_k^{(i)}.$$

We can also implement the EM algorithm using particle smoothing. Recall that to implement the EM algorithm, we need to evaluate $\mathcal{Q}(\boldsymbol{\theta}, \boldsymbol{\theta}^{(n)})$ via Equation (16.29), which in turn requires us to evaluate the expectations appearing in Equation (16.30). The actual form of the approximation depends on the particle smoother that we are using. In the case of the backward-simulation particle smoother, we have the following simple algorithm (Wills et al., 2013).

Algorithm 16.18 (Evaluation of $\mathcal{Q}$ via the backward-simulation particle smoother). *Assume that we have simulated S trajectories $\{\tilde{\mathbf{x}}_{0:T}^{(i)} : i = 1, \ldots, S\}$ using the backward-simulation particle smoother in Algorithm 15.2 with parameter values fixed to $\boldsymbol{\theta}^{(n)}$. Then the integrals* *in Equation* (16.30) *can be approximated as*

$$
\begin{aligned}
I_1(\boldsymbol{\theta}, \boldsymbol{\theta}^{(n)}) &\approx \frac{1}{S} \sum_{i=1}^{S} \log p(\tilde{\mathbf{x}}_0^{(i)} \mid \boldsymbol{\theta}), \\
I_2(\boldsymbol{\theta}, \boldsymbol{\theta}^{(n)}) &\approx \sum_{k=0}^{T-1} \frac{1}{S} \sum_{i=1}^{S} \log p(\tilde{\mathbf{x}}_{k+1}^{(i)} \mid \tilde{\mathbf{x}}_k^{(i)}, \boldsymbol{\theta}), \\
I_3(\boldsymbol{\theta}, \boldsymbol{\theta}^{(n)}) &\approx \sum_{k=1}^{T} \frac{1}{S} \sum_{i=1}^{S} \log p(\mathbf{y}_k \mid \tilde{\mathbf{x}}_k^{(i)}, \boldsymbol{\theta}).
\end{aligned}
\tag{16.66}
$$

If we are using the reweighting (or marginal) particle smoother in Algorithm 15.7, the corresponding expectations can be approximated as follows (Schön et al., 2011).

Algorithm 16.19 (Evaluation of $\mathcal{Q}$ via the reweighting smoother). *Assume that we have the set of particles* $\{\mathbf{x}_k^{(i)} : k = 0, \ldots, T; i = 1, \ldots, N\}$ *representing the filtering distribution and we have calculated the weights* $\{w_{k|T}^{(i)} : k = 0, \ldots, T; i = 1, \ldots, N\}$ *using Algorithm 15.7. Then we can approximate the integrals in Equation* (16.30) *as follows:*

$$
\begin{aligned}
I_1(\boldsymbol{\theta}, \boldsymbol{\theta}^{(n)}) &\approx \sum_i w_{0|T}^{(i)} \log p(\mathbf{x}_0^{(i)} \mid \boldsymbol{\theta}), \\
I_2(\boldsymbol{\theta}, \boldsymbol{\theta}^{(n)}) &\approx \sum_{k=0}^{T-1} \sum_i \sum_j \frac{w_{k+1|T}^{(j)} \, w_k^{(i)} \, p(\mathbf{x}_{k+1}^{(j)} \mid \mathbf{x}_k^{(i)}, \boldsymbol{\theta}^{(n)})}{\left[\sum_l w_k^{(l)} \, p(\mathbf{x}_{k+1}^{(j)} \mid \mathbf{x}_k^{(l)}, \boldsymbol{\theta}^{(n)})\right]} \\
&\qquad \times \log p(\mathbf{x}_{k+1}^{(j)} \mid \mathbf{x}_k^{(i)}, \boldsymbol{\theta}), \\
I_3(\boldsymbol{\theta}, \boldsymbol{\theta}^{(n)}) &\approx \sum_{k=1}^{T} \sum_i w_{k|T}^{(i)} \log p(\mathbf{y}_k \mid \mathbf{x}_k^{(i)}, \boldsymbol{\theta}).
\end{aligned}
\tag{16.67}
$$

By differentiating the expressions in the above algorithms with respect to $\boldsymbol{\theta}$, we can also get an approximation for $\partial \mathcal{Q}(\boldsymbol{\theta}, \boldsymbol{\theta}^{(n)}) / \partial \boldsymbol{\theta}$. This approximation can be further used in Fisher's identity in Equation (16.32) to give approximations for the gradients of the energy function (Ninness et al., 2010). For more information on this kind of approach as well as other methods for particle filtering-based parameter estimation, the reader is referred to Andrieu et al. (2004), Kantas et al. (2009), and Poyiadjis et al. (2011).

16.3.6 Rao–Blackwellization of Parameters

In this section we show how Rao–Blackwellization can sometimes be used for marginalizing out the static parameters in state space models. Let's start by considering the following generalized version of the pendulum example used in Särkkä (2006) and Särkkä and Sottinen (2008):

$$\begin{aligned} \mathbf{x}_k &= \mathbf{f}(\mathbf{x}_{k-1}) + \mathbf{q}_{k-1}, \\ y_k &= h(\mathbf{x}_k) + r_k, \\ r_k &\sim \mathrm{N}(0, R), \\ R &\sim \mathrm{Inv}\text{-}\chi^2(\nu_0, R_0), \end{aligned} \tag{16.68}$$

where $\mathbf{q}_{k-1} \sim \mathrm{N}(\mathbf{0}, \mathbf{Q})$. This is thus the same kind of non-linear state space model that we have already considered in this book, except that here the measurement noise variance R is considered as unknown and given an inverse-chi-squared prior distribution $\mathrm{Inv}\text{-}\chi^2(\nu_0, R_0)$.

It turns out that we can do sequential Monte Carlo sampling in this model such that we do not need to sample the values of R. Instead, it is enough to sample the state values and then carry the parameters of the distribution of R, conditional on the previous measurements and the histories of samples. The idea is the following.

1. Assume that we have generated a set of particle histories $\{w_k^{(i)}, \mathbf{x}_{0:k}^{(i)} : i = 1, \ldots, N\}$ that approximate the full distribution of the states as follows:

$$p(\mathbf{x}_{0:k-1} \mid y_{1:k-1}) \approx \sum_i w_{k-1}^{(i)} \, \delta(\mathbf{x}_{0:k-1} - \mathbf{x}_{0:k-1}^{(i)}), \tag{16.69}$$

which is thus the conventional SIR filter approximation when we store the full sample histories (see Section 15.1). Further assume that the conditional distribution of R given measurements $y_{1:k-1}$ and the sampled state history $\mathbf{x}_{0:k-1}^{(i)}$ is

$$p(R \mid \mathbf{x}_{0:k-1}^{(i)}, y_{1:k-1}) = \mathrm{Inv}\text{-}\chi^2(R \mid \nu_{k-1}^{(i)}, R_{k-1}^{(i)}), \tag{16.70}$$

where $\nu_{k-1}^{(i)}, R_{k-1}^{(i)}$ have already been computed for each i. Then we have the following approximation for the full distribution of states and parameters:

$$\begin{aligned} &p(\mathbf{x}_{0:k-1}, R \mid y_{1:k-1}) \\ &= p(R \mid \mathbf{x}_{0:k-1}, y_{1:k-1}) \, p(\mathbf{x}_{0:k-1} \mid y_{1:k-1}) \\ &\approx \sum_i w_{k-1}^{(i)} \, \mathrm{Inv}\text{-}\chi^2(R \mid \nu_{k-1}^{(i)}, R_{k-1}^{(i)}) \, \delta(\mathbf{x}_{0:k-1} - \mathbf{x}_{0:k-1}^{(i)}). \end{aligned} \tag{16.71}$$

2. Let us now draw samples from an importance distribution:

$$\mathbf{x}_k^{(i)} \sim \pi(\mathbf{x}_k^{(i)} \mid \mathbf{x}_{k-1}^{(i)}, y_k). \tag{16.72}$$

3. We can now evaluate the likelihood of the measurement given $\mathbf{x}_k^{(i)}$ and the previous measurements as follows:

$$\begin{aligned} &p(y_k \mid \mathbf{x}_k^{(i)}, y_{1:k-1}) \\ &= \int \mathrm{N}(y_k \mid h(\mathbf{x}_k^{(i)}), R) \, \text{Inv-}\chi^2(R \mid \nu_{k-1}^{(i)}, R_{k-1}^{(i)}) \, \mathrm{d}R \\ &= t_{\nu_k^{(i)}}(y_k \mid h(\mathbf{x}_k^{(i)}), R_k^{(i)}), \end{aligned} \tag{16.73}$$

where the parameters of the Student's t-distribution above are

$$\begin{aligned} \nu_k^{(i)} &= \nu_{k-1}^{(i)} + 1, \\ R_k^{(i)} &= \frac{\nu_{k-1}^{(i)} R_{k-1}^{(i)} + (y_k - h(\mathbf{x}_k^{(i)}))^2}{\nu_{k-1}^{(i)} + 1}. \end{aligned} \tag{16.74}$$

This allows us to compute the next step importance weights for the SIR algorithm as follows:

$$w_k^{(i)} \propto \frac{p(y_k \mid \mathbf{x}_k^{(i)}, y_{1:k-1}) \, \mathrm{N}(\mathbf{x}_k^{(i)} \mid \mathbf{f}(\mathbf{x}_{k-1}^{(i)}), \mathbf{Q})}{\pi(\mathbf{x}_k^{(i)} \mid \mathbf{x}_{k-1}^{(i)}, y_k)}. \tag{16.75}$$

4. Given the measurement and the state, we can further compute the conditional distribution of R given $y_{1:k}$ and $\mathbf{x}_{0:k}^{(i)}$:

$$p(R \mid \mathbf{x}_{0:k}^{(i)}, y_{1:k}) = \text{Inv-}\chi^2(R \mid \nu_k^{(i)}, R_k^{(i)}). \tag{16.76}$$

5. Now we again have a similar representation for the filtering distribution as in step 1, and we can set $k \leftarrow k+1$ and go back to step 1. But before that we can also do resampling jointly for the state histories $\mathbf{x}_{0:k}^{(i)}$ and the parameters $\nu_k^{(i)}, R_k^{(i)}$ as in the conventional SIR algorithm.

In the above algorithm, we would not actually need to carry the whole state histories in the filter, but theoretically the parameters of the inverse chi squared distributions are indeed conditioned on the full state histories.

The procedure above is a Rao–Blackwellized particle filter where the static parameter R has been marginalized out and is carried via its sufficient statistics $\nu_k^{(i)}, R_k^{(i)}$ conditioned on the particle histories $\mathbf{x}_{0:k}^{(i)}$ and measurements. In another example given in Särkkä (2006) and Särkkä and Sottinen

(2008), this procedure is used for marginalizing out the unknown population size in a Poisson measurement model. The same idea can be used in various other types of model.

In an abstract sense the method can be applied to a class of models of the form

$$\begin{aligned} \mathbf{x}_k &\sim p(\mathbf{x}_k \mid \mathbf{x}_{k-1}, \boldsymbol{\theta}), \\ \mathbf{y}_k &\sim p(\mathbf{y}_k \mid \mathbf{x}_k, \boldsymbol{\theta}), \\ \boldsymbol{\theta} &\sim p(\boldsymbol{\theta}), \end{aligned} \tag{16.77}$$

where the vector $\boldsymbol{\theta}$ contains the unknown static parameters. Now if the posterior distribution of the parameters $\boldsymbol{\theta}$ depends only on some sufficient statistics

$$\mathcal{T}_k = \mathcal{T}_k(\mathbf{x}_{1:k}, \mathbf{y}_{1:k}), \tag{16.78}$$

and if the sufficient statistics are easy to update recursively, $\mathcal{T}_k \leftarrow \mathcal{T}_{k-1}$, then sampling of the state and parameters can be performed by recursively computing the sufficient statistics conditionally on the sampled states and the measurements analogously to the example above. The original idea of the method seems to have appeared quite independently in Storvik (2002), Fearnhead (2002), and Djuric and Miguez (2002), and it has been applied to estimation of full noise covariances in state space models by Saha et al. (2010).

A particularly useful special case, which includes the example above, is obtained when the dynamic model is independent of the parameters $\boldsymbol{\theta}$. In this case, if conditionally to the state $\mathbf{x}_k$ the prior $p(\boldsymbol{\theta})$ belongs to the conjugate family of the likelihood $p(\mathbf{y}_k \mid \mathbf{x}_k, \boldsymbol{\theta})$, the static parameters $\boldsymbol{\theta}$ can be marginalized out, and only the states need to be sampled. This idea can be extended to the time-varying case if the dynamic model has a form that keeps the predicted distribution of the parameter within the conjugate family (cf. Särkkä and Nummenmaa, 2009).

When the static parameter appears linearly in the model, we recover a noise-free version of the conditionally Gaussian Rao–Blackwellization considered in Section 11.7 (see Schön and Gustafsson, 2003). The Rao–Blackwellized particle filter can then be seen as a time-varying extension of this method in the conditionally linear Gaussian case.

16.4 Exercises

16.1 Implement the EM algorithm for ML estimation of the measurement noise variance in the Gaussian random walk model considered in Examples 6.4, 6.7, and 12.3. Test the algorithm with simulated data.

16.2 Implement the algorithm for computing the energy function for the Gaussian random walk model as well as its derivative with respect to the noise variance (via the sensitivity equations given in Section A.10). Generate some simulated data, and use a gradient-based optimization method to find the ML estimate of the parameter.

16.3 With the Gaussian random walk model, find the expression for the Fisher's identity-based derivative with respect to the noise parameter. Check numerically that it matches the expression obtained with the sensitivity equations.

16.4 Implement a random walk Metropolis-based MCMC method for estimating the noise variance in the Gaussian random walk model. Use the Kalman filter to evaluate the energy function. For simplicity, you can assume that the prior for the parameter has the form $p(R) \propto 1$.

16.5 Derive the sensitivity equations for the first order extended Kalman filter.

16.6 Derive the equation for the derivative of the energy function resulting from differentiating the Gaussian smoothing-based approximation in Algorithm 16.15 and using Fisher's identity (Equation (16.32)).

16.7 Compute and plot the approximate energy function obtained for the noise variance in the model given in Exercise 7.1 by using a non-linear Kalman filter-based estimate of the energy function. You can select the filter freely.

16.8 Implement a random walk Metropolis-based MCMC method for estimating the noise variance in the model given in Exercise 7.1. Use one of the non-linear Kalman filters to approximate the energy function.

16.9 Implement a random walk Metropolis-based particle MCMC method for estimating the noise variance in the Gaussian random walk model. Use a simple bootstrap filter as the particle filter.

16.10 Implement a random walk Metropolis-based particle MCMC method for estimating the noise variance in the model given in Exercise 7.1.

17

Epilogue

17.1 Which Method Should I Choose?

An important question when preparing to solve a specific filtering, smoothing, or parameter estimation problem for state space models is: *Which of the numerous methods should I choose for a particular application*? Obviously if the problem is linear, then the Kalman filter and RTS smoother are natural choices – also for evaluating the quantities needed for parameter estimation. But if the system is non-linear/non-Gaussian, the question is harder.

When the noises in the system can be modeled as Gaussian, and the model is of the form

$$\begin{aligned} \mathbf{x}_k &= \mathbf{f}(\mathbf{x}_{k-1}) + \mathbf{q}_{k-1}, \\ \mathbf{y}_k &= \mathbf{h}(\mathbf{x}_k) + \mathbf{r}_k, \end{aligned} \tag{17.1}$$

where $\mathbf{f}$ and $\mathbf{h}$ are somewhat well-behaved functions, then the first choice would be one of the Gaussian approximation-based filters and smoothers – provided that we are working on an application, and the theoretical exactness of the solution is not important per se, but we are interested in getting good estimates of the state and parameters. If theoretical exactness is needed, then the only option is to use particle filters and smoothers (or grid-based solutions).

Among the Gaussian approximation-based filters and smoothers, it is always a good idea to start with an EKF and an ERTSS. These are the only algorithms that have been used for over half a century in practical applications, and there are good reasons for that – they simply work. Iterated versions of the EKF and ERTSS can then be used to improve the results – in particular, the iterated ERTSS is a powerful algorithm that often leads to good results, due to its convergence to the MAP estimate.

With some models the EKF and ERTSS do not work well or at all, and in that case we can move to the sigma point methods. The spherical cubature and unscented methods have the advantage of being computationally quite light, but they still tend to produce very good results. However, these methods have the problem that their error estimates might not always be consistent with the actual errors, a problem which the EKF/ERTSS methods also tend to have. The unscented transform has more parameters to tune for a particular problem than the spherical cubature method, which can be an advantage or a disadvantage (recall that the spherical cubature method is an unscented transform with a certain selection of parameters). The Gauss–Hermite-based methods tend to be more consistent in errors and are thus more robust approximations, but they have the disadvantage of having high computational complexity. The higher order cubature/unscented rules also have a moderate computational complexity while having quite high accuracy, but their problem is that they can lead to negative weights. However, there are various other higher integration rules that do not have (all of) these problems but which have not been explicitly discussed in this book.

The iterated sigma point methods can also be used to improve the performance of the non-iterated versions, and sometimes Monte Carlo integration can also be used to replace the sigma point-based numerical integration when using Gaussian approximations. One should always remember that there is no guarantee that using more complicated filtering and smoothing algorithms would actually improve the results; therefore it is a good idea to always test the EKF and ERTSS first. The bootstrap filter has the advantage that it is very easy to implement, and thus it can sometimes be used as a reference solution when debugging and testing the performance of Gaussian approximation-based filters and smoothers.

If the problem has a more complicated form that cannot be fitted into the non-linear Gaussian framework or when the Gaussian approximations do not work for other reasons, we need to go to particle-based solutions. Because the bootstrap filter is very easy to implement, it (and probably one of the particle smoothers) should be the first option to test with a sufficiently large number of particles. However, the clear disadvantage of particle methods is the high computational load, and thus it is a good idea to check at quite an early stage if any of the states or parameters can be marginalized out ("Rao–Blackwellized") exactly or approximately. If this is the case, then one should always prefer marginalization to sampling.[1]

[1] The rule of thumb is: Use Monte Carlo sampling only as a last resort when all the other options have failed.

The other thing to check is if it is possible to use the optimal or almost optimal importance distribution in the particle filter. In principle, non-linear Gaussian approximation-based filters can be used to form such importance distributions, but this may also lead to overly heavy computational methods as well as to convergence problems. If they are used, then it might be advisable to artificially increase the filter covariances somewhat or to use Student's t-distributions instead of using the Gaussian approximations as such.

When it comes to parameter estimation, it is generally a good idea to use the same approximations in the parameter estimation method as will be used in the final application, assuming that the parameter estimation results will later be used in filters and smoothers to solve a state estimation problem. Furthermore, if a single parameter estimation result (point estimate) is needed anyway, ML and MAP estimates are not bad choices, but it might be useful to check the spread of the posterior distribution of parameters using an MCMC method. But if the true values of the parameters are of interest, then the combination of particle filtering and MCMC is probably the safest bet. However, one should remember that estimating the true parameters of the system is possible only in simulated scenarios and in real applications the models used will be more or less wrong anyway. On the other hand, we should be careful not to ruin already probably inaccurate models with bad approximations of filters and smoothers.

17.2 Further Topics

This book is mainly concerned with non-linear Kalman filtering and smoothing as well as particle filtering and smoothing approaches to Bayesian filtering and smoothing, but numerous other methods exist as well. It is impossible to list all of them, but below we try to give pointers to some of the methods. Regarding filters and smoothers themselves, there are also various subareas that we did not discuss, and we try to give some pointers to them as well.

First of all, one huge area that we have not mentioned at all is continuous time filters and smoothers. In these methods the dynamics of the state and sometimes the measurements as well are modeled using stochastic differential equations (SDEs, Øksendal, 2003; Särkkä and Solin, 2019). The full theory of Bayesian filtering in such models can be found in the classic book of Jazwinski (1970), and the smoothing theory is due to Striebel

(1965) and Leondes et al. (1970). The extended Kalman type of filter approximation can be found in the abovementioned references as well. Extensions of unscented Kalman filters and smoothers to the continuous-time setting can be found in Särkkä (2006, 2007, 2010). Extensions of Gaussian filters and smoothers to continuous-time setting have been discussed in Singer (2008), Arasaratnam et al. (2010), Singer (2011), Särkkä and Solin (2012), and Särkkä and Sarmavuori (2013). Extensions of particle filters and smoothers to continuous-time setting can be found, for example, in Crisan and Rozovskii (2011), Särkkä and Sottinen (2008), Murray and Storkey (2011), and references therein.

There also exist various other kinds of Gaussian integration methods that we have not presented here that could be used for constructing new kinds of Gaussian filters and smoothers (see, e.g., O'Hagan, 1991; Nørgaard et al., 2000; Lefebvre et al., 2002; Wu et al., 2006; Särkkä and Hartikainen, 2010b; Sandblom and Svensson, 2012). One particularly interesting approach is to approximate the non-linear functions with a Gaussian process-based non-parametric model that is fitted using a finite number of sample points (Deisenroth et al., 2009, 2012). These methods are related to so-called kernel or Gaussian process quadratures (see, e.g., O'Hagan, 1991; Särkkä et al., 2016; Karvonen and Särkkä, 2018; Karvonen et al., 2019; Prüher et al., 2021; Hennig et al., 2022), which are numerical integration methods based on Gaussian process regression.

One useful class of discrete-time methods is the multiple model approaches, such as the generalized pseudo-Bayesian methods (GPB1 and GPB2) as well as the interacting multiple model (IMM) algorithm (Bar-Shalom et al., 2001). These methods can be used for approximating the Bayesian solutions to problems with a fixed number of models or modes of operation. The active mode of the system is described by a discrete latent variable, which is modeled as a discrete-state Markov chain. Given the value of the latent variable, the system is (approximately) Gaussian. The GPB1, GPB2, and IMM algorithms are based on an approximation that is a mixture of Gaussians (a bank of Kalman or extended Kalman filters) to the Bayesian filtering solutions by using moment matching.

The abovementioned multiple model methods are also closely related to so-called expectation correction (EC, Barber, 2006) and expectation propagation (EP, Zoeter and Heskes, 2011) methods, which can also be used for Bayesian filtering and smoothing in switching linear dynamic systems (SLDS), which is another term used for multiple mode/model problems. These models can also be considered as special cases of the conditionally Gaussian models, and the history of similar approximations dates back to the works of Alspach and Sorenson (1972) and Akashi and Kumamoto (1977). The relationship between various methods for this type of model has also been analyzed by Barber (2011).

When the measurement model is non-Gaussian (e.g., Student's t), it is sometimes possible to use variational Bayes approximations (Agamennoni et al., 2011; Piché et al., 2012) to yield to tractable inference. The expectation propagation (EP) algorithm (Ypma and Heskes, 2005) can also be used for approximate inference in non-linear and non-Gaussian dynamic systems. Both of these approaches are also closely related to the Gaussian filters and smoothers considered in this book. Variational Bayesian approximations can also be used for estimation of unknown time-varying parameters in state space models (Särkkä and Nummenmaa, 2009).

In the multiple target tracking context, there exist a number of specific methods to cope with the problems arising there, namely, the data association problem and an unknown numbers of targets. The classical approaches to the multiple target tracking can be found in the books of Bar-Shalom and Li (1995) and Blackman and Popoli (1999). For further approaches, the reader is referred to Ristic et al. (2004), Challa et al. (2011), and Mahler (2014), along with Särkkä et al. (2007b), Svensson et al. (2011), García-Fernández et al. (2020); Granström et al. (2020), and García-Fernández et al. (2021).

There are a number of other important topics that have had to be omitted here. For example, the Cramér–Rao lower bounds (CRLB, see, e.g., Ristic et al., 2004) are theoretical lower bounds for the mean-squared error that can be achieved with any non-linear filter in a given filtering problem. We have also omitted the observability and controllability questions of linear and non-linear systems (see, e.g., Kailath et al., 2000; Bar-Shalom et al., 2001), which are related to the question whether it is possible to determine the state of a given system from the available measurements at all. A somewhat related issue is the question of under what conditions does a particle filter converge to the true distribution. For more details on this topic, the reader is referred to the works of Crisan and Doucet (2002) and Hu et al. (2008, 2011). The numerical stability of linear and non-linear Kalman filters and RTS smoothers can sometimes also be enhanced by using square root versions of them. This kind of methods can be found, for example, in the works of Bierman (1977), Van der Merwe and Wan (2001), Arasaratnam and Haykin (2009, 2011), and Grewal and Andrews (2015).

There has also been some recent interest in using graphics processing units (GPUs, Barlas, 2015) for parallel computations, and special parallel filtering and smoothing methods that allow for parallelization can be found, for example, in Särkkä and García-Fernández (2021), Hassan et al. (2021), Yaghoobi et al. (2021), and Corenflos et al. (2022).

Appendix

Additional Material

A.1 Properties of Gaussian Distribution

Definition A.1 (Gaussian distribution). *A random variable* $\mathbf{x} \in \mathbb{R}^n$ *has a Gaussian distribution with mean* $\mathbf{m} \in \mathbb{R}^n$ *and covariance* $\mathbf{P} \in \mathbb{R}^{n \times n}$ *if its probability density has the form*

$$\mathrm{N}(\mathbf{x} \mid \mathbf{m}, \mathbf{P}) = \frac{1}{(2\,\pi)^{n/2}\,|\mathbf{P}|^{1/2}} \exp\left(-\frac{1}{2}(\mathbf{x}-\mathbf{m})^{\mathsf{T}}\,\mathbf{P}^{-1}\,(\mathbf{x}-\mathbf{m})\right), \tag{A.1}$$

where $|\mathbf{P}|$ *is the determinant of the matrix* $\mathbf{P}$.

Lemma A.2 (Joint distribution of Gaussian variables). *If random variables* $\mathbf{x} \in \mathbb{R}^n$ *and* $\mathbf{y} \in \mathbb{R}^m$ *have the Gaussian probability distributions*

$$\begin{aligned} \mathbf{x} &\sim \mathrm{N}(\mathbf{m}, \mathbf{P}), \\ \mathbf{y} \mid \mathbf{x} &\sim \mathrm{N}(\mathbf{H}\,\mathbf{x} + \mathbf{u}, \mathbf{R}), \end{aligned} \tag{A.2}$$

then the joint distribution of $\mathbf{x}, \mathbf{y}$ *and the marginal distribution of* $\mathbf{y}$ *are given as*

$$\begin{aligned} \begin{pmatrix} \mathbf{x} \\ \mathbf{y} \end{pmatrix} &\sim \mathrm{N}\left(\begin{pmatrix} \mathbf{m} \\ \mathbf{H}\,\mathbf{m} + \mathbf{u} \end{pmatrix}, \begin{pmatrix} \mathbf{P} & \mathbf{P}\,\mathbf{H}^{\mathsf{T}} \\ \mathbf{H}\,\mathbf{P} & \mathbf{H}\,\mathbf{P}\,\mathbf{H}^{\mathsf{T}} + \mathbf{R} \end{pmatrix}\right), \\ \mathbf{y} &\sim \mathrm{N}(\mathbf{H}\,\mathbf{m} + \mathbf{u}, \mathbf{H}\,\mathbf{P}\,\mathbf{H}^{\mathsf{T}} + \mathbf{R}). \end{aligned} \tag{A.3}$$

Lemma A.3 (Conditional distribution of Gaussian variables). *If the random variables* $\mathbf{x}$ *and* $\mathbf{y}$ *have the joint Gaussian probability distribution*

$$\begin{pmatrix} \mathbf{x} \\ \mathbf{y} \end{pmatrix} \sim \mathrm{N}\left(\begin{pmatrix} \mathbf{a} \\ \mathbf{b} \end{pmatrix}, \begin{pmatrix} \mathbf{A} & \mathbf{C} \\ \mathbf{C}^{\mathsf{T}} & \mathbf{B} \end{pmatrix}\right), \tag{A.4}$$

then the marginal and conditional distributions of $\mathbf{x}$ *and* $\mathbf{y}$ *are given as follows:*

$$\begin{aligned}
\mathbf{x} &\sim \mathrm{N}(\mathbf{a}, \mathbf{A}), \\
\mathbf{y} &\sim \mathrm{N}(\mathbf{b}, \mathbf{B}), \\
\mathbf{x} \mid \mathbf{y} &\sim \mathrm{N}(\mathbf{a} + \mathbf{C}\,\mathbf{B}^{-1}\,(\mathbf{y} - \mathbf{b}), \mathbf{A} - \mathbf{C}\,\mathbf{B}^{-1}\mathbf{C}^{\mathsf{T}}), \\
\mathbf{y} \mid \mathbf{x} &\sim \mathrm{N}(\mathbf{b} + \mathbf{C}^{\mathsf{T}}\,\mathbf{A}^{-1}\,(\mathbf{x} - \mathbf{a}), \mathbf{B} - \mathbf{C}^{\mathsf{T}}\,\mathbf{A}^{-1}\,\mathbf{C}).
\end{aligned} \tag{A.5}$$

A.2 Block Matrix Inverses and Matrix Inversion Formulas

Block matrix inversion formulas (see, e.g., Lütkepohl, 1996) and matrix inversion formulas (sometimes called Woodbury's identities) are useful in simplifying various expressions involving Gaussian distributions. The block matrix inversion formulas are the following, and they are easy to verify by direct calculation.

Theorem A.4 (Block matrix inverses). *When* $\mathbf{A}$ *and* $\mathbf{D} - \mathbf{C}\,\mathbf{A}^{-1}\,\mathbf{B}$ *are invertible, then*

$$\begin{aligned}
&\begin{pmatrix} \mathbf{A} & \mathbf{B} \\ \mathbf{C} & \mathbf{D} \end{pmatrix}^{-1} \\
&= \begin{pmatrix} \mathbf{A}^{-1} + \mathbf{A}^{-1}\,\mathbf{B}\,(\mathbf{D} - \mathbf{C}\mathbf{A}^{-1}\,\mathbf{B})^{-1}\mathbf{C}\mathbf{A}^{-1} & -\mathbf{A}^{-1}\,\mathbf{B}\,(\mathbf{D} - \mathbf{C}\mathbf{A}^{-1}\,\mathbf{B})^{-1} \\ -(\mathbf{D} - \mathbf{C}\mathbf{A}^{-1}\,\mathbf{B})^{-1}\,\mathbf{C}\mathbf{A}^{-1} & (\mathbf{D} - \mathbf{C}\mathbf{A}^{-1}\,\mathbf{B})^{-1} \end{pmatrix}.
\end{aligned} \tag{A.6}$$

When $\mathbf{D}$ *and* $\mathbf{A} - \mathbf{B}\,\mathbf{D}^{-1}\,\mathbf{C}$ *are invertible, then*

$$\begin{aligned}
&\begin{pmatrix} \mathbf{A} & \mathbf{B} \\ \mathbf{C} & \mathbf{D} \end{pmatrix}^{-1} \\
&= \begin{pmatrix} (\mathbf{A} - \mathbf{B}\,\mathbf{D}^{-1}\,\mathbf{C})^{-1} & -(\mathbf{A} - \mathbf{B}\,\mathbf{D}^{-1}\,\mathbf{C})^{-1}\,\mathbf{B}\,\mathbf{D}^{-1} \\ -\mathbf{D}^{-1}\mathbf{C}(\mathbf{A} - \mathbf{B}\,\mathbf{D}^{-1}\mathbf{C})^{-1} & \mathbf{D}^{-1} + \mathbf{D}^{-1}\mathbf{C}\,(\mathbf{A} - \mathbf{B}\,\mathbf{D}^{-1}\mathbf{C})^{-1}\mathbf{B}\,\mathbf{D}^{-1} \end{pmatrix}.
\end{aligned} \tag{A.7}$$

We can derive various useful formulas for matrices by matching the blocks in the representations in Equations (A.6) and (A.7). For example, the following formulas are sometimes useful in deriving Gaussian conditioning, and Kalman filtering and smoothing equations.

Corollary A.5 (Matrix inversion lemmas). *By putting* $\mathbf{A} = \mathbf{P}^{-1}$, $\mathbf{D} = \mathbf{R}$, $\mathbf{B} = -\mathbf{H}^\mathsf{T}$, *and* $\mathbf{C} = \mathbf{H}$, *and matching the top left blocks in Equations* (A.6) *and* (A.7), *we get*

$$\left(\mathbf{P}^{-1} + \mathbf{H}^\mathsf{T}\,\mathbf{R}^{-1}\,\mathbf{H}\right)^{-1} = \mathbf{P} - \mathbf{P}\mathbf{H}^\mathsf{T}\,\left(\mathbf{R} + \mathbf{H}\,\mathbf{P}\,\mathbf{H}^\mathsf{T}\right)^{-1}\mathbf{H}\,\mathbf{P}, \tag{A.8}$$

and from the top right block, we get

$$\left(\mathbf{P}^{-1} + \mathbf{H}^\mathsf{T}\,\mathbf{R}^{-1}\,\mathbf{H}\right)^{-1}\,\mathbf{H}^\mathsf{T}\,\mathbf{R}^{-1} = \mathbf{P}\mathbf{H}^\mathsf{T}\,\left(\mathbf{R} + \mathbf{H}\,\mathbf{P}\,\mathbf{H}^\mathsf{T}\right)^{-1}. \tag{A.9}$$

A.3 Cholesky Factorization and Its Derivative

The Cholesky factor of the symmetric positive definite matrix $\mathbf{P}$ is a lower triangular matrix $\mathbf{A}$ such that

$$\mathbf{P} = \mathbf{A}\,\mathbf{A}^\mathsf{T}. \tag{A.10}$$

The matrix $\mathbf{A}$ can be computed by the Cholesky factorization algorithm (see, e.g., Golub and van Loan, 1996) presented below.

Algorithm A.6 (Cholesky factorization). *The Cholesky factor* $\mathbf{A}$ *of the matrix* $\mathbf{P} \in \mathbb{R}^{n \times n}$ *can be computed as follows:*

procedure CHOL($\mathbf{P}$)
  for $i \leftarrow 1 \ldots n$ ***do***
    $A_{ii} = \sqrt{P_{ii} - \sum_{k<i} A_{ik}^2}$
    for $j \leftarrow i + 1 \ldots n$ ***do***
      $A_{ji} = \left(P_{ji} - \sum_{k<i} A_{jk}\,A_{ik}\right) / A_{ii}$
    end for
  end for
  return $\mathbf{A}$
end procedure

The partial derivative of the Cholesky factor $\partial\mathbf{A}/\partial\theta$ with respect to a scalar parameters θ can be computed using the following algorithm once $\mathbf{P}$ and $\partial\mathbf{P}/\partial\theta$ are known. The algorithm can be derived by formally differentiating the Cholesky factorization algorithm equations.

Algorithm A.7 (Partial derivative of Cholesky factorization I). *The partial derivative* $\mathbf{D} = \partial\mathbf{A}/\partial\theta$ *of the Cholesky factor of* $\mathbf{P} \in \mathbb{R}^{n \times n}$ *with respect to a scalar parameter* θ *can be computed as follows:*

procedure DCHOL($\mathbf{P}$, $\partial\mathbf{P}/\partial\theta$)
  for $i \leftarrow 1 \ldots n$ ***do***

$A_{ii} \leftarrow \sqrt{P_{ii} - \sum_{k<i} A_{ik}^2}$
$D_{ii} \leftarrow (\partial P_{ii}/\partial\theta - \sum_{k<i} 2 D_{ik}\, A_{ik})/A_{ii}/2$
for $j \leftarrow i+1 \ldots n$ ***do***
  $A_{ji} \leftarrow (P_{ji} - \sum_{k<i} A_{jk}\, A_{ik})/A_{ii}$
  $\text{temp} \leftarrow \partial P_{ji}/\partial\theta - \sum_{k<i}(D_{jk}\, A_{ik} + A_{jk}\, D_{ik})$
  $D_{ji} \leftarrow \text{temp}/A_{ii} - (D_{ii}/A_{ii})\, A_{ji}$
end for
end for
return $\mathbf{D}$
end procedure

Another way to compute the same derivative is via the following theorem.

Theorem A.8 (Partial derivative of Cholesky factorization II). *The partial derivative* $\partial \mathbf{A}/\partial\theta$ *of the lower triangular Cholesky factor* $\mathbf{A}$ *such that* $\mathbf{P} = \mathbf{A}\,\mathbf{A}^\mathsf{T}$ *with respect to a scalar parameter* θ *can be computed as*

$$\frac{\partial \mathbf{A}}{\partial\theta} = \mathbf{A}\,\Phi\left(\mathbf{A}^{-1}\,\frac{\partial \mathbf{P}}{\partial\theta}\,\mathbf{A}^{-\mathsf{T}}\right), \tag{A.11}$$

where $\Phi(\cdot)$ *is a function returning the lower triangular part and half the diagonal of the argument as follows:*

$$\Phi_{ij}(\mathbf{M}) = \begin{cases} M_{ij}, & \text{if } i > j, \\ \frac{1}{2} M_{ij}, & \text{if } i = j, \\ 0, & \text{if } i < j. \end{cases} \tag{A.12}$$

Proof We use a similar trick that was used in the derivation of the time derivative of the Cholesky factor in Morf et al. (1978). We have

$$\mathbf{P} = \mathbf{A}\,\mathbf{A}^\mathsf{T}. \tag{A.13}$$

By taking the derivative with respect to θ, we get

$$\frac{\partial \mathbf{P}}{\partial\theta} = \frac{\partial \mathbf{A}}{\partial\theta}\,\mathbf{A}^\mathsf{T} + \mathbf{A}\,\frac{\partial \mathbf{A}^\mathsf{T}}{\partial\theta}. \tag{A.14}$$

Multiplying from the left with $\mathbf{A}^{-1}$ and from the right with $\mathbf{A}^{-\mathsf{T}}$ gives

$$\mathbf{A}^{-1}\,\frac{\partial \mathbf{P}}{\partial\theta}\,\mathbf{A}^{-\mathsf{T}} = \mathbf{A}^{-1}\,\frac{\partial \mathbf{A}}{\partial\theta} + \frac{\partial \mathbf{A}^\mathsf{T}}{\partial\theta}\,\mathbf{A}^{-\mathsf{T}}. \tag{A.15}$$

Now the right-hand side is the sum of a lower triangular matrix and an upper triangular matrix with identical diagonals. Thus we can recover

$\mathbf{A}^{-1}\ \partial\mathbf{A}/\partial\theta$ via

$$\mathbf{A}^{-1}\,\frac{\partial \mathbf{A}}{\partial \theta} = \Phi\left(\mathbf{A}^{-1}\,\frac{\partial \mathbf{P}}{\partial \theta}\,\mathbf{A}^{-\mathsf{T}}\right), \tag{A.16}$$

where the function $\Phi(\cdot)$ returns the (strictly) lower triangular part of the argument and half of the diagonal. Multiplying from the left by $\mathbf{A}$ gives the result. □

A.4 Affine Stochastic Differential Equations

Many of the stochastic differential equations (SDEs) that we study in Chapter 4 are affine, and the following lemma can be used to derive several of the key results in that chapter. Note that a more rigorous derivation of these results can be obtained using Itô calculus (see, e.g., Särkkä and Solin, 2019).

Lemma A.9 (Analytical solution to affine SDE). *Suppose $\mathbf{x}(t)$ obeys the stochastic differential equation (SDE)*

$$\frac{\mathrm{d}\mathbf{x}(t)}{\mathrm{d}t} = \mathbf{F}\,\mathbf{x}(t) + \mathbf{b}(t) + \mathbf{L}\,\mathbf{w}(t), \tag{A.17}$$

where $\mathbf{F}$ is constant, $\mathbf{b}(t)$ is a deterministic function of time, and $\mathbf{w}(t)$ is white Gaussian noise with the moments (4.4). *It then holds that*

$$\begin{aligned}\mathbf{x}(t+\tau) = \exp(\mathbf{F}\,\tau)\mathbf{x}(t) &+ \int_t^{t+\tau} \exp(\mathbf{F}\,(t+\tau-v))\,\mathbf{b}(v)\,\mathrm{d}v \\ &+ \int_t^{t+\tau} \exp(\mathbf{F}\,(t+\tau-v))\,\mathbf{L}\,\mathbf{w}(v)\,\mathrm{d}v,\end{aligned} \tag{A.18}$$

where $\int_t^{t+\tau} \exp(\mathbf{F}\,(t+\tau-v))\,\mathbf{L}\,\mathbf{w}(v)\,\mathrm{d}v$ is a Gaussian random variable with zero mean and covariance

$$\int_0^{\tau} \exp(\mathbf{F}\,v)\,\mathbf{L}\,\mathbf{Q}^{\mathrm{c}}\,\mathbf{L}^{\mathsf{T}}\,\exp(\mathbf{F}\,v)^{\mathsf{T}}\,\mathrm{d}v. \tag{A.19}$$

Consequently, $\mathbf{x}(t+\tau) \mid \mathbf{x}(t)$ is Gaussian with the moments

$$\begin{aligned}\mathrm{E}\left[\mathbf{x}(t+\tau) \mid \mathbf{x}(t)\right] &= \exp(\mathbf{F}\,\tau)\,\mathbf{x}(t) \\ &\quad + \int_t^{t+\tau} \exp(\mathbf{F}\,(t+\tau-v))\,\mathbf{b}(v)\,\mathrm{d}v, \\ \mathrm{Cov}\left[\mathbf{x}(t+\tau) \mid \mathbf{x}(t)\right] &= \int_0^{\tau} \exp(\mathbf{F}\,v)\,\mathbf{L}\,\mathbf{Q}^{\mathrm{c}}\,\mathbf{L}^{\mathsf{T}}\exp(\mathbf{F}\,v)^{\mathsf{T}}\,\mathrm{d}v.\end{aligned} \tag{A.20}$$

Furthermore, if $\mathbf{b}(t) = \mathbf{b}$ *is a constant, it holds that*

$$\int_t^{t+\tau} \exp(\mathbf{F}\,(t+\tau-v))\,\mathbf{b}\,\mathrm{d}v = \int_0^{\tau} \exp(\mathbf{F}\,v)\,\mathrm{d}v\,\mathbf{b} = \left(\sum_{i=1}^{\infty} \frac{\tau^i\,\mathbf{F}^{i-1}}{i!}\right)\mathbf{b}, \tag{A.21}$$

which further simplifies the expressions for $\mathbf{x}(t+\tau)$ *and* $\mathrm{E}\left[\mathbf{x}(t+\tau) \mid \mathbf{x}(t)\right]$.

Proof To solve (A.17), we move the term $\mathbf{F}\,\mathbf{x}(t)$ to the left-hand side and multiply both sides of the equation with the integrating factor $\exp(-\mathbf{F}\,t)$, which gives

$$\exp(-\mathbf{F}\,t)\frac{\mathrm{d}\mathbf{x}(t)}{\mathrm{d}t} - \exp(-\mathbf{F}\,t)\,\mathbf{F}\,\mathbf{x}(t) = \exp(-\mathbf{F}\,t)\,(\mathbf{b}(t) + \mathbf{L}\,\mathbf{w}(t))\,. \tag{A.22}$$

From the definition of the matrix exponential,

$$\exp(\mathbf{F}\,t) = \sum_{i=0}^{\infty} \frac{\mathbf{F}^i\,t^i}{i!} = \mathbf{I} + \mathbf{F}\,t + \frac{\mathbf{F}^2\,t^2}{2!} + \cdots, \tag{A.23}$$

it follows that $\frac{\mathrm{d}}{\mathrm{d}t}\exp(-\mathbf{F}\,t) = -\mathbf{F}\,\exp(-\mathbf{F}\,t)$, and we can use this relation to identify that the left-hand side in (A.22) is $\frac{\mathrm{d}}{\mathrm{d}t}(\exp(-\mathbf{F}\,t)\,\mathbf{x}(t))$, which implies that

$$\frac{\mathrm{d}}{\mathrm{d}t}(\exp(-\mathbf{F}\,t)\,\mathbf{x}(t)) = \exp(-\mathbf{F}\,t)\;(\mathbf{b}(t) + \mathbf{L}\,\mathbf{w}(t))\,. \tag{A.24}$$

Integrating (A.24) from t to $t+\tau$ therefore gives

$$\begin{aligned} &\exp(-\mathbf{F}\,(t+\tau))\,\mathbf{x}(t+\tau) - \exp(-\mathbf{F}\,t)\,\mathbf{x}(t) \\ &\qquad = \int_t^{t+\tau} \exp(-\mathbf{F}\,v)\;(\mathbf{b}(t) + \mathbf{L}\,\mathbf{w}(v))\,\mathrm{d}v. \end{aligned} \tag{A.25}$$

By multiplying both sides of (A.25) by $\exp(\mathbf{F}\,(t+\tau))$, we obtain (A.18).

Let us now analyze the two remaining integrals in (A.18) one at a time. If $\mathbf{b}(t)$ is a constant, it holds that

$$\int_t^{t+\tau} \exp(\mathbf{F}\,(t+\tau-v))\,\mathbf{b}\,\mathrm{d}v = \int_t^{t+\tau} \exp(\mathbf{F}\,(t+\tau-v))\,\mathrm{d}v\,\mathbf{b}. \tag{A.26}$$

Using appropriate changes of integration variables, combined with the definition of the matrix exponential, we get

$$\begin{aligned}\int_t^{t+\tau} \exp(\mathbf{F}\,(t+\tau-v))\;dv &= \int_0^{\tau} \exp(\mathbf{F}\,v)\;dv \\ &= \int_0^{\tau} \sum_{i=0}^{\infty} \frac{\mathbf{F}^i\,v^i}{i!}\;dv \\ &= \sum_{i=0}^{\infty} \frac{\mathbf{F}^i\,\tau^{i+1}}{(i+1)!} \\ &= \mathbf{I}\tau + \frac{\mathbf{F}\,\tau^2}{2} + \frac{\mathbf{F}^2\,\tau^3}{3!} + \cdots,\end{aligned} \tag{A.27}$$

which confirms (A.21).

The last integral in (A.18) contains $\mathbf{w}(t)$, which is a Gaussian process with zero mean, and $\int_t^{t+\tau} \exp(\mathbf{F}\,(t+\tau-v))\,\mathbf{L}\,\mathbf{w}(v)\;dv$ is therefore also a zero mean Gaussian random variable. Two results are central for deriving its covariance matrix. First, that the covariance matrix of a zero mean random vector is the expected value of the vector times the transpose of the vector. Second, that the definition of the Dirac delta function $\delta(\cdot)$ that appears in (4.4) is that $\int \delta(v-v_0)\,f(v)\;dv = f(v_0)$. Using these results we get

$$\begin{aligned}&\mathrm{E}\Big[\int_t^{t+\tau} \exp(\mathbf{F}\,(t+\tau-v_1))\,\mathbf{L}\,\mathbf{w}(v_1)\;dv_1 \\ &\qquad \times \int_t^{t+\tau} \mathbf{w}^{\mathsf{T}}(v_2)\,\mathbf{L}^{\mathsf{T}} \exp(\mathbf{F}\,(t+\tau-v_2))^{\mathsf{T}}\;dv_2\Big] \\ &= \int_t^{t+\tau}\int_t^{t+\tau} \exp(\mathbf{F}\,(t+\tau-v_1))\,\mathbf{L}\,\mathrm{E}\big[\mathbf{w}(v_1)\,\mathbf{w}^{\mathsf{T}}(v_2)\big]\,\mathbf{L}^{\mathsf{T}} \\ &\qquad \times \exp(\mathbf{F}\,(t+\tau-v_2))^{\mathsf{T}}\;dv_1\,dv_2 \\ &= \int_t^{t+\tau}\int_t^{t+\tau} \exp(\mathbf{F}\,(t+\tau-v_1))\,\mathbf{L}\,\mathbf{Q}^{\mathrm{c}}\,\delta(v_1-v_2)\,\mathbf{L}^{\mathsf{T}} \\ &\qquad \times \exp(\mathbf{F}\,(t+\tau-v_2))^{\mathsf{T}}\;dv_1\,dv_2 \\ &= \int_t^{t+\tau} \exp(\mathbf{F}\,(t+\tau-v_1))\,\mathbf{L}\,\mathbf{Q}^{\mathrm{c}}\,\mathbf{L}^{\mathsf{T}} \exp(\mathbf{F}\,(t+\tau-v_1))^{\mathsf{T}}\;dv_1 \\ &= \int_0^{\tau} \exp(\mathbf{F}\,v)\,\mathbf{L}\,\mathbf{Q}^{\mathrm{c}}\,\mathbf{L}^{\mathsf{T}} \exp(\mathbf{F}v)^{\mathsf{T}}\;dv.\end{aligned} \tag{A.28}$$

The above equation confirms (A.19). Finally, (A.20) follows from the earlier results. □

A.5 Time Derivative of Covariance in Theorem 4.13

In this appendix, we derive the expression for the time derivative of $\mathbf{P}(t)$ presented in Theorem 4.13. Recall that $\mathbf{P}(t) = \mathrm{E}[(\mathbf{x}(t) - \mathbf{m}(t))(\mathbf{x}(t) - \mathbf{m}(t))^T]$ and that

$$\frac{\mathrm{d}\mathbf{P}(t)}{\mathrm{d}t} = \lim_{h \to \infty} \frac{\mathbf{P}(t+h) - \mathbf{P}(t)}{h}. \tag{A.29}$$

We therefore seek an approximation to $\mathbf{P}(t+h)$ where $o(h)$ terms are omitted, that is, we ignore terms that go to zero faster than h as $h \to 0$. First of all we have

$$\begin{aligned}\mathbf{x}(t+h) - \mathbf{x}(t) &= \int_t^{t+h} \mathbf{a}(\mathbf{m}(s))\,\mathrm{d}s \\ &+ \int_t^{t+h} \mathbf{A}_{\mathbf{x}}(\mathbf{m}(s))\,(\mathbf{x}(s) - \mathbf{m}(s))\,\mathrm{d}s + \int_t^{t+h} \mathbf{L}\,\mathbf{w}(s)\,\mathrm{d}s.\end{aligned} \tag{A.30}$$

On the other hand,

$$\mathbf{m}(t+h) - \mathbf{m}(t) = \int_t^{t+h} \mathbf{a}(\mathbf{m}(s))\,\mathrm{d}s. \tag{A.31}$$

We can now write

$$\begin{aligned}&\mathbf{x}(t+h) - \mathbf{m}(t+h) \\ &= \mathbf{x}(t+h) - \mathbf{m}(t+h) + \mathbf{x}(t) - \mathbf{x}(t) + \mathbf{m}(t) - \mathbf{m}(t) \\ &= [\mathbf{x}(t) - \mathbf{m}(t)] + [\mathbf{x}(t+h) - \mathbf{x}(t)] - [\mathbf{m}(t+h) - \mathbf{m}(t)] \\ &= [\mathbf{x}(t) - \mathbf{m}(t)] + \int_t^{t+h} \mathbf{A}_{\mathbf{x}}(\mathbf{m}(s))\,(\mathbf{x}(s) - \mathbf{m}(s))\,\mathrm{d}s + \int_t^{t+h} \mathbf{L}\,\mathbf{w}(s)\,\mathrm{d}s \\ &= [\mathbf{x}(t) - \mathbf{m}(t)] + \mathbf{A}_{\mathbf{x}}(\mathbf{m}(t))\,(\mathbf{x}(t) - \mathbf{m}(t))\,h + o(h) + \int_t^{t+h} \mathbf{L}\,\mathbf{w}(s)\,\mathrm{d}s,\end{aligned} \tag{A.32}$$

where we have left the integral involving white noise $\mathbf{w}(t)$ intact because $\int_t^{t+h} \mathbf{L}\,\mathbf{w}(s)\,\mathrm{d}s \neq \mathbf{L}\,\mathbf{w}(t)\,h + o(h)$ (cf. Särkkä and Solin, 2019), but instead we need to explicitly use Equation (4.4). Because $\int_t^{t+h} \mathbf{L}\,\mathbf{w}(s)\,\mathrm{d}s$ is independent of the other terms and $\mathrm{E}\left[\int_t^{t+h} \mathbf{L}\,\mathbf{w}(s)\,\mathrm{d}s\right] = 0$, we then get

$$
\begin{aligned}
\mathbf{P}(t+h) &= \mathrm{E}\left[(\mathbf{x}(t+h) - \mathbf{m}(t+h))\,(\mathbf{x}(t+h) - \mathbf{m}(t+h))^\mathsf{T}\right] \\
&= \mathrm{E}\left[(\mathbf{x}(t) - \mathbf{m}(t))\,(\mathbf{x}(t) - \mathbf{m}(t))^\mathsf{T}\right] \\
&\quad + \mathbf{A}_\mathbf{x}(\mathbf{m}(t))\;\mathrm{E}\left[(\mathbf{x}(t) - \mathbf{m}(t))\,(\mathbf{x}(t) - \mathbf{m}(t))^\mathsf{T}\right] h \\
&\quad + \mathrm{E}\left[(\mathbf{x}(t) - \mathbf{m}(t))\,(\mathbf{x}(t) - \mathbf{m}(t))^\mathsf{T}\right] \mathbf{A}_\mathbf{x}^\mathsf{T}(\mathbf{m}(t))\,h \\
&\quad + \mathrm{E}\left[\int_t^{t+h} \mathbf{L}\,\mathbf{w}(s)\,\mathbf{w}^\mathsf{T}(s')\,\mathbf{L}^\mathsf{T}\,\mathrm{d}s\,\mathrm{d}s'\right] + o(h) \\
&= \mathbf{P}(t) + \mathbf{A}_\mathbf{x}(\mathbf{m}(t))\,\mathbf{P}(t)\,h + \mathbf{P}(t)\,\mathbf{A}_\mathbf{x}^\mathsf{T}(\mathbf{m}(t))\,h \\
&\quad + \mathbf{L}\,\mathbf{Q}^\mathrm{c}\,\mathbf{L}^\mathsf{T}\,h + o(h).
\end{aligned}
\tag{A.33}
$$

Rearranging and taking the limit $h \to 0$ then yields Equation (4.53).

A.6 Derivation of Mean for Bicycle Model

In this section, the aim is to derive the solution to the mean differential equation in Example 4.17. Our aim is now to solve the differential equation

$$
\frac{\mathrm{d}}{\mathrm{d}t}\begin{pmatrix} x_1(t) \\ x_2(t) \\ s(t) \\ \theta(t) \\ \delta(t) \end{pmatrix} = \begin{pmatrix} s(t)\cos(\theta(t)) \\ s(t)\sin(\theta(t)) \\ 0 \\ s(t)/r(t) \\ 0 \end{pmatrix}
\tag{A.34}
$$

where $r(t) = L/\tan(\delta(t))$ for $\mathbf{x}(t) = (x_1(t), x_2(t), s(t), \theta(t), \delta(t))$ given the initial condition $\mathbf{x}(t_{k-1}) = \mathbf{x}_{k-1}$. It holds that $s(t) = s_{k-1}$, $\delta(t) = \delta_{k-1}$, and thus $r(t) = r_{k-1} = L/\tan(\delta_{k-1})$, which implies that $\theta(t) = \theta_{k-1} + (t - t_{k-1})\,s_{k-1}/r_{k-1}$. We can now solve for

$$
\begin{aligned}
x_1(t) &= x_{1,k-1} + \int_{t_{k-1}}^{t} s(v)\cos(\theta(v))\,\mathrm{d}v \\
&= x_{1,k-1} + \int_0^{t-t_{k-1}} s_{k-1}\cos(\theta_{k-1} + v\,s_{k-1}/r_{k-1})\,\mathrm{d}v \\
&= x_{1,k-1} + r_{k-1}\left(\sin(\theta_{k-1} + (t - t_{k-1})\,s_{k-1}/r_{k-1}) - \sin(\theta_{k-1})\right).
\end{aligned}
\tag{A.35}
$$

Using a similar derivation for $x_2(t)$, we finally get

$$\begin{pmatrix} x_1(t) \\ x_2(t) \\ s(t) \\ \theta(t) \\ \delta(t) \end{pmatrix} = \begin{pmatrix} x_{1,k-1} + r_{k-1}\,(\sin(\theta_{k-1} + \beta(t - t_{k-1})) - \sin(\theta_{k-1})) \\ x_{2,k-1} + r_{k-1}\,(\cos(\theta_{k-1}) - \cos(\theta_{k-1} + \beta(t - t_{k-1}))) \\ s_{k-1} \\ \theta_{k-1} + \beta(t - t_{k-1}) \\ \delta_{k-1} \end{pmatrix}, \tag{A.36}$$

where for brevity, we have defined $\beta(t) = t\, s_{k-1}/r_{k-1}$. When the above solution is evaluated at $t = t_k$, the corresponding discretized model then becomes Equation (4.66).

A.7 Mean Discretization for the Polar Coordinated Turn Model

Let us now derive the solution to the differential equation (4.56) that corresponds to the mean discretization of the polar coordinated turn model. The state is given as $\mathbf{x}(t) = (x_1(t), x_2(t), s(t), \varphi(t), \omega(t))$, and the initial condition is $\mathbf{x}(t_{k-1}) = \mathbf{x}_{k-1}$. Because the time derivatives of $s(t)$ and $\omega(t)$ are zero, it holds that

$$s(t) = s_{k-1}, \quad \omega(t) = \omega_{k-1}. \tag{A.37}$$

Because $\omega(t)$ is the time derivative of $\varphi(t)$, it follows that

$$\varphi(t) = \varphi(t_{k-1}) + (t - t_{k-1})\,\omega(t_{k-1}). \tag{A.38}$$

Given (A.37) and (A.38), we get

$$\begin{aligned} x_1(t) &= x_{1,k-1} + \int_{t_{k-1}}^{t} s(v)\,\cos(\varphi(v))\,\mathrm{d}v \\ &= x_{1,k-1} + \int_{t_{k-1}}^{t} s_{k-1}\,\cos(\varphi_{k-1} + (v - t_{k-1})\,\omega_{k-1})\,\mathrm{d}v \\ &= x_{1,k-1} + \frac{s_{k-1}}{\omega_{k-1}}\,(\sin(\varphi_{k-1} + (t - t_{k-1})\,\omega_{k-1}) - \sin(\varphi_{k-1})) \\ &= x_{1,k-1} + \frac{2 s_{k-1}}{\omega(t_{k-1})}\,\sin\left(\frac{(t - t_{k-1})\,\omega_{k-1}}{2}\right) \\ &\quad \times \cos\left(\varphi_{k-1} + \frac{(t - t_{k-1})\,\omega_{k-1}}{2}\right) \end{aligned} \tag{A.39}$$

and, using similar calculations,

$$x_2(t) = x_{2,k-1} + \frac{2s_{k-1}}{\omega_{k-1}} \sin\left(\frac{(t - t_{k-1})\,\omega(t_{k-1})}{2}\right) \times \sin\left(\varphi_{k-1} + \frac{(t - t_{k-1})\,\omega(t_{k-1})}{2}\right). \tag{A.40}$$

The final discretization is obtained by evaluation at $t = t_k$.

A.8 Approximating $\mathbf{Q}_{k-1}$ in the Polar Coordinated Turn Model

Let us now derive a constant gradient-based covariance approximation as given by Algorithm 4.19 for the polar coordinated turn model, whose mean discretization was considered in Example 4.15. The derivation of the mean discretization is given above in Section A.7. The covariance expression is thus given as

$$\mathbf{Q}_{k-1} = \int_0^{\Delta t_{k-1}} \exp(\mathbf{A}_{\mathbf{x}}(\mathbf{x}_{k-1})\,v)\,\mathbf{L}\,\mathbf{Q}^{\mathrm{c}}\,\mathbf{L}^{\mathsf{T}} \exp(\mathbf{A}_{\mathbf{x}}(\mathbf{x}_{k-1})\,v)^{\mathsf{T}}\,\mathrm{d}v. \tag{A.41}$$

For brevity, we write

$$\mathbf{x}_{k-1} = \begin{pmatrix} x_1 & x_2 & s & \varphi & \omega \end{pmatrix}^{\mathsf{T}}. \tag{A.42}$$

It follows from (4.56) that

$$\mathbf{A}_{\mathbf{x}}(\mathbf{x}_{k-1}) = \begin{pmatrix} 0 & 0 & \cos(\varphi) & -s\sin(\varphi) & 0 \\ 0 & 0 & \sin(\varphi) & s\cos(\varphi) & 0 \\ 0 & 0 & 0 & 0 & 0 \\ 0 & 0 & 0 & 0 & 1 \\ 0 & 0 & 0 & 0 & 0 \end{pmatrix}, \tag{A.43}$$

and it is easy to verify that

$$(\mathbf{A}_{\mathbf{x}}(\mathbf{x}_{k-1}))^2 = \begin{pmatrix} 0 & 0 & 0 & 0 & -s\sin(\varphi) \\ 0 & 0 & 0 & 0 & s\cos(\varphi) \\ 0 & 0 & 0 & 0 & 0 \\ 0 & 0 & 0 & 0 & 0 \\ 0 & 0 & 0 & 0 & 0 \end{pmatrix}, \tag{A.44}$$

whereas $(\mathbf{A_x}(\mathbf{x}_{k-1}))^n = \mathbf{0}$ for $n \geq 3$. The definition of the matrix exponential therefore gives

$$\begin{aligned}
&\exp(\mathbf{A_x}(\mathbf{x}_{k-1})\, v) \\
&= \mathbf{I} + v\, \mathbf{A_x}(\mathbf{x}_{k-1}) + \frac{v^2\, (\mathbf{A_x}(\mathbf{x}_{k-1}))^2}{2} \\
&= \begin{pmatrix} 1 & 0 & v\cos(\varphi) & -vs\sin(\varphi) & -\frac{v^2}{2}s\sin(\varphi) \\ 0 & 1 & v\sin(\varphi) & vs\cos(\varphi) & \frac{v^2}{2}s\cos(\varphi) \\ 0 & 0 & 1 & 0 & 0 \\ 0 & 0 & 0 & 1 & v \\ 0 & 0 & 0 & 0 & 1 \end{pmatrix},
\end{aligned} \tag{A.45}$$

and from the expression for $\mathbf{L}$ in (4.43) we now obtain

$$\exp(\mathbf{A_x}(\mathbf{x}_{k-1})\, v)\, \mathbf{L} = \begin{pmatrix} v\cos(\varphi) & -\frac{v^2}{2}s\sin(\varphi) \\ v\sin(\varphi) & \frac{v^2}{2}s\cos(\varphi) \\ 1 & 0 \\ 0 & v \\ 0 & 1 \end{pmatrix}. \tag{A.46}$$

For simplicity, we assume that $\mathbf{Q}^{\mathrm{c}} = \mathrm{diag}(q_1, q_2)$. To save space, we use the shorthand notations $\mathrm{si} = \sin(\varphi)$, $\mathrm{co} = \cos(\varphi)$ and $\mathrm{cs} = \sin(\varphi)\cos(\varphi)$ and we only express the lower triangle of the following symmetric matrices. Equation (A.46) yields

$$\begin{aligned}
&\exp(\mathbf{A_x}(\mathbf{x}_{k-1}))\, v)\, \mathbf{L}\, \mathbf{Q}^{\mathrm{c}} \mathbf{L}^{\mathsf{T}} \exp(\mathbf{A_x}(\mathbf{x}_{k-1})\, v)^{\mathsf{T}} \\
&= \begin{pmatrix} q_1 v^2 \mathrm{co}^2 + q_2 \frac{v^4}{4} s^2 \mathrm{si}^2 & & & & \\ q_1 v^2 \mathrm{cs} - q_2 \frac{v^4}{4} s^2 \mathrm{cs} & q_1 v^2 \mathrm{si}^2 + q_2 \frac{v^4}{4} s^2 \mathrm{co}^2 & & & \\ q_1 v \mathrm{co} & q_1 v \mathrm{si} & q_1 & & \\ -q_2 s \frac{v^3}{2} \mathrm{si} & q_2 s \frac{v^3}{2} \mathrm{co} & 0 & q_2 v^2 & \\ -q_2 s \frac{v^2}{2} \mathrm{si} & q_2 s \frac{v^2}{2} \mathrm{co} & 0 & q_2 v & q_2 \end{pmatrix}.
\end{aligned} \tag{A.47}$$

Finally, we integrate over v to obtain

$$
\mathbf{P}(\tau + t_{k-1}) = \begin{pmatrix}
q_1 \frac{\tau^3}{3}\mathrm{co}^2 + q_2 \frac{\tau^5}{20} s^2 \mathrm{si}^2 & & & & \\
q_1 \frac{\tau^3}{3}\mathrm{cs} - q_2 \frac{\tau^5}{20} s^2 \mathrm{cs} & q_1 \frac{\tau^3}{3}\mathrm{si}^2 + q_2 \frac{\tau^5}{20} s^2 \mathrm{co}^2 & & & \\
q_1 \frac{\tau^2}{2}\mathrm{co} & q_1 \frac{\tau^2}{2}\mathrm{si} & q_1 \tau & & \\
-q_2 s \frac{\tau^4}{8}\mathrm{si} & q_2 s \frac{v^4}{8}\mathrm{co} & 0 & q_2 \frac{\tau^3}{3} & \\
-q_2 s \frac{\tau^3}{6}\mathrm{si} & q_2 s \frac{\tau^3}{6}\mathrm{co} & 0 & q_2 \frac{\tau^2}{2} & q_2 \tau
\end{pmatrix}, \tag{A.48}
$$

which gives $\mathbf{Q}_{k-1}$ by setting $\tau = \Delta t_{k-1}$.

A.9 Conditional Moments Used in SLR

We can use statistical linear regression (SLR) to approximate the dynamic and measurement models in both filtering and smoothing. Theorems 9.12 and 9.16, combined with Corollaries 9.14, 9.17, and 9.18, provide five different expressions for moments that we need to compute to perform SLR; one expression for the additive noise models in (9.1) and two expressions each for the non-additive noise models in (9.9) and for the conditional distribution models in (9.10). In this section, we present the explicit forms that these equations take for the dynamic and measurement models, respectively.

Using SLR, the dynamic model is approximated as

$$
\mathbf{x}_k \simeq \mathbf{A}_{k-1}\,\mathbf{x}_{k-1} + \mathbf{a}_{k-1} + \mathbf{e}_{k-1}, \tag{A.49}
$$

where $\mathbf{e}_{k-1} \sim \mathrm{N}(\mathbf{0}, \mathbf{\Lambda}_{k-1})$. If[1] $\mathbf{x}_{k-1} \sim \mathrm{N}(\tilde{\mathbf{m}}_{k-1}, \tilde{\mathbf{P}}_{k-1})$, it holds that

$$
\begin{aligned}
\mathbf{A}_{k-1} &= (\mathbf{P}_k^{\mathrm{xx}})^\mathsf{T}\,\tilde{\mathbf{P}}_{k-1}^{-1}, \\
\mathbf{a}_{k-1} &= \boldsymbol{\mu}_k^- - \mathbf{A}_{k-1}\,\tilde{\mathbf{m}}_{k-1}, \\
\mathbf{\Lambda}_{k-1} &= \mathbf{P}_k^{\mathrm{x}} - \mathbf{A}_{k-1}\,\tilde{\mathbf{P}}_{k-1}\mathbf{A}_{k-1}^\mathsf{T},
\end{aligned} \tag{A.50}
$$

where Theorem 9.16 and Corollaries 9.14, 9.17, and 9.18 provide the following five expressions for $\boldsymbol{\mu}_k^- = \mathrm{E}[\mathbf{x}_k]$, $\mathbf{P}_k^{\mathrm{xx}} = \mathrm{Cov}[\mathbf{x}_{k-1}, \mathbf{x}_k]$, and $\mathbf{P}_k^{\mathrm{x}} =$

[1] We use tildes to indicate that the equations are valid for any mean and covariance, but note that the choice of these moments influences the resulting SLR approximation.

$\text{Cov}[\mathbf{x}_k]$. For additive noise models, Corollary 9.14 implies that

$$\begin{aligned}
\boldsymbol{\mu}_k^- &= \int \mathbf{f}(\mathbf{x}_{k-1})\, \mathrm{N}(\mathbf{x}_{k-1} \mid \tilde{\mathbf{m}}_{k-1}, \tilde{\mathbf{P}}_{k-1})\, \mathrm{d}\mathbf{x}_{k-1}, \\
\mathbf{P}_k^{\mathrm{xx}} &= \int (\mathbf{x}_{k-1} - \tilde{\mathbf{m}}_{k-1})\, (\mathbf{f}(\mathbf{x}_{k-1}) - \boldsymbol{\mu}_k^-)^\mathsf{T}\, \mathrm{N}(\mathbf{x}_{k-1} \mid \tilde{\mathbf{m}}_{k-1}, \tilde{\mathbf{P}}_{k-1})\, \mathrm{d}\mathbf{x}_{k-1}, \\
\mathbf{P}_k^{\mathrm{x}} &= \int (\mathbf{f}(\mathbf{x}_{k-1}) - \boldsymbol{\mu}_k^-)\, (\mathbf{f}(\mathbf{x}_{k-1}) - \boldsymbol{\mu}_k^-)^\mathsf{T}\, \mathrm{N}(\mathbf{x}_{k-1} \mid \tilde{\mathbf{m}}_{k-1}, \tilde{\mathbf{P}}_{k-1})\, \mathrm{d}\mathbf{x}_{k-1} \\
&\quad + \mathbf{Q}_{k-1};
\end{aligned} \tag{A.51}$$

for non-additive noise models, Equations (9.50) in Corollary 9.17 give

$$\begin{aligned}
\boldsymbol{\mu}_k^- &= \int \mathbf{f}(\mathbf{x}_{k-1}, \mathbf{q}_{k-1}) \\
&\qquad \times \mathrm{N}(\mathbf{x}_{k-1} \mid \tilde{\mathbf{m}}_{k-1}, \tilde{\mathbf{P}}_{k-1})\, \mathrm{N}(\mathbf{q}_{k-1} \mid \mathbf{0}, \mathbf{Q}_{k-1})\, \mathrm{d}\mathbf{x}_{k-1}\, \mathrm{d}\mathbf{q}_{k-1}, \\
\mathbf{P}_k^{\mathrm{xx}} &= \int (\mathbf{x}_{k-1} - \tilde{\mathbf{m}}_{k-1})\, (\mathbf{f}(\mathbf{x}_{k-1}, \mathbf{q}_{k-1}) - \boldsymbol{\mu}_k^-)^\mathsf{T} \\
&\qquad \times \mathrm{N}(\mathbf{x}_{k-1} \mid \tilde{\mathbf{m}}_{k-1}, \tilde{\mathbf{P}}_{k-1})\, \mathrm{N}(\mathbf{q}_{k-1} \mid \mathbf{0}, \mathbf{Q}_{k-1})\, \mathrm{d}\mathbf{x}_{k-1}\, \mathrm{d}\mathbf{q}_{k-1}, \\
\mathbf{P}_k^{\mathrm{x}} &= \int (\mathbf{f}(\mathbf{x}_{k-1}, \mathbf{q}_{k-1}) - \boldsymbol{\mu}_k^-)\, (\mathbf{f}(\mathbf{x}_{k-1}, \mathbf{q}_{k-1}) - \boldsymbol{\mu}_k^-)^\mathsf{T} \\
&\qquad \times \mathrm{N}(\mathbf{x}_{k-1} \mid \tilde{\mathbf{m}}_{k-1}, \tilde{\mathbf{P}}_{k-1})\, \mathrm{N}(\mathbf{q}_{k-1} \mid \mathbf{0}, \mathbf{Q}_{k-1})\, \mathrm{d}\mathbf{x}_{k-1}\, \mathrm{d}\mathbf{q}_{k-1};
\end{aligned} \tag{A.52}$$

and for conditional distribution models, Equations (9.53) in Corollary 9.18 give

$$\begin{aligned}
\boldsymbol{\mu}_k^- &= \int \mathbf{x}_k\, p(\mathbf{x}_k \mid \mathbf{x}_{k-1})\, \mathrm{N}(\mathbf{x}_{k-1} \mid \tilde{\mathbf{m}}_{k-1}, \tilde{\mathbf{P}}_{k-1})\, \mathrm{d}\mathbf{x}_k\, \mathrm{d}\mathbf{x}_{k-1}, \\
\mathbf{P}_k^{\mathrm{xx}} &= \int (\mathbf{x}_{k-1} - \tilde{\mathbf{m}}_{k-1})\, (\mathbf{x}_k - \boldsymbol{\mu}_k^-)^\mathsf{T} \\
&\qquad \times p(\mathbf{x}_k \mid \mathbf{x}_{k-1})\, \mathrm{N}(\mathbf{x}_{k-1} \mid \tilde{\mathbf{m}}_{k-1}, \tilde{\mathbf{P}}_{k-1})\, \mathrm{d}\mathbf{x}_k\, \mathrm{d}\mathbf{x}_{k-1}, \\
\mathbf{P}_k^{\mathrm{x}} &= \int (\mathbf{x}_k - \boldsymbol{\mu}_k^-)\, (\mathbf{x}_k - \boldsymbol{\mu}_k^-)^\mathsf{T} \\
&\qquad \times p(\mathbf{x}_k \mid \mathbf{x}_{k-1})\, \mathrm{N}(\mathbf{x}_{k-1} \mid \tilde{\mathbf{m}}_{k-1}, \tilde{\mathbf{P}}_{k-1})\, \mathrm{d}\mathbf{x}_k\, \mathrm{d}\mathbf{x}_{k-1}.
\end{aligned} \tag{A.53}$$

Theorem 9.16 also implies that we can use the conditional moments formulation to obtain alternative expressions for $\boldsymbol{\mu}_k^-$, $\mathbf{P}_k^{\mathrm{xx}}$, and $\mathbf{P}_k^{\mathrm{x}}$. Introducing

$\boldsymbol{\mu}_k^-(\mathbf{x}_{k-1}) = \mathrm{E}[\mathbf{x}_k \mid \mathbf{x}_{k-1}]$ and $\mathbf{P}_k^{\mathrm{x}}(\mathbf{x}_{k-1}) = \mathrm{Cov}[\mathbf{x}_k \mid \mathbf{x}_{k-1}]$, it holds that

$$\begin{aligned}
\boldsymbol{\mu}_k^- &= \int \boldsymbol{\mu}_k^-(\mathbf{x}_{k-1})\, \mathrm{N}(\mathbf{x}_{k-1} \mid \tilde{\mathbf{m}}_{k-1}, \tilde{\mathbf{P}}_{k-1})\, \mathrm{d}\mathbf{x}_{k-1}, \\
\mathbf{P}_k^{\mathrm{xx}} &= \int (\mathbf{x}_{k-1} - \tilde{\mathbf{m}}_{k-1})\, (\boldsymbol{\mu}_k^-(\mathbf{x}_{k-1}) - \boldsymbol{\mu}_k^-)^\mathsf{T} \\
&\quad \times \mathrm{N}(\mathbf{x}_{k-1} \mid \tilde{\mathbf{m}}_{k-1}, \tilde{\mathbf{P}}_{k-1})\, \mathrm{d}\mathbf{x}_{k-1}, \\
\mathbf{P}_k^{\mathrm{x}} &= \int (\boldsymbol{\mu}_k^-(\mathbf{x}_{k-1}) - \boldsymbol{\mu}_k^-)\, (\boldsymbol{\mu}_k^-(\mathbf{x}_{k-1}) - \boldsymbol{\mu}_k^-)^\mathsf{T} \\
&\quad \times \mathrm{N}(\mathbf{x}_{k-1} \mid \tilde{\mathbf{m}}_{k-1}, \tilde{\mathbf{P}}_{k-1})\, \mathrm{d}\mathbf{x}_{k-1} \\
&\quad + \int \mathbf{P}_k^{\mathrm{x}}(\mathbf{x}_{k-1})\, \mathrm{N}(\mathbf{x}_{k-1} \mid \tilde{\mathbf{m}}_{k-1}, \tilde{\mathbf{P}}_{k-1})\, \mathrm{d}\mathbf{x}_{k-1}.
\end{aligned} \tag{A.54}$$

For additive noise models, $\boldsymbol{\mu}_k^-(\mathbf{x}_{k-1}) = \mathbf{f}(\mathbf{x}_{k-1})$ and $\mathbf{P}_k^{\mathrm{x}}(\mathbf{x}_{k-1}) = \mathbf{Q}_{k-1}$, and (A.54) simplifies to (A.51). For non-additive noise models, Equations (9.51) in Corollary 9.17 imply that

$$\begin{aligned}
\boldsymbol{\mu}_k^-(\mathbf{x}_{k-1}) &= \int \mathbf{f}(\mathbf{x}_{k-1}, \mathbf{q}_{k-1})\, \mathrm{N}(\mathbf{q}_{k-1} \mid \mathbf{0}, \mathbf{Q}_{k-1})\, \mathrm{d}\mathbf{q}_{k-1}, \\
\mathbf{P}_k^{\mathrm{x}}(\mathbf{x}_{k-1}) &= \int (\mathbf{f}(\mathbf{x}_{k-1}, \mathbf{q}_{k-1}) - \boldsymbol{\mu}_k^-(\mathbf{x}_{k-1})) \\
&\quad \times (\mathbf{f}(\mathbf{x}_{k-1}, \mathbf{q}_{k-1}) - \boldsymbol{\mu}_k^-(\mathbf{x}_{k-1}))^\mathsf{T}\, \mathrm{N}(\mathbf{q}_{k-1} \mid \mathbf{0}, \mathbf{Q}_{k-1})\, \mathrm{d}\mathbf{q}_{k-1}.
\end{aligned} \tag{A.55}$$

Finally, for conditional distribution models, Equations (9.54) in Corollary 9.18 give

$$\begin{aligned}
\boldsymbol{\mu}_k^-(\mathbf{x}_{k-1}) &= \int \mathbf{x}_k\, p(\mathbf{x}_k \mid \mathbf{x}_{k-1})\, \mathrm{d}\mathbf{x}_k, \\
\mathbf{P}_k^{\mathrm{x}}(\mathbf{x}_{k-1}) &= \int (\mathbf{x}_k - \boldsymbol{\mu}_k^-(\mathbf{x}_{k-1}))\, (\mathbf{x}_k - \boldsymbol{\mu}_k^-(\mathbf{x}_{k-1}))^\mathsf{T}\, p(\mathbf{x}_k \mid \mathbf{x}_{k-1})\, \mathrm{d}\mathbf{x}_k.
\end{aligned} \tag{A.56}$$

In total, the above equations present five different alternatives to compute the moments $\boldsymbol{\mu}_k^-$, $\mathbf{P}_k^{\mathrm{xx}}$, and $\mathbf{P}_k^{\mathrm{x}}$ that we need to approximate the dynamic model using SLR.

Similarly, we can use SLR to approximate the measurement model as

$$\mathbf{y}_k \simeq \mathbf{H}_k\, \mathbf{x}_k + \mathbf{b}_k + \mathbf{v}_k, \tag{A.57}$$

where $\mathbf{v}_k \sim \mathrm{N}(\mathbf{0}, \boldsymbol{\Omega}_k)$. If $\mathbf{x}_k \sim \mathrm{N}(\tilde{\mathbf{m}}_k, \tilde{\mathbf{P}}_k)$, it holds that

$$\begin{aligned} \mathbf{H}_k &= (\mathbf{P}_k^{\mathrm{xy}})^\mathsf{T} \, \tilde{\mathbf{P}}_k^{-1}, \\ \mathbf{b}_k &= \boldsymbol{\mu}_k^+ - \mathbf{H}_k \, \tilde{\mathbf{m}}_k, \\ \boldsymbol{\Omega}_k &= \mathbf{P}_k^{\mathrm{y}} - \mathbf{H}_k \, \tilde{\mathbf{P}}_k \, \mathbf{H}_k^\mathsf{T}, \end{aligned} \tag{A.58}$$

where we can obtain five different expressions for $\boldsymbol{\mu}_k^+ = \mathrm{E}[\mathbf{y}_k]$, $\mathbf{P}_k^{\mathrm{xy}} = \mathrm{Cov}[\mathbf{x}_k, \mathbf{y}_k]$, and $\mathbf{P}_k^{\mathrm{y}} = \mathrm{Cov}[\mathbf{y}_k]$ from Theorem 9.16 and Corollaries 9.14, 9.17, and 9.18. For additive noise models, Corollary 9.14 implies that

$$\begin{aligned} \boldsymbol{\mu}_k^+ &= \int \mathbf{h}(\mathbf{x}_k) \, \mathrm{N}(\mathbf{x}_k \mid \tilde{\mathbf{m}}_k, \tilde{\mathbf{P}}_k) \, \mathrm{d}\mathbf{x}_k, \\ \mathbf{P}_k^{\mathrm{xy}} &= \int (\mathbf{x}_k - \tilde{\mathbf{m}}_k) \, (\mathbf{h}(\mathbf{x}_k) - \boldsymbol{\mu}_k^+)^\mathsf{T} \, \mathrm{N}(\mathbf{x}_k \mid \tilde{\mathbf{m}}_k, \tilde{\mathbf{P}}_k) \, \mathrm{d}\mathbf{x}_k, \\ \mathbf{P}_k^{\mathrm{y}} &= \int (\mathbf{h}(\mathbf{x}_k) - \boldsymbol{\mu}_k^+) \, (\mathbf{h}(\mathbf{x}_k) - \boldsymbol{\mu}_k^+)^\mathsf{T} \, \mathrm{N}(\mathbf{x}_k \mid \tilde{\mathbf{m}}_k, \tilde{\mathbf{P}}_k) \, \mathrm{d}\mathbf{x}_k + \mathbf{R}_k; \end{aligned} \tag{A.59}$$

for non-additive noise models, Equations (9.50) in Corollary 9.17 give

$$\begin{aligned} \boldsymbol{\mu}_k^+ &= \int \mathbf{h}(\mathbf{x}_k, \mathbf{r}_k) \, \mathrm{N}(\mathbf{x}_k \mid \tilde{\mathbf{m}}_k, \tilde{\mathbf{P}}_k) \, \mathrm{N}(\mathbf{r}_k \mid \mathbf{0}, \mathbf{R}_k) \, \mathrm{d}\mathbf{x}_k \, \mathrm{d}\mathbf{r}_k, \\ \mathbf{P}_k^{\mathrm{xy}} &= \int (\mathbf{x}_k - \tilde{\mathbf{m}}_k) \, (\mathbf{h}(\mathbf{x}_k, \mathbf{r}_k) - \boldsymbol{\mu}_k^+)^\mathsf{T} \\ &\qquad \times \mathrm{N}(\mathbf{x}_k \mid \tilde{\mathbf{m}}_k, \tilde{\mathbf{P}}_k) \, \mathrm{N}(\mathbf{r}_k \mid \mathbf{0}, \mathbf{R}_k) \, \mathrm{d}\mathbf{x}_k \, \mathrm{d}\mathbf{r}_k, \\ \mathbf{P}_k^{\mathrm{y}} &= \int (\mathbf{h}(\mathbf{x}_k, \mathbf{r}_k) - \boldsymbol{\mu}_k^+) \, (\mathbf{h}(\mathbf{x}_k, \mathbf{r}_k) - \boldsymbol{\mu}_k^+)^\mathsf{T} \\ &\qquad \times \mathrm{N}(\mathbf{x}_k \mid \tilde{\mathbf{m}}_k, \tilde{\mathbf{P}}_k) \, \mathrm{N}(\mathbf{r}_k \mid \mathbf{0}, \mathbf{R}_k) \, \mathrm{d}\mathbf{x}_k \, \mathrm{d}\mathbf{r}_k; \end{aligned} \tag{A.60}$$

and for conditional distribution models, Equations (9.53) in Corollary 9.18 give

$$\begin{aligned} \boldsymbol{\mu}_k^+ &= \int \mathbf{y}_k \, p(\mathbf{y}_k \mid \mathbf{x}_k) \, \mathrm{N}(\mathbf{x}_k \mid \tilde{\mathbf{m}}_k, \tilde{\mathbf{P}}_k) \, \mathrm{d}\mathbf{y}_k \, \mathrm{d}\mathbf{x}_k, \\ \mathbf{P}_k^{\mathrm{xy}} &= \int (\mathbf{x}_k - \tilde{\mathbf{m}}_k) \, (\mathbf{y}_k - \boldsymbol{\mu}_k^+)^\mathsf{T} \, p(\mathbf{y}_k \mid \mathbf{x}_k) \, \mathrm{N}(\mathbf{x}_k \mid \tilde{\mathbf{m}}_k, \tilde{\mathbf{P}}_k) \, \mathrm{d}\mathbf{y}_k \, \mathrm{d}\mathbf{x}_k, \\ \mathbf{P}_k^{\mathrm{y}} &= \int (\mathbf{y}_k - \boldsymbol{\mu}_k^+) \, (\mathbf{y}_k - \boldsymbol{\mu}_k^+)^\mathsf{T} \, p(\mathbf{y}_k \mid \mathbf{x}_k) \, \mathrm{N}(\mathbf{x}_k \mid \tilde{\mathbf{m}}_k, \tilde{\mathbf{P}}_k) \, \mathrm{d}\mathbf{y}_k \, \mathrm{d}\mathbf{x}_k. \end{aligned} \tag{A.61}$$

The conditional moments formulation provides two more expressions for the required moments. Introducing $\boldsymbol{\mu}_k^+(\mathbf{x}_k) = \mathrm{E}[\mathbf{y}_k \mid \mathbf{x}_k]$ and

$\mathbf{P}_k^{\mathrm{y}}(\mathbf{x}_k) = \mathrm{Cov}[\mathbf{y}_k \mid \mathbf{x}_k]$, Theorem 9.16 gives that

$$\begin{aligned}
\boldsymbol{\mu}_k^+ &= \int \boldsymbol{\mu}_k^+(\mathbf{x}_k)\, \mathrm{N}(\mathbf{x}_k \mid \tilde{\mathbf{m}}_k, \tilde{\mathbf{P}}_k)\, \mathrm{d}\mathbf{x}_k, \\
\mathbf{P}_k^{\mathrm{xy}} &= \int (\mathbf{x}_k - \tilde{\mathbf{m}}_k)\, (\boldsymbol{\mu}_k^+(\mathbf{x}_k) - \boldsymbol{\mu}_k^+)^\mathsf{T}\, \mathrm{N}(\mathbf{x}_k \mid \tilde{\mathbf{m}}_k, \tilde{\mathbf{P}}_k)\, \mathrm{d}\mathbf{x}_k, \\
\mathbf{P}_k^{\mathrm{y}} &= \int (\boldsymbol{\mu}_k^+(\mathbf{x}_k) - \boldsymbol{\mu}_k^+)\, (\boldsymbol{\mu}_k^+(\mathbf{x}_k) - \boldsymbol{\mu}_k^+)^\mathsf{T}\, \mathrm{N}(\mathbf{x}_k \mid \tilde{\mathbf{m}}_k, \tilde{\mathbf{P}}_k)\, \mathrm{d}\mathbf{x}_k \\
&\quad + \int \mathbf{P}_k^{\mathrm{y}}(\mathbf{x}_k)\, \mathrm{N}(\mathbf{x}_k \mid \tilde{\mathbf{m}}_k, \tilde{\mathbf{P}}_k)\, \mathrm{d}\mathbf{x}_k.
\end{aligned} \tag{A.62}$$

For additive noise models, $\boldsymbol{\mu}_k^+(\mathbf{x}_k) = \mathbf{h}(\mathbf{x}_k)$ and $\mathbf{P}_k^{\mathrm{y}}(\mathbf{x}_k) = \mathbf{R}_k$, which means that (A.62) simplifies to (A.59). For non-additive noise models, Equations (9.51) in Corollary 9.17 imply that

$$\begin{aligned}
\boldsymbol{\mu}_k^+(\mathbf{x}_k) &= \int \mathbf{h}(\mathbf{x}_k, \mathbf{r}_k)\, \mathrm{N}(\mathbf{r}_k \mid \mathbf{0}, \mathbf{R}_k)\, \mathrm{d}\mathbf{r}_k, \\
\mathbf{P}_k^{\mathrm{y}}(\mathbf{x}_k) &= \int (\mathbf{h}(\mathbf{x}_k, \mathbf{r}_k) - \boldsymbol{\mu}_k^+(\mathbf{x}_k))\, (\mathbf{h}(\mathbf{x}_k, \mathbf{r}_k) - \boldsymbol{\mu}_k^+(\mathbf{x}_k))^\mathsf{T} \\
&\qquad \times \mathrm{N}(\mathbf{r}_k \mid \mathbf{0}, \mathbf{R}_k)\, \mathrm{d}\mathbf{r}_k;
\end{aligned} \tag{A.63}$$

and, finally, for conditional distribution models, Equations (9.54) in Corollary 9.18 give

$$\begin{aligned}
\boldsymbol{\mu}_k^+(\mathbf{x}_k) &= \int \mathbf{y}_k\, p(\mathbf{y}_k \mid \mathbf{x}_k)\, \mathrm{d}\mathbf{y}_k, \\
\mathbf{P}_k^{\mathrm{y}}(\mathbf{x}_k) &= \int (\mathbf{y}_k - \boldsymbol{\mu}_k^+(\mathbf{x}_k))\, (\mathbf{y}_k - \boldsymbol{\mu}_k^+(\mathbf{x}_k))^\mathsf{T}\, p(\mathbf{y}_k \mid \mathbf{x}_k)\, \mathrm{d}\mathbf{y}_k.
\end{aligned} \tag{A.64}$$

We have now presented five different alternatives for computing the moments $\boldsymbol{\mu}_k^+$, $\mathbf{P}_k^{\mathrm{xy}}$, and $\mathbf{P}_k^{\mathrm{y}}$, which are required to approximate the measurement model using SLR.

A.10 Parameter Derivatives for the Kalman Filter

Theorem 16.9 gives the energy function (i.e., the negative logarithm of the unnormalized posterior distribution) of the parameters for the following linear Gaussian model:

$$\begin{aligned}
\mathbf{x}_k &= \mathbf{A}(\boldsymbol{\theta})\, \mathbf{x}_{k-1} + \mathbf{q}_{k-1}, \\
\mathbf{y}_k &= \mathbf{H}(\boldsymbol{\theta})\, \mathbf{x}_k + \mathbf{r}_k,
\end{aligned} \tag{A.65}$$

where $\mathbf{q}_{k-1} \sim \mathrm{N}(\mathbf{0}, \mathbf{Q}(\boldsymbol{\theta}))$, $\mathbf{r}_k \sim \mathrm{N}(\mathbf{0}, \mathbf{R}(\boldsymbol{\theta}))$, and $\mathbf{x}_0 \sim \mathrm{N}(\mathbf{m}_0(\boldsymbol{\theta}), \mathbf{P}_0(\boldsymbol{\theta}))$. The parameters' derivatives, which are needed, for example, for implementing a gradient-based optimization method for finding ML or MAP estimates, can be evaluated via the following *sensitivity equations* (Gupta and Mehra, 1974), which can be derived by termwise differentiation of the energy function and Kalman filter equations in Theorem 16.9.

Theorem A.10 (Energy function derivative for linear Gaussian model I). *The derivative of the energy function given in Theorem 16.9 can be computed via the following recursion along with Kalman filtering:*

$$\begin{aligned}\frac{\partial \varphi_k(\boldsymbol{\theta})}{\partial \theta_i} &= \frac{\partial \varphi_{k-1}(\boldsymbol{\theta})}{\partial \theta_i} + \frac{1}{2}\operatorname{tr}\left(\mathbf{S}_k^{-1}(\boldsymbol{\theta})\,\frac{\partial \mathbf{S}_k(\boldsymbol{\theta})}{\partial \theta_i}\right) + \mathbf{v}_k^{\mathsf{T}}(\boldsymbol{\theta})\,\mathbf{S}_k^{-1}(\boldsymbol{\theta})\,\frac{\partial \mathbf{v}_k(\boldsymbol{\theta})}{\partial \theta_i} \\ &\quad - \frac{1}{2}\mathbf{v}_k^{\mathsf{T}}(\boldsymbol{\theta})\,\mathbf{S}_k^{-1}(\boldsymbol{\theta})\,\frac{\partial \mathbf{S}_k(\boldsymbol{\theta})}{\partial \theta_i}\,\mathbf{S}_k^{-1}(\boldsymbol{\theta})\,\mathbf{v}_k(\boldsymbol{\theta}),\end{aligned} \tag{A.66}$$

where on the Kalman filter prediction step, we compute

$$\begin{aligned}\frac{\partial \mathbf{m}_k^-(\boldsymbol{\theta})}{\partial \theta_i} &= \frac{\partial \mathbf{A}(\boldsymbol{\theta})}{\partial \theta_i}\,\mathbf{m}_{k-1}(\boldsymbol{\theta}) + \mathbf{A}(\boldsymbol{\theta})\,\frac{\partial \mathbf{m}_{k-1}(\boldsymbol{\theta})}{\partial \theta_i}, \\ \frac{\partial \mathbf{P}_k^-(\boldsymbol{\theta})}{\partial \theta_i} &= \frac{\partial \mathbf{A}(\boldsymbol{\theta})}{\partial \theta_i}\,\mathbf{P}_{k-1}(\boldsymbol{\theta})\,\mathbf{A}^{\mathsf{T}}(\boldsymbol{\theta}) + \mathbf{A}(\boldsymbol{\theta})\,\frac{\partial \mathbf{P}_{k-1}(\boldsymbol{\theta})}{\partial \theta_i}\,\mathbf{A}^{\mathsf{T}}(\boldsymbol{\theta}) \\ &\quad + \mathbf{A}(\boldsymbol{\theta})\,\mathbf{P}_{k-1}(\boldsymbol{\theta})\,\frac{\partial \mathbf{A}^{\mathsf{T}}(\boldsymbol{\theta})}{\partial \theta_i} + \frac{\partial \mathbf{Q}(\boldsymbol{\theta})}{\partial \theta_i},\end{aligned} \tag{A.67}$$

and on the Kalman filter update step, we compute

$$
\begin{aligned}
\frac{\partial \mathbf{v}_k(\boldsymbol{\theta})}{\partial \theta_i} &= -\frac{\partial \mathbf{H}(\boldsymbol{\theta})}{\partial \theta_i} \mathbf{m}_k^-(\boldsymbol{\theta}) - \mathbf{H}(\boldsymbol{\theta}) \frac{\partial \mathbf{m}_k^-(\boldsymbol{\theta})}{\partial \theta_i}, \\
\frac{\partial \mathbf{S}_k(\boldsymbol{\theta})}{\partial \theta_i} &= \frac{\partial \mathbf{H}(\boldsymbol{\theta})}{\partial \theta_i} \mathbf{P}_k^-(\boldsymbol{\theta}) \mathbf{H}^\top(\boldsymbol{\theta}) + \mathbf{H}(\boldsymbol{\theta}) \frac{\partial \mathbf{P}_k^-(\boldsymbol{\theta})}{\partial \theta_i} \mathbf{H}^\top(\boldsymbol{\theta}) \\
&\quad + \mathbf{H}(\boldsymbol{\theta}) \mathbf{P}_k^-(\boldsymbol{\theta}) \frac{\partial \mathbf{H}^\top(\boldsymbol{\theta})}{\partial \theta_i} + \frac{\partial \mathbf{R}(\boldsymbol{\theta})}{\partial \theta_i}, \\
\frac{\partial \mathbf{K}_k(\boldsymbol{\theta})}{\partial \theta_i} &= \frac{\partial \mathbf{P}_k^-(\boldsymbol{\theta})}{\partial \theta_i} \mathbf{H}^\top(\boldsymbol{\theta}) \mathbf{S}_k^{-1}(\boldsymbol{\theta}) + \mathbf{P}_k^-(\boldsymbol{\theta}) \frac{\partial \mathbf{H}^\top(\boldsymbol{\theta})}{\partial \theta_i} \mathbf{S}_k^{-1}(\boldsymbol{\theta}) \\
&\quad - \mathbf{P}_k^-(\boldsymbol{\theta}) \mathbf{H}^\top(\boldsymbol{\theta}) \mathbf{S}_k^{-1}(\boldsymbol{\theta}) \frac{\partial \mathbf{S}_k(\boldsymbol{\theta})}{\partial \theta_i} \mathbf{S}_k^{-1}(\boldsymbol{\theta}), \\
\frac{\partial \mathbf{m}_k(\boldsymbol{\theta})}{\partial \theta_i} &= \frac{\partial \mathbf{m}_k^-(\boldsymbol{\theta})}{\partial \theta_i} + \frac{\partial \mathbf{K}_k(\boldsymbol{\theta})}{\partial \theta_i} \mathbf{v}_k(\boldsymbol{\theta}) + \mathbf{K}_k(\boldsymbol{\theta}) \frac{\partial \mathbf{v}_k(\boldsymbol{\theta})}{\partial \theta_i}, \\
\frac{\partial \mathbf{P}_k(\boldsymbol{\theta})}{\partial \theta_i} &= \frac{\partial \mathbf{P}_k^-(\boldsymbol{\theta})}{\partial \theta_i} - \frac{\partial \mathbf{K}_k(\boldsymbol{\theta})}{\partial \theta_i} \mathbf{S}_k(\boldsymbol{\theta}) \mathbf{K}_k^\top(\boldsymbol{\theta}) \\
&\quad - \mathbf{K}_k(\boldsymbol{\theta}) \frac{\partial \mathbf{S}_k(\boldsymbol{\theta})}{\partial \theta_i} \mathbf{K}_k^\top(\boldsymbol{\theta}) - \mathbf{K}_k(\boldsymbol{\theta}) \mathbf{S}_k(\boldsymbol{\theta}) \frac{\partial \mathbf{K}_k^\top(\boldsymbol{\theta})}{\partial \theta_i}.
\end{aligned}
\tag{A.68}
$$

The recursion should be started from the initial condition $\partial \varphi_0(\boldsymbol{\theta})/\partial \boldsymbol{\theta} = -\,\partial \log p(\boldsymbol{\theta})/\partial \boldsymbol{\theta}$.

Another way to compute the same derivative is by using Fisher's identity (Equation 16.32) together with the expression for $\mathcal{Q}$ in Theorem 16.11. The result is the following.

Theorem A.11 (Energy function derivative for linear Gaussian model II). *The derivative of the energy function given in Theorem 16.9 can be computed as*

$$
\frac{\partial \varphi_T(\boldsymbol{\theta})}{\partial \boldsymbol{\theta}} = -\frac{\partial \log p(\boldsymbol{\theta})}{\partial \boldsymbol{\theta}} - \left. \frac{\partial \mathcal{Q}(\boldsymbol{\theta}, \boldsymbol{\theta}^{(n)})}{\partial \boldsymbol{\theta}} \right|_{\boldsymbol{\theta}^{(n)} = \boldsymbol{\theta}}, \tag{A.69}
$$

where

$$
\begin{aligned}
&\left.\frac{\partial \mathcal{Q}(\boldsymbol{\theta}, \boldsymbol{\theta}^{(n)})}{\partial \theta_i}\right|_{\boldsymbol{\theta}^{(n)} = \boldsymbol{\theta}} \\
&= -\frac{1}{2} \operatorname{tr}\left(\mathbf{P}_0^{-1} \frac{\partial \mathbf{P}_0}{\partial \theta_i}\right) - \frac{T}{2} \operatorname{tr}\left(\mathbf{Q}^{-1} \frac{\partial \mathbf{Q}}{\partial \theta_i}\right) - \frac{T}{2} \operatorname{tr}\left(\mathbf{R}^{-1} \frac{\partial \mathbf{R}}{\partial \theta_i}\right) \\
&+ \frac{1}{2} \operatorname{tr}\left\{\mathbf{P}_0^{-1} \frac{\partial \mathbf{P}_0}{\partial \theta_i} \mathbf{P}_0^{-1} \left[\mathbf{P}_0^{\mathrm{s}} + (\mathbf{m}_0^{\mathrm{s}} - \mathbf{m}_0)(\mathbf{m}_0^{\mathrm{s}} - \mathbf{m}_0)^{\mathsf{T}}\right]\right\} \\
&+ \frac{1}{2} \operatorname{tr}\left\{\mathbf{P}_0^{-1} \left[\frac{\partial \mathbf{m}_0}{\partial \theta_i} (\mathbf{m}_0^{\mathrm{s}} - \mathbf{m}_0)^{\mathsf{T}} + (\mathbf{m}_0^{\mathrm{s}} - \mathbf{m}_0) \frac{\partial \mathbf{m}_0^{\mathsf{T}}}{\partial \theta_i}\right]\right\} \\
&+ \frac{T}{2} \operatorname{tr}\left\{\mathbf{Q}^{-1} \frac{\partial \mathbf{Q}}{\partial \theta_i} \mathbf{Q}^{-1} \left[\boldsymbol{\Sigma} - \mathbf{C}\,\mathbf{A}^{\mathsf{T}} - \mathbf{A}\,\mathbf{C}^{\mathsf{T}} + \mathbf{A}\,\boldsymbol{\Phi}\,\mathbf{A}^{\mathsf{T}}\right]\right\} \\
&- \frac{T}{2} \operatorname{tr}\left\{\mathbf{Q}^{-1} \left[-\mathbf{C} \frac{\partial \mathbf{A}^{\mathsf{T}}}{\partial \theta_i} - \frac{\partial \mathbf{A}}{\partial \theta_i} \mathbf{C}^{\mathsf{T}} + \frac{\partial \mathbf{A}}{\partial \theta_i} \boldsymbol{\Phi}\,\mathbf{A}^{\mathsf{T}} + \mathbf{A}\,\boldsymbol{\Phi} \frac{\partial \mathbf{A}^{\mathsf{T}}}{\partial \theta_i}\right]\right\} \\
&+ \frac{T}{2} \operatorname{tr}\left\{\mathbf{R}^{-1} \frac{\partial \mathbf{R}}{\partial \theta_i} \mathbf{R}^{-1} \left[\mathbf{D} - \mathbf{B}\,\mathbf{H}^{\mathsf{T}} - \mathbf{H}\,\mathbf{B}^{\mathsf{T}} + \mathbf{H}\,\boldsymbol{\Sigma}\,\mathbf{H}^{\mathsf{T}}\right]\right\} \\
&- \frac{T}{2} \operatorname{tr}\left\{\mathbf{R}^{-1} \left[-\mathbf{B} \frac{\partial \mathbf{H}^{\mathsf{T}}}{\partial \theta_i} - \frac{\partial \mathbf{H}}{\partial \theta_i} \mathbf{B}^{\mathsf{T}} + \frac{\partial \mathbf{H}}{\partial \theta_i} \boldsymbol{\Sigma}\,\mathbf{H}^{\mathsf{T}} + \mathbf{H}\,\boldsymbol{\Sigma} \frac{\partial \mathbf{H}^{\mathsf{T}}}{\partial \theta_i}\right]\right\},
\end{aligned}
\tag{A.70}
$$

where all the terms are evaluated at $\boldsymbol{\theta}$.

A.11 Parameter Derivatives for the Gaussian Filter

In this section we consider the computation of the gradient of the Gaussian filtering-based energy function in Algorithm 16.14, which was considered with models of the form

$$
\begin{aligned}
\mathbf{x}_k &= \mathbf{f}(\mathbf{x}_{k-1}, \boldsymbol{\theta}) + \mathbf{q}_{k-1}, \\
\mathbf{y}_k &= \mathbf{h}(\mathbf{x}_k, \boldsymbol{\theta}) + \mathbf{r}_k.
\end{aligned}
\tag{A.71}
$$

In order to compute the derivative, it is convenient to first rewrite the expectations as expectations over unit Gaussian distributions as follows:

$$
\begin{aligned}
\mathbf{m}_k^-(\boldsymbol{\theta}) &= \int \mathbf{f}(\mathbf{x}_{k-1}, \boldsymbol{\theta})\, \mathrm{N}(\mathbf{x}_{k-1} \mid \mathbf{m}_{k-1}(\boldsymbol{\theta}), \mathbf{P}_{k-1}(\boldsymbol{\theta}))\, \mathrm{d}\mathbf{x}_{k-1} \\
&= \int \mathbf{f}(\mathbf{m}_{k-1}(\boldsymbol{\theta}) + \sqrt{\mathbf{P}_{k-1}(\boldsymbol{\theta})}\, \boldsymbol{\xi}, \boldsymbol{\theta})\, \mathrm{N}(\boldsymbol{\xi} \mid \mathbf{0}, \mathbf{I})\, \mathrm{d}\boldsymbol{\xi}.
\end{aligned}
\tag{A.72}
$$

The derivative of this expression can now be computed as

$$\frac{\partial \mathbf{m}_k^-(\boldsymbol{\theta})}{\partial \theta_i} = \int \Bigg[\mathbf{F}_x(\mathbf{m}_{k-1}(\boldsymbol{\theta}) + \sqrt{\mathbf{P}_{k-1}(\boldsymbol{\theta})}\,\boldsymbol{\xi}, \boldsymbol{\theta}) \times \left(\frac{\partial \mathbf{m}_{k-1}(\boldsymbol{\theta})}{\partial \theta_i} + \frac{\partial \sqrt{\mathbf{P}_{k-1}(\boldsymbol{\theta})}}{\partial \theta_i}\,\boldsymbol{\xi} \right) + \frac{\partial \mathbf{f}}{\partial \theta_i}(\mathbf{m}_{k-1}(\boldsymbol{\theta}) + \sqrt{\mathbf{P}_{k-1}(\boldsymbol{\theta})}\,\boldsymbol{\xi}, \boldsymbol{\theta}) \Bigg] \mathrm{N}(\boldsymbol{\xi} \mid \mathbf{0}, \mathbf{I})\, \mathrm{d}\boldsymbol{\xi}.$$

Assuming that we are using the Cholesky factorization-based matrix square root, the derivative $\partial \sqrt{\mathbf{P}_{k-1}(\boldsymbol{\theta})} / \partial \theta_i$ can be computed with Algorithm A.7 or Theorem A.8 given in Section A.3.

The above form is directly suitable for sigma point methods, because they are based on the same change of variables. That is, the corresponding derivative of the sigma point approximation will be

$$\frac{\partial \mathbf{m}_k^-(\boldsymbol{\theta})}{\partial \theta_i} \approx \sum_j W_j \Bigg[\mathbf{F}_x(\mathbf{m}_{k-1}(\boldsymbol{\theta}) + \sqrt{\mathbf{P}_{k-1}(\boldsymbol{\theta})}\,\boldsymbol{\xi}^{(j)}, \boldsymbol{\theta}) \times \left(\frac{\partial \mathbf{m}_{k-1}(\boldsymbol{\theta})}{\partial \theta_i} + \frac{\partial \sqrt{\mathbf{P}_{k-1}(\boldsymbol{\theta})}}{\partial \theta_i}\,\boldsymbol{\xi}^{(j)} \right) + \frac{\partial \mathbf{f}}{\partial \theta_i}(\mathbf{m}_{k-1}(\boldsymbol{\theta}) + \sqrt{\mathbf{P}_{k-1}(\boldsymbol{\theta})}\,\boldsymbol{\xi}^{(j)}, \boldsymbol{\theta}) \Bigg], \tag{A.73}$$

where W_j and $\boldsymbol{\xi}^{(j)}$ are the weights and unit sigma points of the used sigma point method. A nice thing in the above expression is that it is exactly the derivative of the sigma point approximation to $\mathbf{m}_k^-(\boldsymbol{\theta})$.

The derivatives of the Gaussian filtering-based energy function can now be expressed as follows.

Algorithm A.12 (Derivatives of Gaussian filtering-based energy function). *The recursion for the derivative of the approximate energy function is*

$$\frac{\partial \varphi_k(\boldsymbol{\theta})}{\partial \theta_i} \simeq \frac{\partial \varphi_{k-1}(\boldsymbol{\theta})}{\partial \theta_i} + \frac{1}{2} \operatorname{tr}\left(\mathbf{S}_k^{-1}(\boldsymbol{\theta}) \frac{\partial \mathbf{S}_k(\boldsymbol{\theta})}{\partial \theta_i} \right) + \mathbf{v}_k^\mathsf{T}(\boldsymbol{\theta})\, \mathbf{S}_k^{-1}(\boldsymbol{\theta}) \frac{\partial \mathbf{v}_k(\boldsymbol{\theta})}{\partial \theta_i} - \frac{1}{2} \mathbf{v}_k^\mathsf{T}(\boldsymbol{\theta})\, \mathbf{S}_k^{-1}(\boldsymbol{\theta}) \frac{\partial \mathbf{S}_k(\boldsymbol{\theta})}{\partial \theta_i} \mathbf{S}_k^{-1}(\boldsymbol{\theta})\, \mathbf{v}_k(\boldsymbol{\theta}). \tag{A.74}$$

The derivatives of the prediction step are

$$\begin{aligned}
\frac{\partial \mathbf{m}_k^-}{\partial \theta_i} &= \int \Bigg[\mathbf{F}_x \left(\mathbf{m}_{k-1} + \sqrt{\mathbf{P}_{k-1}}\, \boldsymbol{\xi}, \boldsymbol{\theta} \right) \left(\frac{\partial \mathbf{m}_{k-1}}{\partial \theta_i} + \frac{\partial \sqrt{\mathbf{P}_{k-1}}}{\partial \theta_i}\, \boldsymbol{\xi} \right) \\
&\quad + \frac{\partial \mathbf{f}}{\partial \theta_i} \left(\mathbf{m}_{k-1} + \sqrt{\mathbf{P}_{k-1}}\, \boldsymbol{\xi}, \boldsymbol{\theta} \right) \Bigg] \mathrm{N}(\boldsymbol{\xi} \mid \mathbf{0}, \mathbf{I})\, \mathrm{d}\boldsymbol{\xi}, \\
\frac{\partial \mathbf{P}_k^-}{\partial \theta_i} &= \int \Bigg[\mathbf{F}_x \left(\mathbf{m}_{k-1} + \sqrt{\mathbf{P}_{k-1}}\, \boldsymbol{\xi}, \boldsymbol{\theta} \right) \left(\frac{\partial \mathbf{m}_{k-1}}{\partial \theta_i} + \frac{\partial \sqrt{\mathbf{P}_{k-1}}}{\partial \theta_i}\, \boldsymbol{\xi} \right) \\
&\quad + \frac{\partial \mathbf{f}}{\partial \theta_i} \left(\mathbf{m}_{k-1} + \sqrt{\mathbf{P}_{k-1}}\, \boldsymbol{\xi}, \boldsymbol{\theta} \right) - \frac{\partial \mathbf{m}_k^-}{\partial \theta_i} \Bigg] \\
&\quad \times \left[\mathbf{f} \left(\mathbf{m}_{k-1} + \sqrt{\mathbf{P}_{k-1}}\, \boldsymbol{\xi}, \boldsymbol{\theta} \right) - \mathbf{m}_k^- \right]^\mathsf{T} \mathrm{N}(\boldsymbol{\xi} \mid \mathbf{0}, \mathbf{I})\, \mathrm{d}\boldsymbol{\xi} \\
&\quad + \int \left[\mathbf{f} \left(\mathbf{m}_{k-1} + \sqrt{\mathbf{P}_{k-1}}\, \boldsymbol{\xi}, \boldsymbol{\theta} \right) - \mathbf{m}_k^- \right] \\
&\quad \times \Bigg[\mathbf{F}_x \left(\mathbf{m}_{k-1} + \sqrt{\mathbf{P}_{k-1}}\, \boldsymbol{\xi}, \boldsymbol{\theta} \right) \left(\frac{\partial \mathbf{m}_{k-1}}{\partial \theta_i} + \frac{\partial \sqrt{\mathbf{P}_{k-1}}}{\partial \theta_i}\, \boldsymbol{\xi} \right) \\
&\quad + \frac{\partial \mathbf{f}}{\partial \theta_i} \left(\mathbf{m}_{k-1} + \sqrt{\mathbf{P}_{k-1}}\, \boldsymbol{\xi}, \boldsymbol{\theta} \right) - \frac{\partial \mathbf{m}_k^-}{\partial \theta_i} \Bigg]^\mathsf{T} \mathrm{N}(\boldsymbol{\xi} \mid \mathbf{0}, \mathbf{I})\, \mathrm{d}\boldsymbol{\xi} + \frac{\partial \mathbf{Q}_{k-1}}{\partial \theta_i},
\end{aligned} \tag{A.75}$$

where $\mathbf{F}_x$ is the Jacobian matrix of $\mathbf{x} \mapsto \mathbf{f}(\mathbf{x}, \boldsymbol{\theta})$. In the above expressions as well as in the following, we have dropped the dependencies of various terms on the parameters $\boldsymbol{\theta}$ to simplify the notation. The derivatives of the update step are

$$\begin{aligned}
\frac{\partial \boldsymbol{\mu}_k}{\partial \theta_i} &= \int \Bigg[\mathbf{H}_x \left(\mathbf{m}_k^- + \sqrt{\mathbf{P}_k^-}\, \boldsymbol{\xi}, \boldsymbol{\theta} \right) \left(\frac{\partial \mathbf{m}_k^-}{\partial \theta_i} + \frac{\partial \sqrt{\mathbf{P}_k^-}}{\partial \theta_i}\, \boldsymbol{\xi} \right) \\
&\quad + \frac{\partial \mathbf{h}}{\partial \theta_i} \left(\mathbf{m}_k^- + \sqrt{\mathbf{P}_k^-}\, \boldsymbol{\xi}, \boldsymbol{\theta} \right) \Bigg] \mathrm{N}(\boldsymbol{\xi} \mid \mathbf{0}, \mathbf{I})\, \mathrm{d}\boldsymbol{\xi}, \\
\frac{\partial \mathbf{v}_k}{\partial \theta_i} &= -\frac{\partial \boldsymbol{\mu}_k}{\partial \theta_i}, \\
\frac{\partial \mathbf{S}_k}{\partial \theta_i} &= \int \Bigg[\mathbf{H}_x \left(\mathbf{m}_k^- + \sqrt{\mathbf{P}_k^-}\, \boldsymbol{\xi}, \boldsymbol{\theta} \right) \left(\frac{\partial \mathbf{m}_k^-}{\partial \theta_i} + \frac{\partial \sqrt{\mathbf{P}_k^-}}{\partial \theta_i}\, \boldsymbol{\xi} \right)
\end{aligned}$$

$$
\begin{aligned}
&\qquad + \frac{\partial \mathbf{h}}{\partial \theta_i}\left(\mathbf{m}_k^- + \sqrt{\mathbf{P}_k^-}\,\boldsymbol{\xi}, \boldsymbol{\theta}\right) - \frac{\partial \boldsymbol{\mu}_k}{\partial \theta_i}\Bigg] \\
&\qquad \times \left[\mathbf{h}\left(\mathbf{m}_k^- + \sqrt{\mathbf{P}_k^-}\,\boldsymbol{\xi}, \boldsymbol{\theta}\right) - \boldsymbol{\mu}_k\right]^{\mathsf{T}} \mathrm{N}(\boldsymbol{\xi} \mid \mathbf{0}, \mathbf{I})\, \mathrm{d}\boldsymbol{\xi} \\
&+ \int \left[\mathbf{h}\left(\mathbf{m}_k^- + \sqrt{\mathbf{P}_k^-}\,\boldsymbol{\xi}, \boldsymbol{\theta}\right) - \boldsymbol{\mu}_k\right] \\
&\qquad \times \Bigg[\mathbf{H}_x\left(\mathbf{m}_k^- + \sqrt{\mathbf{P}_k^-}\,\boldsymbol{\xi}, \boldsymbol{\theta}\right)\left(\frac{\partial \mathbf{m}_k^-}{\partial \theta_i} + \frac{\partial \sqrt{\mathbf{P}_k^-}}{\partial \theta_i}\,\boldsymbol{\xi}\right) \\
&\qquad + \frac{\partial \mathbf{h}}{\partial \theta_i}\left(\mathbf{m}_k^- + \sqrt{\mathbf{P}_k^-}\,\boldsymbol{\xi}, \boldsymbol{\theta}\right) - \frac{\partial \boldsymbol{\mu}_k}{\partial \theta_i}\Bigg]^{\mathsf{T}} \mathrm{N}(\boldsymbol{\xi} \mid \mathbf{0}, \mathbf{I})\, \mathrm{d}\boldsymbol{\xi} + \frac{\partial \mathbf{R}_k}{\partial \theta_i}, \\
\frac{\partial \mathbf{C}_k}{\partial \theta_i} &= \int \Bigg\{ \frac{\partial \sqrt{\mathbf{P}_k^-}}{\partial \theta_i}\,\boldsymbol{\xi}\left(\mathbf{h}\left(\mathbf{m}_k^- + \sqrt{\mathbf{P}_k^-}\,\boldsymbol{\xi}, \boldsymbol{\theta}\right) - \boldsymbol{\mu}_k\right)^{\mathsf{T}} \\
&\qquad + \sqrt{\mathbf{P}_k^-}\,\boldsymbol{\xi}\Bigg[\mathbf{H}_x\left(\mathbf{m}_k^- + \sqrt{\mathbf{P}_k^-}\,\boldsymbol{\xi}, \boldsymbol{\theta}\right)\left(\frac{\partial \mathbf{m}_k^-}{\partial \theta_i} + \frac{\partial \sqrt{\mathbf{P}_k^-}}{\partial \theta_i}\,\boldsymbol{\xi}\right) \\
&\qquad + \frac{\partial \mathbf{h}}{\partial \theta_i}\left(\mathbf{m}_k^- + \sqrt{\mathbf{P}_k^-}\,\boldsymbol{\xi}, \boldsymbol{\theta}\right) - \frac{\partial \boldsymbol{\mu}_k}{\partial \theta_i}\Bigg]^{\mathsf{T}}\Bigg\} \mathrm{N}(\boldsymbol{\xi} \mid \mathbf{0}, \mathbf{I})\, \mathrm{d}\boldsymbol{\xi}, \\
\frac{\partial \mathbf{K}_k}{\partial \theta_i} &= \frac{\partial \mathbf{C}_k}{\partial \theta_i}\,\mathbf{S}_k^{-1} - \mathbf{C}_k\,\mathbf{S}_k^{-1}\,\frac{\partial \mathbf{S}_k}{\partial \theta_i}\,\mathbf{S}_k^{-1}, \\
\frac{\partial \mathbf{m}_k}{\partial \theta_i} &= \frac{\partial \mathbf{m}_k^-}{\partial \theta_i} + \frac{\partial \mathbf{K}_k}{\partial \theta_i}\,\mathbf{v}_k + \mathbf{K}_k\,\frac{\partial \mathbf{v}_k}{\partial \theta_i}, \\
\frac{\partial \mathbf{P}_k}{\partial \theta_i} &= \frac{\partial \mathbf{P}_k^-}{\partial \theta_i} - \frac{\partial \mathbf{K}_k}{\partial \theta_i}\,\mathbf{S}_k\,\mathbf{K}_k^{\mathsf{T}} - \mathbf{K}_k\,\frac{\partial \mathbf{S}_k}{\partial \theta_i}\,\mathbf{K}_k^{\mathsf{T}} - \mathbf{K}_k\,\mathbf{S}_k\,\frac{\partial \mathbf{K}_k^{\mathsf{T}}}{\partial \theta_i},
\end{aligned}
\tag{A.76}
$$

where $\mathbf{H}_x$ is the Jacobian matrix of $\mathbf{x} \mapsto \mathbf{h}(\mathbf{x}, \boldsymbol{\theta})$. The derivatives of the Cholesky factors can be computed with Algorithm A.7 or Theorem A.8 given in Section A.3.

The corresponding sigma point approximations can be obtained by approximating the integrals analogously to Equation (A.73). The resulting derivative will be exact in the sense that it is the exact derivative of the corresponding sigma point-based approximation to the energy function.

We could also change back to the original variable, which gives, for example, the following representation for the derivative of the predicted mean:

$$\frac{\partial \mathbf{m}_k^-}{\partial \theta_i} = \int \left[\mathbf{F}_x(\mathbf{x}_{k-1}, \boldsymbol{\theta})\, \mathbf{g}(\mathbf{x}_{k-1}, \boldsymbol{\theta}) + \frac{\partial \mathbf{f}\,(\mathbf{x}_{k-1}, \boldsymbol{\theta})}{\partial \theta_i} \right] \times \mathrm{N}(\mathbf{x}_{k-1} \mid \mathbf{m}_{k-1}, \mathbf{P}_{k-1})\, \mathrm{d}\mathbf{x}_{k-1}, \tag{A.77}$$

where

$$\mathbf{g}(\mathbf{x}_{k-1}, \boldsymbol{\theta}) = \frac{\partial \mathbf{m}_{k-1}}{\partial \theta_i} + \frac{\partial \sqrt{\mathbf{P}_{k-1}}}{\partial \theta_i} \left(\sqrt{\mathbf{P}_{k-1}}\right)^{-1} (\mathbf{x}_{k-1} - \mathbf{m}_{k-1}). \tag{A.78}$$

The derivation of the full set of equations is left as an exercise for the reader.

References

Agamennoni, G., Nieto, J., and Nebot, E. 2011. An outlier-robust Kalman filter. Pages 1551–1558 of: *IEEE International Conference on Robotics and Automation (ICRA)*. (Cited on page 353.)

Akashi, H., and Kumamoto, H. 1977. Random sampling approach to state estimation in switching environments. *Automatica*, **13**(4), 429–434. (Cited on pages 247 and 353.)

Alspach, D. L., and Sorenson, H. W. 1972. Nonlinear Bayesian estimation using Gaussian sum approximations. *IEEE Transactions on Automatic Control*, **17**(4). (Cited on page 353.)

Anderson, B. D. O., and Moore, J. B. 1979. *Optimal Filtering*. Prentice-Hall. (Cited on page 102.)

Andrieu, C., De Freitas, N., and Doucet, A. 2002. Rao-Blackwellised particle filtering via data augmentation. In: Dietterich, T. G., Becker, S., and Ghahramani, Z. (eds), *Advances in Neural Information Processing Systems 14*. MIT Press. (Cited on page 7.)

Andrieu, C., Doucet, A., Singh, S., and Tadic, V. 2004. Particle methods for change detection, system identification, and control. *Proceedings of the IEEE*, **92**(3), 423–438. (Cited on pages 341 and 344.)

Andrieu, C., and Thoms, J. 2008. A tutorial on adaptive MCMC. *Statistics and Computing*, **18**(4), 343–373. (Cited on page 325.)

Andrieu, C., Doucet, A., and Holenstein, R. 2010. Particle Markov chain Monte Carlo methods. *The Royal Statistical Society: Series B (Statistical Methodology)*, **72**(3), 269–342. (Cited on pages 324 and 342.)

Aoki, M. 1967. *Optimization of Stochastic Systems*. Academic Press. (Cited on page 6.)

Arasaratnam, I., and Haykin, S. 2009. Cubature Kalman filters. *IEEE Transactions on Automatic Control*, **54**(6), 1254–1269. (Cited on pages 131, 135, 142, 143, 145, 149, 165, 280, 289, and 353.)

Arasaratnam, I., and Haykin, S. 2011. Cubature Kalman smoothers. *Automatica*, **47**(10), 2245–2250. (Cited on pages 282 and 353.)

Arasaratnam, I., Haykin, S., and Elliott, R. J. 2007. Discrete-time nonlinear filtering algorithms using Gauss–Hermite quadrature. *Proceedings of the IEEE*, **95**(5), 953–977. (Cited on pages 140, 168, 179, and 187.)

Arasaratnam, I., Haykin, S., and Hurd, T. R. 2010. Cubature Kalman filtering for continuous-discrete systems: Theory and simulations. *IEEE Transactions on Signal Processing*, **58**(10), 4977–4993. (Cited on page 352.)

Arulampalam, M. S., Maskell, S., Gordon, N., and Clapp, T. 2002. A tutorial on particle filters for online nonlinear/non-Gaussian Bayesian tracking. *IEEE Transactions on Signal Processing*, **50(2)**, 174–188. (Cited on page 244.)

Axelsson, P., and Gustafsson, F. 2015. Discrete-time solutions to the continuous-time differential Lyapunov equation with applications to Kalman filtering. *IEEE Transactions on Automatic Control*, **60(3)**, 632–643. (Cited on page 52.)

Bar-Shalom, Y., and Li, X.-R. 1995. *Multitarget-Multisensor Tracking: Principles and Techniques*. YBS Publishing. (Cited on pages 3 and 353.)

Bar-Shalom, Y., Li, X.-R., and Kirubarajan, T. 2001. *Estimation with Applications to Tracking and Navigation*. Wiley. (Cited on pages 2, 4, 8, 44, 113, 201, 352, and 353.)

Barber, D. 2006. Expectation correction for smoothed inference in switching linear dynamical systems. *The Journal of Machine Learning Research*, **7**, 2515–2540. (Cited on pages 317 and 352.)

Barber, D. 2011. Approximate inference in switching linear dynamical systems using Gaussian mixtures. Chap. 8, pages 166–181 of: Barber, D., Cemgil, A. T., and Chiappa, S. (eds), *Bayesian Time Series Models*. Cambridge University Press. (Cited on page 353.)

Barfoot, T. D. 2017. *State Estimation for Robotics*. Cambridge University Press. (Cited on page 5.)

Barlas, G. 2015. *Multicore and GPU Programming: An Integrated Approach*. Morgan Kaufmann Publishers Inc. (Cited on page 353.)

Bell, B. M. 1994. The iterated Kalman smoother as a Gauss–Newton method. *SIAM Journal on Optimization*, **4**(3), 626–636. (Cited on pages 271 and 275.)

Bell, B. M., and Cathey, F. W. 1993. The iterated Kalman filter update as a Gauss–Newton method. *IEEE Transactions on Automatic Control*, **38**(2), 294–297. (Cited on page 121.)

Berger, J. O. 1985. *Statistical Decision Theory and Bayesian Analysis*. Springer. (Cited on pages 20, 21, and 247.)

Bernardo, J. M., and Smith, A. F. M. 1994. *Bayesian Theory*. Wiley. (Cited on pages 17 and 20.)

Bierman, G. J. 1977. *Factorization Methods for Discrete Sequential Estimation*. Academic Press. (Cited on page 353.)

Bishop, C. M. 2006. *Pattern Recognition and Machine Learning*. Springer. (Cited on pages 7, 81, and 84.)

Blackman, S., and Popoli, R. 1999. *Design and Analysis of Modern Tracking Systems*. Artech House Radar Library. (Cited on pages 3 and 353.)

Brooks, S., Gelman, A., Jones, G. L., and Meng, X.-L. 2011. *Handbook of Markov Chain Monte Carlo*. Chapman & Hall/CRC. (Cited on pages 23, 324, 325, and 326.)

Cappé, O., Moulines, E., and Rydén, T. 2005. *Inference in Hidden Markov Models*. Springer. (Cited on pages 87, 241, 262, 323, 330, and 334.)

Challa, S., Morelande, M. R., Mušicki, D., and Evans, R. J. 2011. *Fundamentals of Object Tracking*. Cambridge University Press. (Cited on pages 2 and 353.)

Chen, R., and Liu, J. S. 2000. Mixture Kalman filters. *Journal of the Royal Statistical Society: Series B (Statistical Methodology)*, **62**(3), 493–508. (Cited on page 247.)

Chopin, N., and Papaspiliopoulos, O. 2020. *An Introduction to Sequential Monte Carlo*. Springer. (Cited on pages 238, 239, 240, 312, and 314.)

Christianson, B. 1994. Reverse accumulation and attractive fixed points. *Optimization Methods and Software*, **3**(4), 311–326. (Cited on page 340.)

Corenflos, A., Chopin, N., and Särkkä, S. 2022. De-Sequentialized Monte Carlo: a parallel-in-time particle smoother. *Journal of Machine Learning Research*, **23**(283), 1–39. (Cited on page 354.)

Cox, H. 1964. On the estimation of state variables and parameters for noisy dynamic systems. *IEEE Transactions on Automatic Control*, **9**(1), 5–12. (Cited on page 267.)

Crassidis, J. L., and Junkins, J. L. 2004. *Optimal Estimation of Dynamic Systems*. Chapman & Hall/CRC. (Cited on page 2.)

Creal, D. 2012. A survey of sequential Monte Carlo methods for economics and finance. *Econometric Reviews*, **31**(3), 245–296. (Cited on page 341.)

Crisan, D., and Doucet, A. 2002. A survey of convergence results on particle filtering for practitioners. *IEEE Transactions on Signal Processing*, **50**(3), 736–746. (Cited on page 353.)

Crisan, D., and Rozovskii, B. (eds). 2011. *The Oxford Handbook of Nonlinear Filtering*. Oxford University Press. (Cited on page 352.)

Daum, F., and Huang, J. 2003. Curse of dimensionality and particle filters. Pages 1979–1993 of: *Proceedings of the IEEE Aerospace Conference*, vol. 4. (Cited on page 230.)

Deisenroth, M. P., Huber, M. F., and Hanebeck, U. D. 2009. Analytic moment-based Gaussian process filtering. Pages 225–232 of: *Proceedings of the 26th International Conference on Machine Learning*. (Cited on pages 135 and 352.)

Deisenroth, M., Turner, R., Huber, M., Hanebeck, U., and Rasmussen, C. 2012. Robust filtering and smoothing with Gaussian processes. *IEEE Transactions on Automatic Control*, **57**(7), 1865–1871. (Cited on page 352.)

Dempster, A., Laird, N., and Rubin, D. 1977. Maximum likelihood from incomplete data via the EM algorithm. *Journal of the Royal Statistical Society: Series B (Methodological)*, **39**(1), 1–38. (Cited on page 327.)

Djuric, P., and Miguez, J. 2002. Sequential particle filtering in the presence of additive Gaussian noise with unknown parameters. Pages 1621–1624 of: *IEEE International Conference on Acoustics, Speech, and Signal Processing (ICASSP)*, vol. 2. (Cited on page 347.)

Douc, R., Garivier, A., Moulines, E., and Olsson, J. 2011. Sequential Monte Carlo smoothing for general state space hidden Markov models. *Annals of Applied Probability*, **21**(6), 2109–2145. (Cited on pages 312 and 314.)

Doucet, A., and Johansen, A. M. 2011. A tutorial on particle filtering and smoothing: Fifteen years later. Chap. 24, pages 656–704 of: Crisan, D., and Rozovskiĭ, B. (eds), *Handbook of Nonlinear Filtering*. Oxford University Press. (Cited on page 244.)

Doucet, A., Godsill, S. J., and Andrieu, C. 2000. On sequential Monte Carlo sampling methods for Bayesian filtering. *Statistics and Computing*, **10**(3), 197–208. (Cited on pages 241, 309, and 314.)

Doucet, A., De Freitas, N., and Gordon, N. 2001. *Sequential Monte Carlo Methods in Practice*. Springer. (Cited on pages 234, 239, 241, 247, and 248.)

Duane, S., Kennedy, A. D., Pendleton, B. J., and Roweth, D. 1987. Hybrid Monte Carlo. *Physics Letters B*, **195**(2), 216–222. (Cited on page 326.)

Fatemi, M., Svensson, L., Hammarstrand, L., and Morelande, M. 2012. A study of MAP estimation techniques for nonlinear filtering. Pages 1058–1065 of: *15th International Conference on Information Fusion (FUSION)*. (Cited on page 126.)

Fearnhead, P. 2002. Markov chain Monte Carlo, sufficient statistics, and particle filters. *Journal of Computational and Graphical Statistics*, **11**(4), 848–862. (Cited on page 347.)

Fearnhead, P., and Clifford, P. 2003. On-line inference for Hidden Markov models via particle filters. *Journal of the Royal Statistical Society: Series B (Statistical Methodology)*, **65**(4), 887–899. (Cited on page 241.)

Fong, W., Godsill, S. J., Doucet, A., and West, M. 2002. Monte Carlo smoothing with application to audio signal enhancement. *IEEE Transactions on Signal Processing*, **50**(2), 438–449. (Cited on pages 6 and 317.)

Fraser, D., and Potter, J. 1969. The optimum linear smoother as a combination of two optimum linear filters. *IEEE Transactions on Automatic Control*, **14**(4), 387–390. (Cited on page 255.)

García-Fernández, Á. F., Svensson, L., Morelande, M., and Särkkä, S. 2015. Posterior linearization filter: Principles and implementation using sigma points. *IEEE Transactions on Signal Processing*, **63**(20), 5561–5573. (Cited on pages 204, 208, and 226.)

García-Fernández, Á. F., Svensson, L., and Särkkä, S. 2017. Iterated posterior linearization smoother. *IEEE Transactions on Automatic Control*, **62**(4), 2056–2063. (Cited on page 297.)

García-Fernández, Á. F., Tronarp, F., and Särkkä, S. 2019. Gaussian process classification using posterior linearisation. *IEEE Signal Processing Letters*, **26**(5), 735–739. (Cited on page 82.)

García-Fernández, Á. F., Svensson, L., and Morelande, M. R. 2020. Multiple target tracking based on sets of trajectories. *IEEE Transactions on Aerospace and Electronic Systems*, **56**(3), 1685–1707. (Cited on page 353.)

García-Fernández, Á. F., Williams, J. L., Svensson, L., and Xia, Y. 2021. A Poisson multi-Bernoulli mixture filter for coexisting point and extended targets. *IEEE Transactions on Signal Processing*, **69**, 2600–2610. (Cited on page 353.)

Gelb, A. 1974. *Applied Optimal Estimation*. MIT Press. (Cited on pages 119, 121, 171, and 176.)

Gelb, A., and Vander Velde, W. 1968. *Multiple-Input Describing Functions and Nonlinear System Design*. McGraw-Hill. (Cited on page 178.)

Gelman, A., Carlin, J. B., Stern, H. S., Dunson, D. B., Vehtari, A., and Rubin, D. B. 2013. *Bayesian Data Analysis*. Third edn. Chapman & Hall/CRC. (Cited on pages 17, 19, 22, 40, 108, and 323.)

Gilks, W., Richardson, S., and Spiegelhalter, D. (eds). 1996. *Markov Chain Monte Carlo in Practice*. Chapman & Hall. (Cited on page 23.)

Godsill, S. J., and Rayner, P. J. 1998. *Digital Audio Restoration: A Statistical Model Based Approach*. Springer-Verlag. (Cited on page 6.)

Godsill, S. J., Doucet, A., and West, M. 2004. Monte Carlo smoothing for nonlinear time series. *Journal of the American Statistical Association*, **99**(465), 156–168. (Cited on page 310.)

Golub, G. H., and van Loan, C. F. 1996. *Matrix Computations*. Third edn. The Johns Hopkins University Press. (Cited on page 357.)

Golub, G. H., and Welsch, J. H. 1969. Calculation of Gauss quadrature rules. *Mathematics of Computation*, **23**(106), 221–230. (Cited on page 138.)

Gonzalez, R. C., and Woods, R. E. 2008. *Digital Image Processing*. Third edn. Prentice Hall. (Cited on page 7.)

Gordon, N. J., Salmond, D. J., and Smith, A. F. M. 1993. Novel approach to nonlinear/non-Gaussian Bayesian state estimation. Pages 107–113 of: *IEEE Proceedings on Radar and Signal Processing*, vol. 140. (Cited on pages 239 and 242.)

Granström, K., Fatemi, M., and Svensson, L. 2020. Poisson multi-Bernoulli mixture conjugate prior for multiple extended target filtering. *IEEE Transactions on Aerospace and Electronic Systems*, **56**(1), 208–225. (Cited on page 353.)

Grewal, M. S., and Andrews, A. P. 2015. *Kalman Filtering: Theory and Practice Using MATLAB*. 4th edn. Wiley. (Cited on pages 98, 113, and 353.)

Grewal, M. S., Weill, L. R., and Andrews, A. P. 2001. *Global Positioning Systems, Inertial Navigation and Integration*. Wiley. (Cited on pages 3 and 4.)

Griewank, A., and Walther, A. 2008. *Evaluating Derivatives: Principles and Techniques of Algorithmic Differentiation*. SIAM. (Cited on pages 117 and 127.)

Gupta, N., and Mehra, R. 1974. Computational aspects of maximum likelihood estimation and reduction in sensitivity function calculations. *IEEE Transactions on Automatic Control*, **19**(6), 774–783. (Cited on pages 334 and 372.)

Haario, H., Saksman, E., and Tamminen, J. 1999. Adaptive proposal distribution for random walk Metropolis algorithm. *Computational Statistics*, **14**(3), 375–395. (Cited on page 325.)

Haario, H., Saksman, E., and Tamminen, J. 2001. An adaptive Metropolis algorithm. *Bernoulli*, **7**(2), 223–242. (Cited on page 325.)

Hairer, E., Nørsett, S. P., and Wanner, G. 2008. *Solving Ordinary Differential Equations I: Nonstiff Problems*. Springer Series in Computational Mathematics, vol. 1. Springer Science & Business. (Cited on page 60.)

Hartikainen, J., and Särkkä, S. 2010. Kalman filtering and smoothing solutions to temporal Gaussian process regression models. Pages 379–384 of: *Proceedings of IEEE International Workshop on Machine Learning for Signal Processing (MLSP)*. (Cited on page 7.)

Hassan, S., Särkkä, S., and García-Fernández, A. F. 2021. Temporal parallelization of inference in hidden Markov models. *IEEE Transactions on Signal Processing*, **69**, 4875–4887. (Cited on page 354.)

Hauk, O. 2004. Keep it simple: A case for using classical minimum norm estimation in the analysis of EEG and MEG data. *NeuroImage*, **21**(4), 1612–1621. (Cited on page 5.)

Hayes, M. H. 1996. *Statistical Digital Signal Processing and Modeling*. Wiley. (Cited on pages 7 and 86.)

Haykin, S. 2001. *Kalman Filtering and Neural Networks*. Wiley. (Cited on pages 7 and 85.)

Hennig, P., Osborne, M. A., and Kersting, H. P. 2022. *Probabilistic Numerics*. Cambridge University Press. (Cited on page 352.)

Hiltunen, P., Särkkä, S., Nissilä, I., Lajunen, A., and Lampinen, J. 2011. State space regularization in the nonstationary inverse problem for diffuse optical tomography. *Inverse Problems*, **27**, 025009. (Cited on page 5.)

Ho, Y. C., and Lee, R. C. K. 1964. A Bayesian approach to problems in stochastic estimation and control. *IEEE Transactions on Automatic Control*, **9**(4), 333–339. (Cited on page 8.)

Hostettler, R., Tronarp, F., García-Fernández, Á. F., and Särkkä, S. 2020. Importance densities for particle filtering using iterated conditional expectations. *IEEE Signal Processing Letters*, **27**, 211–215. (Cited on page 241.)

Hu, X., Schön, T., and Ljung, L. 2008. A basic convergence result for particle filtering. *IEEE Transactions on Signal Processing*, **56**(4), 1337–1348. (Cited on page 353.)

Hu, X., Schön, T., and Ljung, L. 2011. A general convergence result for particle filtering. *IEEE Transactions on Signal Processing*, **59**(7), 3424–3429. (Cited on page 353.)

Hürzeler, M., and Kunsch, H. R. 1998. Monte Carlo approximations for general state-space models. *Journal of Computational and Graphical Statistics*, **7**(2), 175–193. (Cited on page 314.)

Ito, K., and Xiong, K. 2000. Gaussian filters for nonlinear filtering problems. *IEEE Transactions on Automatic Control*, **45**(5), 910–927. (Cited on pages 131, 133, 135, 140, 165, 280, and 289.)

Jazwinski, A. H. 1966. Filtering for nonlinear dynamical systems. *IEEE Transactions on Automatic Control*, **11**(4), 765–766. (Cited on page 8.)

Jazwinski, A. H. 1970. *Stochastic Processes and Filtering Theory*. Academic Press. (Cited on pages xi, 8, 113, and 351.)

Johansen, A. M., and Doucet, A. 2008. A note on auxiliary particle filters. *Statistics & Probability Letters*, **78**(12), 1498–1504. (Cited on pages 244, 246, and 247.)

Julier, S. J. 2002. The scaled unscented transformation. Pages 4555–4559 of: *Proceedings of the 2002 American Control Conference*, vol. 6. (Cited on pages 150 and 152.)

Julier, S. J., and Uhlmann, J. K. 1995. *A General Method of Approximating Nonlinear Transformations of Probability Distributions*. Tech. rept. Robotics Research Group, Department of Engineering Science, University of Oxford. (Cited on pages 131, 135, 150, and 152.)

Julier, S. J., and Uhlmann, J. K. 2004. Unscented filtering and nonlinear estimation. *Proceedings of the IEEE*, **92**(3), 401–422. (Cited on pages 150, 155, 157, and 159.)

Julier, S. J., Uhlmann, J. K., and Durrant-Whyte, H. F. 1995. A new approach for filtering nonlinear systems. Pages 1628–1632 of: *Proceedings of the 1995 American Control, Conference, Seattle, Washington*. (Cited on pages 149, 155, and 159.)

Julier, S. J., Uhlmann, J. K., and Durrant-Whyte, H. F. 2000. A new method for the nonlinear transformation of means and covariances in filters and estimators. *IEEE Transactions on Automatic Control*, **45**(3), 477–482. (Cited on pages 131, 135, 150, 165, 187, and 289.)

Kailath, T., Sayed, A. H., and Hassibi, B. 2000. *Linear Estimation*. Prentice Hall. (Cited on pages 102 and 353.)

Kaipio, J., and Somersalo, E. 2005. *Statistical and Computational Inverse Problems*. Applied Mathematical Sciences, no. 160. Springer. (Cited on pages 5 and 7.)

Kalman, R. E. 1960a. Contributions to the theory of optimal control. *Boletin de la Sociedad Matematica Mexicana*, **5**(1), 102–119. (Cited on page 8.)

Kalman, R. E. 1960b. A new approach to linear filtering and prediction problems. *Transactions of the ASME, Journal of Basic Engineering*, **82**(1), 35–45. (Cited on pages 8, 16, 96, and 101.)

Kalman, R. E., and Bucy, R. S. 1961. New results in linear filtering and prediction theory. *Transactions of the ASME, Journal of Basic Engineering*, **83**(3), 95–108. (Cited on page 8.)

Kantas, N., Doucet, A., Singh, S., and Maciejowski, J. 2009. An overview of sequential Monte Carlo methods for parameter estimation in general state-space models. In: *Proceedings IFAC Symposium on System Identification (SYSID)*. (Cited on pages 343 and 344.)

Kaplan, E. D. 1996. *Understanding GPS, Principles and Applications*. Artech House. (Cited on page 2.)

Karvonen, T., and Särkkä, S. 2018. Fully symmetric kernel quadrature. *SIAM Journal on Scientific Computing*, **40**(2), A697–A720. (Cited on page 352.)

Karvonen, T., Särkkä, S., and Oates, C. J. 2019. Symmetry exploits for Bayesian cubature methods. *Statistics and Computing*, **29**, 1231–1248. (Cited on pages 135, 165, 290, and 352.)

Keeling, M., and Rohani, P. 2007. *Modeling Infectious Diseases in Humans and Animals*. Princeton University Press. (Cited on page 5.)

Kim, C.-J. 1994. Dynamic linear models with Markov-switching. *Journal of Econometrics*, **60**, 1–22. (Cited on page 317.)

Kitagawa, G. 1987. Non-Gaussian state-space modeling of nonstationary time series. *Journal of the American Statistical Association*, **82**(400), 1032–1041. (Cited on page 254.)

Kitagawa, G. 1994. The two-filter formula for smoothing and an implementation of the Gaussian-sum smoother. *Annals of the Institute of Statistical Mathematics*, **46**(4), 605–623. (Cited on page 255.)

Kitagawa, G. 1996. Monte Carlo filter and smoother for non-Gaussian nonlinear state space models. *Journal of Computational and Graphical Statistics*, **5**(1), 1–25. (Cited on pages 239, 308, 309, and 311.)

Kloeden, P. E., and Platen, E. 1999. *Numerical Solution to Stochastic Differential Equations*. Springer. (Cited on pages 57, 58, and 69.)

Kokkala, J., Solin, A., and Särkkä, S. 2016. Sigma-Point Filtering and Smoothing Based Parameter Estimation in Nonlinear Dynamic Systems. *Journal of Advances in Information Fusion*, **11**(1), 15–30. (Cited on page 327.)

Koller, D., and Friedman, N. 2009. *Probabilistic Graphical Models: Principles and Techniques*. MIT Press. (Cited on pages 255 and 264.)

Kotz, S., and Nadarajah, S. 2004. *Multivariate T-Distributions and Their Applications*. Cambridge University Press. (Cited on page 78.)

Larson, R. E., and Peschon, J. 1966. A dynamic programming approach to trajectory estimation. *IEEE Transactions on Automatic Control*, **11**(3), 537–540. (Cited on page 262.)

Lee, R. C. K. 1964. *Optimal Estimation, Identification and Control*. MIT Press. (Cited on page 8.)

Lefebvre, T., Bruyninckx, H., and Schutter, J. D. 2002. Comment on "A new method for the nonlinear transformation of means and covariances in filters and estimators" [and authors' reply]. *IEEE Transactions on Automatic Control*, **47**(8), 1406–1409. (Cited on pages 168, 187, and 352.)

Leondes, C. T., Peller, J. B., and Stear, E. B. 1970. Nonlinear smoothing theory. *IEEE Transactions on Systems Science and Cybernetics*, **6**(1), 63–71. (Cited on pages 8 and 352.)

Lin, F.-H., Wald, L. L., Ahlfors, S. P., Hämäläinen, M. S., Kwong, K. K., and Belliveau, J. W. 2006. Dynamic magnetic resonance inverse imaging of human brain function. *Magnetic Resonance in Medicine*, **56**(4), 787–802. (Cited on page 5.)

Lindsten, F. 2011. *Rao–Blackwellised Particle Methods for Inference and Identification*. Licentiate's thesis, Linköping University. (Cited on page 317.)

Lindsten, F., and Schön, T. B. 2013. Backward simulation methods for Monte Carlo statistical inference. *Foundations and Trends in Machine Learning*, **6**(1), 1–143. (Cited on page 310.)

Lindsten, F., Bunch, P., Särkkä, S., Schön, T., and Godsill, S. 2016. Rao-Blackwellized particle smoothers for conditionally linear Gaussian models. *IEEE Journal of Selected Topics in Signal Processing*, **10**(2), 353–365. (Cited on page 317.)

Liu, J. S. 2001. *Monte Carlo Strategies in Scientific Computing*. Springer. (Cited on pages 23, 230, 231, 324, 325, and 326.)

Liu, J. S., and Chen, R. 1995. Blind deconvolution via sequential imputations. *Journal of the American Statistical Association*, **90**(430), 567–576. (Cited on pages 239 and 240.)

Luenberger, D. G., and Ye, Y. 2008. *Linear and Nonlinear Programming*. Third edn. Springer. (Cited on pages 323 and 329.)

Luengo, D., Martino, L., Bugallo, M., Elvira, V., and Särkkä, S. 2020. A survey of Monte Carlo methods for parameter estimation. *EURASIP Journal on Advances in Signal Processing*, **2020**(25), 1–62. (Cited on page 313.)

Lütkepohl, H. 1996. *Handbook of Matrices*. Wiley. (Cited on page 356.)

Mahler, R. P. S. 2014. *Advances in Statistical Multisource-Multitarget Information Fusion*. Artech House. (Cited on pages 3 and 353.)

Maybeck, P. 1982a. *Stochastic Models, Estimation and Control*. Vol. 3. Academic Press. (Cited on pages 6 and 8.)

Maybeck, P. 1982b. *Stochastic Models, Estimation and Control*. Vol. 2. Academic Press. (Cited on pages 113 and 133.)

Mbalawata, I. S., Särkkä, S., and Haario, H. 2013. Parameter estimation in stochastic differential equations with Markov chain Monte Carlo and non-linear Kalman filtering. *Computational Statistics*, **28**(3), 1195–1223. (Cited on pages 326 and 338.)

McNamee, J., and Stenger, F. 1967. Construction of fully symmetric numerical integration formulas. *Numerische Mathematik*, **10**, 327–344. (Cited on pages 135, 143, 162, 165, 280, 287, and 289.)

Milton, J. S., and Arnold, J. C. 1995. *Introduction to Probability and Statistics, Principles and Applications for Engineering and the Computing Sciences*. McGraw-Hill. (Cited on page 17.)

Morelande, M., and García-Fernández, Á. F. 2013. Analysis of Kalman filter approximations for nonlinear measurements. *IEEE Transactions on Signal Processing*, **61**(22), 5477–5484. (Cited on page 124.)

Morf, M., Lévy, B., and Kailath, T. 1978. Square-root algorithms for the continuous-time linear least-square estimation problem. *IEEE Transactions on Automatic Control*, **23**(5), 907–911. (Cited on page 358.)

Murray, J. D. 1993. *Mathematical Biology*. Springer. (Cited on page 5.)

Murray, L., and Storkey, A. 2011. Particle smoothing in continuous time: A fast approach via density estimation. *IEEE Transactions on Signal Processing*, **59**(3), 1017–1026. (Cited on page 352.)

Neal, R. M. 2011. MCMC using Hamiltonian dynamics. Chap. 5 of: Brooks, S., Gelman, A., Jones, G. L., and Meng, X.-L. (eds), *Handbook of Markov Chain Monte Carlo*. Chapman & Hall/CRC. (Cited on page 326.)

Neal, R., and Hinton, G. 1999. A view of the EM algorithm that justifies incremental, sparse, and other variants. Pages 355–370 of: Jordan, M. I. (ed), *Learning in Graphical Models*. MIT Press. (Cited on page 327.)

Ninness, B., and Henriksen, S. 2010. Bayesian system identification via Markov chain Monte Carlo techniques. *Automatica*, **46**(1), 40–51. (Cited on page 324.)

Ninness, B., Wills, A., and Schön, T. B. 2010. Estimation of general nonlinear state-space systems. Pages 6371–6376 of: *Proceedings of the 49th IEEE Conference on Decision and Control (CDC), Atlanta, USA*. (Cited on page 344.)

Nørgaard, M., Poulsen, N. K., and Ravn, O. 2000. New developments in state estimation for nonlinear systems. *Automatica*, **36**(11), 1627–1638. (Cited on page 352.)

Nocedal, J., and Wright, S. J. 2006. *Numerical Optimization*. 2nd edn. Springer. (Cited on pages 122 and 126.)

O'Hagan, A. 1991. Bayes-Hermite quadrature. *Journal of Statistical Planning and Inference*, **29**(3), 245–260. (Cited on pages 135 and 352.)

Øksendal, B. 2003. *Stochastic Differential Equations: An Introduction with Applications*. Sixth edn. Springer–Verlag. (Cited on page 351.)

Olsson, R., Petersen, K., and Lehn-Schiøler, T. 2007. State-space models: From the EM algorithm to a gradient approach. *Neural Computation*, **19**(4), 1097–1111. (Cited on pages 330 and 337.)

Piché, R., Särkkä, S., and Hartikainen, J. 2012. Recursive outlier-robust filtering and smoothing for nonlinear systems using the multivariate Student-t distribution. In: *Proceedings of IEEE International Workshop on Machine Learning for Signal Processing (MLSP)*. (Cited on page 353.)

Pitt, M. K., and Shephard, N. 1999. Filtering via simulation: Auxiliary particle filters. *Journal of the American Statistical Association*, **94**(446), 590–599. (Cited on pages 244, 245, and 246.)

Poyiadjis, G., Doucet, A., and Singh, S. 2011. Particle approximations of the score and observed information matrix in state space models with application to parameter estimation. *Biometrika*, **98**(1), 65–80. (Cited on page 344.)

Proakis, J. G. 2001. *Digital Communications*. Fourth edn. McGraw-Hill. (Cited on pages 6 and 7.)

Prüher, J., Karvonen, T., Oates, C. J., Straka, O., and Särkkä, S. 2021. Improved calibration of numerical integration error in sigma-point filters. *IEEE Transactions on Automatic Control*, **66**(3), 1286–1292. (Cited on pages 135 and 352.)

Punskaya, E., Doucet, A., and Fitzgerald, W. J. 2002. On the use and misuse of particle filtering in digital communications. In: *Proceedings of EUSIPCO*. (Cited on page 242.)

Rabiner, L. R. 1989. A tutorial on hidden Markov models and selected applications in speech recognition. *Proceedings of the IEEE*, **77**(2), 257–286. (Cited on pages 87, 103, and 105.)

Raiffa, H., and Schlaifer, R. 2000. *Applied Statistical Decision Theory*. Wiley. (Cited on page 20.)

Rasmussen, C. E., and Williams, C. K. I. 2006. *Gaussian Processes for Machine Learning*. MIT Press. (Cited on pages 5, 7, and 81.)

Rauch, H. E. 1963. Solutions to the linear smoothing problem. *IEEE Transactions on Automatic Control*, **8**(4), 371–372. (Cited on page 8.)

Rauch, H. E., Tung, F., and Striebel, C. T. 1965. Maximum likelihood estimates of linear dynamic systems. *AIAA Journal*, **3**(8), 1445–1450. (Cited on pages 8 and 255.)

Ristic, B., Arulampalam, S., and Gordon, N. 2004. *Beyond the Kalman Filter*. Artech House. (Cited on pages 239, 241, 242, 247, and 353.)

Roberts, G. O., and Rosenthal, J. S. 2001. Optimal scaling for various Metropolis–Hastings algorithms. *Statistical Science*, **16**(4), 351–367. (Cited on page 325.)

Roth, M. 2012. *On the multivariate t distribution*. Tech. rept. LiTH-ISY-R-3059. Linköping University Electronic Press. (Cited on pages 78 and 79.)

Roth, M., and Gustafsson, F. 2011. An efficient implementation of the second order extended Kalman filter. In: *Proceedings of the 14th International Conference on Information Fusion*. (Cited on page 201.)

Roth, M., Özkan, E., and Gustafsson, F. 2013. A Student's t filter for heavy tailed process and measurement noise. Pages 5770–5774 of: *Proceedings of the 2013 IEEE International Conference on Acoustics, Speech and Signal Processing*. (Cited on page 78.)

Roweis, S., and Ghahramani, Z. 2001. Learning nonlinear dynamical systems using the expectation–maximization algorithm. Chap. 6, pages 175–220 of: Haykin, S. (ed), *Kalman Filtering and Neural Networks*. Wiley-Interscience. (Cited on page 327.)

Sage, A. P., and Melsa, J. L. 1971. *Estimation Theory with Applications to Communications and Control*. McGraw-Hill. (Cited on page 267.)

Saha, S., Ozkan, E., Gustafsson, F., and Smidl, V. 2010. Marginalized particle filters for Bayesian estimation of Gaussian noise parameters. Pages 1–8 of: *13th Conference on Information Fusion (FUSION)*. (Cited on page 347.)

Sandblom, F., and Svensson, L. 2012. Moment estimation using a marginalized transform. *IEEE Transactions on Signal Processing*, **60**(12), 6138–6150. (Cited on page 352.)

Särkkä, S., and García-Fernández, A. F. 2021. Temporal parallelization of Bayesian smoothers. *IEEE Transactions on Automatic Control*, **66**, 299–306. (Cited on page 354.)

Särkkä, S. 2006. *Recursive Bayesian Inference on Stochastic Differential Equations*. Doctoral dissertation, Helsinki University of Technology. (Cited on pages 285, 345, 346, and 352.)

Särkkä, S. 2007. On unscented Kalman filtering for state estimation of continuous-time nonlinear systems. *IEEE Transactions on Automatic Control*, **52**(9), 1631–1641. (Cited on page 352.)

Särkkä, S. 2008. Unscented Rauch-Tung-Striebel smoother. *IEEE Transactions on Automatic Control*, **53**(3), 845–849. (Cited on pages 285 and 286.)

Särkkä, S. 2010. Continuous-time and continuous-discrete-time unscented Rauch-Tung-Striebel smoothers. *Signal Processing*, **90**(1), 225–235. (Cited on page 352.)

Särkkä, S. 2011. Linear operators and stochastic partial differential equations in Gaussian process regression. In: *Proceedings of ICANN*. (Cited on page 5.)

Särkkä, S., and Hartikainen, J. 2010a. On Gaussian optimal smoothing of non-linear state space models. *IEEE Transactions on Automatic Control*, **55**(8), 1938–1941. (Cited on pages 278 and 280.)

Särkkä, S., and Hartikainen, J. 2010b. Sigma point methods in optimal smoothing of non-linear stochastic state space models. Pages 184–189 of: *Proceedings of IEEE International Workshop on Machine Learning for Signal Processing (MLSP)*. (Cited on page 352.)

Särkkä, S., and Nummenmaa, A. 2009. Recursive noise adaptive Kalman filtering by variational Bayesian approximations. *IEEE Transactions on Automatic Control*, **54**(3), 596–600. (Cited on pages 79, 347, and 353.)

Särkkä, S., and Sarmavuori, J. 2013. Gaussian filtering and smoothing for continuous-discrete dynamic systems. *Signal Processing*, **93**(2), 500–510. (Cited on page 352.)

Särkkä, S., and Solin, A. 2012. On continuous-discrete cubature Kalman filtering. Pages 1210–1215 of: *Proceedings of SYSID 2012*. (Cited on page 352.)

Särkkä, S., and Solin, A. 2019. *Applied Stochastic Differential Equations*. Cambridge University Press. (Cited on pages 7, 44, 46, 52, 53, 57, 58, 59, 61, 69, 351, 359, and 362.)

Särkkä, S., and Sottinen, T. 2008. Application of Girsanov theorem to particle filtering of discretely observed continuous-time non-linear systems. *Bayesian Analysis*, **3**(3), 555–584. (Cited on pages 5, 345, 346, and 352.)

Särkkä, S., and Svensson, L. 2020. Levenberg–Marquardt and line-search extended Kalman smoothers. In: *Proceedings of 45th International Conference on Acoustics, Speech, and Signal Processing (ICASSP 2020)*. (Cited on pages 127, 276, and 277.)

Särkkä, S., Vehtari, A., and Lampinen, J. 2007a. CATS benchmark time series prediction by Kalman smoother with cross-validated noise density. *Neurocomputing*, **70**(13–15), 2331–2341. (Cited on page 7.)

Särkkä, S., Vehtari, A., and Lampinen, J. 2007b. Rao-Blackwellized particle filter for multiple target tracking. *Information Fusion Journal*, **8**(1), 2–15. (Cited on pages 250 and 353.)

Särkkä, S., Bunch, P., and Godsill, S. J. 2012a. A backward-simulation based Rao-Blackwellized particle smoother for conditionally linear Gaussian models. Pages 506–511 of: *Proceedings of SYSID 2012*. (Cited on page 317.)

Särkkä, S., Solin, A., Nummenmaa, A., Vehtari, A., Auranen, T., Vanni, S., and Lin, F.-H. 2012b. Dynamic retrospective filtering of physiological noise in BOLD fMRI: DRIFTER. *NeuroImage*, **60**(2), 1517–1527. (Cited on page 5.)

Särkkä, S., Solin, A., and Hartikainen, J. 2013. Spatiotemporal learning via infinite-dimensional Bayesian filtering and smoothing. *IEEE Signal Processing Magazine*, **30**(4), 51–61. (Cited on page 7.)

Särkkä, S., Hartikainen, J., Svensson, L., and Sandblom, F. 2016. On the relation between Gaussian process quadratures and sigma-point methods. *Journal of Advances in Information Fusion*, **11**(1), 31–46. (Cited on pages 135, 165, 201, 290, and 352.)

Sarmavuori, J., and Särkkä, S. 2012. Fourier-Hermite Kalman filter. *IEEE Transactions on Automatic Control*, **57**(6), 1511–1515. (Cited on pages 176 and 201.)

Schön, T., and Gustafsson, F. 2003. Particle filters for system identification of state-space models linear in either parameters or states. Pages 1287–1292 of: *Proceedings of the 13th IFAC Symposium on System Identification, Rotterdam, The Netherlands.* (Cited on page 347.)

Schön, T., Gustafsson, F., and Nordlund, P.-J. 2005. Marginalized particle filters for mixed linear/nonlinear state-space models. *IEEE Transactions on Signal Processing*, **53**(7), 2279–2289. (Cited on page 250.)

Schön, T., Wills, A., and Ninness, B. 2011. System identification of nonlinear state-space models. *Automatica*, **47**(1), 39–49. (Cited on pages 327, 328, and 344.)

Segal, M., and Weinstein, E. 1989. A new method for evaluating the log-likelihood gradient, the Hessian, and the Fisher information matrix for linear dynamic systems. *IEEE Transactions on Information Theory*, **35**(3), 682–687. (Cited on page 330.)

Shiryaev, A. N. 1996. *Probability*. Springer. (Cited on page 17.)

Shumway, R., and Stoffer, D. 1982. An approach to time series smoothing and forecasting using the EM algorithm. *Journal of Time Series Analysis*, **3**(4), 253–264. (Cited on pages 327 and 334.)

Šimandl, M., and Duník, J. 2006. Design of derivative-free smoothers and predictors. Pages 991–996 of: *Preprints of the 14th IFAC Symposium on System Identification.* (Cited on page 285.)

Singer, H. 2008. Nonlinear continuous time modeling approaches in panel research. *Statistica Neerlandica*, **62**(1), 29–57. (Cited on page 352.)

Singer, H. 2011. Continuous-discrete state-space modeling of panel data with nonlinear filter algorithms. *AStA Advances in Statistical Analysis*, **95**(4), 375–413. (Cited on page 352.)

Skoglund, M., Hendeby, G., and Axehill, D. 2015. Extended Kalman filter modifications based on an optimization view point. Pages 1856–1861 of: *18th International Conference on Information Fusion (FUSION).* (Cited on pages 126 and 127.)

Snyder, C., Bengtsson, T., Bickel, P., and Anderson, J. 2008. Obstacles to high-dimensional particle filtering. *Monthly Weather Review*, **136**(12), 4629–4640. (Cited on page 230.)

Stengel, R. F. 1994. *Optimal Control and Estimation*. Dover. (Cited on pages 6, 8, and 176.)

Stone, L. D., Barlow, C. A., and Corwin, T. L. 2014. *Bayesian Multiple Target Tracking*. 2nd edn. Artech House. (Cited on page 3.)

Storvik, G. 2002. Particle filters in state space models with the presence of unknown static parameters. *IEEE Transactions on Signal Processing*, **50**(2), 281–289. (Cited on page 347.)

Stratonovich, R. L. 1968. *Conditional Markov Processes and Their Application to the Theory of Optimal Control*. Elsevier. (Cited on page 8.)

Striebel, C. T. 1965. Partial differential equations for the conditional distribution of a Markov process given noisy observations. *Journal of Mathematical Analysis and Applications*, **11**, 151–159. (Cited on page 351.)

Svensson, L., Svensson, D., Guerriero, M., and Willett, P. 2011. Set JPDA filter for multitarget tracking. *IEEE Transactions on Signal Processing*, **59**(10), 4677–4691. (Cited on page 353.)

Taghavi, E., Lindsten, F., Svensson, L., and Schön, T. B. 2013. Adaptive stopping for fast particle smoothing. Pages 6293–6297 of: *IEEE International Conference on Acoustics, Speech and Signal Processing (ICASSP)*. (Cited on page 314.)

Tarantola, A. 2004. *Inverse Problem Theory and Methods for Model Parameter Estimation*. SIAM. (Cited on pages 5 and 7.)

Thrun, S., Burgard, W., and Fox, D. 2005. *Probabilistic Robotics*. MIT Press. (Cited on page 5.)

Titterton, D. H., and Weston, J. L. 1997. *Strapdown Inertial Navigation Technology*. Peter Pregrinus Ltd. (Cited on page 3.)

Tronarp, F., García-Fernández, A. F., and Särkkä, S. 2018. Iterative filtering and smoothing in nonlinear and non-Gaussian systems using conditional moments. *IEEE Signal Processing Letters*, **25**(3), 408–412. (Cited on pages 168, 182, 187, 204, and 297.)

Väänänen, V. 2012. *Gaussian Filtering and Smoothing Based Parameter Estimation in Nonlinear Models for Sequential Data*. Master's Thesis, Aalto University. (Cited on page 327.)

Van der Merwe, R., and Wan, E. A. 2001. The square-root unscented Kalman filter for state and parameter estimation. Pages 3461–3464 of: *International Conference on Acoustics, Speech, and Signal Processing*. (Cited on page 353.)

Van der Merwe, R., De Freitas, N., Doucet, A., and Wan, E. 2001. The unscented particle filter. Pages 584–590 of: *Advances in Neural Information Processing Systems 13*. (Cited on page 241.)

Van Trees, H. L. 1968. *Detection, Estimation, and Modulation Theory Part I*. Wiley. (Cited on page 6.)

Van Trees, H. L. 1971. *Detection, Estimation, and Modulation Theory Part II*. Wiley. (Cited on page 6.)

Vihola, M. 2012. Robust adaptive Metropolis algorithm with coerced acceptance rate. *Statistics and Computing*, **22**(5), 997–1008. (Cited on pages 325 and 326.)

Viterbi, A. J. 1967. Error bounds for convolutional codes and an asymptotically optimum decoding algorithm. *IEEE Transactions on Information Theory*, **13**(2). (Cited on pages 6 and 262.)

Wan, E. A., and Van der Merwe, R. 2001. The unscented Kalman filter. Chap. 7 of: Haykin, S. (ed), *Kalman Filtering and Neural Networks*. Wiley. (Cited on pages 150 and 155.)

West, M., and Harrison, J. 1997. *Bayesian Forecasting and Dynamic Models*. Springer-Verlag. (Cited on page 8.)

Wiener, N. 1950. *Extrapolation, Interpolation and Smoothing of Stationary Time Series with Engineering Applications*. Wiley. (Cited on page 7.)

Wills, A., Schön, T. B., Ljung, L., and Ninness, B. 2013. Identification of Hammerstein–Wiener models. *Automatica*, **49**(1), 70–81. (Cited on page 343.)

Wu, Y., Hu, D., Wu, M., and Hu, X. 2005. Unscented Kalman filtering for additive noise case: Augmented versus nonaugmented. *IEEE Signal Processing Letters*, **12**(5), 357–360. (Cited on page 157.)

Wu, Y., Hu, D., Wu, M., and Hu, X. 2006. A numerical-integration perspective on Gaussian filters. *IEEE Transactions on Signal Processing*, **54**(8), 2910–2921. (Cited on pages 131, 133, 135, 142, 143, 145, 162, 165, 280, 287, 289, and 352.)

Yaghoobi, F., Corenflos, A., Hassan, S., and Särkkä, S. 2021. Parallel iterated extended and sigma-point Kalman smoothers. In: *Proceedings of IEEE International Conference on Acoustics, Speech and Signal Processing (ICASSP)*. (Cited on page 354.)

Ypma, A., and Heskes, T. 2005. Novel approximations for inference in nonlinear dynamical systems using expectation propagation. *Neurocomputing*, **69**(1), 85–99. (Cited on page 353.)

Zoeter, O., and Heskes, T. 2011. Expectation propagation and generalized EP methods for inference in switching linear dynamical systems. Chap. 7, pages 141–165 of: *Bayesian Time Series Models*. Cambridge University Press. (Cited on page 352.)

List of Examples

List of Theorems, Corollaries, and Algorithms

Index